# Life at High Pressure

Alister Macdonald

# Life at High Pressure

In the Deep Sea and Other Environments

 Springer

Alister Macdonald
Emeritus Reader in Physiology
University of Aberdeen, Kings College
Aberdeen, UK

ISBN 978-3-030-67589-9        ISBN 978-3-030-67587-5    (eBook)
https://doi.org/10.1007/978-3-030-67587-5

This Springer imprint is published by the registered company Springer Nature Switzerland AG.
The registered company address is: Gewerbestrasse 11, 6330 Cham, Switzerland

*To*
*Florence Telfer and Joseph Macdonald*

# Foreword

For approximately 50 years, Alister Macdonald has been the most prominent scientist in the field of high pressure physiology. I first met Macdonald in the early 1970s at a meeting on high pressure aquarium systems. He presented the pioneering work he and Ian Gilchrist were doing on retrieving water and deep-sea animals in pressure-retaining devices. Their work was foundational for my subsequent work and should still be for anyone choosing to recover living organisms from great ocean depths. I have followed his research ever since. Shortly after this meeting Macdonald wrote the influential *Physiological Aspects of Deep Sea Biology* (Cambridge University Press, 1975) and after that was an editor of several conference proceedings relating to high pressure research in biology.

Since then, Macdonald has maintained an admirable expansive and scholarly command of the field of high pressure physiology not only by organizing international conferences and editing their proceedings, but also by teaching and, in particular, by engaging in research. In this book, Macdonald shares with the reader his vast knowledge, experience, and enthusiasm for studies of life at high pressure. His expertise is wide-ranging from deep-sea studies to laboratory experiments at high pressure. This breadth can be seen in the book by noting that he has original research publications on the subjects of each chapter, except for the one on hydrothermal vent biology. The topics covered include the processes influenced by high pressures in joints of animals, in diving mammals, in deep ocean inhabitants, and in microorganisms found in the deep biosphere.

Macdonald has written this book for a general reader and has succeeded in that objective. He takes time to explain certain biological and physical concepts perhaps unfamiliar to a general reader. Nevertheless, beginning and advanced students and researchers can also benefit from reading this well-referenced book. The field of life at high pressures is of interest to earth scientists, physicists, chemists, and human physiologists as well as to biologists. This book with supplemental reading would form the basis for an advanced course on biological processes at high pressure.

When looking at aspects of organisms living at high pressure, the question Macdonald asks is "How do they do it"? This book provides some answers and introduces the reader to research areas where answers are still wanting. Anyone entering the field of high pressure biology will benefit from the insights shared in this book and gained by Alister Macdonald over a long and productive career.

Professor of Biophysics Emeritus                                     A. Aristides Yayanos
Scripps Institution of Oceanography
University of California San Diego
San Diego, CA, USA

# Preface

It seemed a good idea at the time. Well, we have all been there, but in this case the time spanned many years. It was only on retirement that I felt up to the task of writing about high pressure environments and the organisms and cells which occupy them, with the hope that they would become more widely recognised in biology. The amazing process of natural selection has produced molecular adaptations to high pressure and they are of fundamental interest. And the high pressure adaptation of complex physiological systems is equally striking and beautiful to behold.

Extreme environments are currently well recognised as niche environments, but two on a global scale is hardly consistent with a niche. High pressure is fundamental in physics and chemistry, so it surely means high pressure biology should be centre stage in the teaching of biology, and not a minor speciality. The scope of this book will convey that point, which is not, of course, an original one. Perhaps the attempt to deal with all grades of life in three high pressure environments will make the point, rather than the actual delivery, which I am aware is patchy. The book is intended to interest a wide range of readers, and it certainly lacks the rigour of an advanced textbook. Plenty of references to original papers are given in the text, but to accommodate a broad readership there are supplementary sources, not cited in the text, in the lists at the end of each chapter, which can be spotted easily enough.

My own experience in this area has been reasonably broad, and it has naturally involved numerous collaborations. I want to acknowledge them, starting with my student days at Bristol University, UK. Jack Kitching, otherwise Professor J. A. Kitching, FRS, was my PhD supervisor who I accompanied when he was appointed to a foundation Chair at the new University of East Anglia, Norwich, UK. During that time I embarked on a long-term collaboration with Ian Gilchrist, an engineer then working at Queens University, Belfast. For him working with a background of the "troubles" in Northern Ireland was in marked contrast to the lively, stimulating atmosphere of a new University which I enjoyed. On moving to the University of Aberdeen, Scotland, UK, I entered a much bigger and older institution, about 500 years older, where I found an impressive community of scientists. Collaboration with Ian Gilchrist continued, but new projects with scientists in the University and the associated research institutes proved rewarding: John Dewry, Ken Wann and Sandy Harper, electrophysiologists; Roger Pertwee, pharmacologist; Klaus Wahle, Charles Earl and John Sargent, biochemists; Clem Wardle, R. Johnstone and Peter

Fraser, animal physiologists; and Paul Wraight and Mike Player, physicists. Collaboration outwith Aberdeen was also important: Alec Bangham, Cambridge; Andrew Cossins, Liverpool; and Peter Usherwood, Bob Ramsey and Mark Sansom, Nottingham. Contacts with scientists abroad were similarly friendly and rewarding: Karel Heremans and Andre Pequeux (Belgium), Philippe Sebert (France), Stefan Heineman and Horst Ludwig (Germany) and Alex Vjotosh (Russia); in the USA John Teal (Woods Hole), Keith Miller (Harvard), Bob Marquis (Rochester), Perry Hogan (Buffalo), Art Yayanos (Scripps Institution) and Ralph Brauer (North Carolina); in Bermuda, Tony Knap; in Japan, Fumiyoshi Abe; in Canada, Art Zimmerman; and in Australia, Boris Martinac (with Doug Bartlett (USA) playing a role in one project).

Students, working for their first degree or a higher degree, provided another type of collaboration from which I benefitted as much as I hope they did. That applies particularly to PhD students. When I switched from full-time research to the traditional University mix of teaching and research, I came to appreciate the old adage that to teach (well) is to learn.

This book is mildly advocatory, a contribution to a re-balancing of the way we regard high pressure in biology. In presenting the case I have tried to avoid disastrous errors and in that I have been helped by John Huntley, Vivien Thain, John Thain, Ian Booth, Horst Ludwig and Ian Gilchrist, who have variously commented on sections of the book, and for that I am grateful. I am also grateful to the library staff of Aberdeen University and to the authors who directly and indirectly gave permission for their work to be reproduced in the book. That said, at a more personal level, I am deeply indebted to my wife for the support she has provided at a time when she expected life to be more relaxed.

The customary acknowledgement that the author is responsible for all shortcomings cannot be evaded. Here is hoping they do not detract from the central idea of the book. It seemed a good idea at the start of writing and it seems a good idea at the end.

Aberdeen, Scotland                                                      Alister Macdonald, D Sc
2020

# Contents

# High Pressure and High-Pressure Environments

1

It is well known that there are major environments in which life thrives, freshwater, the sea, and tropical and temperate lands. There are very hot and cold regions where life may not appear to thrive, but nevertheless exists in surprising abundance. This book is about the little known high-pressure environments and the life which exists there.

Pressure, like temperature, determines the structure of molecules and their ability to react with others. In this context, pressure means hydrostatic pressure, a force that acts in all directions. It is the pressure at the bottom of a tank of water or a deep ocean, of which more soon. The world of biology which is familiar to us and medically important is assumed to work at normal atmospheric pressure. That is reflected in many biochemistry and physiology textbooks which either ignore pressure or mention it hardly at all. Textbooks concerned with physical and chemical processes deal with hydrostatic pressure as it contributes to their energetics or thermodynamics, as we shall see. And as we shall also see, there is a great deal of biology going on at high hydrostatic pressure.

## 1.1 High-Pressure Environments

The environments into which life has evolved include a wide range of hydrostatic pressure. Our terrestrial environment is actually a gaseous one whose atmospheric pressure decreases markedly with altitude, but although that affects the pressure at which oxygen is available for us to breathe, it is quite distinct from the hydrostatic pressure with which this book is concerned. Because water is dense, aquatic environments provide a wide range of hydrostatic pressures which have a profound effect on all the basic life processes. The sea extends to a depth of nearly 11,000 m creating a hydrostatic pressure of more than 100 MPa. This is a very high pressure. (The unit of pressure, Mega Pascal, MPa, will be explained soon.) Even the pressure at the average depth of the oceans, 3800 m, is very significant. So, the deep ocean,

© Springer Nature Switzerland AG 2021
A. Macdonald, *Life at High Pressure*, https://doi.org/10.1007/978-3-030-67587-5_1

until recently thought to be the largest biological environment on our planet, is characterised by high hydrostatic pressures.

Deep-sea pressures disrupt basic life processes in ordinary "surface" organisms. They prevent cell division and cause shallow water animals to convulse and then become paralysed. In the longer term, they are lethal. Thus the wide variety of organisms living in the deep sea, microorganisms, exotic squid and huge whales (on a brief visit, holding their breath), must have evolved adaptations to high pressure, and the question is, how?

On the ocean floor, there are sites from which superheated waters gush. These are the hydrothermal vents that are colonised by special organisms, many of which cope both with high temperatures and high pressures. Additionally, there are a few freshwater lakes, such as Lake Baikal, which are deep enough to provide pressures sufficient to elicit evolutionary adaptations to pressure, in splendid isolation. But these aquatic environments are not the only places where life exists at pressures much higher than normal atmospheric pressure.

The second high-pressure environment to be considered could hardly be more different from the deep sea. It is often called the Deep Biosphere. Its scale and significance have only recently been appreciated and its full extent has yet to be established. It seems to be the largest of all biological environments, but one that only accommodates microorganisms. The earth's crust, which lies beneath the oceans and the continents, contains water in which microorganisms live. Parts of it beneath the oceans seem to be connected to the hydrothermal vents just mentioned, but other regions are quite separate. The fluids which are present throughout the earth's crust support chemical reactions and microbial life on an amazing scale. The hydrostatic pressures present are notable, in the range of those in the deep sea and perhaps greater. The earth's crust is "infected" with bacteria and archaea, of which more below (Bar-On et al. 2018). Some scientists have argued that life may have originated in the Deep Biosphere and its alien conditions also give us ideas about how life may have evolved on planets other than Earth.

The third high-pressure environment could also hardly be more different from the previous two. When vertebrates evolved onto land they relinquished their S-shaped body movements which propelled them through water and instead supported their body weight on four legs, articulating them to move about. Such a mechanical solution to the problem of moving whilst counteracting gravity created load-bearing sites in the articulating joints. The problem of coping with the ensuing stress was solved by evolving shock absorbers and ways of lubricating the joints. We know that in the healthy development of load-bearing bones and joints the mechanical stresses actually influence the growth of the detailed structure, to achieve a light and strong system. One of those stresses is hydrostatic pressure, which occurs in pulses during movement. The pressures are significant, similar to those which occur at intermediate depths in the deep sea. Hydrostatic pressure is attracting much interest in orthopaedic medicine and the broader subject of Biomechanics. If it seems odd to regard joints as a distinct biological environment then reflect on the environments occupied by animal parasites, infective bacteria or the cellular environments of viruses. Perhaps micro-environment is a better term, but nevertheless evolution takes place there. Also, in bones and other organs and tissues, there are small hydrostatic pressures that influence the metabolism of cells in ways not well understood. Some of the smallest pressure match those which aquatic animals detect in their environment, so some pressures act on sensory systems. There is a serious view that holds that the smaller the pressure the harder it is to understand how it works in nature.

So, this book is concerned with how the organisms and cells which occupy the three very different high-pressure environments are adapted to high pressure. The adaptations are necessarily either molecular in nature or in the case of certain complex physiological systems they are multicellular. The adaptations are invariably plausible but by no means firmly proved. "Ecologically significant" is one phrase physiologists and biochemists use to imply an adaptive role in a particular feature. The natural selection of chance mutations, which drives adaptation, complies with physical chemical laws, and this book attempts to describe high-pressure adaptations, putative or definite, in that light.

The effects on organisms of very high pressures, often higher than those which occur in the environments just mentioned, are also described, albeit briefly. Such pressures are used to investigate molecules and their reactions and are also much used in food technology, for example to inactivate bacteria without spoiling flavour. Astrobiologists devise experiments to test the tolerance and adaptability of organisms to pressures ten or more times those in the deepest oceans and the fleeting high pressures (and temperatures) which accompany meteorite impacts (Meersman et al. 2013; Winter 2013; Akasaka and Matsuki 2015; Merino et al. 2019).

### 1.1.1 Measuring Pressure

The way hydrostatic pressure is measured and the unit used need an explanation. The units of pressure are explained in Box 1.1 and the actual devices used to measure pressure are described in Chapter 13.

A picture of what pressure actually is may be helpful, and the one provided by Hesketh in his physics book for the uninitiated is that atoms and molecules move around and bounce off each other. We perceive this motion as pressure and temperature, thus more bouncing gives higher pressure and faster bouncing higher temperature.

---

**Box 1.1 Units Used to Measure Pressure**

Pressure is force per area and its basic unit of measurement is the Pascal. A Pascal is one Newton (force) per $m^2$, a derived unit of the International System of units. Pascal (1623–1662) and Newton (1643–1727) were, respectively, French and English mathematician-physicists.

---

(continued)

**Box 1.1**  (continued)

A depth of 10 m of water generates a pressure equal to the normal atmospheric pressure at the surface, thus descending 10 m doubles the pressure. The pressure at a depth of 10 m is also one-tenth of that at 100 m and in fact pressure increases in proportion to depth all the way to the greatest depths in the ocean, nearly 11,000 m, as already mentioned. The pressure at any depth may be calculated by multiplying the water density in units of kg per cubic metre, the depth in m, and the acceleration due to gravity, which is approximately 10 in the equation. This gives 100,000 Pascals per 10 m. A Pascal (Pa) is a very small unit of pressure, so 1000 Pa or 1 kPa is used. The above ignores variations in seawater salinity and temperature, which affect its density, and local variations in gravity, all of which are very small and unimportant here. So, the pressure at a depth of 10 m is 100 kPa additional to atmospheric pressure which, by convention, is ignored unless otherwise specified. A deep-sea pressure, say at 1000 m depth, is 10,000 kPa, usually expressed as 10 Mega Pascals, 10 MPa, and at a depth of 5000 m, typical of the ocean floor, it is 50 MPa.

To complete the picture, normal or standard atmospheric pressure (dry air at 0 °C and at a latitude of 45 degrees North), supports a column of mercury 760 mm high and is equal to 101.325 kPa, or in the delightful units favoured in some parts of the world, 14.696 pounds per $inch^2$. 100 kPa is also equal to 1 bar, a long-established unit of pressure superseded by the System International.

## 1.1.2   Negative Pressure

Liquids, particularly water, have strong cohesive forces and can be subjected to tension. Imagine a narrow bore glass tube containing a column of water. If suction is applied to the ends of the tube the pressure in the water will decrease to zero and then become negative. At some point cavitation, the collapse of the water structure and the appearance of a bubble will mark the limit of the tensile strength of the water column. The bubble will contain water vapour and any previously dissolved gases. During the decrease in pressure, the water is in a metastable (i.e., unstable) state and is susceptible to micronuclei triggering cavitation, analogous to supercooled water suddenly freezing solid when nucleated with an ice crystal. In special and rather rare experiments water has been shown to cavitate at negative pressures of several hundred MPa. Ordinary seawater generally cavitates at a few kPa, negative. What has this got to do with a book on life at high pressure? Well, Cephalopods such as squid and octopus have effective suckers that can create powerful "tenacity" (a technical term). One of the limiting factors in tenacity is cavitation, which is decreased by the positive hydrostatic pressure which occurs at depth, so deep living Cephalopods have suckers with great tenacity. The effect is also apparent in limpets

(Smith 1991; Smith et al. 1993). This is probably one of the most obscure effects of deep-sea pressures, but not the least interesting one, you will encounter in this book. Negative pressure and the tensile strength of water come into its own in understanding how tall trees manage to get sap to rise sufficiently high, but that is for another book.

## 1.2 The Deep Ocean

The bulk of the sea is, in fact, deep ocean. Pressure, increasing in proportion to depth and reaching 10 MPa at 1000 m, a hundred times atmospheric pressure, is physiologically significant. The shallow continental shelf, at a depth of 200 m, abruptly slopes down to the abyssal plain which provides most of the ocean floor. The very restricted deep-sea trenches, regions where the tectonic plates are being subducted, moving down into the mantle, are the deepest zones. The deepest of these is the Challenger Deep in the Mariana Trench which reaches a depth of nearly 11,000 m. The average ocean depth of 3800 m generates a pressure of 38 MPa (Fig. 1.1). This is reasonably described as a high pressure but in engineering terms, it is pretty ordinary. Hydraulic machinery and industrial processes use similar pressures. Chemists, geologists and physicists are involved with much higher pressures and astrophysicists with pressures orders of magnitude higher.

The relative bulk of the oceanic environment is huge compared to the shallow seas. According to Roberts (2012) the total area of the sea is 360 million $km^2$ and its

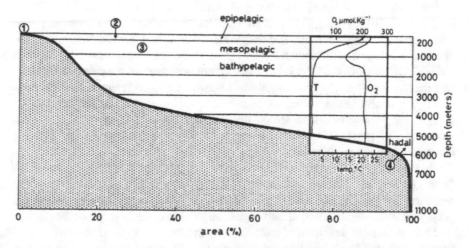

**Fig. 1.1** The hypsographic curve. The average profile of the oceans. (Macdonald et al. 1980). The distribution of depth in the ocean is plotted against area. The insets show typical and approximate vertical distributions of temperature and dissolved oxygen. Reproduced with permission from Macdonald et al. (1980)

volume is 1.3 billion km$^3$, but the deep sea is very thinly populated. This arises from its remoteness from the sun lit waters where photosynthesis takes place, driving the growth of microscopic plants (phytoplankton) and thus generating the primary source of food in the marine food chain. The deep sea is dark because sunlight is both scattered and absorbed by the water column. The green-blue part of the light spectrum penetrates furthest, such that at a depth of 500 m in clear oceanic water the human eye can just detect the residual sunlight. The eyes of animals living in the "twilight" zone have a photopigment that is sensitive to the appropriate wavelengths. Those living beyond the range of sunlight typically have eyes sensitive to the wavelength of the bioluminescence which is emitted by a variety of animals. Generally, the deep sea is cold with the temperature of deep waters around 4 °C and relatively constant. There are local exceptions to the deep and cold rule, for example the western Mediterranean Sea has deep water at a temperature of 15 °C and in the Red sea, there are even higher temperatures. Seawater is slightly compressible and at a depth of 10,000 m, it is about 4% denser than surface water. High pressure also lowers the freezing point of water, at 100 MPa pure water freezes at −9 °C.

Water is an effective medium for transmitting sound, which travels slightly faster through deep, compressed seawater than through shallow water. To some extent, the perception of sound replaces sight in many deep-sea animals. The deep sea is not stagnant however and the complex circulating currents ensure that deep seawater contains dissolved air and salts similar to those in surface waters. They also ensure that pollutants, especially microscopic particles derived from degraded plastic, are present throughout the water column. The deep sea is also susceptible to pollution from mining, drilling and other industrial operations, and there is also evidence that global warming is affecting the depths (Levitus et al. 2000; Devine et al. 2006; Fischer et al. 2015; Petersen et al. 2016; Harden-Davies 2017). The combination of low temperature, high pressure and darkness creates a vast, alien environment, but one which is populated throughout by all grades of organism, although it was once speculated to be devoid of life. It has a significant genetic resource (Harden-Davies). The deep ocean floor supports a significant benthic fauna and microbiological flora which can penetrate the sediments to a considerable extent. The environment is notable for its low oxidative stress, due to the lack of sunlight and generally moderate concentrations of dissolved oxygen. There are exceptions to this, for example when oxygen is present at elevated pressure, of which more later.

Hydrothermal vents, mentioned earlier, create highly localised oases on the ocean floor, and will be described in Chapter 9. Here it is sufficient to point out that the geothermally superheated waters which pour into the ocean bottom water are prevented from boiling (turning into vapour) by the ambient high pressure.

There is no rigid definition of the deep sea, and there is no need for one, at least for the biologist. Deep has to have a context, so deep-sea diving in the case of air-breathing mammals such as whales and seals generally implies lesser depths and pressures than experienced by many deep-sea fish. Some diving mammals descend, briefly, to depths of several 1000 m and provide physiologists with some fascinating problems. Avoiding the word deep is not difficult and the following terms are generally recognised. Hadal refers to ocean floor depths greater than 6000 m, whilst abyssal applies to depths of 6000–3000 m, (4) in Fig. 1.1. Bathyal refers to 3000 m up to 200 m, (3) in Fig. 1.1 and sublittoral to depths of less than 200 m, (2) in Fig. 1.1, typically on the continental shelf. Littoral means the seashore and the intertidal zone, (1) in Fig. 1.1. Mid water or pelagic zones are correspondingly, hadopelagic (>6000 m), abyssopelagic (3–6000 m), bathypelagic (1–3000 m) and mesopelagic (200–1000 m). The commonly fished depths of less than 200 m is epipelagic.

## 1.2.1 The Ocean Depths and Their Inhabitants

The history of the exploration of the deep oceans has a romance and excitement not unlike that of Polar exploration. The historical paths taken in learning about the oceans, their extent, depth and inhabitants, are many and varied. There is a good chronological account in the classic "The Depths of the Ocean" by Murray and Hjort (published in Murray and Hjort 1912). It is part of the history of the subject and well worth the effort of finding a specialised library to see a copy. A full historical record is to be found in Rozwadowski (2008) and more scientific accounts include; Idyll (1971), Menzies et al. (1973), Marshall (1979), Kunzig (2000) and Herring (2001). Contemporary work, including that being pursued in the deepest trenches, is reported in numerous publications which will be mentioned in various sections of this book, starting here with Jamieson (2015, 2018).

The modern phase of deep-sea biology began in the early nineteenth century and reached a critical stage in the early 1950s, after which the traditional expeditionary approach gave way to more focussed, and more frequent, investigations. In the 1950s two large scale, nation-based research cruises were undertaken, one by the Russians using the ship Vityaz in the northwest Pacific in 1949–1953 and the other by the Danes using the ship Galathea, which operated all round the world in 1950–1952. The Galathea expedition was led by Anton Bruun and it established that animals and bacteria lived in the sediments of the greatest depths, in the deep-sea trenches. Specifically, animals were recovered from a depth of 10,190 m in the Philippine Trench. They included sea anemones (phylum Coelenterata, class Anthozoa) and Amphipods(phylum Arthropoda, class Crustacea), specifically *Galatheanthemum hadale* and *Padaliscoides longicaudatus*, respectively). This settled, once and for all, the question of whether high hydrostatic pressure prevented life from colonising the full vertical range of the oceans. All the trenches with a depth in excess of 10,000 m are in the Pacific Ocean (Schrope 2014). The Philippine Trench, whose maximum depth is 10,497 m, is not the deepest. It is exceeded by the Kurile-Kamchatka Trench (10,542 m), the Japan Trench (10,554 m), the Tonga Trench (10,882 m) and the Mariana Trench (10,915 m) (Menzies et al. 1973). The reasonable assumption that the extra 400 m depth of the Mariana Trench would not preclude life has been subsequently confirmed. The deepest part of the Mariana Trench, the Challenger Deep, was established in 1951 by the British survey ship HMS Challenger, named after the illustrious nineteenth century Challenger, of which more below (Nakanishi and Hashimoto 2011). The greatest depth in the

Challenger Deep has been variously reported: 10,920 m (Gvirtzman and Strern 2004), 10,924 m (Akimoto et al. 2001; Taira et al. 2004) and 10,984 ± 25 m (Gardner et al. 2014). The last has a strong claim to be the deepest point. Since then small animals have been collected from a depth of 10,929 m (see Lan et al. 2017 and Yayanos, Section 5.2.2). Temperatures from 1 °C to 2.6 °C have been recorded.

The Galathea expedition trawled five deep-sea trenches, including the Tonga and Philippine Trenches, using acoustic soundings to assist the passage of the trawl along the steep-sided topography. It collected fish from trench depths, including Bassogigas from 7160 m in the Sunda Trench. Quantitative sampling was also a feature of much of the work. The results were published in ten volumes, from 1959 to 1969, in the English language, and had an important impact on both Biology and Oceanography. For example a particularly interesting creature called Neopilina was described. It looks like a limpet which it is not, and its formal classification is phylum Mollusca, class Amphineura, order Monoplacophora. It is a relict species, a living fossil, with a segmented organisation of the foot musculature and internal organs. Neopilina galatheae was collected from a seafloor depth of 3600 m off the west coast of Mexico and is closely related to fossil forms such as Pilina, known from the Cambrian–Devonian eras. Its discovery stimulated a notion that deep oceans might be a haven for living fossils, but this romantic idea is not supported by the evidence. However, the results of the Galathea Expedition led Anton Bruun to the concept of a hadal fauna, one which is endemic to the deep trenches, which is romantic enough and generally accepted. The deep-sea fauna also includes giant species of various major groups of Crustacea, Molluscs and even Protists, but so do other environments. The Russian Vityaz expedition also established the hadal concept and its results are generally less well known in the west. Quantitative sampling was also a feature of the work on Vityaz, and a set of data it produced which did become well known in the west was the vertical distribution of plankton biomass, the mixed assembly of organisms, all so small their geographical distribution is dependent on currents. Irrespective of the density of the biomass in the surface waters, it declines steadily over the full range of ocean depths.

The early, erratic accumulation of knowledge about deep-sea life is illustrated by a number of significant observations. In 1818 John Ross, exploring the Arctic for a Northwest passage to the Pacific, retrieved starfish, hydroids and worms from depths of 1600–1900 m. This was followed by the collection of animals from 1800 m in Antarctic waters during cruises of the Erebus and the Terror. At about the same time an academic, Edward Forbes, extrapolating from the limited results he obtained whilst dredging in the eastern Mediterranean Sea, suggested that animals would not be found deeper than 600 m. This highly speculative "azoic" concept at least had the merit of being provocative and testable, although unknown to him those tests had already been accomplished. Also, at the same time, Charles Darwin was aboard HMS Beagle, participating in a cruise that was not primarily concerned with life in the ocean depths but which was to lead to the Theory of Evolution by Natural Selection. This theory would provide an understanding of how the deep oceans might have been colonised, from cold polar regions or directly from warm shallow depths. It influenced the outlook of Charles Wyville Thomson who led expeditions on HMS Lightning and HMS Porcupine in 1868–1870 during which dredging was extended to depths greater than 4000 m off the Spanish coast and north and west of Britain, which yielded a variety of invertebrates. This was a precursor to the great HMS Challenger expedition of 1872–1876, which Wyville Thomson organised and led. The Challenger Expedition was a British venture, comparable in its prestige to the twentieth century space programmes of the USA and Russia. It encompassed deep-sea biology, oceanography and much else besides. Abundant evidence was secured to formally refute the azoic hypothesis, which was dead in the water anyway, see below. The expedition's results were published in 34 volumes and are still consulted. Other nations were not to be outdone as the following list of expeditions, each named after the lead ship, shows Hirorondell, Prince Alice I (Monaco), Ingolf, (Denmark), Michael Sars (Norway), Valdivia (Germany) and Blake, Albatross (USA). The French expeditionary cruises of the Travailleur and one cruise of the Talisman in 1882–1883 were particularly significant as they provided sediment samples from 5000 m depths. Certes, based as a part-time amateur naturalist in the Pasteur Institute in Paris, established that bacteria in the samples could be cultivated in the laboratory. (The word bacterium entered the English language in 1847, according to Bragg 2003). Microbiologists like Certes were capable of using the sterile techniques developed by Pasteur and had embarked on experiments at high pressure (Certes 1884a, b; Certes and Cochin 1884; Follonier et al. 2012). The excellent historical account provided by Adler and Ducke (2018) explains that Regard, a contemporary of Certes and not a microbiologist but more a general physiologist, undertook high-pressure experiments with small aquatic animals, of which more later. The high-pressure equipment he shared with Certes derived from the work of the French chemist—metallurgist Cailletet, a pioneer in high-pressure chemistry and low-temperature physics. He is best known for liquefying oxygen in 1877 (Papanelopoulou 2013).

One of the objectives of nineteenth century deep-sea exploration was to find routes along which telegraph cables might be laid. During the soundings made for this purpose deep living animals were accidentally retrieved. But could starfish, clinging to the end of a sounding cable, be reliably regarded as coming from the ocean floor? The uncertainty was obvious but completely set aside when telegraph cables, hauled from a depth of more than 2000 m for repair, were found to be encrusted with coral, polyzoa (small colonial animals, commonly seen on rocky beaches), bivalve and snail-like molluscs. This splendid accidental experiment took place between Sardinia and the African coast in 1860. A twentieth century sequel to this demonstration was the discovery that deep-diving sperm whales became entangled in cables lying on the ocean floor, sometimes at depths in excess of 1000 m (Heezen 1957). The entangled whales were not intent on a record dive, they were probably feeding, which makes the 10 MPa pressure they experienced remarkably interesting, as will be explained in Chapter 11.

In "The Silent Deep" Koslow provides us with further evidence of the erratic development of deep-sea biology by pointing out that deep-sea fisheries had existed for a very long time. We normally think that fisheries are situated on the continental shelf but Koslow describes traditional "artisanal" fisheries in the South Pacific, the

Azores and the west Greenland fjords where, for several 100 years, fish had been caught at depths of 700–1000 m. It appears that these were unknown to the European explorers and scientists, particularly Edward Forbes. The contemporary exploitation of conventional fishing grounds has led to depleted returns such that deeper water fish have been sought. In turn this has led to the depletion of, for example the blue hake (*Antimora rostrata*) and the roundnose grenadier (*Coryphaenoides rupestris*), which are regarded by Devine and his colleagues as endangered species.

Exploring the depths of the ocean by means of underwater vehicles has contributed greatly to our knowledge of life there. In the 1930s William Beebe, Director of the Department of Tropical Research of the New York Zoological Society, collaborated with an engineer, Otis Barton, to develop a pressure-resistant steel sphere which could accommodate two people. The life support system was primitive, but adequate, and the bathysphere was equipped with three quartz windows and a secure attachment point for a cable from which it was suspended from a surface barge. "Dives" were carried out off Bermuda to depths of 610 m. Many observations were made of the creatures in the mid water, twilight zone, a region which had never been seen before. In 1932 Beebe even involved the National Broadcasting Company (of the USA) and, from a depth of 610 m, broadcast a radio commentary of his observations (Beebe 1935). In 2013 The Japanese agency JAMSTEC organised an Internet broadcast from the submersible Shinkai 6500 operating at a depth of 5000 m in the Cayman Trench. Viewers saw the deepest hydrothermal vent and much else besides (Kawama 2014). The parallel with the public's involvement with the manned landing on the moon in 1969 is striking. Beebe's second sphere was used in 1934 to descend to a depth of 914 m. His account of what he and his team accomplished is a fascinating example of brave adventure and objective scientific reporting, contributing to a maritime tradition that continues to the present day.

In the mid-1950s Piccard's bathyscaph Trieste began operating. This vessel comprised a very large buoyancy tank below which was mounted a pressure sphere, similar to a bathysphere, which accommodated a pilot and an observer. It was an aquatic equivalent of an airship, slow and cumbersome, but it proved capable of descending to the greatest ocean depths. In 1960 Piccard and Walsh, aboard the Trieste, descended to a depth of 10,916 m in the Mariana Trench. Note that this is consistent with the depths mentioned earlier. They saw a large flatfish swimming along the seafloor (Piccard and Dietz 1957). However, in 1961 a fish expert, Torben Wolff, suggested their observations were unreliable and argued that the creature was probably a holothurian (sea cucumber). This was thoroughly supported by Jamieson and Yancy in a 2012 paper. The trench depths are more frequently sampled these days (Jamieson 2015: 208). Notable progress is the case of the amphipod *Hirondellea gigas*, numerous specimens of which were collected in a simple baited trap from a depth of 10,929 m in the Challenger Deep in the Mariana Trench, Fig. 1.2 (Lan et al. 2017).

Of this animal, from the deepest part of the ocean, several points can be made. The specimens were dead due to the rise in temperature and decrease in pressure experienced during collection. The specimens were promptly frozen for subsequent

**Fig. 1.2** The Amphipod *Hirondellea gigas* from the Challenger Deep in the Mariana Trench. Scale bar 5 mm. Reproduced with permission from Lan et al. (2017)

analysis. Note the individual shown in Fig. 1.2 is about 8 cm long, a giant species of this type of Crustacean, but giantism is not uncommon in the deep-sea fauna. However, its size has nothing to do with its most interesting feature, namely its ability to live at a pressure of around 1100 MPa. Deep-sea animals provide no visible clue about their pressure adaptions, which are on a molecular scale. And although the transcriptome (the nucleic acid RNA and associated genes) of this animal has been investigated, it cannot, at the present state of knowledge and perhaps never will, provide the necessary detail we need to understand the molecular mechanisms underlying adaptation to high pressure. For that we need to investigate the enzymes, structural proteins and other molecules making up the membranes, nerves, muscles and how they coordinate in metabolism. The same applies to single cell organisms. This book attempts to explain this problem fully and show what progress has been made in understanding such adaptation.

Returning to the exploration of the deeps, modern deep sea vehicles comprise manned and unmanned submersible craft, many of which have a modest depth range. An example of an unmanned submersible is Nereus, described by Schrope in Scientific American (2014). Very few manned submersibles can enter the hadal zone. The remarkable Deepsea Challenger, built in Australia, was piloted to the bottom of the Mariana Trench by its creator, James Cameron, in 2012. A depth of 10,898 m was recorded by this one-man submersible (Lutz and Falkowski 2012). In 2019 Victor Vescovo piloted his American built, two-person submersible, the Limiting Factor, to a depth of 10,928 m in the Challenger Deep. For good measure he has also visited other deep trenches, returning to the Challenger Deep in 2020, to descend with a passenger (Stokstad 2018). Collaboration with the British deep-sea biologist Jamieson is planned, so in due course, we shall hear of scientific results from their exploits (Jamieson 2020). Appropriate Websites, including that of the National Geographic, provide plenty of information and drama.

A number of publications provide well-illustrated accounts of general oceanic research activity: Ballard (2000), Nouvain (2007), Boyle (2009) and Byatt et al. (2001), the last is also available in video form, "The Blue Planet," originally broadcast by the BBC. The reader will also find lots of related information on various websites, some of which should be taken with a pinch of salt.

In conclusion, we should think of marine organisms in four categories; mega-fauna, macrofauna, meiofauna (miniature and microscopic animals typically living in sediments and retained by a 40 μm mesh sieve) and microorganisms. Generally, the first three decreases in biomass with increase in depth. Microorganisms are present throughout the entire water column and in the sediments on the seafloor, where their number, per unit volume, is many thousand times greater than in the bulk deep seawater. The term microorganism includes Bacteria, Archaea (of which more soon), Protists, and Fungi. The subject of Microbiology is a major player in Deep Sea Biology.

## 1.2.2   Freshwater Lakes and Ice

Some lakes are deep enough to create a high-pressure environment isolated from the oceans. Lake Baikal arose over 20 million years ago. It is 630 km long, its greatest width extends to 80 km, and its maximum depth of 1600 m (16 MPa) means it is the deepest of any lake on Earth. Although it is situated in Siberia it is not particularly remote and a research institute has been created to investigate its remarkable endemic flora and fauna. Some of the high-pressure work carried out there will be described in Chapter 5. Other deep lakes, such as Lake Tanganyika, which is 1470 m deep and the brackish water Caspian Sea, 1000 m deep, present other isolated high-pressure environments well worth investigating.

In the 1970s airborne radio-echo soundings detected a large lake beneath the Antarctic Ice cap which has proved to be a remarkable, high-pressure biological environment. The lake was named Lake Vostok, being in the Russian sector. It is 250 km long, 50 km wide and more than 1200 m deep. The lake water is liquid because of the upward flux of geothermal heat. The hydrostatic pressure has been estimated to reach about 40 MPa due to the 4000 m depth of ice sheet covering it. The lake originated over 35 million years ago in a rift valley before the ice arrived. It has been isolated by ice for 14 million years as a dark, deep, cold water mass, similar to the deep sea. The questions of whether it sustains or merely preserves life and the source of that life, are proving difficult to investigate (Siegert et al. 2001, 2003; Shtarkman et al. 2013; Chapter 8). The ice cap covering the lake moves across the width of the lake at 3–4 m per year. Contact between ice and liquid water produces freshly frozen ice called accretionary ice, which accumulates to a thickness of 230 m between the water and the ice cap, downstream of the glacial flow. The only way to investigate this distinct layer of ice and the water below is by drilling through the ice cap, which is a controversial issue. Drilling risks contaminating the pristine environment, especially as, so far, a traditional "dirty" technique has been used and not a cleaner hot water drilling technique. Drilling conditions on the ice cap are harsh and can only proceed in summer, and at great expense. Great care has been taken to control contamination and the deep borehole which has been drilled stopped in the accretionary layer and did not enter the water. Ices core thus obtained have been analysed. The upper layer of accretionary ice contains salts, organic carbon and, from traces of RNA (ribonucleic acid, a marker chemical of life). Gene sequences were identified, indicating the presence of several thousand different bacteria species. Microbial cell counts have been carried out and the results are compared with similar counts in other subglacial ice and subglacial waters. The lower layer of accretionary ice held much less biomass. Much of the biological material in the upper layer was presumably derived from the ice cap and had been trapped in ice for a million or so years, but some were likely to have come from the lake water. One identified organism, Hydrogenophilus thermoluteolus, is a bacterium able to use hydrogen as an energy source, live in quite hot water, and is known to be present in faulted bedrock at the bottom of the lake (Karl et al. 1999; Priscu et al. 1999). A lot of work remains to be carried out before we will have a clear understanding of exactly what is going on in and below Lake Vostok, and the role of high pressure there, 40 MPa. It has attracted interest in the Astrobiological community who see similarities between Lake Vostok and deep water on a moon of Jupiter, Europa (Carr et al. 1998; Bulat et al. 2011; Chyba and Phillips 2001; Muir 2002). There, ice covers water a hundred km deep, providing a significant pressure in which life may have evolved.

In 1977 our ideas about the oceans and their inhabitants were revolutionised by the discovery, made by scientists on board the American submersible Alvin, of hydrothermal vents in the deep ocean floor. The geothermally heated water which flows from the vents supports a special flora and fauna which will be described in Chapter 9. Understanding how such organisms survive and thrive has revolutionised bio- and geochemistry (Suttle 2007; Anderson et al. 2011, 2013; Forterre 2013; Koonin and Dolja 2014). A few years later the less spectacular Cold seeps were discovered (Fig. 1.3); some in shallow water but others as deep as 3500 m (Bradley 2009).

Somewhat earlier, In the 1950s, a new facet of deep marine life became recognised, the presence of viruses. As viruses only come to life when they enter a living cell and hijack its metabolism, they can be regarded as non-living. However, they are the most abundant and genetically diverse life forms in the ocean. Typical ocean water contains ten million virus particles per mL (millilitre) and even deep seawater contains a million or so. The deep sediments contain comparable numbers. Viruses with other "Mobile Genetic Elements" infect all grades of the organism, from bacteria to marine mammals and kill many. They play a role in geochemical cycles and their gene pool interacts with that of marine organisms. Their susceptibility to high pressure in seawater has yet to be investigated but well-known "laboratory" viruses have been the subject of some excellent high-pressure studies.

## 1.3  The Deep Biosphere

The term Deep Biosphere used to refer to depths greater than 1000 m but now it refers to the earth's crust, both continental and oceanic. Water in the Earth's crust, in pores and especially in fractures, contains dissolved salts and gases partly derived from the surrounding rock although some of the water may have originated from meteors or from ancient oceans. The main features of this crustal environment can be

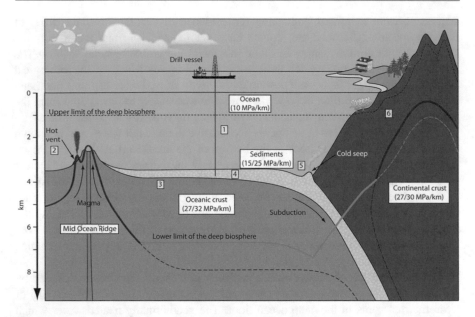

**Fig. 1.3** A diagrammatic cross-section of the Deep Biosphere in and beneath the oceanic and continental crust. The depth of the ocean is shown as close to the average depth of the world oceans. The Deep Biosphere is estimated to achieve a similar depth in the earth's crust (blue line). The horizontal dashed red line in the ocean, labelled upper limit, reflects past terminology. The red lines, dashed and continuous, indicate the isotherm (approximately 122 °C) which might limit the distribution of life. The blue line indicates the 150 MPa isobar, which might also limit life. Whatever the uncertainties, the Deep Biosphere is a global environment. Note the pressure-depth relationships, in boxes, and the specific zones; *1*, deep sea; *2*, hydrothermal vents; *3*, oceanic crust; *4*, deep sediments; *5*, cold seep; *6*, continental crust. Reproduced with permission from Oger and Jebbar (2010)

readily summarised, but the chemical aspects are very complex (Van der Pluijm et al. 2004).

The hydrostatic gradient describing the pressure in the ocean water column extends into sediments and crustal rock when the porosity is sufficient. Porosity is the percentage of a rock that is void space, and which can be filled with water, oil or gas. Coarse gravel has a porosity of 24–36% and shale and dense crystalline rock have a lower porosity of 0–10%. Surprisingly, basalt's porosity can be as high as 35%. If there are microorganisms living in porous rocks beneath the Mariana Trench, they will be at a higher hydrostatic pressure than those in the Trench sediments. There is also a lithostatic pressure gradient, which can be calculated using the simplified equation mentioned earlier, in which the density term is much higher than that of water. Taking the density of sedimentary rock as 2.7 g per cubic cm instead of the 1.0 for water, we can see that the pressure generated by a column of rock is proportionately greater. The pressure under a 10,000 m depth of such rock is 270 MPa. Rocks with a density of 20% greater exist, producing an even steeper pressure gradient. The water in the rock has to be unconnected to the surface for the lithostatic gradient to apply to microorganisms living there, and in many circumstances, it is unclear if that is the case. Generally, we can assume that in the Deep Biosphere the hydrostatic pressure at a given depth will be equal to or greater than the equivalent water depth.

In complete contrast to the oceans, the temperature in the crust increases steadily with depth. The gradient varies according to geological conditions, from 5 °C per km depth to 50 °C per km. The heat comes from deeper down and is generated by radioactivity and compression. The Deep Biosphere is a very warm place and in certain localities, for example the Nankai Trough in the Pacific, the temperature at a depth of 1000 m below the seafloor is in excess of 100 °C. It has been suggested that life exists in any region of the earth's crust where there is sufficient water and the temperature is below a critical limit. Exactly what that limit has yet to be resolved, but as life exists at 120 °C at high pressure in hydrothermal vent fluids, it is probably around that figure and unlikely to be as high as, say, 150 °C. (Note that hydrostatic pressure raises the boiling point of water, keeping it liquid at 120 °C and it may also influence the thermal limit of life by affecting protein structure). In the Deep Biosphere pressure is unlikely to limit the depth to which life extends as high temperature intervenes.

Another physical feature of the Deep Biosphere is its size. The crustal environment is well defined but the biomass therein is difficult to quantify. It may be divided into two categories; (a) pelagic, meaning in suspension in water which can be sampled and (b) attached to the surface or to the interior of rock, a biofilm, which can be examined microscopically if samples can be extracted. A feature that is so obvious it can easily be overlooked is that the Deep Biosphere is essentially static, although it moves on a geological time scale and water passes through it on a somewhat shorter time scale. Samples from deep sediments, which accumulate on a geological time scale, or from rock can be from indigenous cultures that have been isolated for millions of years, a claim first made by Lipman in 1931. Today 250 million years seems to be the current record age. Then there is the issue of metabolic turnover and how that compares with, say, that of the deep sea. The Deep Biosphere plays a unique role in both the biological and geochemical cycling of elements and compounds. According to Gold, hydrocarbons are intrinsically present in the depths and not the product of photosynthetically produced organic material. The Deep Biosphere, or regions of it, may however acquire organic compounds produced by photosynthesis which are carried by percolating water. Other regions are self-supporting, where microorganisms are sustained by non-biological (abiotic) sources of energy, such as hydrogen and methane, and are thus independent of sun driven photosynthesis, (Trevors 2002). These chemolithoautotrophic organisms exist in syntrophic communities of cells. That is, they feed on metabolic products such as acetate, formate and hydrogen such that these substances do not accumulate to a toxic level. The infected crust of the Earth is complicated.

Estimates of the size of the biomass in the Deep Biosphere have proved controversial but for present purposes, it is sufficient to recognise it as global in scale (Chapter 8). More important in the present context is its adaption to high pressure as well as other distinctive factors. Much of it may also reproduce and metabolise at an exceptionally low rate. This has been calculated by independent methods (Hoehler and Jorgensen 2013) producing mean generation times of 1000 years or more and the concept of a "basal power requirement" to sustain cells in a viable state, of which more in Chapter 8. The origin, evolution, and energy sources of life in the Deep Biosphere are of considerable theoretical interest and the environment is also of economic importance. For example it contains oil wells, which provided the first scientific clue, published in 1926, that bacteria could survive in alien crustal fluids. More recently oil wells have reminded us of their size when the drilling rig Deepwater Horizon exploded and sank in the Gulf of Mexico in 2010. The ruptured riser pipe released oil and gas through the damaged well head which was on the seafloor, at a depth of 1544 m, for 83 days. The release of the vast quantity of oil and gas provided an unplanned experimental perturbation of the region's ecology. The Deep Biosphere also includes regions where radioactive waste is stored in containers that are susceptible to microbially driven corrosion. It may become a store for significant amount of carbon dioxide, extracted from the atmosphere to minimise climate change and last but not least important, it is one of our main sources of water.

The Deep Biosphere consists of two main sectors.

## 1.3.1   Oceanic Crust

The oceanic crust is characterised by its slow and steady eruption along mid ocean ridges, from which the newly formed crust spreads laterally, expanding the ocean floor. It is about 6–10 km thick. The upper few 100 m of newly formed crust is much fractured basalt, a rock rich in iron-bearing silicates. On average the Earth produces about 3 $km^2$ of crust each year. Away from the ridges the flanks are up to 20 million years old and have accumulated sediment, perhaps as much as 100 m thick. The older crust can be up to 70 million years old and like the flanks will probably consist of crystalline rock down to a depth of 5–6 km. At its margins, the crust descends in the subduction zones creating deep-sea trenches and volcanic activity.

The ocean water penetrates the crust, driven as slow convection currents through the porous, fractured rock, dissolving salts as it goes. The "residence" time of ocean water passing through is up to 10,000 years and the amount of dissolved material it delivers to the bulk of the oceans is considerable. For example half of the iron present in seawater comes from the oceanic crust in this way, and the other half comes from runoff from the land. Iron is an essential trace element required for the photosynthesis carried out in the surface waters. The circulation of water suggests that microorganisms could be introduced to, or removed from, the Deep Biosphere, but some crustal regions are known to be geologically isolated from the surface water for millions of years, and yet sustain microorganisms. Presumably such regions experience the full lithostatic pressure gradient.

The Juan de Fuca Ridge provides a good example of the Deep Biosphere in oceanic crust. It lies on the flank of the Mid Ocean Ridge in the eastern Pacific and has been investigated by deep drilling techniques, part of the Integrated Ocean Drilling Project (IODP) and Deep Sea Drilling Programs. Special research ships were used, capable of drilling into the sea bed in a seawater depth of several km. Two boreholes, coded as IODP301 and 1026, produced much interesting data, over a period of years. They were drilled through the sediment and just into the underlying basalt crust. The age of the crust was estimated at 3.5 million years.

The boreholes were lined with steel casing to ensure that samples taken from the top came from the crustal water and not from the sediment. The top of each borehole was capped with a special device that sealed the hole, preventing crustal water from flowing into the sea. Crustal water could also be intermittently withdrawn by means of a valve which could be opened to fill a container. This required a submersible and the samples were subsequently analysed on the mother ship or in a land laboratory.

The crustal water, heated to 65 °C, returning to the ocean was sampled from the top of the borehole. It was enriched with ammonium, silicate, calcium and strontium. It contained less sulphate and magnesium than bottom water. The changes in sulphate and ammonium in particular are good evidence for the existence of a microbial population actively metabolising in the crust. More direct information about that population was obtained from filtered samples from the top of the borehole. The number of living cells, Bacteria and Archaea, was counted by various established techniques, producing counts similar to those found in bottom water. Microscopic examination revealed a wide variety of bacterial forms; spheres, rods, crescents and chains. Extracting nucleic acid from the cells established the identity of some of the cells. For example one was a bacterium closely related to *Ammonifex degensii* which is known to convert (reduce) nitrate to ammonia. The reaction releases energy for subsequent use by the cell and is an example of how this type of bacterium is independent of photosynthesis, deriving its energy from inorganic sources. As previously mentioned chemolithoautotrophic is the formal term for this and the energy source is referred to as "dark energy," not to be confused with the cosmological term.

Bacterial-like cells known as Archaea (to be explained in the last part of this chapter), were also detected. They are responsible for reducing the sulphate concentration of the crustal water by combining the sulphate with hydrogen to produce hydrogen sulphide and water. How high pressure affects these reactions seems to be unknown.

The hydrogen involved in these reactions can be produced inorganically, for example by "serpentinisation." This is a reaction in which water and a mineral such as olivine react to produce the minerals serpentine, magnetite and brucite plus hydrogen. The production of hydrogen by this type of reaction in the crust, and in the deeper mantle, is well established. It provides an energy source for indigenous microorganisms, as above, and it may have been an important energy source in the origin of life.

In more recent work on crustal water from the Juan de Fuca Ridge, nucleic acids were isolated from which genes responsible for the synthesis of certain parts of enzymes were identified. One was part of the enzyme essential in the conversion (oxidation) of methane and another was part of the enzyme involved in converting (reducing) sulphate. These are examples of using "biomarkers," to detect organisms living in these inaccessible high-pressure (about 30 MPa) regions.

Another example of life in the Oceanic crust, beneath the Atlantic Ocean, is at a site north of the Lost City hydrothermal vent (Chapter 9), The IODP Hole 1309D penetrated deep into the gabbro of the crust and detected RNA and bacteria, one species of which was closely related to *Ralstonia pickettii*, known to grow on hydrocarbons such as benzene. The sample depth was 1313 m below the seafloor in a similar depth of water, so the pressure was approximately 26 MPa. The temperature was 79 °C. How such conditions affect the particular organisms present is unknown, but similar examples have been studied and will be described in future chapters.

The hydrostatic pressure of the crustal fluids is often given little emphasis or is not mentioned at all in the published reports. The exploratory phase will eventually give rise to more analytical work, but specialist microbiological skills are needed to study the effects of pressure on these remarkable organisms.

## 1.3.2   Continental Crust

The continental crust is thicker than the Oceanic crust, averaging 35–40 km. It is more heterogeneous but smaller in area than oceanic crust. Geologists say it is less mafic, meaning it has less manganese and iron in its minerals, and therefore less favourable for serpentinisation. It is also rather more convenient to explore than oceanic crust.

The following world-wide selection of sites gives us a sense of the scale of the high-pressure environments which the continental crust provides. Estimates of its biomass have been a source of discussion for years and a 2014 paper concluded that it amounts to 2–19% of the Earth's biomass (McMahon and Parnell 2014).

In Sweden, borehole samples from a granite aquifer (the Fennoscandian Shield) have produced autotrophic bacteria from depths of several 1000 m at temperatures of 60–70 °C. This study site was one of several which led to the concept of hydrogen being an energy source for microbial populations, independent of surface energy sources, sunlight and oxygen. Some, perhaps a lot, of the hydrogen is produced by the radiolysis of water, a process that harnesses the energy released from the radioactive decay of certain elements. It was found that certain Bacteria and Archaea produce methane, water and energy for the cell, by the following reaction. $4H_2 + CO_2 = CH_4 + 2H_2O$. They are called methanogens. Similarly, acetogens, typically bacteria, produce acetic acid, water and energy by $4H_2 + 2CO_2 = CH_3COOH + 2H_2O$. The "fixed" carbon in acetic acid and methane can be subsequently converted into more complicated molecules. Methane, for example can be polymerised to produce long chain hydrocarbons.

In China, in the Dabie-Sulu mountain building region, a borehole yielded plenty of evidence of bacterial life at depths of 530–2030 m, where the temperature was 38 °C and 95 °C, respectively. The borehole was in hard, coarse-grained metamorphic rock (largely amphibolite), which had been depressed to a depth of 100 km but raised to the surface 220 million years ago. The origin of the water in the rock, which had a porosity of from 1% to 3.5%, may have been an ancient ocean. The source of the organisms living there is also unknown, but the nature of the rock would provide an effective barrier to microbial invasion, suggesting that they too may have been derived from the ancient ocean. Nucleic acids recovered from the fluids indicated the dominant bacteria belonged to a major group called Proteobacteria. Methane was probably the major source of energy for many. Some grew in the culture at 37 °C at normal atmospheric pressure, demonstrating the presence of living populations in the crust. The hydrostatic pressure was not mentioned but we can assume the deepest sample came from at least 20 MPa.

The Great Artesian Basin of Australia, sampled by a borehole to 2000 m depth, has temperatures of 40–95 °C, and produced bacteria that were metabolically active mobile rods, closely related to *Bacillus infernus* and *B. firmus*. As the region is an aquifer we can assume they were living at a minimal hydrostatic pressure of 20 MPa.

Drilling boreholes in deep mines has obvious advantages and in the Witwatersrand Basin in South Africa, deep gold mines have been used in this way. Starting at a depth of 2716 m a borehole was drilled in the Tona Tau mine to provide water samples from depths of 125–648 m below the start depth (i.e., a total depth of 3354 m). The water temperatures were moderate, 40–50 °C, and its original source was probably a meteor. The nucleic acids recovered indicated that both Bacteria and Archaea were present, and among the latter there were some unique Archaea. They probably used hydrogen as an energy source, at an assumed hydrostatic pressure of slightly more than 33 MPa. The Driefontein Consolidated Mine nearby has provided access to 3200 m depths where many aspects of scientific drilling have been studied, including the prevention of contamination and enrichment, the process of providing nutrients to boost the growth of indigenous populations of microorganisms.

The Taylorsville Basin in Virginia, the USA, has natural gas-bearing rock which was one of the first major deep microbiological sites to be investigated. The deepest samples came from a depth of 2800 m and a temperature of 70–80 °C. The lithostatic pressure gradient contributed to the hydrostatic pressure in a complicated way, but the pressure at the sample depth of 2800 m was 32 MPa. Some of the sampled bacteria, probably a very small proportion, grew in culture. One, *Bacillus infernus*, grew anaerobically at 61 °C, at normal atmospheric pressure and its isolation is described in Chapter 8.

So far only single celled organisms, bacteria and archaea, have been mentioned, but the Earth's crust is also inhabited by multicellular organisms. Although rare, fungi have been isolated from deep oceanic crust, for example an Exophiala-related form from below the Mid Atlantic Ridge. Worms of a semi microscopic size (0.5 mm long) have been collected from deep borehole water in the South African mines just mentioned (Borgonie et al. 2011). They came from depths of 900–3600 m, 24–48 °C respectively. The worms belong to the phylum Nematoda, otherwise known as roundworms. They are far removed from familiar earthworms or lugworms used as bait or wireworms, which are Arthropods. They are, however, complicated multicellular animals, with nerve and muscle cells and celebrated in Zoological circles for being able to live everywhere. So, the discovery of the Nematode *Halicephalobus mephisto* in the Deep Biosphere about 10 years ago was both interesting and credible. It apparently grazes on bacterial films adhering to rock. 48 °C is a record high temperature for a Nematode. As a phylum Nematodes are the most common animal in deep-sea sediments, which of course are cold, and their abundance extends to trench sediments where the pressures are much greater than the 36 MPa reported in the S. African mine boreholes.

Viruses, it is said, exist wherever there are living organisms, and the Deep Biosphere is no exception, where both Bacteria and Archaea are hosts to viruses. For example viruses are present in the Deep Biosphere groundwater 450 m deep in the Swedish granites mentioned earlier. Active viruses have been isolated from bacteria collected from the eastern Pacific and from the Peru Margin. The deepest sample was from a sediment depth of 268 m below the seafloor, which in turn was 5086 m below sea level, and thus at a pressure of more than 50 MPa. The report included pictures of viruses inside and attached to the outside of bacteria. Viruses play a significant role in the turnover of biological material, as they lyse (rupture) their host cell. They also contribute to the genetics of the populations in the Deep Biosphere, just as they do in other more familiar environments. Their pressure tolerance must be comparable to that of deep-sea viruses, which awaits investigation.

The Deep Biosphere is still a relatively recently discovered environment, but it has attracted particular interest because of the autotrophic nature of some of its inhabitants. Their biochemistry is seen as similar to that which probably existed in the early stages of the evolution of life on Earth, and perhaps on other planets. The transition of geochemistry to biochemistry is beyond the scope of this book, but it is worth noting that papers have been written arguing that life could have evolved in the Deep Biosphere implying, among other things, at high pressure. The arguments have two distinct implications. One is that life could have evolved early in the development of the Earth and if present in the depths would have influenced the evolution of geochemistry. This is a view very different from the traditional one of the developed earth providing Darwin's "warm pond" or the primordial soup of others, in which life originated. The second implication is extraterrestrial. The search for life on other planets should not overlook deep, high-pressure environments, such as occur on Europa, mentioned earlier (and Chapter 9; Box 9.1).

Both the Deep Biosphere and the Deep Sea are open to commercial exploitation, and in particular to mining. The International Seabed Authority, set up by the United Nations, is the body responsible for managing such activity beyond the limits of national jurisdiction (Lodge et al. 2014; Petersen et al. 2016; Van Dover 2014).

## 1.4   High-Pressure Joints

When, in the course of animal evolution, vertebrates colonised the land, they lost the buoyancy of water and evolved strong bones and load-bearing joints. These structures cope with a variety of mechanical stresses which can, in certain localised sites within joints, generate a significant, pure, hydrostatic pressure.

For example the complex microstructure of bone contains fluid-filled cavities in which hydrostatic pressure is generated. The layer of cartilage cells covering the

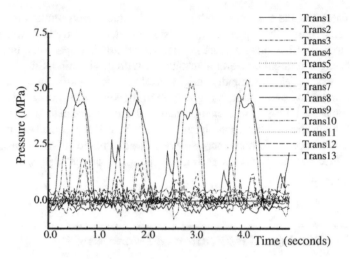

**Fig. 1.4** The hydrostatic pressures experienced at the surface of the head of a femur in a freely walking man. Each line is the record from a specific transmitter in the head and clearly two such sites report the largest pulses of pressure (Details in Fig. 13.8). Reproduced with permission from Morrell et al. (2005)

load-bearing surfaces of bones experiences cyclically fluctuating hydrostatic pressure during walking and running. It was only when a cleverly instrumented artificial hip joint was devised that the magnitude and time course of the pressures were actually measured in a moving human (Fig. 1.4).

The details will be described in Chapter 3 and here it is sufficient to note that cyclic hydrostatic pressures in the range of 2–20 MPa (5 MPa in Fig. 1.4), are experienced by load-bearing cartilage in humans, and doubtless other vertebrates. Remarkably, they are comparable to the pressure at depths of 200–2000 m in the ocean.

In the joint, the hydrostatically pressurised layer of cartilage is thought to be lubricated by a film of liquid moving over its surface, so-called "weeping" lubrication. The fact that the hydrostatic pressure is cyclic is important and interesting, but its duration is long on a molecular time scale and does not affect the way it affects chemical reactions. Both bone and cartilage respond to the mechanical stress they experience and within certain limits, the response is one that improves the performance of the structure. In the knee and hip joints of humans, for example the layer of cartilage experiencing cyclic loading becomes thicker with increased mechanical stress and thus performs better as a shock absorber and self-lubricating surface. Shear stress is certainly one of the factors involved, but hydrostatic pressure is important and by using isolated cartilage its specific effects can be investigated (Rushfeldt et al. 1981; Hodge et al. 1986; Van Rossom et al. 2017). In evolutionary terms, the adaptation to hydrostatic pressure in joints should be seen as part of the adaptation of vertebrates to life without buoyancy and to the force of gravity. The health of human cartilage is obviously medically very significant and has prompted much research but the extent of the high-pressure micro-environment in vertebrate joints has not been fully explored. The tension created by muscle power might be sufficient to pressurise joints that are not carrying a gravitational load. For example fast swimming fish might be worth investigating with this in mind. Small insects that accelerate and decelerate rapidly might well create significant point loading pressures, and what, one wonders, is going on in the joints of stampeding elephants?

To summarise high-pressure environments; the deep ocean is a vast and cold, with local hydrothermal hot spots. In the equally vast Deep Biosphere hydrostatic pressure and temperature both increase with depth. In the micro-environment of vertebrate load-bearing joints lesser, but significant, hydrostatic pressures fluctuate at body temperature.

## 1.5    The Modern Classification of Organisms

In the title of this book the word Life means all types of life and for many readers that will mean animals, plants, bacteria, perhaps some other single celled organisms and perhaps viruses. Here we have to be clear about the categories of organisms we are dealing with because a term such as bacteria, which is often used in a non-technical sense (usually correctly) has acquired a precise meaning due to major developments in our understanding of evolution.

We classify organisms in a way that reflects their evolutionary relationships, a phylogenetic classification. Historically these relationships were worked out on the basis of structure, and that worked pretty well for a time. It enabled Darwin to produce his Theory of Evolution. However about 50 years ago the understanding of the genetic code set in motion a new way of assessing relationships between organisms, led by Carl Woese (see Box 1.2).

**Box 1.2**
The essential feature of the new classification is that it is based on the structure of a special nucleic acid, found in all cells. It is generally known that DNA (deoxyribonucleic acid) is the "blueprint for life," because the linear sequence of its four constituent subunits (called bases) codes for the manufacture of proteins in the cell. That process uses copies of the DNA sequence in the form of ribonucleic acid molecules, and in particular, the form known as ribosomal RNA, rRNA. Woese 's insight was the rRNA conserves the primary DNA sequence in a convenient form and that differences between the rRNAs of different organisms could reflect their evolutionary separation. His initial studies utilised fragments of 16 s rRNA, made of about 1500 bases. He found organisms thought to be closely related by traditional methods had, in fact, very similar rRNAs. When he and his colleagues studied methane-producing microorganisms they found novel sequences, which were also

(continued)

**Box 1.2** (continued)

present in other microorganisms from extreme environments. Thus, the Archaea were discovered and established as one of the three domains of life; Archaea, Bacteria and Eukarya. This was a highly controversial result. However, the new classification, described below, is withstanding the test of time and is now, along with Natural selection, fundamental to modern biology.

The modern phylogenetic classification shows that all organisms belong in one of three domains; Archaea, Bacteria and Eukarya. The Bacteria are somewhat apart from the Archaea and the Eukaryotes, but all three Domains originated early in the evolution of life.

Do familiar bacteria belong in the Bacteria Domain? The answer is they do. What is special about Archaea? (The singular is archaeon). Both Archaea and Bacteria are single celled organisms, prokaryotes, which lack a nucleus, the structure which, in Eukaryote cells, contains the cell's DNA. Superficially Bacteria and Archaea resemble each other. They live in the same environments, in much the same way, so how is this coexistence sustained and how is it they are significantly different? It seems very odd.

Archaea differ from Bacteria in numerous basic features. Their enclosing plasma membranes are made of distinctive lipids, quite unlike those in Bacterial membranes, although the membranes are superficially structurally similar. The core phospholipid in both has a "backbone" known chemically as a glycerophosphate, but the structure of that in Bacterial phospholipids is the mirror image of the glycerophosphate in Archaeal phospholipids. Such mirror image structures are known as enantiomers. The enzymes which synthesise them are radically different, implying the evolutionary divergence is ancient and occurred close to the origin of cells. Wachterhauser has argued that Bacteria diverged from a universal common ancestral "precell" and Archaea diverged subsequently. Other experts hold different views about this key evolutionary issue called the Lipid Divide (Koga 2011), which will be further explained in Chapter 7. Other significant differences between Archaea and Bacteria include the outer cell wall, the S layer, analogous to the outer layer of Gram Positive Bacteria, which is made of quite different compounds. They also differ in their RNAs and in the way they respond to antibiotics. It is obvious that, in evolutionary terms, they are very different, yet they arose soon after the origin of life.

Bacteria have evolved about 15 times more genera than Archaea and in a general way, the experts regard Archaea as strong at coping with a shortage of energy from nutrients and the Bacteria better at diversifying and exploiting opportunities. The Archaea are notable for their ability to colonise extreme environments. There are six main groups. Halophiles are tolerant of extremely concentrated salt solutions. Thermophiles are able to grow at temperatures above 80 °C and up to 122 °C in one case, which of course need high pressure to keep the water liquid. Acidophiles are able to grow in strong acids and Nitrifiers can oxidise ammonia. Methanogens we have met. They do not need oxygen (anaerobic) and generate methane. Finally, some

Archaea are anaerobic methane oxidisers. Like Bacteria the Archaea occupy high-pressure environments.

The single cell design of Bacteria and Archaea and their small size means their internal biochemistry proceeds without the complexity of the internal membranes and the nuclear compartment seen in the larger Eukaryote cells. It is thought that the complexity and large size of the Eukaryote cell (a thousand times the size of a typical prokaryote) evolved by an archaeon-like cell engulfing a bacterium, to create an effective symbiosis. Intracellular structures called mitochondria, it is thought, arose from symbiotic bacteria. They are described in Chapter 2. The Eukaryotes are thought to be the last Domain to evolve and now comprise five main groups, with strange names, especially for those who, like the author, learned their biology before the phylogenomic revolution.

Although their names will not appear much in the rest of the book, the Eukaryote groups are listed as follows. The Plantae comprise familiar land plants and green and red algae. Their chloroplasts which contain chlorophyll, evolved like mitochondria, from symbiotic bacteria. The Rhizarians are well established by molecular data but have no characteristic form. The microscopic Foraminifera, formerly regarded as Protozoa, now a defunct term, belong here. The Chromalveolates include diatoms, dinoflagellates and ciliates, all single celled and previously lumped as Protozoa. The Excavates are described as a diverse group in which the Trypanosomes (also formerly Protozoa) are perhaps the most familiar, being pathogenic parasites. The fifth group is the Unikonts which include animals, fungi and some amoebae. Like the author, many readers will be surprised to learn they are Unikonts!

The familiar term species needs special comment. It is the only taxonomic unit that can be defined in phylogenetic terms, according to Wayne and colleagues (Wayne et al. 1987). In animal biology, there is a rule of thumb that one species cannot breed with another species, or if they can it is rare and can lead to sterile offspring. It gives some weight to the word species, which we cannot avoid when we need to be accurate. But what about bacteria and archaea? According to Wayne et al, a bacterial species, and presumably an archaeal species, includes a population of cells with approximately 70% DNA–DNA relatedness and other features in common. Thus the term species, in the context of Bacteria and Archaea, implies a lot of DNA in common, giving real meaning to the word.

Finally, although the new classification of Life is soundly based, it is incomplete (Keeling et al. 2006; Williams et al. 2013). Some special cases await their final placement. One such organism, formerly the Helizoan Actinosphaerium, is mentioned in Chapter 2 as it played a significant role in our understanding the effects of high pressure on cells. But most of the organisms we shall encounter will be familiar enough, and quite unmoved by the revolution.

We also have to consider a classification of "non-living entities" which lie outside the above scheme. They exert considerable influence on life in the marine environment and its deep regions, hot and cold, and in the Deep Biosphere. "Mobile genetic elements" (MGE), is an unappealing term that includes viruses and a host of simpler multimolecular structures such as virus-like particles, plasmids, vesicles (in this context the word has a specific meaning), transposons and others. They are found in all environments on Earth, most certainly in the marine environment, in prodigious numbers, conveniently summarised by an expert as "everything everywhere." MGEs lack the RNA gene and are thus excluded from Woese's classification.

The collective term "mobilome" is used for the mobile gene pool generated by MGEs, analogous to the genome for an individual organism. A virus is the most familiar example of an MGE. As mentioned earlier a virus is deemed to be non-living because it only comes to life when it infects a living cell, takes over its metabolism and replicates. It is not self-sufficient. A virus is essentially a protein shell, which encloses DNA, or RNA. Sometimes a virus can incorporate a host cell gene and may then pass it on when it infects another cell. Thus, genes can move from one cell to another, an important process called horizontal gene transfer, which drives significant evolution. As viruses infect Eukaryotes, Bacteria and Archaea this happens across the whole spectrum of life.

## References[1]

Adler A, Ducke E (2018) When Pasteurian science went to sea: the birth of marine microbiology. J Hist Biol 51:107–133

Akasaka K, Matsuki H (2015) High pressure bioscience. Springer, New York

Akimoto K, Hattori M, Uematsu K, Kato C (2001) The deepest living foraminifera, challenger deep, mariana Trench. Mar Micropaleontol 42:95–97

Anderson RE, Brazelton W, Baross JA (2011) Is the genetic landscape of the deep subsurface biosphere affected by viruses? Front Microbiol 219:1–15

Anderson RE, Brazelton W, Baross JA (2013) The deep viriosphere: assessing the viral impact on microbial community dynamics in the deep sub surface. Rev Mineral Geochem 75:649–675

Ballard RD (2000) The eternal darkness. Princeton University Press, Princeton, NJ

Bar-On YM, Phillips R, Milo R (2018) The biomass distribution on Earth. Proc. Nat. Acad. Sci USA 115:6506–6511

Beebe W (1935) Half mile down. John Lane. The Bodley Head, London

Bomberg M, Ahonen L (2017) Geomicrobes: life in terrestrial deep subsurface. Front Microbiol. 8:103. *

Borgonie G et al (2011) Nematoda from the terrestrial deep subsurface of South Africa. Nature 474:79–82

Boyle P (2009) Life in the mid atlantic. Bergen Museum Press, Bergen

Bradley AS (2009) Expanding the limits of life. Sci Am 301:62–67

Bulat SA, Alekhina IA, Mariem D, Martins J, Petit JR (2011) Searching for life in extreme environments relevant to Jovian's Europa: Lessons from subglacial ice studies at Lake Vostok (East Antarctica). Adv Space Res 48:697–701

Byatt A, Fothergill A, Holmes M (2001) The blue planet. BBC, UK

Carr MH et al (1998) Evidence for a subsurface ocean on Europa. Nature 391:363–365

Certes A (1884a) Sur la culture a l'abri des germes atmospheriques, des eaux et des sediments rapports par les expeditions du Travailleur et du Talisman: 1882—1883. Comptes Rendus des Hebdomadaires. Sean Acad Sci 98:690–693

Certes A (1884b) Note relative a l'action des hautes pressions sur la vitalite des micro-organisms d'eau douce at d'eau de mar. C R Soc Biol 36:220–222

---

[1]Those marked * are further reading.

Certes A, Cochin D (1884) Actions des hautes pressions sur la vitalite de la levure et sur les phenomens de la fermentation. C R Soc Biologie 36:639–640

Chyba CF, Phillips CB (2001) Possible ecosystems and the search for life on Europa. Proc Natl Acad Sci U S A 98:801–804

Colwell FS, D'Hondt S (2013) Nature and extent of the deep biosphere. Rev Mineral Geochem 75:547–574. *

Devine JA, Baker KD, Haedrich RL (2006) Deep sea fishes qualify as endangered. Nature 439:29

Edwards KJ, Fisher AT, Wheat CG (2012) The deep subsurface biosphere in igneous ocean crust: frontier habitats for microbiological exploration. Front Microbiol 3:8. *

Engelen B et al (2008) Fluids from the oceanic crust support microbial activities within the deep biosphere. Geomicrobiol J 25:56–66. *

Fischer V, Elsner NO, Brenke N, Escwabe E, Brandt A (2015) Plastic pollution of the Kurile-Kamchatka Trench area (NW Pacific). Deep-Sea Res II 111:399–405

Follonier S, Panke S, Zinn M (2012) Pressure to kill or pressure to boost: a review of the various effects and applications of hydrostatic pressure in bacterial biotechnology. Appl Microbiol Biotechnol 93:1805–1815

Forterre P (2013) The virocell concept and environmental microbiology. ISME J 7:233–236

Gardner JV, Armstrong AA, Calder BR, Beaudoin J (2014) So, how deep is the mariana trench? Mar Geod 37:1–13

Gihring TM et al (2006) The distribution of microbial taxa in the subsurface water of the Kalahari Shield, South Africa. Geomicrobiol J 23:415–430. *

Gold T (1992) The deep hot biosphere. Proc Natl Acad Sci U S A 89:6045–6049. *

Gold T (2001) The deep hot biosphere. The myth of fossil fuels. Copernicus Books, Torun. *

Gvirtzman Z, Strern RJ (2004) Bathymetry of Mariana Trench-arc-system and formation of the challenger deep as a consequence of weak plate coupling. Tectonics 23:TC2011

Harden-Davies H (2017) Deep sea genetic resources: new frontiers for science and stewardship in areas beyond national jurisdiction. Deep-Sea Res II 137:504–513

Heezen B (1957) Whales entangled in deep sea cables. Deep-Sea Res 4:105–115

Herring PJ (2001) Biology of the deep ocean. Oxford University Press, Oxford, UK

Hinrich K-U, Inagaki F (2012) Downsizing the deep biosphere. Science 338:204–205. *

Hodge WA et al (1986) Contact pressures in the human hip joint measured in vivo. Proc Natl Acad Sci U S A 83:2879–2883

Hoehler TM, Jorgensen BB (2013) Microbial life under extreme energy limitation. Nat Rev Microbiol 11:83–94

Huber JA (2015) Making methane down deep. Science 349:376–377. *

Idyll CP (1971) Abyss. The deep sea and the creatures that live in it. Thomas C. Crowal Company, New York, NY

Inagaki F et al (2015) Exploring deep microbial life in coal bearing sediments down to 2.5 km below the ocean floor. Science 349:420–424. *

Ivarsson M, Schnurer A, Bengston S, Neubeck A (2016) Anaerobic fungi: a potential source of biological H2 in the oceanic crust. Front Microbiol 7:674. *

Jamieson AJ (2015) The hadal zone. Cambridge University Press, Cambridge, UK

Jamieson AJ (2018) A contemporary perspective on hadal science. Deep-Sea Res II 155:4–10

Jamieson AJ (2020) The five deeps expedition and an update of full ocean depth exploration and explorers. Mar Technol Soc J 54:6–12

Jamieson AJ, Yancy PH (2012) On the validity of the Trieste flatfish: dispelling the myth. Biol Bull 222:171–175. *

Karl DM et al (1999) Microorganisms in the accreted ice of Lake Vostok, Antarctica. Science 286:2144–2147

Kawama I (2014) Live broadcasting of deep sea hydrothermal vents. Seal Technol 55:39–42

Keeling PJ et al (2006) The tree of eukaryotes. Trends Ecol Evol 20:670–676

Kimes NE et al (2013) Metagenomic analysis and metabolite profiling of deep sea sediments from the Gulf of Mexico following the Deepwater Horizon oil spill. Front Microbiol 4:50. *

Koga Y (2011) Early evolution of membrane lipids: how did the lipid divide occur. J Mol Evol 72:274–282

Koga Y, Kyuragi T, Nishihara M, Sone N (1998) Did Archaeal and Bacterial cells arise from noncellular precursors? J Mol Evol 47:54–63. *

Koonin EV (2010) The origin and early evolution of eukaryotes in the light of phylogenomics. Genome Biol 11:209–220. *

Koonin EV, Dolja VV (2014) Virus world as an evolutionary network of viruses and capsidless selfish elements. Microbiol Mol Biol Rev 78:278–303

Koslow T (2000) The silent deep. University of Chicago Press, Chicago, USA. *

Kunzig R (2000) Mapping the deep. The extraordinary story of ocean science. Sort of Books, New York

Lan Y et al (2017) Molecular adaptation in the world's deepest-living animals: Insights from transcriptome sequencing of the hadal amphipod, Hirondellea gigas. Mol Ecol 26:3732–3743

Lau MCY et al (2016) An oligotrophic deep-subsurface community dependent on syntrophy is dominated by sulfur-driven autotrophic denitrifiers. Proc Natl Acad Sci U S A 21:E7927–E7936. *

Levitus S, Aonov JI, Boyer TP, Stephens C (2000) Warming of the world ocean. Science 287:2225–2229

Lipman CB (1931) Living microorganisms in ancient rock. J Bacteriol 22:183–198. *

Lodge M et al (2014) Seabed mining: International Seabed Authority environmental management plan for the Clarion–Clipperton Zone. A partnership approach. Mar Policy 49:66–72

Lutz RA, Falkowski PG (2012) A dive to challenger deep. Science 336:301–302

Macdonald AG, Gilchrist I, Wann KT, Wilcock SE (1980) The tolerance of animals to pressure. In: Gilles R (ed) Animals and environmental fitness. Pergamon Press, Pergamon, pp 395–403

Marshall NB (1979) Developments in deep sea biology. Blandford Press, Blandford, UK

Mason OU et al (2010) First investigation of the microbiology of the deepest layer of ocean crust. PLoS One e15399:5. *

McMahon S, Parnell J (2014) Weighing the deep continental biosphere. FEMS Microbiol Ecol 87:113–120

Meersman F, Daniel I, Bartlett DH, Winter R, Hazael R, McMillan PF (2013) High pressure biochemistry and biophysics. Rev Mineral Geochem 75:607–648

Menzies RJ, George RY, Rowe GT (1973) Abyssal environment and ecology of the world's oceans. Wiley Interscience, New York

Merino N et al (2019) Living at the extremes: extremophiles and the limits of life in a planetary context. Front Microbiol 10:780

Morrell KC, Hodge WA, Krebs DE, Mann RW (2005) Corroboration of in vivo cartilage pressures with implications for synovial joint tribology and osteoarthritis causation. Proc Natl Acad Sci U S A 102:14819–14824

Muir H (2002) Europa has raw ingredients for life. New Scientist 22

Murray J, Hjort J (1912) The depths of the Ocean. Macmillan, London

Nakanishi M, Hashimoto J (2011) A precise bathymetric map of the world's deepest sea floor, challenger deep, in the mariana trench. Mar Geophys Res 32:455–463

Nouvain C (2007) The deep. University of Chicago Press, Chicago

Oger PM, Jebbar MJ (2010) The many ways of coping with pressure. Res Microbiol 161:799–809

Onstott TC et al (1998) Observations pertaining to the origin and ecology of microorganisms recovered from the deep subsurface of Taylorsville Basin, Virginia. Geomicrobiol J 15:353–385. *

Onstott TC et al (2003) Indigenous and contaminant microbes in ultradeep mines. Environ Microbiol 5:1168–1191. *

Orsi WD, Edgcomb VP, Chrsitman GD, Biddle JF (2013) Gene expression in the deep biosphere. Nature 499:205–208. *

Papanelopoulou F (2013) Louis Paul Cailletet: the liquefaction of oxygen and the emergence of low temperature research. Notes Rec R Soc Lond 67:355–373

Parnell J, McMahon S (2016) Physical and chemical controls on habitats for life in the deep subsurface beneath continents and ice. Phil Trans R Soc A 374:20140293. *

Petersen M et al (2016) News from the seabed—geological characteristics and resource potential of deep sea mineral resources. Mar Policy 70:175–187

Piccard J, Dietz RS (1957) Oceanographic observations by the Barhyscaph Trieste (1953–1956). Deep-Sea Res 4:221–229

Priscu JC et al (1999) Geomicrobiology of subglacial ice above Lake Vostok, Antarctica. Science 286:2141–2144

Rex MA et al (2006) Global bathymetric patterns of standing stock and body size in the deep sea benthos. Mar Ecol Prog Ser 317:1–8. *

Roberts C (2012) Ocean of life. Penguin Books, Allen Lane

Rozwadowski HM (2008) Fathoming the ocean. The discovery and exploration of the deep sea. The Belknap Press of Harvard University Press, London, UK

Rushfeldt PD, Mann RW, Harris WH (1981) Improved techniques for measuring in vitro the geometry and pressure distribution in the human acetabulum–II, Instrumented endoprosthesis measurement of articular surface pressure distribution. J Biomech 14:315–323

Schrope M (2014) Journey to the bottom of the sea. Sci Am:60–69

Shtarkman YM et al (2013) Subglacial Lake Vostok (Antarctica) accretion ice contains a diverse set of sequences from aquatic , marine and sediment–inhabiting bacteria and Eukarya. PLoS One 8: e67221

Siegert MJ et al (2001) Physical, chemical and biological processes in Lake Vostok and other Antarctic subglacial lakes. Nature 414:603–609

Siegert MJ, Tranater M, Ellis-Evans JC, Priscu JC, Lyons WB (2003) The hydrochemistry of Lake Vostok and the potential for life in Antarctic subglacial lakes. Hydrol Process 17:795–814

Smith AM (1991) Negative pressure generated by octopus suckers: a study of the tensile strength of water in nature. J Exp Biol 157:257–271

Smith AM, Kier WM, Johnsen S (1993) The effect of depth on the attachment force of limpets. Biol Bull 184:338–341

Stokstad E (2018) About Vescovo and the limiting factor. Science 362:1342–1343

Suttle CA (2007) Marine viruses–major players in the global ecosystem. Nat Rev Microbiol 5:801–812

Swewzyk U, Szewzyk R, Stenstrom TA (2004) Thermophilic, anaerobic bacteria isolated from a deep borehole in granite in Sweden. Proc Natl Acad Sci U S A 91:1810–1813. *

Taira K, Kitagawa S, Yamashiro T, Yanagimoto D (2004) Deep and bottom currents in the challenger deep, Mariana Trench, measured with super deep current meters. J Oceanogr 60:919–926

Takai K, Moser DP, Deflaun M, Onstott TC, Frederickson JK (2001) Archaea diversity in waters from deep South African gold mines. Appl Environ Microbiol 67:5750–5760. *

Trevors JT (2002) The subsurface origin of microbial life on earth. Res Microbiol 153:487–491

Van der Pluijm B, Marshak S, Allmendinger RW (2004) Earth structure. W.W. Norton & Co, New York, NY

Van Dover CL (2014) Impacts of anthropogenic disturbances at deep sea hydrothermal vent ecosystems: a review. Mar Environ Res 102:59–72

Van Rossom S et al (2017) Knee cartilage thickness, Tlp and T2 relaxation times are related to articular cartilage loading in healthy adults. PLoS One 12:e170002

Vescovo V (2020.) The Times, London, 8 and 9 June 2020 *

Wachterhauser G (2003) From pre-cells to Eukarya–a tale of two lipids. Mol Microbiol 47:13–22. *

Wayne LG et al (1987) Report of the ad hoc committee on reconciliation of approaches to bacterial systematics. Int J Syst Bacteriol 37:463–464

Williams TA, Foster PG, Cox CJ, Embley TM (2013) An archaeal origin of eukaryotes supports only two primary domains of life. Nature 504:231–236

Winter R (2013) Biomolecular systems under extreme environmental conditions. Biophys Chem 183:1–2; and following papers

Woese CR (2004a) A new biology for a new century. Microbiol Mol Biol Rev 68:173–186. *
Woese R (2004b) The archaea concept and the world it lives in: a retrospective. Photosynth Res
    80:361–372. *
Woese CR, Kandler O, Wheelis ML (1990) Towards a natural system of organsisms: proposals for
    the domains of Archaea, Bacteria and Eucarya. Proc Natl Acad Sci U S A 87:4576–4579. *
Wolff T (1961) The deepest recorded fishes. Nature 190:283. *
Zhang G, Dong H, Xu Z, Zhao D, Zhang C (2005) Microbial diversity in ultra high pressure rocks
    and fluids from the Chinese Continental Scientific Drilling Project in China. Appl Environ
    Microbiol 71:3213–3227. *
Zimmer C (2008) Microcosm, *E.coli* and the new science of life. Heinemann, London, p 243. *
ZoBell CE, Johnson FH (1949) The influence of hydrostatic pressure on the growth and viability of
    terrestrial and marine bacteria. J Bacteriol 57:179–189. *

# High Pressure: Molecules, Chemical Process and Cellular Structures

**2**

A first stage in understanding how organisms can live in high-pressure environments is to understand how high pressure affects ordinary organisms and to understand that we need to know how pressure affects their cellular processes and the molecules involved. We begin with water.

## 2.1  Water, Dissolved Salts and Proteins

There are good reasons for thinking that water is not a homogeneous liquid, composed of identical units. The structure of the water molecule, two hydrogen atoms combined with a single oxygen atom, forms a *V* shape. A cloud of electrons surrounding the atomic nuclei defines the radius of each atom, the Van der Waals radius. It is 1.4 and 1.2 Å for oxygen and hydrogen, respectively (an Angstrom is $10^{-8}$ cm). Another shape attributed to the water molecule is a tetrahedron, thought to be its shape in ice but possibly one which persists in liquid water. However, these model shapes should be imagined as the average the molecule might occupy as, like all molecules, water vibrates at a very high frequency.

Within the vibrating shape, there is a persistent asymmetry in the distribution of electric charge. Water is weakly polar, a dipole, and it electrostatically attracts negatively charged atoms such as the oxygen of another water molecule. This attraction between the weakly positive hydrogen and the negative oxygen in a separate water molecule is called a hydrogen bond and creates cohesion in water (Jha 2016; Ben-Naim 2009). As the temperature of water increases so the hydrogen bonds are broken. Pure bulk water has its maximum density at 3.98 °C and, at 0 °C, it freezes forming ice, with lots of bonds, in equilibrium with the water. As water forms ice it increases in volume, becoming less dense than liquid water, a process which is opposed by high pressure. This provides us with a good example of Le Chatelier's Principle, which is an important thermodynamic principle which states that a system at equilibrium responds to an imposed change by adopting the lesser energy state. The molecular volume increase in the formation of ice is opposed by high pressure,

© Springer Nature Switzerland AG 2021
A. Macdonald, *Life at High Pressure*, https://doi.org/10.1007/978-3-030-67587-5_2

which causes melting. High pressure opposes molecular processes which involve a volume increase. Skaters unknowingly demonstrate this. The contact between their skates and the ice creates a local region of hydrostatic pressure which melts the ice, creating a lubricating film of water.

At atmospheric pressure and a temperature of $0\,°C$, the freezing point of water, ice and water co-exist, and the liquid water has four nearest neighbour molecules, a small number which implies structure. At a temperature of less than $10\,°C$, pressures in the deep-sea range reduce the viscosity of water (see Box 2.1), which is also inconsistent with a homogeneous liquid. Pure water can be imagined as a mixture of "flickering clusters" of water molecules, with the clusters in equilibrium with smaller groupings and individual molecules, all buzzing randomly about and colliding at a speed determined by the temperature. The density (packing) of these clusters differ. If we now add a sodium ion, positively charged, it becomes closely surrounded by water molecules (it is hydrated). The water molecules nearest the ion are electro-restricted, more densely packed than the bulk water further away, and out of range of the electrical forces. If we introduce an uncharged (apolar, or hydrophobic) mole-cule, such as the hydrocarbon benzene, structural changes also occur in the water. The volume of water the benzene molecule displaces is less than the volume which it occupies in pure benzene. Its displacement volume is called its partial molar volume. The reason is that the water molecules collapse around the benzene molecule, become more tightly packed, so the water–benzene mixture shrinks. Hydrophobic hydration is a strange phenomenon; water can pack more closely around certain hydrophobic groups, causing a volume decrease, but that does not occur with all hydrophobic molecules. In contrast, ionic hydration just mentioned, always leads to the electro-restriction of water and a molar volume decrease. A good example of this is the ionisation of magnesium sulphate in seawater. In surface waters, this salt exists as separate magnesium and sulphate ions in equilibrium with undissociated magne-sium sulphate. In deeper water the high pressure favours an increase in the dissocia-tion to separate ions, because they cause electro-restriction and a decrease in molar volume. Another example is calcium carbonate which, either as crystalline aragonite or calcite, forms the skeleton of various unicellular creatures. Pressure favours the solvation (dissolving) of calcium carbonate as it involves a molecular volume decrease. It therefore tends to oppose skeleton formation. When the creatures die their remains sink and form part of the ocean sediments. A depth of 4000 m, hence a pressure of 40 MPa, is generally sufficient to ensure that calcium carbonate is dissolved and absent from the sediments.

**Box 2.1: Viscosity**
Viscosity is resistance to flow or, in culinary terms, thickness, and it is a property of fluids much affected by hydrostatic pressure. The resistance to flow reflects the friction between the constituent molecules; small round ones create less friction than long angular molecules. Water's viscosity changes

(continued)

**Box 2.1** (continued)

with temperature and pressure in a complicated way because of the variable structural packing its molecules adopt. Viscosity is just one of water's many unusual physical properties which have attracted the interest of distinguished chemists and physicists. One such was Sir Francis Crick who, at an early stage in his career, was involved in the design of an apparatus for measuring the viscosity of water at high temperatures and pressures. This intriguing fact is recorded by his equally distinguished collaborator in establishing the structure of DNA, James Watson, in his book, "Avoid boring people". The viscosity of cytoplasm is important in high-pressure cell biology, because, as we shall see, high pressure reduces it, which is rather unexpected. As with water, the explanation lies in the underlying molecular structures.

Now what about the behaviour of proteins in water, under high pressure? Proteins are very large, being made of a string of amino acid molecules, often a very long string, containing hundreds of them. There are 20 amino acids in nature, and each has its special properties. Some carry an electric charge and bind well with water molecules (i.e., they are hydrophilic), and thus occur on the outer face of the protein. Others are hydrophobic, such as alanine and valine, and their favourable energy state is to be buried in the interior of the protein, well away from water. A dissolved protein interacts with its "atmosphere" of water molecules, which are virtually part of its structure.

The amino acids which make up a protein are connected by covalent bonds, which are very strong. The string of amino acids is the primary structure of the protein (more in Section 2.4.3). It becomes arranged in various ways, for example as a helix, a pleated sheet or a helical coil. This is the secondary structure. In many cases, the whole is then folded in an apparently tangled fashion, the tertiary structure. The folding process is critical in providing the precise shape for the protein to perform properly. Molecular models may appear to show a protein to be a tangled mess but in fact the structure is precisely evolved. Molecular models are static whereas in reality proteins are vibrant, adopting, fleetingly, a variety of microstates, the average of which is frozen in the model. Their complicated higher order structure is held together by weak bonds, such as ionic bonds which arise from the attraction between positive and negative charged groups. Hydrogen bonds also play a part, arising from the weak attraction between polar groups. The weakest bonds of all are Van der Waals interactions, which we can imagine as the attraction between atoms that happen to be close together. Finally, for the sake of completeness, there are hydrophobic interactions, the mutual attraction of groups seeking to minimise their contact with water, or, as Chandler puts it, water is more strongly attracted to itself than it is to hydrophobic groups. Hydrophobic interactions are sometimes incorrectly referred to as hydrophobic bonds. These weak bonds and hydrophobic interactions contribute a volume to the structure of proteins and are thus susceptible to high pressure.

Some proteins, for example those involved in muscle contraction, are very orderly, reflecting the linear motion their function entails. Many proteins are multimeric, i.e., polymers, made up of similar subunits. However, in water they have a partial molar volume, just like small molecules, best imagined as a displacement volume. It consists of three parts; the volume of the amino acids, the volume of any voids in the folded structure which water does not normally penetrate and the "interfacial volume", that is the water layer intimately associated with the surface of the protein. The last and particularly the second of these components of the volume of a protein are affected by environmental pressures. The voids in proteins are thought to render them flexible. A significant volume decrease occurs when high pressure forces water molecules into the structural voids and thus affect the flexible structure of the molecule. High pressure also favours unfolding as the increased exposure of charged molecules to the solvent water causes a volume decrease. Uncharged, hydrophobic regions of the protein also alter their hydration, incurring a volume change which may be an increase or a decrease (Gross and Jaenicke 1994; Meersman et al. 2006; Roche et al. 2012; Akasaka et al. 2013; Meersman and McMillan 2014; Chen and Makhatadze 2017).

Generally, proteins pressurised in water open up and unfold, simultaneously becoming more hydrated. Multimeric proteins dissociate into their subunits, of which more soon. At very high pressures protein unfolding is more drastic and called denaturation. In the laboratory, protein denaturation is often more conveniently achieved by heating or by exposure to special, destabilising molecules.

Understanding the adaptation of organisms to high pressure involves understanding the pressure adaptation of proteins (and much else besides), so the work of chemists, who use high pressure as an informative probe of protein structure, provides an essential platform from which to launch studies of adaptation. For years the general expectation has been that adapting proteins to high pressure entails a succession of small changes, arising from mutations and retained by natural selection. This view was strengthened by a classic paper published in 1966 by Murayama, on the effect of high pressure on a special protein, an abnormal haemoglobin known as Sickle cell haemoglobin (HbS). It differs from normal human haemoglobin in just one amino acid. Normal haemoglobin comprises two pairs of polypeptide chains, alpha and beta. In the case of HbS, valine, a hydrophobic amino acid, replaces glutamate at position 6 in each of the two beta chains (the abbreviation for this is beta E6V, E being the one letter symbol for glutamate). Glutamate is negatively charged and hydrophilic. This seemingly small change has significant consequences. In the deoxygenated state HbS aggregates into rods, distorting the usual doughnut-shaped blood cell into an angular (sickle) shape. The cells readily rupture, causing anaemia. However, this pathological condition confers some protection against malaria, and so in regions of the world where the disease is rife, sickle cell haemoglobin provides some compensation. The mortality from sickle cell anaemia is less than that inflicted by malaria in people with normal haemoglobin, so HbS sustains itself in the population. Remarkably, in 1966, Murayama predicted, and experimentally confirmed, that high hydrostatic pressure (5–30 MPa according to conditions) causes sickle cells to revert to a normal shape.

The effect of the same pressure on normal blood cells is negligible. Murayama reasoned that in the deoxygenated state the substituted valine in HbS hydrophobically interacts with a "normal" valine, forming a close association, each minimising its contact with water. This causes the quaternary structure of HbS to aggregate into rods, creating angular packing of the semi-solid haemoglobin, which damages the blood cell. High pressure disaggregates the HbS, separating the valines by causing a molecular volume decrease. Thus, the sickle cell shaped red cell is restored to the normal round shape. Furthermore, cooling disaggregates HbS by affecting the hydrophobic hydration of valine, just as it disaggregates a special synthetic protein called poly valyl RNase, of which more soon.

Essentially then, normal haemoglobin is pressure—stable whereas HbS is pressure—labile. The message is clear. A single change in the sequence of amino acids can drastically affect a protein's response to high pressure. Adaptation of proteins to function at high pressure by a single amino acid substitution is probably rare, but the notion of a succession of small changes (mutations) in a protein's primary sequence, rendering it fit for purpose at high pressure is a valid way of imagining how evolution takes place. As we shall see it is not the only way.

Returning to the way high pressure opens up proteins and very high pressure denatures them, heat is a very effective way of denaturing proteins, which explains how cooking makes food more digestible (Kaur et al. 2016). The author attended a High Pressure Biochemistry meeting at which a speaker distributed small pots of delicious jam to members of the audience. The jam was made, not by boiling fruit, but by pressurising it. This excellent episode of scientific showmanship reminded some of us of the story of Bridgman's "boiled" egg. Percy Bridgman was a professor of physics at Harvard University during which time he was awarded, in 1946, the Nobel Prize for Physics, in recognition of his fundamental high-pressure studies (see Chapter 13). His culinary contribution, probably apocryphal, was to pressurise an egg, making it similar to a boiled egg. In fact, in 1914 he published a paper on the "coagulation of albumen by pressure," which was the first demonstration of structural changes in proteins caused by high pressure (for example 700 MPa for 30 min at 20 °C). He was at pains to point out that the pressure coagulation (denaturation) might not be exactly the same as that produced by heat. He was right, and after a delay of many years, chemists, notably Japanese but soon an international assembly, started serious work on the phenomenon of high-pressure denaturation.

What is meant by opening up proteins? They have a radius of gyration which can be measured by X-ray methods. One particularly small protein, SNase, has a radius of gyration of 17 Å at room pressure which increases to 35 Å at 300 MPa (Krywka et al. 2008; Section 6.3).

## 2.2    Chemical Reactions in Solution, at High Pressure

### 2.2.1    Equilibria

Multimeric proteins are dissociated to their subunits when pressurised. Decompression leads them to spontaneously re-assemble, so it is a reversible process. Good examples are the structural proteins actin and tubulin. Actin is ubiquitous, involved in muscle contraction and the structure of many unspecialised cells, and tubulin is involved in intracellular movements and both are discussed later in the chapter. Many enzymes are multimers, for example lactate dehydrogenase and so are oxygen-carrying proteins such as haemoglobin, as we have seen. Giant supramolecular structures such as viruses also disaggregate under high pressure. In all these cases pressure acts by increasing the hydration of the proteins, causing them to occupy a state of least volume, in accordance with Le Chatelier's Principle. Box 2.2 explains the relationship between the amount of pressure and the degree of dissociation in formal terms, but the main point is the bigger the volume change, the more sensitive the equilibrium is to pressure. The following account is highly simplified, but provides a basis for understanding the importance $\Delta V$, the volume change (the triangle is the Greek letter $D$, delta, meaning change in, or difference).

---

**Box 2.2: Effect of Pressure on Dissociation**
The dissociation of a multimeric protein to subunits is reversible, an equilibrium process.

$$\text{Multimer} - \text{Sub units} \tag{2.1}$$

This is equilibrium is like the dissociation of a salt in water to its constituent ions, mentioned earlier.

$$AB = (A^+)(B^-) \tag{2.2}$$

The degree of dissociation or ionisation is expressed as a ratio: $(A^+)$ $(B^-)/$ $(AB)$ and is called an Equilibrium constant, conventionally designated $K$. The energy involved in equilibrium shifts is, for historical reasons, called the Gibbs free energy change, $\Delta G$. It is made up of $\Delta H$, the heat content or enthalpy minus the product of $T$, temperature and $\Delta S$, the entropy change.

$$\Delta G = \Delta H - \Delta(TS) \tag{2.3}$$

It is rarely explained in biological textbooks that $\Delta H$ is made up of two terms: $\Delta U$, the internal energy and the product of $P$ and $\Delta V$, the pressure and volume change in the system.

---

(continued)

**Box 2.2** (continued)

As much of biology is not involved with significant pressure it is understandable that many textbooks ignore the $\Delta V$ term in $\Delta H$, but for understanding how high pressure acts, $\Delta V$ is crucial. And, of course, for the purist, a rigorous understanding of the molecular reactions requires a full thermodynamic treatment.

Returning to $K$, the equilibrium constant, it changes with the conditions; RT $\ln K = -\Delta G$. In the present case, it changes under pressure, at constant temperature, according to:

$$d \ln K/dP = -\Delta V/R \times T \qquad (2.4)$$

This states that a small change in $K$, per small change in pressure is given by the change in volume of the system divided by the product of $R$, a special constant, and the temperature. The change in volume can be an increase (+) or a decrease ($-$). It is expressed as ml per mol and for readers unfamiliar with the term mol it will be helpful to divide the volume by Avogadro's number $(6.023 \times 10^{23})$ because then the volume, conveniently expressed in units of cubic Angstroms (Å), relates to a single molecule. This is often done when considering a large protein whose dimensions and hydrated structure is known. Here we are thinking of pressure favouring ionisation or the above multimeric protein equilibrium shifting to the right, with $\Delta V$ being negative. The larger the $\Delta V$ the more the equilibrium is shifted by a given pressure. In other words, the sensitivity of the equilibrium to high pressure is proportional to the volume change.

In passing it should be noted that the effect of temperature on an equilibrium at constant pressure is treated in analogous fashion

$$d \ln K/dT = \Delta H/R\ T^2 \qquad (2.5)$$

Here is a celebrated example that illustrates both hydrophobic interactions and the relation between the volume change and the effect of pressure on an equilibrium. It features the enzyme ribonuclease, RNase, which, many years ago was modified by a group of American chemists (Kettman et al. 1966), whose work influenced Murayama's understanding of sickle cell haemoglobin, just mentioned. They added the hydrophobic amino acid valine to the surface of RNase. The poly valyl RNase became an unusual protein with hydrophobic groups on its surface. The result was that it spontaneously aggregates at a temperature above 30 °C, as the hydrophobic groups huddled together, trying to minimise contact with the solvent water. As the hydrophobic groups come together as a result of random collisions, their tightly bound water molecules are released, causing a volume increase in the system. Conversely, an increase in pressure drives the dissociation of poly valyl RNase because water molecules interact closely with the exposed hydrophobic groups, creating a volume decrease.

$$\text{Poly RNase}, \; -\Delta V \rightarrow \text{Sub units}$$

This is a good example of an equilibrium process, in which the position of the equilibrium, that is the proportion of the system which is in the aggregated state, is the main interest. How fast the equilibrium shifts at a given temperature and pressure is dealt with in the next section, on reaction rates (Incidentally, at low temperature the hydration of the hydrophobic groups is stable and aggregation does not proceed).

## 2.2.2  Rates of Reaction

Although understanding how pressure affects an equilibrium is important, understanding how pressure affects the rate of chemical reactions is far more important. It is also rather more complicated, especially in reactions catalysed by enzymes, which are our main focus.

However, we can start with two simple generalisations. In most cases, a change in temperature has a predictable effect on the rate of a biological reaction. Between well-understood limits, increasing the temperature by 10 °C usually doubles the rate of reaction. This is the result of an increase in the rate at which the molecules are "bouncing about" (Chapter 1), and colliding more frequently. No such generalisation can be made of a change in pressure.

The second generalisation reinforces the point; it is very difficult to predict how a change in pressure might change the rate of a reaction. It is certainly not related to the overall volume change which occurs when a substrate is converted to a product, which unfortunately, is sometimes stated.

### 2.2.2.1 Simple Reactions, Not Involving an Enzyme

The thermally driven aggregation of poly valyl-RNase, above, does not involve an enzyme and provides us with an example of a simple reaction.

Rate process theory, also called transition state theory, devised by Eyring, is the way chemists deal with reaction rates in general. Its essential feature is activation. In the case of a simple reaction in which substance **A** simply converts to **B**, or poly valyl RNase dissociates to subunits, a two-stage process is envisaged. First, **A** becomes activated, by gaining energy. This is called Gibbs energy ($G^*$) for historical reasons and is also known as the activation energy. Formally it is given by the Eq. (2.6) which is similar to Eq. (2.3),

$$\Delta G^* = \Delta H^* - \Delta(TS)^* \tag{2.6}$$

The symbol $^*$ denotes activation, a transitional high energy state in which the bonds in the molecule can be imagined to be poised ready for change. $\Delta V^*$ is very important. By analogy with $\Delta V$ in Eq. (2.4), $\Delta V^*$ is the change in molecular volume (displacement volume) which arises when a molecule changes from the ground state to the activated state. If it is positive, i.e., it undergoes an increase, then high pressure will slow the rate of reaction. Conversely if it is negative high pressure will accelerate the reaction.

The activated stage in reaction $A - B$ is designated $A^*$. It has a fleeting existence during which the bonds holding the atomic components of the molecule together, becoming strained. Either $A^*$ decays to the product $B$, or back to $A$. Activation is thus treated as a reversible equilibrium process.

$$A \leftrightarrow A^* \leftrightarrow B \qquad (2.7)$$

For a simple reaction, the rate (also called the rate constant $k$), at high pressures and constant temperature, is given by an Eq. (2.8), similar to Eq. (2.4) above. Note the rate constant $k$ is in the lower case whereas the equilibrium constant $K$ is a capital.

$$d \ln k/dP = -\Delta V^*/R \times T \qquad (2.8)$$

The equation says that the rate changes with pressure according to the $\Delta V^*$ term. The rate of aggregation of poly valyl RNase is reduced at pressure. $\Delta V^*$ for the activation process entails a transient volume increase, which is actually quite large making the reaction sensitive to pressure. The reaction is not mediated by an enzyme so activation can be thought of as the necessary preliminary alignment of the subunits before they come closely together.

Whilst the volume change across an equilibrium tells us how much the equilibrium will shift when pressurised by a certain amount (thermodynamics), even in a simple reaction like the aggregation of poly valyl RNase, the volume change between the aggregated and disaggregated state tells us nothing about the rate at which that reaction proceeds (kinetics). It is the $\Delta V^*$, the transitional activation volume, which is the key parameter which determines the rate of the reaction at high pressure.

## 2.2.2.2   Enzyme Catalysed Reactions

Enzymes are proteins with a very specific structure for the particular reaction they catalyse. They catalyse reversible reactions in both directions, the actual direction, forward or reverse, is determined by the reaction conditions. They enormously accelerate biochemical reactions that would otherwise hardly proceed at all by effectively reducing the energy barrier to the reaction. How that is achieved is something of a mystery but it involves the way the enzyme binds to the substrate (i.e., the substance which is to be converted into something else) at a special (active) site in the enzyme's structure. The enzyme must also accommodate, at different sites, regulatory molecules, as enzymes are not much use if the rate of the reaction they catalyse cannot be controlled to suit the cells' needs.

The first step in enzyme catalysis is thus a binding equilibrium in which weak bonds secure the substrate in the active site of the enzyme.

$$E + S = ES^* = ES \text{ complex.} \qquad (2.9)$$

$E$ and $S$ represent the concentration of enzyme and substrate respectively. This reversible binding reaction proceeds via an activation step such that $ES^*$ fleetingly exists, and then either decays to the product $ES$ or to the reactants, $E$ and $S$. A second, catalytic, stage follows, in which $ES$ forms the product $P$, and releases the unaltered enzyme, $E$, ready to go through the cycle again. This too proceeds through a transition step, $ES^*$.

$$ES = (ES)^* = EP = E + P \qquad (2.10)$$

To summarise:

$$E + S = ES = ES^* = EP = E + P \qquad (2.11)$$

In this sort of multi-stage process the slowest stage, the bottleneck, is the important rate-determining stage and its $\Delta V^*$ will determine how pressure affects the overall rate, either slowing or speeding the reaction, as rigorously explained by Laidler (1951), in a classic paper (see also Morild (1981) and Mozhaev et al. (1996). Reactions in cells usually run on a low substrate concentration, so experiments seeking to measure the effect of pressure in cellular conditions but in vitro should use low substrate concentrations (Box 2.3).

---

**Box 2.3: Measuring an Activation Volume ($\Delta V^*$) and an Equilibrium Volume ($\Delta V$)**

A convenient equation to calculate the Activation volume. $V^*$, in a reaction is

$$\Delta V^* = 2.303 \, RT \log_{10}(kp_1 - kp_2)/(p_2 - p_1) \qquad (2.12)$$

$p_1$ is atmospheric pressure, $p_2$ is high pressure, $k$ is the rate constant for the reaction and other terms are as in Section 2.2.

As $\Delta V^*$ has a fleeting existence it can only be quantified in this way (Johnson and Eyring 1970).

The volume change across an equilibrium ($\Delta V$) can be calculated from an analogous equation, using $K$ the equilibrium constant. However, in many cases, $\Delta V$ can also be directly measured.

## 2.3    The Affinity of an Enzyme for Its Substrate

For any organism to live at high pressure it must have enzymes capable of functioning adequately at the required pressure. To measure the sensitivity of an enzyme to high pressure, and hence its adaptation, we need to measure its catalytic power at pressure. One way of doing that is to measure the ease with which it binds to its substrate, i.e., its affinity for its substrate (Box 2.4). This has the symbol Km and it is measured in units of concentration. If an enzyme's Km for a particular substrate is increased by high pressure that means its affinity is reduced by pressure and it probably is not well adapted to function at that pressure. An enzyme that has a high affinity for its substrate at high pressure (a low Km value) is well adapted to function at the pressure in question. There are numerous other ways in which chemists assess the fitness of an enzyme to function at high pressure and we will meet some of them in later chapters.

> **Box 2.4: Measuring How High Pressure Affects the Affinity of an Enzyme for Its Substrate**
>
> Imagine an experiment in which the product ($P$) of an enzyme reaction can be monitored easily using an optical method, such as a colour change. Start the experiment by mixing the enzyme with a certain concentration of its substrate at a controlled temperature. As the reaction proceeds, it slows down as the concentration of the substrate decreases. It is the initial rate at which the colour changes which indicates the formation of P, which is instructive. It is found that this initial rate of reaction increases with the increase in the initial concentration of the substrate, but it reaches a limiting value, the maximum reaction velocity or Vmax.
>
> Now this Vmax is useful, because by reasoning which we can ignore here, it can be shown that the quantity half Vmax provides a good measure of the ease with which the enzyme binds to its substrate. The concentration of substrate at which the half $V$ max occurs is called the Michaelis Constant, Km. A low value means the enzyme has a strong affinity for its substrate, i.e., it binds easily with the enzyme. Using a high-pressure optical apparatus to measure the rate of colour change (Chapter 13) it might be shown that high pressure increases a Km value, which means it reduces the affinity of the enzyme for its substrate. The enzyme–substrate binding process is thus affected by pressure, probably because pressure directly affects the structure of the enzyme.
>
> Affinity implies adaptation, which is one of the main features of the enzymes we need to understand. The approach of using the Km can be used with enzymes from any organism; bacteria, archaea, and eukaryotes such as fungi or deep-sea fish. An important difficulty in doing so is recreating the physiological conditions in vitro.

## 2.4 Cells Under High Pressure

Here we are primarily concerned with the effects of pressures of less than 100 MPa on cellular structures. This range of pressures approximates to that of the known high-pressure environments. Higher pressures are of interest, particularly to chemists, microbiologists and food scientists and are discussed throughout the book. A paper by Frey et al. (2008) describes the scope of a fourth group, medical cellular biochemists. These groups are, of course, merely a way of organising knowledge and we shall see that the unity of nature ignores them. For example proteins adapted to high pressure in deep-sea organisms and other proteins responsible for human disease are studied with the same techniques (Meersman et al. 2006), and sometimes by the same scientists.

### 2.4.1 Cell Membranes

Membranes enclose all cells and in the Eukaryotes they are also present within the cell, enclosing the DNA in the nucleus and providing other compartments and surfaces on which reactions can take place. They are all based on an assembly of lipid molecules which is intrinsically stable, typically only two molecules thick, hence the term lipid bilayer. The individual structure of the lipids provides the stability: a hydrophobic chain with a polar, hydrophilic group at the end (Fig. 2.1a). This lipid bilayer structure is characteristic of Eukaryotes and Bacteria, although in the latter the Gram-Negative bacteria have an inner and an outer membrane. The inner membrane has an orthodox composition and structure but the outer membrane is rather different, see Chapter 8. The Archaea have a membrane that is composed of very different lipids (Chapters 7 and 9), some of which traverse the membrane, forming a thick monolayer. Both kinds of membrane contain proteins, organised on and in the lipid structure. In a crude summary, you could say that the lipids provide the cell with an impermeable barrier whilst the proteins do the clever stuff, transporting molecules across the barrier, recognising foreign surfaces outside the cell and sticking it to other cells. However, the bilayer has special properties too. It can transmit tension to open special proteins (channels) mounted in the bilayer which then allow solutes and water to pass through the membrane. Other channel proteins are opened by voltage changes or by binding to special molecules. The performance of ion channels under high pressure is discussed in Chapter 4. The bilayer lipids provide a special environment for channels and many other proteins. Some can be detected as "rafts" floating in the membrane and indeed clusters of distinct lipids occur as rafts too, in a fluid mosaic. When looking at diagrams of such structures it is important to remember that the molecules are in continuous motion, which can be measured. They wobble, flex and laterally diffuse across the membrane. Special techniques exist which can visualise and measure their molecular motion.

Many of the lipids are involved in clever stuff too, such as signalling and organising bulk transport of vesicular traffic, both inwards and outwards (endocytosis and exocytosis).

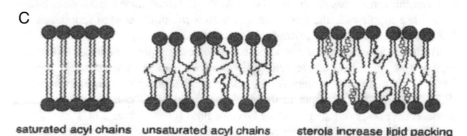

**A**

**Bacteria**

*fatty acid*

*ester bond*

*glycerol-3-phosphate*

**B**

**C**

saturated acyl chains
non-fluid

unsaturated acyl chains
fluid

sterols increase lipid packing
and membrane thickness

**Fig. 2.1** (**a**) The elementary structure of membrane lipids in Bacteria and Eukaryotes consists of a glycerol backbone, with carbon atoms 1, 2 and 3, linked to a head group and two fatty acid hydrocarbon chains. (From Caforio and Dreissen 2017). (**b**) The headgroup and hydrocarbon chain confer a distinct shape on individual phospholipids (right) and the presence of a double bond in the hydrocarbon chain creates a marked kink which disrupts the otherwise close packing in a bilayer (left). Palmitic acid has a chain of 16 carbon atoms and no double bonds. Below it, vaccenic acid has an 18 carbon chain with one double bond at a specific position. Often a phospholipid (**a**) has one saturated and one unsaturated hydrocarbon chain. (From Ernst et al. 2016). (**c**) Lipid bilayers present a significant permeability barrier when highly ordered (left) or more disordered when containing some unsaturated chains, or when heated, or both (middle). The presence of sterols orders and thickens the bilayer (right). Reproduced with permission from Caforio and Dreissen (2017), Ernst et al. (2016)

Simplified membranes are studied in order to focus on particular properties of natural membranes. There are planar bilayers, which are lipid membranes covering a small hole in a plastic plate, effectively separating two compartments. The diffusion or transport of dissolved substances across the planar bilayer can be studied at high hydrostatic pressure by mounting such a plate within a pressure vessel. A very different type of artificial membrane is provided in the form of lipid vesicles, freely suspended in solution. These are called liposomes and have taught us a great deal about the properties of natural membranes.

The introduction of the author to the study of liposome membranes arose at the 50th Anniversary meeting of the UK Society for Experimental Biology, held at Cambridge. There, Alec Bangham, then head of Biophysics at the Babraham Institute, pressed the case for some high-pressure experiments with liposomes. He kindly extended an invitation to his lab to enable me to learn how to work with them.

Liposomes are microscopic lipid vesicles. They assemble themselves, spontaneously, when certain lipids and water are brought into contact in the right way, for example by exposing lipids deposited on a clean substrate to an aqueous solution of salts. The vesicles which spontaneously form have an onion-like structure, consisting of concentric layers of lipid bilayer membranes. If such a suspension is subjected to ultrasound, the multi-layered vesicles break up into vesicles bounded by just one bilayer, enclosing the aqueous solution in which they were formed. They resemble a simple cell, and this, combined with their ability to form spontaneously, has prompted ideas about the origins of cellular life.

Bacterial membranes are composed of phospholipids whereas eukaryotic membranes may also contain sphingolipids and sterols which are rarely seen in bacteria. Of main interest here are the phospholipids of which there can be many different types. For example analysis of one strain of E. coli recently revealed 31 different phospholipids plus other types of lipid (Oursel et al. 2007).

When subjected to high hydrostatic pressure such bilayers are compressed in an asymmetrical way. Pressure causes the hydrocarbon chains to straighten out and lengthen as they pack more closely laterally. Hence the thickness of the bilayer slightly increases at high pressure; its lateral compressibility exceeds its transverse compressibility. This has been shown by X-ray crystallography (Czeslik et al. 1998), and also by ingenious measurements of the diameter of liposomes in suspension. When pressurised, up to 300 MPa, the diameter of the vesicles slightly decreases, because the lateral compression of the bilayer reduces the surface area of the vesicle. Liposome bilayers enriched with cholesterol have a reduced lateral compressibility as there is less free volume in their packing, and their diameter is little changed at high pressure (Beney et al. 1997).

Liposome bilayers made of s single type of phospholipid manifest a phase transition when heated. The lipid bilayer abruptly changes from an orderly stack of lipid molecules to a more disorderly (fluid) bilayer, at a characteristic temperature. This "fluid" state retains a coherent structure and is a barrier to both water and dissolved ions (Fig. 2.1c). The phase transition is similar to a melting point but is better described as a gel to liquid crystalline phase transition. A familiar phase transition, the water–ice transition mentioned earlier, involves an increase in volume, as the molecules of water rearrange themselves on freezing to form rigid ice. As already noted, high pressure melts ice, forcing the water (ice) molecules to adopt the lesser volume. In the case of the bilayer phase transition, high pressure favours the ordered gel state, which occupies lesser volume. Thus, the volume change in the bilayer gel to liquid crystalline transition is positive, the more fluid, disorderly isotropic packing of the bilayer occupies more volume than the orderly gel state. In the case of liposomes made of the phospholipid dipalmitoyl phosphatidylcholine, DPPC, the transition occurs abruptly at a temperature close to 41 °C with a volume increase of 25 ml per mol (meaning per standard number of molecules). The transition is opposed by high pressure; a higher temperature is required to drive the gel to liquid transition at high pressure. The increase in the transition temperature of DPPC (and other) liposomes at high pressure has been measured by a variety of techniques, one of which is dilatometry which entails the direct measurement of the volume change, $\Delta V$, which occurs at the transition. The dilatometer measures the expansion of a very dense suspension of liposomes much as a thermometer measures the thermal expansion of mercury. In the case of DPPC liposomes the increase in transition temperature arising from an increase in pressure is 0.2 °C per 1 MPa (Macdonald 1978; Bangham 1983). The significance of the numbers will emerge in due course. This example of Le Chatelier's Principle is quantitatively treated by elementary thermodynamics in Bangham's engaging book, "Liposome Letters," which also deals with many other facets of liposomes including their role in medicine.

What has this got to do with real biological cell membranes? The fluid state of liposome bilayers is comparable to the state of bilayers in natural membranes, which require fluid molecular motion to function. All organisms regulate the composition of their membranes to provide the "correct" fluidity (see Chapter 7). Being composed of a mixture of lipids, natural membranes generally do not manifest pure phase transitions, but there are rare examples of phase transition-like behaviour in a natural membrane. Consider an enzyme incorporated in such a membrane and surrounded by lipids (boundary lipids) whose fluid state determines the flexing of the enzyme and hence the speed at which it catalyses a reaction. The rate of reaction is decreased by cooling. However, if a critical temperature is reached when the liquid to gel transition takes place, the effect on the enzyme is abrupt and considerable. Its catalytic activity is greatly reduced. Examples showing this behaviour are certain enzymes, embedded in the structure of the membranes in muscle cells and prokaryotes, which hydrolyse adenosine triphosphate (ATP). The energy so released drives the transport of ions across the membrane. How pressure affects such structures is taken up in Chapter 7.

Phase transitions are instructive but the graded fluid state in the lipid bilayer of natural membranes is of more widespread physiological importance and is significantly affected by pressure. It can be measured by a variety of techniques, some of which are complicated and require very expensive equipment, but it can also be measured by an optical technique using equipment which is not especially expensive, namely fluorescence spectroscopy. It has been widely used in biological laboratories and is readily adapted to work at high pressure. Box 2.5 describes the

technique which provides a numerical measure of bilayer fluidity. Readers wishing
to skip this Box can proceed on the basis that fluidity can be simply scored.

**Box 2.5: Measuring the Fluidity of Lipid Bilayers at High Pressure**

About 3 mL of a suspension of bilayers, liposomes or fragments of natural
membranes, looking like dilute milk, is held in a cuvette, a special transparent
container. A fluorescent molecule is incorporated (dissolved) in the bilayer.
The most commonly used fluorescent "probe" molecule is DPH
(diphenylhexatriene). It has a rod-like shape and dissolves in lipid bilayers
such that it can only oscillate within the confines of the hydrocarbon chains.
Very little is present between the lipid bilayers and in water, it does not
fluoresce. When DPH is incorporated in the bilayers and excited by light at a
wavelength of 365 nm, it emits light (i.e., it fluoresces) at a wavelength of
425 nm, and, most important, the light waves are emitted in the same plane as
the excited DPH molecule, confined by the hydrocarbon chains in the bilayer.
Ordinary light can be imagined as travelling as waves lying in all planes
whereas polarised light waves lie in a single plane. The large number of cell
membrane vesicles in suspension will, of course, be randomly orientated, and
will emit light polarised in all planes possible, but the intensity of that light is
measured in one plane only.

Imagine a lipid bilayer which is cold, with little thermal motion. The DPH
will be tightly constrained, emitting its polarised light similarly constrained in
a single plane. When measured in the corresponding plane the light will be of
high intensity. In contrast, when the temperature is increased there is more
motion in the bilayer and the hydrocarbon chains oscillate. The DPH will
similarly oscillate between the confining chains much more (wobble is used
here as a technical term), emitting polarised light with a wider range of planes.
The emitted light is said to be depolarised. When its intensity is measured in
the exactly the same plane as before, it will be less intense. In this way
the DPH "reports" the fluidity (wobbliness) of its lipid environment and
hence the fluidity of the bilayer. That difference is numerically expressed.
Adapting the technique to work at high pressure introduces no fundamental
problems (Chapters 7 and 13).

This version of the technique is called the steady-state depolarisation of
DPH, or more formally, the steady-state fluorescence anisotropy of DPH.
There is a more complicated version of the technique called time resolved
fluorescence spectroscopy, which provides more refined measurements which
can be related to those obtained by other spectroscopic techniques such as
electron spin resonance spectroscopy (ESR) and nuclear magnetic resonance
spectroscopy (NMR).

See also Earl (1986); Heerklotz and Seelig (2002); Eggeling et al. (2009);
Winter and Jeworrek (2009); Purushothaman et al. (2015) and Ding et al.

(continued)

Bilayers in liposome or in natural membranes are more ordered at high pressure,
mimicking the effect of reducing the temperature. The case of DPPC liposomes, just
mentioned, has proved to be typical of many lipid bilayers, so we have a convenient
generalisation that fluorescence anisotropy measurements of DPH show that an
increase of 1 MPa reduces membrane bilayer fluidity by an amount equal to a
reduction in the temperature of 0.2 °C (Chong and Cossins 1983), a similar equiva-
lence to that seen in the phase transition temperature,

Many cells, eukaryote and bacteria, regulate the lipid composition of their
membranes to maintain the bilayer fluidity within a range suitable for normal
function by regulating the hydrocarbon chain length, degree of saturation and
chain branching. So, in high-pressure environments, and especially cold ones, we
have an interesting situation. At the mid-ocean depth of 4000 m, for example where
the pressure is 40 MPa, the temperature of the water is around 4 °C but for the
fluidity of a hypothetical lipid bilayer, it is effectively at −4 °C, at normal pressure
(40 MPa is equivalent to a reduction of 8 °C. 4−8 = −4). High pressure makes the
deep sea much "colder" for cell membranes than its temperature suggests. The
hydrothermal vent hot spots are clearly exceptions and quite separately so is the
Deep Biosphere, an environment in which the temperature increases with depth and
pressure. In these cases the fluidity of a hypothetical lipid bilayer experiences
mutually opposing effects with increase in depth. Chapter 7 deals with this interesting problem of adapting membrane bilayers to high-pressure environments.

## 2.4.2  Cytoplasm

It is possible to examine cells under the light microscope at high pressure, say
20–30 MPa, using special equipment explained in Chapter 13. When the high
pressure is applied over a few seconds, the cell's appearance changes very little. It
certainly is not squashed or made smaller. What soon becomes apparent, however, is
that if cytoplasmic streaming was evident then it stops, and irregular shaped,
amoeboid cells become rounded. The swimming of ciliated eukaryotic cells is
slowed, and so too is the swimming of bacteria powered by flagella, (see Section
2.5), although flagella are quite different from cilia. If the pressure is then restored to
normal, over a few seconds, then these changes reverse and the cells resume their
normal shape and activity.

The starting point of explaining some of these changes is cytoplasm, which
sounds an old-fashioned term. What exactly is it? In highly differentiated cells,

such as those which make up muscle, there is so much special interior structure that there is little need for the term, but in relatively undifferentiated cells, it is the stuff of which the cell is made. It is conspicuous in amoeboid cells, connective tissue cells called fibroblasts and other cells which are commonly culture in the lab, in eggs, such as those of sea urchins much used in research, and many other cells, perhaps even including bacteria, which are vastly smaller. Seen under the electron microscope, which is much more powerful than the light microscope, cytoplasm has structure. Bray's book, "Cell movement," provides a good description. Typically, the water content of the cytoplasm is about 70% and proteins make up 20% or more by weight. The cytoplasm of the giant axon in the squid, an important experimental cell in electrophysiology, is exceptional being particularly watery with only 5% of its weight being protein. Whatever its protein content, cytoplasm is a mesh of proteins, the main ones being actin, comprising filaments 8 nm in diameter and variously cross-linked, and tubulin, polymerised as microtubules, with a diameter of 25 nm but with a hollow interior 15 nm in diameter. Then there are other proteins making up intermediate filaments and many accessory proteins forming cross bridges, some connecting to proteins embedded in the plasma membrane. Bray vividly describes the dynamic aspect of cytoplasm, stating that every molecule performs a continuous manic dance as it is continuously jostles from one location in the cell to another, and in the case of bacterial cells, he points out that a small molecule visits most locations in the cell within a tenth of a second. Given this teeming picture, we should also note that cytoplasm is crowded, a technical term meaning that much of the water is influenced by hydrophobic and electrically charged solutes and surfaces, Minton (2006). Even bacterial cytoplasm is now thought of in these terms (Fooglia et al. 2019). There is very little unmodified, bulk water structure in a cell.

High pressure appears to liquefy the cytoplasm of undifferentiated cells, by depolymerising protein structures, shifting a sol-gel equilibrium to the left. It causes amoeboid cells to behave as a liquid droplet as they round up. Cytoplasmic streaming stops because the connectivity of the protein mesh is broken. Box 2.6 describes the consequences of this in Eukaryote cell division and mitosis. The simple reversible "equilibrium shift" explanation is a good way to describe the phenomenon, but chemical reactions are probably also involved. Highly differentiated muscle cells, packed with a linear array of contractile proteins, are not affected by the pressure in this way.

**Box 2.6: Cytoplasm Under High Pressure**

The liquefaction of cytoplasmic gels was demonstrated and measured by Brown (1934) and Brown and Marsland (1936), by means of an ingenious centrifugal technique. Granules in the cytoplasm of cells can be displaced by centrifugation, and at high pressure the displacement is greatly accentuated. Other observations showed high-pressure liquefied cytoplasm (Kitching and Pease 1939). Simple gels made of muscle proteins or even methylcellulose

(continued)

**Box 2.6**  (continued)

respond in the same way, becoming more liquid at high pressure, but a gelatine gel is made more firm under pressure. Clearly the bonding of the molecular network differs between these two classes of gel.

Dividing Eukaryote cells require a contractile force to constrict the cell in two, and the cleavage process is inhibited by high pressure, reversibly. The eggs of sea urchins (Echinoderms, a major group of marine animals), are also "liquefied" by high pressure, as shown by Brown's technique, and when fertilised their gel strength increases markedly, before normal cleavage. Marsland, working at the Marine Biological Laboratory at Woods Hole, MA, USA, investigated these phenomena for more than two decades, starting in 1936, using high-pressure light microscopy (Marsland 1958, 1970; Kitching 1954, 1957; Chapter 13). The cleavage furrow, the site where the cell is constricted in two, progresses slowly under a moderate pressure and at higher pressures is caused to reverse, leaving the cell larger than expected but otherwise healthy. In the case of the shallow water sea urchin, *Arbacia punctulata*, the blockade is achieved at 20 °C by 34 MPa and at 10 °C by 13.6 MPa. This blockade is a general phenomenon. The pressure which blocks the cleavage of the eggs of fish and frogs is 34 MPa; of Molluscs (snail), 27 MPa; Polychaetes, marine worms such as Chaetopteris, 27 MPa; Insects (Drosophila the celebrated fruit fly), 40 MPa, and the cell of Amoeba, 40 MPa, all at room temperature. The interesting exception is the cleavage of the egg of the parasitic Nematode worm *Ascaris megalocephala* which, in a 1939 paper, Pease and Marsland reported to cleave at 80 MPa.

The cell cycle, the sequence of growth, DNA duplication and cell division, has been studied at high pressure. A "young" cell passes through a period of time before the synthesis of DNA begins, after which there is a further quiescent period before the chromosomes are replicated and mitosis begins. Mitosis is the elaborate movement of separate chromosomes into the daughter cells formed as the cleavage furrow constricts the "mother" cell in two. Microtubules are involved in this process and they are depolymerised by pressure. This cell cycle is typical of eukaryotes such as yeasts and metazoan cells (Morgan 2007; Section 2.6.2 and Box 2.7).

A simpler cycle can be seen in single celled organisms (formerly Protozoa), such as the ciliated cell *Tetrahymena pyriformis*. It can be grown like a bacterium, in a broth, it has a very short generation time, the complexity of a Eukaryote, but is conveniently large for the microscopy used to follow the cycle at high pressure. A pressure of 25 MPa completely blocks the cleavage of the cell in two, whilst a pressure of 10 MPa has no effect. The intermediate pressure of 17.5 MPa also has no effect on cleavage, but the progress of cells approaching a critical stage prior to cleavage is slowed. On eventually reaching it cells then proceed to cleave at pressure, at 17.5 MPa. Thus, high

(continued)

**Box 2.6** (continued)
pressure affects at least two processes (Macdonald 1967a, b). It inhibits the
constriction of the cell in two and the preparation of the cell for that process,
conclusions which were subsequently confirmed by experiments with cells
growing in suspension and monitored by electronic means, under high
pressure.

### 2.4.3  Cell Organelles

In eukaryote cells, all the organelles are clearly visible under the electron microscope. The nucleus is the largest and ribosomes are the smallest, appearing as
uninteresting granules. They are functionally closely involved with each other.
The nucleus is bounded by a double membrane and contains the cell's genetic
material, DNA, which is an enormously long polymer, packed and supported by
special proteins called histones, such that it is accessible to enzymes which repair
and replicate it. The nucleus is not visibly affected by high pressure unless the cell is
dividing, of which more soon. Recall that Bacteria and Archaea do not have a
nucleus and their DNA is not enclosed.

Ribosomes are supremely important. Their function is to assemble amino acids
into proteins and although this is extremely complicated, it is important to have at
least a simple picture of what goes on as it is sensitive to high pressure. The ribosome
acts as a sort of zipper, aligning and securing amino acids in the sequence required to
make a given protein. There are millions of ribosomes in a typical cell and their
activity consumes a great deal of the cell's energy.

Protein synthesis begins with DNA, in the nucleus. Its sequence of nucleotides
provides coded instructions for the sequence of amino acids in proteins and for
determining when those instructions are to be carried out, i.e., expressed. A segment
of DNA which codes for a given protein or RNA, is called a gene. Whilst DNA is
the archive of the cells' blueprints, RNA molecules are working copies, for use on
the shop floor. The term gene expression is given to the process which results in the
synthesis of protein or RNA, and it consists of two stages. The first is transcription,
the polymerisation of RNA. The reaction involves the double-stranded DNA molecule becoming separated so a single strand can then have its nucleotide sequence
copied to an RNA molecule. There are three types of RNA. rRNA, which was
mentioned in Chapter 1, Box 1.2, is the scaffold for the ribosome. Messenger RNA,
mRNA, encodes a protein and transfer RNA, tRNA, decodes the mRNA, ensuring
the correct amino acid sequence. The tRNA decodes the mRNA using three
nucleotides (i.e., subunits) called an anti-codon that is the complement of the
codon in the mRNA. Each is attached to a single specific amino acid. Remember
there are only 20 amino acids which are used. The ribosome provides the mechanical
support for amino acids attached to the tRNA to be assembled in the correct
sequence by reading the mRNA. Ribosomes look like granules and are, in fact,

multimeric structures of many proteins and several RNA molecules, but they have no membrane. Their gross structure comprises a large and a small subunit with special binding sites. One is a groove in which mRNA binds and as it is a very long molecule it extends sufficiently for other ribosomes to bind to it, making a polyribosome (readers may find this complicated process is well explained with cartoon style illustrations on the Web).

The second stage in gene expression is called translocation. A tRNA molecule carrying its amino acid is base paired adjacent to another tRNA molecule with its different amino acid, both binding to the mRNA on the ribosome. This brings the two amino acids close enough to become joined by a peptide bond. The tRNA molecules diffuse away and a new segment of the ribosomal RNA provides a new codon for a new tRNA molecule with a different amino acid to base pair. A polypeptide chain thus grows in length. Once a string of amino acids has formed the primary structure of a protein it then passes into the lumen of the endoplasmic reticulum (in eukaryotes) where it has to be carefully folded into the native (correct) tertiary structure by chaperone proteins, a process which is followed by a "quality control" check. Faulty proteins are destroyed. Typical proteins have a finite life, hours or days, and are then recycled.

It is difficult to imagine the teeming population of RNA molecules each carrying its specific amino acid, binding (base pairing) and dissociating at high speed. The numbers are impressive. In "The Vital Question", Lane describes the synthesis of the protein actin which comprises a string of 375 amino acids folded in the form of a globular monomer. These are attached to form a polymer or filament. 58 monomers form a 1 μm long filament and a typical cell can synthesise a length of 15 μm per minute, involving 325,380 amino acids assembled per second, in the correct sequence. Specialist cells can work 60 times faster, assembling 19,522,800 amino acids per second, each in the correct sequence. It seems almost inconceivable that such a process works at all. Considering the number and variety of binding and dissociating reactions involved it is not in the least surprising that it is affected by high pressure. Indeed, it has been known for a long time that high pressure dissociates the ribosomes in the bacterium *E. coli* (Hauge 1971 (Chapter 13); Schulz et al. 1976; Infante et al. 1982; Gross et al. 1993), and inhibits protein synthesis, first studied by Landau and by Pollard and Weller (1966). In food technology, a brief exposure of bacteria to very high pressures may fail to inactivate all the cells. A small number of *Bacillus subtilis*, for example may survive, repair their damaged ribosomes and after a delay, start growing (Nguyen et al. 2020).

The faulty proteins, mentioned above, can arise from spontaneous errors in the synthesis or from damage caused by external factors such as heat or radiation. Cells in all three Domains of life are equipped with chaperone proteins, which deal with potentially toxic misfolded proteins. A misfolded protein is liable to have hydrophobic groups on its surface which lead to aggregation and, of course zero function or worse. Additionally, all types of cell possess a special defence capability which deals with proteins and other macromolecules which are damaged by external factors. It is called the cellular stress response. As it was discovered in experiments involving heat shocks, the special proteins whose synthesis is triggered are called heat shock proteins. They can bind to the molecules in question and prevent them from aggregating and, in other cases, act as chaperone proteins. This is now a complex subject but a simple and elegant experiment carried out in the early days of discovery remains illuminating (Ananthan et al. 1986). Proteins injected into a conveniently large cell, known to contain the heat shock genes *hsp*, had no effect, but if the proteins were denatured before injection, then they activated the *hsp* genes and heat shock proteins were synthesised. Even a monomer of a protein polymer was able to activate the *hsp* genes. High pressure appears to act on this system, but the tricky question is, how? Does it create abnormal proteins that activate the *hsp* genes or does it affect the system in various ways, perhaps by acting directly on gene expression, appearing to mimic heat shocks? The former can be regarded as a healthy physiological response and the latter as pathological. The problem resurfaces in the book.

There are other routes by which faulty and unwanted proteins are degraded to their constituent parts and recycled. Bacteria have a system using enzymes called proteases whilst eukaryotes have a separate system called ubiquitination. A protein, ubiquitin, becomes attached to a selected protein which then undergoes degradation by a proteosome, to which the ubiquitinated protein becomes bound. In addition to removing damaged proteins ubiquitination also regulates cell processes, such as cell division (Zhao et al. 2007).

Returning to gene expression, it is important to remember that different cells in an organism have the same DNA but are different because each cell only expresses a fraction of its genetic capability (genome). This point is neatly illustrated in a well-known textbook. Two cells are shown. One, a lymphocyte, is shown as a dot, and the other is a neuron, which looks like a tree, with a main trunk and multiple branches supporting fine dendrite twigs. Different parts of the DNA are expressed to create such different cells. Signals from other cells and from the environment can trigger the transcription of specific proteins, and in many cases, the triggers act by initiating or inhibiting transcription. It is difficult to imagine this amazing sequence of molecular interactions being unaffected by high hydrostatic pressure, and indeed it is not.

The endoplasmic reticulum comprises a system of interconnecting tubular vesicles. Those involved with protein synthesis, mentioned above, have ribosomes associated with them, the rough endoplasmic reticulum as seen by electron microscopy. The smooth endoplasmic reticulum is involved with synthesising and transporting compounds other than proteins. The Golgi complex is another set of vesicles involved with chemically modifying proteins and with the synthesis of lipids and polysaccharides.

Mitochondria are conspicuous in the fine structure of the cell. They are also vesicles enclosed by a double membrane and, as previously mentioned, they arose from bacteria, ingested in the early evolution of cells, which became symbiotic. They divide like bacteria but have lost most of their DNA but what remains, mitochondrial DNA is important in providing for the synthesis of certain proteins. Mitochondrial DNA is significant in eggs (but not sperm), where it plays a role in the inheritance of features through the female line. Mitochondria's two membranes differ. The outer one is permeable, allowing free exchange between the cytoplasm and space between the inner and outer mitochondrial membranes. The inner membrane is much folded and thus has a very large area. It is selectively permeable and is the site of the synthesis of ATP, the energy currency of the cell, of which more at the end of this chapter. It is also the site where oxygen is converted to water in the final stage of cellular respiration. Oxygen consumption, by Eukaryote cells and whole organisms, occurs in mitochondria.

Vertebrate cells possess a primary cilium which may lie within the cell but generally protrudes from it and functions as a small intracellular compartment, about 0.01% of the volume of the cell. It has nine pairs of microtubules common to contractile cilia, with which unicellular organisms such as *Tetrahymena* swim and epithelial cells waft fluid and particles. However, it lacks the central pair. When it was first discovered it was regarded as a relic with no function. However, we now know that its membrane, which is continuous with the cell membrane, carries sensory and signalling molecules. The primary cilium seems to be involved in many signalling pathways (Ruhlen and Marberry 2014; Wheway et al. 2018; Sun et al. 2019), including, in the case of chondrocytes, mechanotransducing molecules such as integrins and mechano-gated ion channels. The latter are described in Chapter 4. Integrins are glycoproteins which cross the cell membrane, connecting the extracellular matrix to the intracellular cytoskeleton, typically actin, usually through a protein called talin. This mechanotransducing system performs a variety of functions in the life of tissue cells. There is evidence that both osteocytes and chondrocytes respond to stretch (strain) and to hydrostatic pressure, of which more in Chapter 3.

Cellular organelles do not show Brownian movement, microscopically visible jiggling movements caused by underlying molecular motion, because they are set in the cytoplasmic gel. This typically has a mesh size of 20–30 nm, which allows small molecules to diffuse through its water-filled vacuities quite rapidly. The cell membrane, enclosing the cytoplasm and hence defining the cell boundary, has a semi-permeable nature, which enables water molecules, but not solutes, to diffuse through it. When a semi-permeable membrane separates a dilute solution from a concentrated one, there is a net diffusion of water into the concentrated solution. This is osmosis, and the net movement of water is an osmotic flux that can generate pressure within the cell which can cause the cell to swell. Countermeasures protect the cell from bursting. Bacteria and plant cells achieve a stable equilibrium by confining the plasma membrane within a porous but rigid cell wall, which confines the pressure generated by osmosis, often referred to as turgor pressure. Some amoeboid cells pump out the osmotically acquired water by special contractile vacuoles. Many animal cells whose environment is a tissue or organ, have a bathing fluid whose composition is regulated by the animal, to osmotically match the cytoplasm and so the cells do not experience a net inflow of water. However, such whole animal osmoregulatory processes are often susceptible to high pressure, in ordinary shallow water animals.

The final cellular process to be considered in this section is exocytosis, the means by which intracellular vesicles, containing chemicals with a specific role, are discharged through the cell's plasma membrane to the extracellular space. In animals, the synapse, the junction between two nerves or between a nerve and a muscle, exhibits one particular type of exocytosis. The propagation of an action potential from one to the other involves the release of a neurotransmitter from the upstream nerve. The neurotransmitter causes the downstream membrane to depolarise and initiate a new action potential, thus continuing the propagation of the nerve impulse. Releasing neurotransmitter by an action potential in this way is a special process to be distinguished from the spontaneous release of individual vesicles containing neurotransmitter from an unstimulated synapse. This latter type of exocytosis is seen in the frog neuromuscular junction, in the absence of electrical stimulation, as very small end-plate current pulses. They arise from the spontaneous exocytosis of vesicle containing the neurotransmitter acetylcholine, which causes the downstream (post-synaptic) membrane to depolarise briefly and allow current to flow. The frequency of release is approximately halved by 5 MPa, making it an unusually sensitive process (Ashford et al. 1982).

The effect of pressure on exocytosis in chromaffin cells, which release epinephrine and norepinephrine from the adrenal medulla in mammals, has been measured by a technique which detects the fusion of intracellular vesicles with the cell membrane. The resultant small increase in the area of the cell membrane is detected as an increase in its capacitance, its ability to store an electrical charge. Stühmer (2015), showed the rate of this exocytosis on command is approximately halved by 10 MPa.

## 2.5    Prokaryotes

With a volume a thousand times smaller than that of Eukaryote cells you might think that there is not much to say about the fine structure of Prokaryotes. In shape they are either spheres about 1 μm across, spirals or rods. However, two intracellular features are important to know. First, the bounding membranes of Archaea and Bacteria are significantly different. The distinctive Archaeal bounding membrane is described in Chapter 9. Bacteria are classified Gram Positive or Gram Negative, because of a basic difference in their external boundary layer. Gram-Positive bacteria, such as Streptococcus, acquire the colour of the Gram stain because their proteoglycan outer cell wall reacts with the dye. Their plasma membrane lies within this tough wall. Gram-Negative bacteria, such as *E. coli*, have an outer membrane whose external surface is coated with a lipopolysaccharide which does not react with the stain. Periplasm lies between the outer membrane and an inner plasma membrane. Within that lies the equivalent of cytoplasm. The molecular structure and composition of bacterial membranes are broadly similar to that of eukaryotes, and so both differ from archaea.

Many bacteria have a flagellum, or several. It is quite different to the Eukaryote flagellum, being made of a distinct protein, flagellin, which is rotated by a special motor. The flagellum acts as a propeller and drives the cell through its aqueous environment, which, because of the cell's high surface to volume ratio, is a very viscous medium to cope with. Viewing the length of the flagellum towards the cell, so to speak, the normal rotation is counter clockwise (CCW). This direction ensures the bundles of filaments cling together. If the rotation changes to clockwise (CW), the flagellum frays and is ineffectual. In this way, the bacterium can respond to a stimulus and proceed in a given direction but can only change it rather haphazardly, by changing the direction of the propeller's rotation. The rotary motor which drives the flagellum is similar to the motor which drives the enzyme ATP Synthase, see Section 2.6.3. It comprises a circular stator, made of many protein units, embedded in the membrane. The base of the flagellum fits in the stator and is rotated by intermolecular forces and powered by an ionic gradient. The change in the direction of rotation requires the presence of a protein component, Che Y, and mutants lacking it are unable to change the CCW rotation. Using a microscopic technique described in Chapter 13, Nishiyama and his colleagues (Nishiyama and Kojima 2012; Nishiyama et al. 2013), studied the effects of pressure on a form of *E. coli* lacking Che y. Increased pressure, at 20 °C, slowed the rate of rotation and at the relatively high pressure of 120 MPa the slow rotation reversed direction, becoming CW. In this respect, the flagellum motor is unlike the ATP Synthase motor, which does not change its direction of rotation at high pressure (Chapter 8, Section 8.9.1). Undoubtedly this reversal is one of the most remarkable effects of high pressure we shall meet in this book.

## 2.6 Proteins

At this juncture, the emphasis shifts from cell biology to biochemistry, and we now consider how some ubiquitous macromolecules, actin and tubulin, respond to high pressure. They are structural proteins that, with other filamentous proteins, create the cytoplasmic gel and carry out a variety of mechanical functions. A particular example of another major group of proteins, enzymes, will be similarly considered at the end of the chapter. In Chapter 4 there is more about the effects of high pressure on other special proteins embedded in the cell membrane, called channels.

### 2.6.1 Actin

Actin, mentioned earlier, is a protein found in all cells. As we have seen, it is made of a string of 375 amino acids, coiled and folded into a bundle, a globular protein. Actin is highly conserved, that is, only small differences exist between the amino acid composition of actin in organisms widely separated in evolution. The monomer, globular or G form, readily polymerises to the F, filamentous form when bound to adenosine triphosphate (ATP), given the right concentration of monomers, salts and temperature. Many years ago Ikkai and Ooi (1966) showed that a decrease in temperature reduced the viscosity of a solution of F-actin and that high pressure likewise reduced viscosity by favouring depolymerisation. They then went on to directly measure the volume change involved in the monomer–polymer equilibrium. Finding it to be very large they attributed it to the release of ordered water from the monomeric subunits as they assembled to form the polymer. Thermodynamically the reaction can be described as endothermic and entropy driven, as in polyvalyl RNase, the gain in entropy coming from the disordering of the previously ordered water molecules clustering around the subunits. The positive $\Delta V$ has implications for the function of actin in deep-sea animals at high pressure, but does actin behave like this in cells?

Actin polymerisation leads to a two stranded helix, forming a gentle spiral, in which each monomer is in contact with four other monomers. The details of the contact between monomers is now known from X-ray crystallographic studies. In the living cell, polymerisation, often called self-assembly, is more complicated and it involves several molecules which bind to actin, including ATP, already mentioned. It is hydrolysed after polymerisation with a release of energy, stored as structural strain in the resultant filament. The form of actin, which is part of the cytoplasmic structure of undifferentiated cells, differs from the actin in muscle cells. In the latter the protein tropomyosin binds in the groove of the filament, stiffening it, with each molecule covering seven actin monomers. A protein called troponin binds to the end of each tropomyosin. The number of other proteins that can also bind to actin is bewildering; there are dozens of them.

Actin, which combines with the muscle protein myosin, is pointed at one end and is barbed at the other. Polymerisation proceeds about ten times more rapidly at the barbed end than at the pointed end, and each end has its characteristic critical concentration of monomers for polymerisation. A critical stage in polymerisation is nucleation, which is sensitive to high pressure, at least in the case of rabbit muscle actin. Detailed studies have been carried out by Winter and his colleagues on actin and other structural proteins (Gao and Winter 2015), and from their work one aspect is selected here to illustrate the effects of macromolecular crowding, mimicking intracellular conditions, on the polymerisation of actin. The experiments were designed to avoid the nucleation stage and to focus on the slow polymerisation at the pointed end of the molecule. In normal laboratory conditions polymerisation, measured by a fluorescence technique, was determined by the concentration of the G form. The "crowding" agent used was Ficoll 70, a polymer of sucrose, and this introduced inert, hydrophobic surfaces to the reaction mixture, and hence significant "excluded volume." The resultant increases in the effective concentration of G actin increased the rate of polymerisation. When the polymerisation reaction was run at high pressure a marked slowing occurred at 80 MPa, but this was abolished in the presence of Ficoll. This means that the positive activation volume is reduced to close to zero by the presence of Ficoll, in "crowded" conditions. The thermodynamics of this is for another day, but for now the point is that if the results of in vitro studies are to reliably guide us in understanding the adaptation of actin to high pressure, in for example deep-sea fish muscle, then due regard for physiological reaction conditions is essential (Chapter 6). A point which, of course, applies generally.

Actin is an important component of the cytoplasmic gel in undifferentiated cells, and close to the cell membrane there are actin filaments lying in the plane of the membrane, creating a distinct cell cortex. An extreme form of this is the contractile ring which constricts a dividing cell in two (Begg et al. 1983).

It will be recalled that the cleavage process in dividing cells is arrested, and indeed the cleavage furrow was seen to relax and reverse, at high pressure. We now know that the furrow consists of a contractile ring of actin filaments which generate force by its interaction with myosin, probably in a muscle like way. However, the actin–myosin system of muscle cells is not pressure labile. In the case of the eggs of the sea urchin (*Arbacia punctulata*) the contraction of the cleavage furrow is arrested by 40 MPa and the bundles of actin filaments in the furrow were shown, by electron microscopy, to disappear. They reappeared within a minute of decompression. This supports the idea that in the living cell high pressure acts on a reaction controlling the self-assembly of the actin filaments rather than on their generation of force.

The shift in the sol-gel equilibrium which high pressure causes in amoeboid cells also involves actin filaments, this time cross-linked to form a gel. Following the initial observations on *Amoebae*, subsequent workers have applied high pressure to a variety of cells in culture, including monkey kidney epithelial cells, chicken heart fibroblasts, rat osteosarcoma and HeLa cells, the last two being cancer cells. The cells grow on the base of a culture dish, adhering with special anchor proteins, which cause the cells to be flat with extended, often thin, processes. Under pressure (30–40 MPa) many cells become rounded up, a condition associated with the disappearance of actin filaments (and other structural proteins). Rounding up, due to surface tension, is what would be expected of a drop of firm gel becoming liquefied. However, the idea that this liquefaction is the result of a simple shift in a physical, sol-gel equilibrium, is inconsistent with the following. In a given popula-tion of cells growing in the same dish, there is variation in the degree of rounding up which occurs. Cells in contact with each other round up much less than isolated cells. Cells in contact with others experience contact inhibition and reduced mobility. They are in a physiologically different state from isolated cells. Quite separately, the protein vimentin is depolymerised in Hela cells by 40 MPa but in rat osteosarcoma it is unaffected. Also, in HeLa cells, the structural protein cytokeratin is readily depolymerised by pressure yet it is extremely resistant to depolymerisation in vitro. These complicating facts led Crenshaw and his colleagues to the view that high-pressure targets reactions that drive the polymerisation of proteins, such as phosphorylation, which also control the state of the proteins in the cell. It does not act on a simple equilibrium.

## 2.6.2  Tubulin

Tubulin is the protein of which microtubules are made. Although microtubules are ubiquitous, dynamic cytoplasmic structures, we only learned of their fine details when glutaraldehyde was introduced as a fixative for electron microscopy (fixation means chemically preserved). One of the first examples of the dynamic properties of microtubules in cells was discovered by Kitching in the 1950s, studying the elegant but obscure Protozoan *Actinophrys sol*, and a multinucleate version *Actinosphaerium*, (previously order Heliozoa but see Chapter 1, Section 1.5). These cells, as their classification suggests, consist of a central body supporting radiating axopods. Each axopod is supported by a double helical array of microtubules, which, when viewed from the cell exterior, are all organised in a clockwise spiral. In normal life, the axopods readily retract and re-grow. Kitching used a microscope pressure chamber to observe that high pressure caused the axopods to collapse and, within 2 min of decompression, to start to grow back, eventually reaching their original size. As with the effects of high pressure on cytoplasmic sol-gel equilibria, the effect of pressure was opposed by an increase in temperature. Thus at 15–20 °C, axopod collapse occurred within the pressure range of 27–34 MPa and at 5–10 °C at 13–20 MPa (Kitching 1957). Pressure-induced regression of microtubular structures has also been studied in *Tetrahymena* and in the flagella of *Polytomella* (Moore 1972; Brown and Rogers 1978).

Microtubules are involved in mitosis, a basic part of the cell cycle (Box 2.7) and they too happen to be sensitive to high pressure.

---

### Box 2.7: Mitosis

Mitosis is an extraordinary process in which the nucleus in a dividing eukaryote cell divides and transfers its replicated chromosomes to the two daughter cells. It is thus part of the cell cycle (Morgan 2007), mentioned earlier, which pressure also affects in ways which are not visibly obvious. The generalised mitotic apparatus comprises three types of microtubular arrays. Websites provide a variety of illustrations. The astral microtubules keep the apparatus in the right position within the cell. The interpolar microtubules run from one centrosome to the other and the kinetochore microtubules are organised with their minus ends in the centrosome and their plus ends attached to the kinetochore, the point of attachment with the chromatid, a recently duplicated chromosome. Much cross-linking between microtubules takes place. The whole apparatus is, after all, a mechanical device for organising the movement of chromatids. Cells vary, but most begin mitosis with the disappearance of the nuclear membrane, exposing the paired chromatids which, miraculously (it seems) separate such that each sister of a pair attaches to microtubules which pull it to opposite spindle poles. What is actually seen is the shortening of the kinetochore microtubules, probably arising from depolymerisation at the plus end attached to the kinetochore. Additionally, depolymerisation at the minus (centrosome), end pulls the microtubules towards the poles. This is just a simplified part of the performance of the mitotic apparatus. Its assembly involves both polymerisation and motor movement. The latter is generated by ingenious motor proteins which, for example in one case bridge a pair of overlapping antiparallel microtubules at the equator and move along them, causing each to move in opposite directions. This results in the microtubules pushing the poles further apart. There is more on motor proteins and their sensitivity to high pressure, below. Websites dealing with mitosis can be very informative.

(continued)

**Box 2.7**  (continued)

What happens to mitosis at high pressure? The eggs of the marine worm *Chaetopteris* were studied by Inoue, Salmon and colleagues using a special high-pressure microscope which revealed the birefringence created by bundles of microtubules in the mitotic array (Chapter 13). The birefringence retardation provides a measure of the degree of polymerisation, in vivo. A pressure of 13.6 MPa was applied to the egg at the metaphase stage of mitosis. This is characterised by the chromatids being aligned at the equator, attached to the microtubules and ready to start their poleward movement. The effect was a marked decrease in spindle birefringence, the quantity of polymerised microtubules, in the light path. Decompression restored the status quo. The length of the spindle decreased with depolymerisation and was restored on decompression.

The $\Delta V$ for the polymerisation equilibrium assumed to be present was calculated from the optical data to be positive and large. Similar experiments showed that cooling the metaphase spindle also depolymerised it, reversibly. Thus, the polymerisation reaction is endothermic and entropy driven, involving the formation of hydrophobic interactions, similar to the aggregation of poly valyl RNase. However, there is so much biochemical complexity involved it is better to conclude that the system responds as if it was a simple equilibrium.

Polymerised tubulin consists of a string of dimers, a beta-tubulin monomer combined with an alpha-tubulin monomer "head to tail." The wall of the microtubule cylinder, circular in cross section, is typically made of an array of 13 (or so) tubulin protofilaments, with neighbouring alpha and beta monomers binding to their like neighbour (homotypic binding). The microtubule has a longitudinal "seam" where alpha monomers bind to beta and vice versa (heterotypic binding). One end of the microtubule ends with an alpha subunit (the minus end) and correspondingly the other end, terminating with a beta subunit, is positive. Monomer subunits can only be added to or subtracted from the ends of the microtubule. The cylindrical body of the tubule is stable and indeed rigid.

Only weak non-covalent bonds contribute to this structure which is so important in the numerous functions which microtubules perform in cells; mechanical support, a framework for motor proteins and generating a motive force of their own. Some microtubular structures are assembled and then disassembled frequently in the life of the cell, as we have seen, which is consistent with the idea that their assembly (polymerisation) arises from a pool of monomers. Other structures, such as the microtubules in cilia, microscopic hair-like processes extending from the surface of cells, are much more permanent.

(continued)

**Box 2.7** (continued)

Experiments with such microtubules in vitro, using the criteria of birefringence, optical density and the number of microtubules seen with electron microscopy, can produce results similar to those obtained with eggs, but generally microtubule activity in vivo is controlled by associated proteins, of which there are many. The ubiquitous distribution of microtubules is illustrated by the fact that the rabbit brain is a good source of tubulin for experiments. Microtubules are associated with the transport of material along the length of neurons which, of course, requires the generation of force. That force arises from special motor proteins, dyneins and kinesins, which move along the monorail provided by microtubules. The squid giant axon was the source of one of the first kinesins to be isolated. It transports membranous vesicles along the axon, away from the cell body, towards the plus end of microtubules (i.e., microtubules have polarity, of which more soon). Dyneins transport in the reverse direction.

The Kinesin 1 motor protein which moves along microtubules is susceptible to high pressure. This was demonstrated in brilliant experiments carried out by Nishiyama and his colleagues (Nishiyama et al. 2009). They used microtubules isolated from the pig brain, and high-pressure fluorescence microscopy (Chapter 13). Microtubules, with their polarity marked, were attached to the surface of the high-pressure glass window so that when the pressure was varied the length of the microtubule could be measured. At atmospheric pressure the lengths were stable, but increasing pressure (up to 140 MPa), reduced their length, but probably not by depolymerisation, because the lengths were not restored on decompression. It was, however, necessary to establish this effect because when the kinesin was attached to the microtubules, two distinct observations were made. First, the attached kinesin did not affect the pressure-induced shortening, but 130 MPa approximately halved the rate at which the kinesin moved towards the plus end of the microtubule. On decompression, the rate of movement returned to normal, so this pressure effect is reversible. So how and where does pressure act?

Kinesin approximates to a $V$ shape, with the free ends able to bind ATP and to a tubulin monomer. Thermal motion allows a free limb of the $V$ to bind to a monomer, which induces conformational changes in the protein structure. The "trailing" limb dissociates and by thermal motion biased by a conformational change, it finds itself ahead of its partner limb and within binding range of another tubulin monomer. The tubulin polymer's directional structure, with plus and minus ends, ensures the Kinesin moves towards the plus end, and the movement is described as walking. The hydrolysis of ATP to ADP and phosphate channels energy into the dissociating stage of the cycle of kinesin binding and dissociation. High pressure inhibits the "stepping" motion, as the authors put it, probably by hydration changes. It does not affect the binding reaction.

Winter and colleagues have applied X-ray and spectroscopic techniques to microtubules extracted from the brain of calves, measuring the effects of pressure on microtubules in different states of organisation (Gao et al. 2018). Very high pressures are required to have much effect on the secondary and quaternary structures of the protein tubulin, but in bundles of microtubules the homotypic lateral interactions between alpha monomers and between beta monomers are more susceptible. Tens of MPa rather than several hundred MPa have a significant effect, causing the separation of protofilaments. Depolymerisation of the head to tail monomers is more pressure resistant. The pressure sensitivity of the intra-dimer, homotypic binding sites, can, in part, be accounted for by the large void volume in the quaternary structures. Other factors may also contribute, such as associated proteins and the excluded volume effect, reducing molar volume changes, as mentioned in connexion with actin.

A theme of this study by Winter's group is that identifying the most pressure-sensitive sites in ordinary microtubules should reveal sites that are likely to require adaptation to function at high pressure. A further point is that the organisational complexity of dynamic instability, mentioned above, suggests that the study of deep-sea microtubules may require the isobaric collection of the cells of interest, whereas the study of deep-sea actins apparently does not (Chapter 5; Section 5.4.3). Finally, microtubules in general show a wide range of pressure resistance which is presumably achieved by diverse ways, some of which might render some microtubules pre-adapted to high pressures.

The last point is illustrated by Yeast cells, for example *Saccharomyces cerevisiae*, which have microtubules within the nucleus where they play a role in its division. Very high pressure (150 MPa, 30 °C, 10 min) was required to depolymerise these microtubules and such damaging high pressures irreversibly inhibit growth and cell division, measured by the ability of the cells to form colonies after decompression, Kobori et al. (1995). In other conditions, certain strains of *S. cerevisiae* have their growth and division reduced at pressures of only 15–25 MPa, which inhibits the essential uptake of the amino acid tryptophan. Interestingly, actin fibres in another yeast, *Schizosaccharomyces pombe*, are also pressure resistant, as 100 MPa is required to disorganise them (Sato et al. 1999).

Bacterial cell division does not entail mitosis as there is no nucleus to divide. However, it has been known for many years that high pressure blocks bacterial division, producing long filaments of cells that have failed to separate. Filamentous forms of *E. coli* arise in this way because pressure affects a protein encoded by the gene *ftsZ*, appropriately called FtsZ. It is structurally similar to tubulin and at the end of the division process previously synthesised monomers of FtsZ undergo polymerisation, forming an equatorial ring at the site where the two daughter cells will separate. The ring constricts the cell in two as a septum grows. High pressure appears to block polymerisation of the Ftsz ring. This was shown in experiments in which *E. coli* was grown at various pressures and "fixed" at high pressure (Ishii et al. 2004). At 40 and 50 MPa cells became filamentous and after fixation they were seen under the electron microscope to lack the characteristic ring of FtsZ protein. If, however, the cells were decompressed from 50 MPa and then fixed a few minutes later, they were subsequently seen to have the protein ring in place. This shows that the protein was probably synthesised under pressure but failed to form its ring, by polymerising. A second line of investigation was to extract the FtsZ protein and follow its polymerisation in vitro. Typically, filaments about 5 μm long are formed, and they also can be fixed and seen with the electron microscope. If the polymerisation reaction is started and allowed to run for 10 min before the reaction mixture is fixed at pressure, the filaments subsequently seen under the electron microscope are greatly affected by the magnitude of the pressure. After 30 MPa treatment the filaments were few in number and shorter than normal. After 50 MPa no filaments were to be seen. The overall conclusion is that pressure interferes with cell division principally by blocking the polymerisation of FtsZ. Other processes, such as DNA synthesis for example may be affected, but they are not responsible for blocking the final stage of division and filament formation.

### 2.6.3  A Special Enzyme, ATP Synthase

The final example of how a macromolecule behaves under high pressure features an enzyme, ATP Synthase, chosen for a host of reasons. As all ready explained, the effects of high pressure on the activity of enzymes are important in understanding life at high pressure, and the subject will be taken further in Chapter 6. ATP Synthase is commonly the source of ATP, adenosine triphosphate, which is generated in the chloroplasts of plants, the plasma membrane of bacteria and the mitochondria of all types of organism. ATP is the compound that provides energy to virtually all the energy-requiring life processes; transporting solutes, synthesising molecules, generating mechanical force, and powering the nervous system of animals Its synthesis is part of the adaptation of organisms to pressure, but here we are simply concerned with how the "standard model" responds to pressure.

Although the general name of the enzyme is ATP Synthase, recognising the forward and reverse directions of the reaction it catalyses, it is also known as Fo F1-ATP synthase and Fo F1-ATPase (Section 2.2.2.2). The former generates ATP by incorporating a high energy phosphate bond (ADP + P = ATP) and the latter uses that energy (ATP = ADP + P plus energy) to pump protons across a membrane and up a concentration gradient.

ATP Synthase is ubiquitous and is thought to have existed at an early stage in the evolution of life. Despite being almost as ancient as life itself its structure is very large and complicated and best thought of as an enzyme complex, sometimes called a nanomachine. It comprises several major units; a hydrophobic unit (Fo) which is normally embedded in a membrane, and hydrophilic parts (F1) which are normally bathed in an aqueous environment. Each of these is made up of many smaller protein subunits. Think of ATP synthesis in mitochondria, usually regarded as the forward reaction of ATP Synthase. It harnesses the energy provided by the gradient of protons across the mitochondrial membrane which has been created by the enzymes in the respiratory chain. In plants photosynthesis does the same. This proton gradient, or proton motive force, drives the synthesis of ATP. This is Chemiosmosis, the term used by Mitchell to describe his theory, for which he was awarded of the Nobel Prize for Chemistry in 1978. The detailed structure and functioning of the ATP Synthase complex were unravelled in subsequent years by numerous workers and two, Boyer and Walker, were jointly awarded the Nobel prize for Chemistry in 1997 for their contribution (sharing it with Skou who worked on a related enzyme).

The following account relates to the inner membrane of mitochondria but similar processes take place in chloroplast membranes in plants and the plasma membrane of bacteria (Adachi et al. 2007).

Cellular respiration, generating the energy required by cells, consists ultimately of a flux of electrons through a series of enzymes in the membrane. Three extract energy to pump protons, thus generating an electrochemical gradient across the membrane which provides the energy for subsequent, separate reactions. Cellular respiration can be measured by the rate at which oxygen is consumed (disappears) in the cell or whole organism. The final stage takes place in cytochrome C oxidase involving the combination of four electrons, four hydrogens and oxygen summarised as:

$$4e + 4H^+ + O_2 = 2H_2O.$$

The size of the Synthase is, in molecular terms, huge. It consists of more than 3500 amino acids organised in various protein subunits. It is a rotary nanomachine which synthesises ATP and its elaborate structure makes rotation possible. We cannot get into the detailed structure here, but we can ask, how is it known that rotation occurs? It is induced by the flow of protons, down their electrochemical gradient through a favourable route provided by the protein structure, releasing energy which is channelled into synthesising ATP from ADP. When first postulated, on structural and other evidence, it was a very original idea. Linear motion but not rotary motion was well established in molecular biology. However, the rotary motion was directly demonstrated by Noji and colleague (Noji et al. 1997), working at the Tokyo Institute of Technology and Keio University, Yokohama, Japan. They used the Synthase from a thermophilic bacterium, modified by mutagenesis to produce an F1 ATPase which (a) had histidine tagged to certain subunits and (b) another subunit was tagged with molecules to which fluorescently labelled actin filaments were attached. The whole structure was then mounted on a specially treated glass plate to which the histidines attached, making it stand upright. A solution of ATP passed through the system caused rotation. This was detected by a CCD camera and image intensifier, focusing on the fluorescently labelled actin. The rotation was 0.5 revolutions per second and, as viewed from the membrane side, the rotation was seen to be anticlockwise. Subsequent work showed that at low concentrations of ATP the rotation was stepwise, proceeding in 120° steps, later resolved into two smaller sub-steps. Other experiments have shown that one step is associated with the binding of ATP and the other step with the release of phosphate, providing the energy for rotation, and in vivo the pumping of protons up their electrochemical gradient.

When the ATP Synthase is working in the forward direction, driven by the proton motive force, a phosphate combines with ADP to form ATP at each of the three 120° steps. Thus, three ATPs are formed each revolution. It takes ten protons to drive one revolution. An often quoted rate of rotation is 8000 revolutions per second, generating around 400 ATP molecules per second, per Synthase complex. A mitochondrion contains many thousands of Synthases and cells generally have hundreds to many thousands of mitochondria, depending on how energetic they are. Rotation is rare in biological structures, the only other example is the motor which drives bacterial flagella.

This account is the minimum required to convey something of the amazing structure of this nanomachine. Readers interested in knowing more should consult Lane's "Why life is the way it is," and for a more formal account the papers listed at the end of the chapter and a good textbook. Its sensitivity to high pressure is considered next.

### 2.6.3.1 The ATP Synthase Under High Pressure

The biological significance of the ATP Synthase has been recognised for a long time and numerous workers have accordingly chosen to study its susceptibility to high hydrostatic pressure. Its multi-subunit structure, embedded in a membrane, provides the prospect of interpreting the results.

The first experiments were carried out by Penniston (1971), who, using mitochondria from the rat liver and the heart of the cow, measured ATPase activity, that is the reaction hydrolysing ATP. The rate of reaction was measured by the rate at which phosphate was detached from ATP, forming ADP, and it was found to be reduced by pressures of up to 140 MPa at 25–27 °C. On decompression to atmospheric pressure, the activity was promptly restored to normal, consistent with the inhibition arising from the reversible dissociation of the enzyme's subunits. Dreyfus and colleagues followed this up, using the same source of enzyme. Submitochondrial particles (membranes) provided ATPase activity which was reduced by high pressure but restored to normal on decompression to atmospheric pressure. This and other evidence also strongly suggested that inhibition was caused by the dissociation of subunits. However, when the enzyme was removed from the membrane, i.e., solubilised and now called the F1 ATPase (see above), pressure inhibited the ATPase activity but in a way which was not restored on decompression. This is consistent with the idea that, in solution the dissociated subunits become more dispersed and re-associate incorrectly on decompression, producing assemblies which, unsurprisingly, do not work as catalysts.

The microbiologists Marquis and Bender found the equivalent enzyme in the plasma membrane of *Streptococcus faecalis* and *E. coli* was inhibited by pressure, and Souza et al. (2004), showed that the same enzyme from plant chloroplasts underwent dissociation to subunits at 200 MPa. Experiments with the Fo F1-ATPase extracted from the heart mitochondria of the cow showed the now familiar dissociation to subunits at 200 MPa, but that the presence of special chemicals, e.g., methylamines, protected the enzyme from the dissociating effects of pressure, meaning the enzyme retained its activity at high pressure (Saad-Nehme et al.

2001). This sort of protection against the effects of high pressure is important and is dealt with in Chapter 7.

More recent experiments have studied the chloroplast ATP Synthase, Souza et al. (2004) and others have measured the rotation of the enzyme at high pressure. This impressive achievement was accomplished by Okuno et al. (2013), with the *Bacillus* synthase expressed in *E. coli*. They used the optical system devised by Nishiyama which comprised an inverted microscope with a long-range objective and an optical pressure chamber connected to a perfusion system (Chapter 13). The F1 ATPase enzyme was essentially the same as previously described but a reflective bead was attached to it in place of the actin filament previously used. Bright-field images were recorded at 100 frames per second. The main effect of high pressure, up to 140 MPa at 22 °C, is to slow the rate of rotation by increasing the length of pauses at the 120° steps and to eliminate the intermediate sub-steps. The direction of rotation and the torque were unaffected. Pressure thus slows the kinetics. In this form the enzyme's multimeric structure appears to be stable at high pressure. A simple interpretation of the experiment is that high pressure increases the Km for ATP, by subtle structural changes in the binding site.

# References[1]

Adachi K et al (2007) Coupling of rotation and catalysis in F1-ATPase revealed by single molecule imaging and manipulation. Cell 130:309–321

Akasaka K, Kitahara R, Kamatari YO (2013) Exploring the folding energy landscape with pressure. Arch Biochem Biophys 531:110–115

Ananthan J, Golberg AL, Voellmy R (1986) Abnormal proteins serve as Eukaryotic stress signals and trigger the activation of heat shock genes. Science 232:522–524

Ashford MLJ, Macdonald AG, Wan KT (1982) The effects of hydrostatic pressure on the spontaneous release of transmitter at the frog neuromuscular junction. J Physiol 333:531–543

Bangham AD (1983) Liposome letters. Academic Press, London

Barshtein G, Bergelson L, Dagan A, Gratton E, Yedgar S (1997) Membrane lipid order of human red blood cells is altered by physiological levels of hydrostatic pressure. Am J Phys 272:H538–H543. *

Begg DA, Salmon ED, Hyatt AH (1983) The changes in structural organization of actin in the sea urchin egg cortex in response to hydrostatic pressure. J Cell Biol 97:1795–1805

Belmont KD, Orlova A, Drubin DG, Egelman EH (1999) A change in actin conformation associated with filament instability after Pi release. Proc Natl Acad Sci U S A 96:29–34. *

Beney L, Perrier-Cornet JM, Wayert M, Gervais P (1997) Modification of phospholipid vesicles induced by high pressure: influence of bilayer compressibility. Biophys J 72:1258–1263

Ben-Naim A (2009) Molecular theory of water and aqueous solutions. Part 1. Understanding water. World Scientific, London

Bray D (2001) Cell movements. From molecules to motility. Garland Publishing, New York. *

Bridgman PW (1914) The coagulation of albumen by pressure. J Biol Chem 19:511–512. *

Brown DES (1934) The pressure coefficient of "viscosity" in the eggs of Arbacia punctulata. J Cell Comp Physiol 5:335–346

---

[1]Those marked * are further reading.

Brown DES, Marsland DA (1936) The viscosity of amoeba at high hydrostatic pressure. J Cell Comp Physiol 8:159–165

Brown DL, Rogers KA (1978) Hydrostatic pressure-induced internalization of flagellar axonemes, disassembly, and reutilization during flagellar regeneration in Polytomella. Exp Cell Res 117:313–324

Bugyi B, Carlier M-F (2010) Control of actin filament treadmilling in cell motility. Annu Rev Biophys 39:449–470. *

Caforio A, Dreissen AJM (2017) Archaeal phospholipids: structural properties and biosynthesis. Biochim Biophys Acta 1862:1325–1339

Chandler D (2002) Two faces of water. Nature 417:491. *

Chen CR, Makhatadze GI (2017) Molecular determinant of the effects of hydrostatic pressure on protein folding stability. Nat Commun 8:ncomms14561

Chong P, Cossins AR (1983) A differential polarised phase fluorometric study of the effects of high hydrostatic pressure upon the fluidity of cellular membranes. Biochemistry 22:409–415

Crenshaw HC, Allen JA, Skeen V, Harris A, Salmon ED (1996) Hydrostatic pressure has different effects on the assembly of tubulin, actin, myosin 11, vinculin, talin, vimentin and cytokeratin in mammalian tissue cells. Exp Cell Res 227:285–297. *

Czeslik C, Reis O, Winter R, Rapp G (1998) Effects of high pressure on the structure of dipalmitoyl phosphatidylcholine bilayer membranes: a synchrotron X-ray diffraction and FT-IR spectroscopy study using the diamond anvil technique. Chem Phys Lipids 91:135–144

Ding W, Palaiokostas M, Shahane G, Wang W, Orsi M (2017) Effects of high pressure on phospholipid bilayers. J Phys Chem B 121:9597–9606

Dreyfus G, Guimaraes-Motta H, Silva JL (1988) Effect of hydrostatic pressure on the mitochondrial ATP Synthase. Biochemist 27:6704–6710. *

Earl CRA (1986) Physical properties of membranes formed of phospholipids having monomethyl branched fatty acyl chains. In: Klein RA, Schmitz B (eds) Topics in lipid research. Royal Society of Chemistry, New York, pp 249–252

Eggeling C et al (2009) Direct observation of the nanoscale dynamics of membrane lipids in a live cell. Nature 457:1159–1163

Ernst R et al (2016) Homeoviscous adaptation and the regulation of membrane lipids. J Mol Biol 428:4776–4791

Fooglia F et al (2019) In vivo water dynamics in Shewanella oneidensis bacteria at high pressure. Sci Report 9:8716–8727

Frey B et al (2008) Cells under pressure–treatment of Eukaryotic cells with high hydrostatic pressure, from physiologic aspects to pressure induced cell death. Curr Med Chem 15:2329–2336

Fujiwara I, Takahashi S, Tadakuma H, Funatsu T, Ishiwata S (2002) Microscopic analysis of polymerization dynamics with individual actin filaments. Nat Cell Biol 4:666–673. *

Gao M, Winter R (2015) Kinetic insights into the elongation reaction of actin filaments as a function of temperature, pressure and macromolecular crowding. ChemPhysChem 16:3681–3686

Gao M et al (2018) On the origin of microtubules' high pressure sensitivity. Biophys J 114:1080–1090

Gromiha MM, Ponnuswamy PK (1993) Relationship between amino acid properties and protein compressibility. J Theor Biol 165:87–100. *

Gross M, Jaenicke R (1994) Proteins under pressure. Eur J Biochem 221:617–630

Gross M, Lehle K, Jaenicke R, Nierhaus KH (1993) Pressure-induced dissociation of ribosomes and elongation cycle intermediates. Eur J Biochem 218:463–468

Gyorgy B et al (2011) Membrane vesicles, current state of the art: emerging role of extracellular vesicle. Cell Mol Life Sci 68:2667–2688. *

Hauge JG (1971) Pressure induced dissociation of ribosomes during ultracentrifugation. FEBS Lett 17:168–172

Heerklotz H, Seelig J (2002) Application of pressure perturbation calorimetry to lipid bilayers. Biophys J 62:1445–1452

Horne RA, Courant RA (1965) Protonic conduction in the water I region. J Phys Chem Ithaca 69:2224–2230. *

Iametti S, Donnizzelli E, Pittia P, Rovere PP, Squarcina N, Bomomi F (1999) Properties of a soluble form of high pressure treated egg albumen. In: Ludwig H (ed) Advances in high pressure bioscience and biotechnology. Springer, Berlin. *

Ikkai T, Ooi T (1966) Actin: volume change on transformation of G-Form to F-Form. Science 152:1756–1757

Infante AA, Demple B, Chaires JB (1982) Analysis of the *Escherichia coli* ribosome-ribosomal subunit equilibriium using pressure -induced dissociation. J Biol Chem 257:80–97

Infrastructure and Activities of Cells (1991) Biotol Series. Butterworh & Heinemann, Heinemann. *

Inoue S, Fuseler J, Salmon ED, Ellis GW (1975) Functional organization of mitotic microtubules. Biophys J 15:725–744. *

Ishii A, Sato T, Wachi M, Nagai K, Kato C (2004) Effects of high hydrostatic pressure on bacterial cytoskeleton FtsZ polymers in vivo and in vitro. Microbiology 150:1965–1972

Jha A (2016) The water book. Headline Publishing Group, UK

Johnson FH, Eyring H (1970) The kinetic basis effects of pressure effects in biology and chemistry. In: Zimmerman AM (ed) High pressure effects on cellular processes. Academic Press, New York, pp 1–44

Kaur L et al (2016) High pressure processing of meat: effects on ultrastructure and protein digestibility. Food Funct 7:2389–2397

Kettman MS, Nishikawa AH, Morita RY, Becker RR (1966) Effect of hydrostatic pressure on the aggregation reaction of poly-L-valyl-ribonuclease. Biochem Biophys Res Commun 22:262–267

Kitching JA (1954) The effects of high pressures on a suctorian. J Exp Biol 31:56–67

Kitching JA (1957) Effects of high hydrostatic pressure on Actinophrys sol (Heliozoa). J Exp Biol 34:511–517

Kitching JA, Pease DC (1939) The liquefaction of the tentacles of suctorian protozoa at high hydrostatic pressure. J Cell Comp Physiol 14:410–412

Kobori H et al (1995) Ultrastructural effects of pressure stress to the nucleus in Saccharomyces cerevisiae: a study by immunoelectron microscopy using frozen thin sections. FEMS Microbiol Lett 132:253–258

Krywka C, Sternemann C, Paulus M, Tolan M, Royer C, Winter R (2008) Effect of osmolytes on pressure-induced unfolding of proteins: a high pressure SAXS study. ChemPhysChem 9:2809–2815

Kuhn JR, Pollard DT (2005) Real-time measurements of actin filament polymerization by total internal reflection fluorescence microscopy. Biophys J 88:1387–1402. *

Kultz D (2005) Molecular and evolutionary basis of the cellular stress response. Annu Rev Physiol 67:225–257. *

Laidler KJ (1951) The influence of pressure on the rate of biological reactions. Arch Biochem Biophys 30:226–236

Landau JV (1966) Protein and nucleic acid synthesis in *Escherichia coli*: pressure and temperature effects. Science 153:1273–1274. *

Lane N (2016) The vital question. Why is life the way it is? Profile Books, London. *

Leberman R, Soper AK (1995) Effect of high salt concentrations on water structure. Nature 378:363–366. *

Lin L-N, Brandts JF, Brandts M, Plotnikov V (2002) Determination of the volumetric properties of proteins and other solutes using pressure perturbation calorimetry. Anal Biochem 302:144–160. *

Macdonald AG (1967a) Effects of high hydrostatic pressure on the cell division and growth of Tetrahymena pyriformis. Exp Cell Res 47:569–580

Macdonald AG (1967b) Delay in the cleavage of Tetrahymena pyriformis exposed to high hydrostatic pressure. J Cell Physiol 70:127–129

Macdonald AG (1978) A dilatometric investigation of the effects of general anaesthetics, alcohols and hydrostatic pressure on the phase transition in smectic mesophases of dipalmitoyl phosphatidylcholine. Biochim Biophys Acta 507:26–37

Macdonald AG (2001) Chapter 59: Effects of high pressure on cellular processes. In: Sperelakis N (ed) Cell physiology sourcebook, a molecular approach. Academic Press, New York, pp 1003–1023. *

Macdonald AC, Kitching JA (1967) Axopodial filaments of Heliozoa. Nature 215:99–100. *

Marenduzzo D, Finan K, Cook PR (2006) The depletion attraction: an underappreciated force driving cellular organization. J Cell Biol 175:681–686. *

Marquis RE, Bender GR (1987) Barophysiology of prokaryotes and proton-translocating ATPases. In: Jannasch HW, Marquis RE, Zimmerman AM (eds) Current perspectives in high pressure biology. Academic Press, New York, pp 65–73. *

Marsland D (1958) Cells at high pressure. Sci Am 199:36–43

Marsland D (1970) Pressure–temperature studies on the mechanisms of cell division. In: Zimmerman A (ed) High pressure effects on cellular processes. Academic Press, New York, pp 259–312

Meersman F, McMillan PF (2014) High hydrostatic pressure: a probing tool and a necessary parameter in biophysical chemistry. Chem Commun 50:766–775

Meersman F, Dobson CM, Heremans K (2006) Protein unfolding, amyloid fibril formation and configurational energy landscapes under high pressure conditions. Chem Soc Rev 35:908–917

Minton A (2006) Macromolecular crowding. Curr Biol 16:R269

Moore KC (1972) Pressure-induced regression of oral apparatus microtubules in synchronised Tetrahymena. J Ultrastruct Res 41:499–518

Morgan DO (2007) The cell cycle. New Science Press, London

Morild E (1981) The theory of pressure effects on enzymes. Adv Protein Chem 34:93–166

Morita T (2003) Structure- based analysis of high pressure adaptation of a-actin. J Biochem 278:28060–28066. *

Mozhaev VV, Heremans K, Frank J, Masson P, Balny C (1996) High pressure effects on protein structure and function. Proteins 34:81–91

Murayama M (1966) Molecular mechanism of red cell "sickling". Science 153:145–149. *

Nguyen HTM et al (2020) Ribosome reconstruction during recovery from high hydrostatic pressure induced injury in Bacillus subtills. Appl Environ Microbiol 80:e01640–e01619

Nishiyama M, Kojima S (2012) Bacterial motility measured by a miniature chamber for high pressure microscopy. Int J Mol Sci 13:9225–9239

Nishiyama M, Sowa Y (2012) Microscopic analysis of bacterial motility at high pressure. Biophys J 102:1872–1880. *

Nishiyama M, Kimura Y, Nishiyama Y, Terazima M (2009) Pressure-induced changes in the structure and function of the kinesin-microtubule complex. Biophys J 96:1142–1150

Nishiyama M et al (2013) High hydrostatic pressure induces counterclockwise to clockwise reversals of the *Escherichia coli* flagellar motor. J Bacteriol 195:1809–1814

Noji H, Yasuda R, Yoshida M, Kinosita K (1997) Direct observation of the rotation of F1-ATPase. Nature 386:299–302

Oda T, Iwasa M, Aihara T, Maeda Y, Narita A (2009) The nature of the globular -to fibrous-actin transition. Nature 457:5441–5445. *

Okuno D, Nishiyama M, Noji H (2013) Single-molecule analysis of the rotation of F1- ATPase under high hydrostatic pressure. Biophys J 105:1635–1642

Olmstead JB, Borisy G (1973) Microtubules. Annu Rev Biochem 42:507–540. *

Oursel D et al (2007) Lipid composition of the membranes of *Escherichia coli* by liquid chromatography/tandem mass spectrometry using negative electrospray ionization. Rapid Commun Mass Spectrom 21:1721–1728

Pease DC, Marsland DA (1939) The cleavage of Ascaris eggs under exceptionally high pressure. J Cell Comp Physiol 14:407–408. *

Penniston JT (1971) High hydrostatic pressure and enzymic activity: inhibition of multimeric enzymes by dissociation. Arch Biochem Biophys 142:322–332

Pollard EC, Weller PK (1966) The effect of hydrostatic pressure on the synthetic processes in bacteria. Biochim Biochem Acta 112:573–580

Pollard TD, Blanchoin L, Mullins RD (2000) Molecular mechanisms controlling actin filament dynamics in non-muscle cells. Annu Rev Biophys Biomol Struct 29:545–576. *

Purushothaman S, Cicuta P, Ces O, Brooks NJ (2015) Influence of high pressure on the bending rigidity of model membranes. J Phys Chem B 119:9805–9810

Raghupathy R et al (2015) Transbilayer lipid interactions mediate nanoclustering of lipid-anchored proteins. Cell 161:581–594. *

Roche J et al (2012) Cavities determine the pressure unfolding of proteins. Proc Natl Acad Sci U S A 109:6945–6950

Rosin C, Erikamp M, von der Ecken J, Raunser S, Winter R (2014) Exploring the stability limits of actin and its suprastructure. Biophys J 107:2982–2992. *

Ruhlen R, Marberry K (2014) The chondrocyte primary cilium. Osteoarthr Cartil 22:1071–1076

Saad-Nehme J, Silva JL, Meyer-Ferandes RJ (2001) Osmolytes protect mitochondrial FoF1 - ATPase complex against pressure inactivation. Biochim Biophys Acta 1546:164–170

Salmon ED (1975a) Pressure induced depolymerization of spindle microtubules. I. Changes in birefringence and spindle length. J Cell Biol 65:603. *

Salmon ED (1975b) Pressure induce depolymerization of spindle microtubules. II Thermodynamics of in vitro spindle assembly. J Cell Biol 66:114–127. *

Salmon ED (1975c) Pressure induced depolymerisation of brain microtubules in vitro. Science 189:884–886. *

Sato M, Hasegawa K, Shimada S, Osumi M (1999) Effects of pressure stress on the fission yeast Schizosaccharomyces pombe cold sensitive mutant nda3. FEMS Microbiol Lett 176:31–38

Schulz E, Ludemann HD, Jaenicke R (1976) High pressure equilibrium studies on the dissociation–association of E. coli ribosomes. FEBS Lett 64:40–43

Silva JL, Oliveira AC, Vieira TCRG, de Oliveira GAP, Suarez MCV, Foguel D (2014) High pressure chemical biology and biotechnology. Chem Rev 114(14):7239–7267. *

Somero GN (1992) Adaptations to high hydrostatic pressure. Annu Rev Physiol 54:557–577. *

Somero GN (2003) Protein adaptations to temperature and pressure: complimentary roles of adaptive changes in amino acid sequence and internal milieu. Comp Biochem Physiol B 136:577–591. *

Souza MO, Creczynski-Pasa TB, Scofano HM, Graber P, Mignaco JA (2004) High hydrostatic pressure perturbs the interactions between CFoF1 subunits and induces a dual effect on activity. J Biochem Cell Biol 36:920–930

Stühmer W (2015) Exocytosis from chromaffin cells: hydrostatic pressure slows vesicle fusion. Phil Trans Roy Soc B 370:20140192

Sun S, Fisher RL, Bowser SS, Pentecost BT, Sui H (2019) Three-dimensional architecture of epithelial primary cilia. Proc Natl Acad Sci U S A 116:9370–9379

Swezey RR, Somero GN (1985) Pressure effects on actin self-assembly: interspecific differences in the equilibrium and kinetics of the G to F transformation. Biochemist 24:852–860. *

Tolgyes FG, Bode C (2010) Chapter 3: Pressure and heat shock proteins. In: Sebert P (ed) Comparative high pressure biology. Science Publishers, New York, pp 43–60. *

Wheway G, Nazlamova L, Hancock JT (2018) Signalling through the primary cilium. Front Cell Devel Biol 6:8

Wiggins P (2008) Life depends on two kinds of water. PLoS One 1:e1406. *

Winter R, Jeworrek C (2009) Effect of pressure on membranes. Soft Matter 5:3157–3173

Wu W, Chong PL-G, Huang C-H (1985) Pressure effect on the rate of crystalline phase formation of L–a—Dipalmitoylphosphatidylcholines in multilamellar dispersions. Biophys J 47:327–242. *

Yoshidi M, Muneyuki E, Hisabori T (2001) ATP Synthase -a marvellous rotary engine of the cell. Nat Rev Mol Cell Biol 2:669–677. *

Zhao J-H, Liu H-L, Lin C-H, Fang H-W, Chan S-S, Ho Y, Tsai W-B, Chan W-Y (2007) Chemical chaperone and inhibitor discovery: potential treatments for protein conformational diseases. Perspect Med Chem 1:39–48

# The High Pressure Micro-environment of Vertebrate Load Bearing Joints

3

## 3.1 Introduction

When, in the course of animal evolution, vertebrates colonised the land, they lost the buoyancy of water and acquired strong bones and load bearing joints. These joints are best known in mammals and in humans in particular. They cope with a variety of mechanical stresses which can, in certain localised sites, generate pure hydrostatic pressure. When human subjects walk, the layer of cartilage cells covering the load bearing surfaces of their bones experience a hydrostatic pressure which fluctuates cyclically. The human hip joint creates hydrostatic pressure up to 18 MPa during normal movement, more fully explained in a following section. This is, of course, a significant "high" hydrostatic pressure, many examples of which are discussed in this book. However, the complex microstructure of bone contains fluid-filled cavities in which small hydrostatic pressure is also generated. Other tissues contain fluid pressure confined in tubes and chambers, such as blood vessels, the bladder, and the vertebrate eye. Their enclosing wall experiences stretch but there is evidence that cells lining the walls experience, and respond to, very small hydrostatic pressures, as distinct from the stretch. Do our blood cells sense the hydrostatic pressure in which they are immersed, otherwise called blood pressure? And do bacteria, whose intra-cellular turgor pressure can be more than 1 MPa, sense this hydrostatic pressure in any way other than by sensing membrane tension?

Quite separately, shallow water animals, for example the planktonic Ctenophore *Pleurobrachia pileus*, the inter-tidal crab *Carcinus* and other invertebrates, respond to changes in hydrostatic pressure of around, or even less than, 10 kPa (0.01 MPa), equivalent to a depth of water of 1 m (Enright 1962; Morgan 1969; Fraser et al. 2003). These animals lack an obvious gas phase or other macroscopic compressible compartment which would convert changes in ambient hydrostatic pressure into a linear stretch, detectable by established receptors. The implication is that these very small changes in hydrostatic pressure, micro-pressures, are detected as such.

To summarise, load bearing joints contain regions which experience significant hydrostatic pressure, and within the bones and other tissues much smaller

© Springer Nature Switzerland AG 2021
A. Macdonald, *Life at High Pressure*, https://doi.org/10.1007/978-3-030-67587-5_3

hydrostatic pressures exist which happen to be similar to the micro-pressures which are sensed by small aquatic animals. This chapter is primarily concerned with the significant hydrostatic pressure in load bearing joints, but we are faced with a range of hydrostatic pressures in tissues. How are we to regard these smaller pressures? The distinction between significant hydrostatic pressure and the micro-pressures detected by aquatic animals has been discussed in a 1999 paper by Macdonald & Fraser. Here it is posed; are micro-pressures detected by means which are fundamentally different from the way significant hydrostatic pressure exerts its effects?

## 3.2    Micro-pressures

Orthodox high hydrostatic pressures are treated in solution chemistry, thermodynamically, in the way set out in Chapter 2. A brief reminder is given here: The volume term, the conjugate of pressure, determines how an equilibrium constant ($K$) in the equation below, responds to pressure, The bigger the volume change the greater the response

$$d \ln K / dP = -\Delta V / R \times T$$

The smaller the volume change the greater the pressure has to be to have much effect. For example 10 MPa at 20 °C causes a 50% increase in the equilibrium constant $K$ when the volume change is −97.4 ml per mol. This is a moderately large volume change and can be accounted for by changes in the hydration and conformation of molecules. For the very small pressure of 0.2 MPa to cause the same change in $K$ the volume change has to be 50-fold greater (−4870 ml per mol). For a reaction in the water this is a very large, perhaps implausibly large, volume change to be caused by hydration and conformation changes.

If we return to the original volume change of −97.4 ml per mol it may be shown that a pressure of 0.2 MPa (200 kPa) causes $K$ to change by 0.8%. If this is regarded as a threshold change sufficient to exert a physiological effect, then 0.2 MPa can be regarded as the smallest pressure which simple solution chemistry can handle. This arbitrarily defines the limit of how very small hydrostatic pressures can be interpreted. An important caveat is that temperature fluctuations cannot be ignored. A 1 °C change can easily affect $K$ by 10%, swamping our pressure effect. Nevertheless, in general we can regard 0.2 MPa as the lower limit for an effect of pressure which can be interpreted in terms of solution chemistry. This simplest of explanations must be applied and if it does not prove satisfactory then different and more complicated interpretations will have to be used.

How, then, does 0.02 MPa, 20 kPa, which is generated by a 2 m depth of water, act to elicit a physiological response (in a gas free system)? It looks as if solution chemistry cannot cope and so a different treatment is required. In Nature there are numerous examples of aquatic animals and cells in tissues responding to pressures as small as 1 kPa (equivalent to a 10 cm depth of water). Bacterial biofilms may also create conditions with a hydrostatic component (Dufrene and Persat 2020). These micro-pressures are discussed in Macdonald and Fraser (1999), and some certainly appear to exert a physiological effect, i.e., they are transduced. The mechanism(s) by which they are transduced appears beyond the scope of solution chemistry. The piston mechanism postulated to function as a sensor in the crab is discussed by Fraser and Macdonald (1994); Fraser et al. (2001); Fraser (2010). The only cellular system for detecting micro-pressures which has been investigated is an ion channel (Olsen et al. 2011).

To summarise, this chapter deals with conventional hydrostatic pressure effects in the unlikely micro-environment of vertebrate load bearing joints and will leave micro-pressures in cells, tissues and in aquatic organisms lacking a gas phase, for another day.

## 3.3    Load Bearing Joints; Cartilage and Bone Cells Exist in Pressurised Environments

In terrestrial vertebrates, and humans in particular, cartilage and bone are pre-eminently structures which develop in response to mechanical stresses. One is shear, which deforms (stretches) cells and another is hydrostatic pressure.

Cartilage arises from cells called fibroblasts forming chondroblasts which produce a composite material, a matrix. It is made of complex macromolecules; glycoproteins, which are fibrous, another group of fibrous proteins, collagens and proteoglycans. These are made of sulphated glycosaminoglycans linked with a protein. For example an important one in cartilage is aggrecan which is present at the level of 7%. These macromolecules are held in a watery fluid, making up 70% of the matrix. The matrix mesh of macromolecules excludes large molecules but not ions and water molecules. The matrix resists compression and the fibres oppose tension. Chondroblasts become embedded in the extracellular matrix, quite sparsely, and are then called chondrocytes and become encapsulated in a chondron. Their largely anaerobic metabolism depends on the diffusional supply of nutrients and the release of lactic acid, as the matrix lacks a blood supply, and indeed nerves. Chondrocytes, like most vertebrate cells, possess a primary cilium which is not particularly obvious in electron micrographs, but is nevertheless important. Hyaline cartilage which coats the load bearing faces of bones is a material well designed to absorb shock, distribute load and minimise friction between the moving parts (joints). It has a limited ability to regenerate following damage. When the cartilage lining a joint surface is loaded, the contact zone hydrostatically pressurises the material. A small quantity of fluid is squeezed out and may function as a lubricant. At rest the matrix has an osmotically generated internal pressure, "turgor pressure," of about 0.2 MPa, which fluctuates during cyclic loading.

We need a more detailed picture of what happens when a load bearing joint experiences a compressive load. Consider the layer of cartilage covering the surface of the joint. Imagine a small part of the interior of the cartilage layer, and in particular a hypothetical small sphere of cartilage. If and when it experiences a purely hydrostatic pressure, the sphere will not change shape, although it might shrink a minute amount, as cartilage is not incompressible (Carter and Wong 2018). When the cartilage experiences a compressive load which causes shear, the sphere will be slightly squashed into the shape of a rugby ball (or an American football if you prefer), and part of the original sphere's surface will experience stretch. Actual examples of this are illustrated in a paper by Pingguan-Murphy et al. (2005). These are the two quite distinct stimuli to which the chondrocyte, and other cells, respond. (In other structures tension is experienced, but we shall not pursue that). Experimentally the two stimuli, hydrostatic pressure and shear, can be separately applied to suitable tissues or cells.

If chondrocytes, or other cells, are hydrostatically pressurised in liquid, in the familiar way, then no compressive load is present and stretch, on a microscopic scale will not take place. (At a molecular level there may be directional effects, of which more later.) The results of such an experiment can be interpreted in the conventional way of solution chemistry, as already mentioned. Stretch and the subsequent strain and shear can be applied to cartilage. Strain can be applied to a cell in the absence of hydrostatic pressure by applying a micropipette and gently sucking out a swelling. The results of these procedures can be interpreted in terms of ion fluxes and structural changes, the currency of cell biology, in which the primary cilium and fibrous proteins are key structures. The conversion of a linear movement on a microscopic scale into a cellular response is called mechanotransduction, one form of which involves mechanoactivated ion channels.

The contact, loading, zone of the joint is where the hydrostatic pressure occurs and there is an adjacent region that experiences shear stresses. As the joint articulates so the zone of pure hydrostatic pressure moves across the cartilage lining. In the case of the head of the human femur, which fits snugly in the cavity of the pelvis, the acetabulum, the pressure experienced by the cartilage lining the head has been directly measured by a femoral head prosthesis. Early in vitro studies by, for example Rushfeldt and his colleagues paved the way. Hodge and his colleagues also at the Massachusetts General Hospital and the Massachusetts Institute of Technology, Boston, USA, designed and fabricated a prosthesis and implanted it in an elderly lady weighing 68 kg and later in an elderly man slightly lighter in weight (Chapter 13, Fig. 13.8). With Morrell and colleagues, they published results in 1986 and 2005, respectively. The device telemetered interarticular pressures during and after the implant operation, and more important, for months afterwards, provided data on the actual pressures experienced by the joint during normal movements. For example climbing stairs typically caused a step increase of 7 MPa over 0.15 s and normal walking produced a pressure wave peaking in the range of 36 MPa with a frequency of 1 s (Chapter 1, Fig. 1.4). Speed walking increased the pressure peak to as much as 25 MPa. The movement of rising from a chair could produce a pressure increase of 18 MPa.

The data established the load bearing joint as a unique high-pressure micro-environment which experiences high hydrostatic pressure (Urban 1994). Other engineering approaches to understanding what goes on in the layer of cartilage seem reluctant to use the term hydrostatic pressure (Karpinski et al. 2019). However, there is a consensus that hydrostatic pressure conditions, cited earlier, are met (Elder and Athanasiou 2009). Looking back, it is interesting to note that this new high-pressure micro-environment was established just a few years after hydrothermal vents became understood as another new and special high-pressure environment.

The doubtless gentle movements of elderly people are unlikely to provide us with the upper limit of the hydrostatic pressures experienced by load bearing joints in humans. Bodyweight can be used as a relative unit of force, for example rising from a sitting position subjects the knee to a force two and a half times bodyweight, and descending stairs exerts three and a half time the bodyweight on the same joint. These figures were obtained from an instrumental knee implant by Kutzner and colleagues. Calculations show that vigorous walking downhill can load the knee joint with seven or even eight times the body weight (familiar to mountaineers). Squatting can be similarly demanding. The peak hydrostatic pressures generated in the knee joint by vigorous movements may be greater than 25 MPa (Smith et al. 2008; Kutzner et al. 2010). And it is worth noting that 25 MPa is the pressure at a depth of 2500 m, a deep-sea pressure, which is being experienced by human tissue.

Surprisingly there is not much information about interarticular joint pressures in animals. Many years ago Simon published data for the compressive stresses experienced by joints in the cow, sheep, dog, rat and mouse. It is entirely possible that the leg joints in ostriches, horses and stampeding elephants generate higher pressures than seen in human joints. As a guide to the pressures to be used in experiments designed to test the response of isolated cartilage to the cyclic hydrostatic pressures generated in humans, we have 25 MPa as an upper value occurring for a limited time, and prolonged exposure to cyclic 3–6 MPa in walking.

### 3.3.1   Bone

Cartilage is just one of five types of connective tissue that derive from an embryonic connective tissue called mesenchyme. The others are, blood cells, adipose (fat) tissue, fibrous connective tissue which, among other things form the deeper layer of our skin, and bone. The difference between bone and cartilage is that bone has a rigid matrix, created by mineral salts crystalized on collagenous fibres. Bone is synthesised by cells called osteoblasts. They form elaborate concentric layers (lamellae) which form the macroscopic structure in which small channels (canaliculi) are occupied by blood vessels supporting the metabolism of the now trapped cells, called osteocytes. Bones sustain mechanical loading, strain, during which the complicated interior structure, largely fluid filled, acquires pockets of increased pressure which induce flow. This lacunar–canalicular system of bones has been studied in engineering style detail to quantify the fluctuating hydrostatic pressures experienced by the cells. One study produced figures of around 50 kPa (0.05 MPa) and others have estimated somewhat higher pressures of around 250 kPa (Henstock et al. 2013; Scheiner et al. 2016). Whitfield's description of what goes on is illuminating, here paraphrased with the units of pressure rounded.

As you are walking and breathing with your lacunar—canalicular fluids pulsing through osteointernets and toggling osteocyte cilia, the compressive forces on the cartilage in your knee shot up from 100–200 kPa to 10–20 MPa when you stood up and are now cycling between 4–5 MPa as you are walking. This strong cyclical vertical compression does not go unnoticed by the chondrocytes.

It is well known that the deposition of mineral crystals in bone is organised along the lines of mechanical stress, producing a nicely engineered, strong yet lightweight, structure. The process continues throughout life; bone is continuously re-modelling. Are osteocytes able to transduce micro-pressures? Loaded joints are an integrated and difficult environment which we know experience pulses of hydrostatic pressure. The purpose of the next section is to consider the evidence that they respond to normal hydrostatic pressures in a way that is meaningful, i.e., adaptive. Then the evidence that higher hydrostatic pressures might cause damage will be examined. The small and micro pressures in bone and other tissues are not considered, as explained earlier.

## 3.4    Experiments with Chondrocytes

Just as there was, and is, confusion about exactly what physical stress, shear or hydrostatic pressure, actually influences the chondrocytes in situ, so there has been some confusion in the design of experiments to subject pieces of cartilage or isolated chondrocytes to hydrostatic pressure, cyclically pulsed or continuous.

Cells, including chondrocytes, are often grown on plastic plates covered by a loose-fitting lid, and incubated at 37 °C in air enriched with carbon dioxide. Some experimenters have used such dishes, stacked in an air-filled vessel, and applied cyclic pulses of air pressure. There is a real chance that the pulses of air pressure flex the base of the dishes, to which the cells adhere, and thereby stretch the cells as well as subjecting them to pulsed changes in temperature, hydrostatic and gas partial pressure. It is generally agreed that this is an unsatisfactory way of subjecting chondrocytes or similar cells to hydrostatic pressure as an independent variable. Surprisingly this approach is seen in recent papers in which sophisticated techniques are used to monitor the cells.

Better experimental designs for hydrostatic pressure experiments use fluid, without a gas phase, to pressurise thin slices of cartilage, or isolated cells growing in a monolayer. An example of the latter are the experiments carried out by Mizuno (2005), who was particularly careful to separate hydrostatic pressure from strain. He used chondrocytes in an elegant microscopic apparatus and a fluorescent method to demonstrate that 0.5 MPa for 5 min caused an influx of calcium ions. It seems reasonable to assume that a train of brief pulses of similar pressure would have had a similar effect.

Cells supported in a matrix of a viscous gel is another attractive preparation. Agarose gels are used, with different levels of stiffness. Special molecules on the chondrocyte surface, called integrins, interact with their extracellular fabric and communicate to the cell interior. The cyclic pressures which joints experience can be used but if the speed of the change in pressure is very rapid then a compressive force may distort, on a microscale, a mass of cells before they equilibrate with the hydrostatic pressure. In this regard, dense cell pellets or aggregates, which have desirable qualities, may be vulnerable. Pressure can also affect a gel matrix supporting growing cells. But this effect can be measured in separate experiments and allowed for. Many experiments use chondrocytes isolated from a cow or some other animal whose joint hydrostatic pressures are unknown. The cells may be subjected to cyclic pressures matching those experienced in human joints, which is not unreasonable but is never acknowledged. The complications of experimenting with chondrocytes and similar cells were not learned overnight! In the following section, some examples in which hydrostatic pressure exerts a physiological effect will be described.

### 3.4.1 The Response of Chondrocytes in Their Matrix to Physiological Pressures

When it became clear that cyclic hydrostatic pressures were at play in load bearing joints, experiments were carried out to see if chondrocytes responded to them. An example from Hall et al. (1991), is the response of slices of bovine metacarpophalangeal cartilage subjected to a 20 s pulse of hydrostatic pressure, 5, 10, or 15 MPa, and subsequently incubated at atmospheric pressure for 2 h. Appropriate nutrient solutions were used, incorporating radiolabelled compounds. The uptake of these indicates the synthesis of matrix protein, which was accelerated in the pressure-pulsed cartilage, as compared to the untreated controls. A 5 min long pressure pulse had a similar effect but when pressure was sustained for the full 2 h it had much less effect. Pressure of 20–50 MPa, similarly applied, was also without effect. An example of more recent work is the experiment of Tatsamura and colleagues (Tatsamura et al. 2013), who used isolated bovine chondrocytes growing in a gel matrix of Type I collagen. The pressure pulse protocol was 0.5 MPa applied at a frequency of 1 pulse per 2 second for 7 days followed by 7 days at atmospheric pressure, and variations thereof. It was found that the gene expression of the proteins aggrecan and type II collagen were enhanced for 4 days (compared to the control cells which were held at atmospheric pressure). Wong et al. used calf head-of-humerus chondrocytes in an alginate gel and subjected them to a train of pressure pulses; 5 MPa each 2 s, for 3 h per day for 3 days. The treatment caused, for example, a decrease in the expression of the enzyme MMP-13 which plays an important role in the "housekeeping" of the chondrocyte. Conversely, it was upregulated by cyclic strain, a pattern seen in other enzymes. Ikenoue and colleagues have used human chondrocytes to show a similar effect. They used non-osteoarthritic cells isolated from the human knee, growing in a dense monolayer on small culture plates. However, these were placed in a fluid-filled bag for pressurisation. The pressure pulse regime comprised 1, 5 or 10 MPa, at a frequency of 1 per second for 4 h on day 1, after which the cells were harvested. A longer regime was also used, 4 h of pulses per day for 4 days, followed by harvesting. The harvested cells were analysed for the level at which aggrecan mRNA was expressed, and similarly for the expression of Collagen II mRNA. The latter was less responsive to the pressure pulses than the former. Collagen II mRNA expression was only increased by the long, 4 day pulse regime of 5 and 10 MPa, whereas aggrecan mRNA expression was increased by just 1 day of 1 MPa pulses. The general conclusion is that pulses of moderate hydrostatic have a meaningful effect on the cells, consistent with a healthy physiological response.

There is evidence that the critical state of chondrocytes can affect their response to hydrostatic pressure pulses. Chondrocytes arise from mesenchymal stem cells whose differentiation is influenced by the pericellular environment of the early stage cells. Steward et al. (2013), have shown that the interaction between the external surface of stem cells and their matrix environment is important in their sensitivity to hydrostatic pressure. Marrow cells from the femur of young pigs were isolated and grown in a stiff agarose gel, which was then "cored" to form 3 mm thick discs, which could then be hydrostatically pulsed in a flexible bag of culture solution, confined in a conventional pressure vessel. These cells were well connected to their agarose gel environment by cell surface integrins which also connect with the cell interior, a plausibly natural and physiological construct. The cyclic pressure treatment consisted of 10 MPa pulsed at 1 per second for 4 h our per day, for 5 days each week, for a total of 3 weeks. This is a serious growth experiment. The synthesis of the matrix molecule sulphated glycosaminoglycan was significantly enhanced by the cyclic pressure treatment (compared to the untreated controls). However, in the presence of a chemical that blocks the binding of integrins to the gel, no increase in synthesis took place.

To further illustrate how the state of the cell, here articular chondrocytes, influences its responsiveness to pressure, consider this result of an experiment carried out by Jortikka et al. (2000). They used patellar surface cells isolated from the femur of a cow and grew them on small dishes (plates). The state of the cytoskeleton was of prime interest. The synthesis of proteoglycan was measured over a period of 20 h during which the cells were cyclically pressure—pulsed 5 MPa each 2 s. Not unexpectedly synthesis was increased by the cyclic treatment (as compared to continuous exposure to 5 MPa or untreated controls), but when the state of the microtubules was upset by specific drugs, no increased synthesis occurred. The drug nocodazole depolymerises the microtubules and causes the Golgi apparatus to become vesiculated. (It is involved in the preparation and secretion of the matrix material.) The drug taxol, which stabilises the turnover of the tubulin structure of microtubules, also abolished the increased synthesis, i.e., the response to pressure.

Ion transport is another pressure-sensitive target. Hall (1999), has shown that the transport of potassium ions by isolated articular chondrocytes from the cow, is variously affected by pressures up to 15 and even 50 MPa, applied for 10 min or 20 s. Transport by the Na/K pump is inhibited markedly and reversibly by pressures of less than 5 MPa, and in a way influenced by the intracellular concentration of Na ions. The other well-defined transport route, the Na/K/Cl co transport system, is more sensitive to pressure. Pressure pulses of 1-s duration, the physiological time-scale, might act similarly and thus influence matrix synthesis by influencing ionic regulation.

There is no shortage of pressure-sensitive sites in joint cells. As there is evidence that the primary cilium is the site where strain caused by mechanical loading is transduced by mechanoactivated ion channels, might hydrostatic pressure also be transduced there? Shao and colleagues have attempted to answer this question by hydrostatically pressurising rat epiphyseal cells grown in aggregated pellet form. The hydrostatic pressure treatment consisted of 1 MPa applied for an hour, alternating with an hour at atmospheric pressure, for various periods of time. A 24 h period stimulated proliferation, demonstrating the health of the cells, but the focus of the experiment was the expression of a certain signalling protein, Ihh (for reasons we have to ignore this stands for Indian Hedgehog (Wu et al. 2001)!), which regulates chondrocyte growth, and which is blocked by a drug called cyclopamine. Periods of pressure treatment as short as half an hour increased the expression of Ihh, and 24 and 48 h of cyclic pressures increased it 7- and 20-fold, respectively. A transcription factor, Gli, induced by Ihh, was increased threefold after 48 h of pressure treatment. This response was blocked by cycloamine. The connection between Ihh signaling and the primary cilium was established by using chloral hydrate. This shortens and deforms the cilium, and presumably damages its membrane. Cells thus treated fail to show an increase in the expression of Ihh when subjected to cyclic pressures. The suggestion is that a hydrostatic pressure of 1 MPa. Cyclically applied, stimulates the primary cilium. This is a small hydrostatic pressure but one which elicits effects in other membranes. For example 1.13 MPa has a significant effect on the exocytosis underlying spontaneous transmitter release in the frog neuromuscular junction. A similar pressure has a drastic effect on the phase transition of DPPC liposomes poised at a critical temperature and caused measurable changes in the membrane fluidity of human red cells. Rather less than 1 MPa affects red cell membranes, causing them to shed a component which is involved in the action of the enzyme phospholipase A2, and 26 kPa can activate human platelets. These effects are caused by hydrostatic pressure with no possibility of a shear force or strain being involved. The references are, respectively (Ashford et al. 1982; Barkai et al. 1983; Barshtein et al. 1997; Halle and Yedgar 1988; Torsellini et al. 1997). The effect on the chondrocyte primary cilium sets no precedent and can, in principle, be interpreted by the thermodynamics of solution chemistry used in interpreting the effects of somewhat greater hydrostatic pressures on, for example ion channel kinetics.

## 3.5  Does High Hydrostatic Pressure Cause Pathological Effects?

Sufficiently high pressure will undoubtedly cause damage and physiologically plausible high pressure can stimulate the presence of a particular class of stress proteins. This was shown by Kaarniranta and colleagues, who found that 30 MPa, continuously applied, caused cultured human chondrocytes to accumulate common stress protein, Hsp 70. The mechanism is unusual; its messenger RNA became stabilised and not, as is the more usual route, by enhancing its gene expression.

Osteoarthritis is commonly associated with a loss of cartilage matrix, the death of chondrocytes and inflammation. Kunimoto and a group of colleagues at Kyoto Prefectural University, Japan (Kunitomo et al. 2009) devised some experiments to see if severe pressure caused changes in cultured cells which resembled these symptoms. Rabbit articular chondrocytes were isolated and cultured in alginate beads to provide two distinct groups of chondrocytes in a quasi-cartilage environment. Cells freshly introduced to the alginate (group 1) would have less matrix surrounding them than cells grown for 2 weeks in the gel (group 2). Both groups were subjected to 50 MPa, judged to be an excessive pressure, for 12 h. In the cells in group 1, 50 MPa induced the enzyme which attacks aggrecan but in group 2 cells it was not induced. Also, in group 1 cells, chondrocyte apoptosis (deliberate cell death) was increased, but not in group 2 cells. Additionally, in the group 1 cells inflammatory cytokines (chemical agents) were induced, but not in group 2 cells. The pathological effects of 50 MPa are significantly affected by the state of the cultured cells. Perhaps the quantity of matrix surrounding the cells is all important.

Another Japanese group, Montagne et al. (2017), have more recently shown that continuous pressure, 25 MPa for 4 or 24 h, modulated (their term) many hundreds of genes in their test cell, a precursor cell for chondrocytes in the mouse. The cells were grown as a monolayer on plastic dishes which were then pressurised in a flexible bag of fluid in a high-pressure vessel. Genes "classified as part of the primary cilium" were down regulated and genes associated with cell death were up regulated. Many of the genes modulated by pressure were the same as those which respond to artificially induced osteoarthritis. The authors conclude that their pressure treatment does not really simulate physiological conditions but it might mimic some of the genetic alterations occurring in osteoarthritis. To that extent, it might be a useful technique.

## 3.6    Conclusion

The physical analysis of moving load bearing joints shows that hydrostatic pressure occurs in a region which passes across the cartilaginous lining of the joint. The magnitude of the pressure fluctuates cyclically from less than 1–25 MPa or more. The biochemistry of the chondrocytes and other cells present respond to similar, experimentally applied hydrostatic pressures in a way which is consistent with adaptation. The load bearing joint is thus a high-pressure micro-environment but little is known about the molecular mechanisms underlying the adaptation of the cells inhabiting it.

In the normal, functional joint, pulses of hydrostatic pressure occur simultaneously with shear stress (strain) and therefore the combination of the two is presumably involved in keeping joints healthy. Pathological effects probably arise when they are present in excess.

# References[1]

Acevedo AD, Bowser SS, Gerritsen ME, Bizios R (1993) Morphological and proliferative responses of endothelial cells to hydrostatic pressure: role of fibroblast growth factor. J Cell Physiol 157:603–614. *

Ashford MLJ, Macdonald AG, Wann KT (1982) The effect of hydrostatic pressure on the spontaneous release of transmitter at the frog neuromuscular junction. J Physiol 333:531–543

Barkai C, Goldman B, Mashiach SM, Shinitzky M (1983) The effect of physiologicaly relevant pressures on the phase transition of liposomes containing dipalmitoyl phosphatidylcholine. J Colloid Interface Sci 94:343–347

Barshtein G, Bergelson L, Dagan A, Gratton E, Yedgar S (1997) Membrane lipid order of human red blood cells is altered by physiological levels of hydrostatic pressure. Am J Phys 272:H538–H543

Becquart P et al (2016) Human mesenchymal stem cell responses to hydrostatic pressure and shear stress. Eur Cell Mater 31:160–173. *

Blaxter JHS, Tytler P (1972) Pressure discrimination in teleost fish. Symp Soc Exp Biol 26:317–444. *

Cao X et al (2015) A mechanical refractory period of chondrocytes after dynamic hydrostatic pressure. Connect Tissue Res 56:212–218. *

Carter DR, Wong M (2018) Modelling cartilage mechanobiology. Philos Trans R Soc Lond B 358:1461–1471

Dufrene YF, Persat A (2020) Mechanomicrobiology: how bacteria sense and respond to forces. Nat Rev Microbiol 18:227–240

Elder BD, Athanasiou KA (2009) Hydrostatic pressure in articular cartilage tissue engineering: from chondrocytes to tissue regeneration. Tissue Eng Part B 15:43–53

Elo MA et al (2000) Differential regulation of stress proteins by high hydrostatic pressure, heat shock, and unbalanced calcium homeostasis in chondrocytic cells. J Cell Biochem 79:610–619. *

Enright JT (1962) Responses of an amphipod to pressure changes. Comp Biochem Physiol 7:131–145

Fraser PJ (2010) Pressure sensing. In: Sebert P (ed) Comparative high pressure biology. Science Publishers, Hauppauge, pp 143–160

Fraser PJ, Macdonald AG (1994) Crab hydrostatic pressure sensors. Nature 3371:383–384

Fraser PJ, Shelmerdine RL (2002) Dogfish hair cells sense hydrostatic pressure. Nature 415:495–496. *

Fraser PJ, Macdonald AG, Cruickshank SF, Schraner MP (2001) Integration of hydrostatic pressure information by identified interneurons in the crab Carcinus maenas (l): long term recordings. J Navig 54:71–79

Fraser PJ, Cruickshank SF, Shelmerdine RL (2003) Hydrostatibular hair cell afferents in fish and crustacea. J Vestib Res 13:235–242

Hall AC (1999) Differential effects of hydrostatic pressure on cation transport pathways of isolated articular chondrocytes. J Cell Physiol 178:197–204

Hall AC, Urban JPG, Gehi KA (1991) Effects of hydrostatic pressure on matrix synthesis in articular cartilage. J Orthop Res 9:1–10

Halle D, Yedgar S (1988) Mild pressure induces resistance of erythrocytes to hemolysis by snake venom phospholipase A2. Biophys J 54:393–396

Haskin CL, Cameron IL, Athanasiou KA (1994) Human osteosarcoma cells alter cytoskeletal and adhesion proteins in response to physiological levels of hydrostatic pressure. In: Bennett PB, Marquis RE (eds) Basic and applied high pressure biology. University of Rochester Press, Rochester, pp 31–40. *

---

[1]Those marked * are further reading.

Henstock JR, Rotherham M, Rose JB, El Haj AJ (2013) Cyclic hydrostatic pressure stimulates enhanced bone development in the foetal chick femur in vivo. Bone 53:468–477

Heupel MR, Simpfendorfer CA, Hueter RE (2003) Running before the storm. Blacktip sharks respond to falling barometric pressure associated with tropical storm Gabrielle. J Fish Biol 63:1357–1363. *

Hodge WA et al (1986) Contact pressures in the human hip joint measured in vivo. Proc Natl Acad Sci U S A 83:2879–2883. *

Ikenoue T et al (2003) Mechanoregulation of human articular chondrocyte aggrecan and Type II collagen expression by intermittent hydrostatic pressure in vitro. J Orthop Res 21:110–116. *

Jortikka MO et al (2000) The role of microtubules in the regulation of proteoglycan synthesis in chondrocytes under hydrostatic pressure. Arch Biochem Biophys 374:172–180

Kaarniranta K et al (1998) Hsp 70 accumulation in chondritic cells exposed to high continuous hydrostatic pressure coincides with mRNA stabilization rather than transcriptional activation. Proc Natl Acad Sci U S A 95:2319–2324. *

Karpinski R, Jaworski J, Jonak J, Krakowski P (2019) Stress distribution in the knee joint in relation to tibiofemoral angle using finite element method. MATEC Web Conf 252:07007

Kunitomo T et al (2009) Influence of extracellular matrix on the expression of inflammatory cytokines, proteases, and apoptosis-related genes induced by hydrostatic pressure in three dimensionally cultured chondrocytes. J Orthop Sci 14:776–783

Kutzner I et al (2010) Loading of the knee joint during activities of daily living measured in vivo in five subjects. J Biomech 43:2164–2173

Macdonald AG, Fraser PJ (1999) The transduction of very small hydrostatic pressures. Comp Biochem Physiol A 122:13–36

Mizuno S (2005) A novel method for assessing effects of hydrostatic fluid pressure on intracellular calcium: a study with bovine articular chondrocytes. Am J Phys Cell Phys 288:C329–C337

Montagne K et al (2017) High hydrostatic pressure induces pro-osteoarthritic changes in cartilage precursor cells: a transcriptome analysis. PLoS One 12:e0183226

Morgan E (1969) The response of nephtys (Polychaeta, Annelida) to changes in hydrostatic prerssure. J Exp Biol 50:501–513

Morrell KC, Hodge WA, Krebs DE, Mann RW (2005) Corroboration of in vivo cartilage pressure with implications for synovial joint tribology and osteoarthritis causation. Proc Natl Acad Sci U S A 102:14819–14824. *

Muir H (1995) The chondrocyte, architect of cartilage. BioEssays 17:1039–1048. *

Olsen SM, Stover JD, Nagatomi J (2011) Examining the role of mechanosensitive ion channels in pressure mechanotransduction in rat bladder urothelial cells. Ann Biomed Eng 39:688–697

Pingguan-Murphy B, Lee DA, Bader DL, Knight MM (2005) Activation of chondrocytes calcium signalling by dynamic compression is independent of number of cycles. Arch Biochem Biophys 444:45–51

Rushfeldt PD, Mann RW, Harris WH (1981) Improved techniques for measuring in vitro the geometry and pressure distribution in the human acetabulum—II. Instrumented endoprothesis measurement of articular surface pressure distribution. J Biomech 14:315–323. *

Sappington RM, Sidorova T, Long DJ, Calkins DJ (2009) TRPV1: contribution to retinal ganglion cell apoptosis and increased intracellular calcium with exposure to hydrostatic pressure. Invest Opthalmol Vis Sci 50:717–728. *

Scheiner S, Pivonka P, Hellmich C (2016) Poromicromechanics reveals that physiological bone strains induce osteocyte-stimulating lacunar pressure. Biomech Model Mechanobiol 15:9–28

Schwartz EA, Bizios R, Medow MS, Gerritsen ME (1999) Exposure of human vascular endothelial cells to sustained hydrostatic pressure stimulates proliferation. Circ Res 84:315–322. *

Shao YY, Wang L, Welter JF, Ballock RT (2012) Primary cilia modulate IHH signal transduction in response to hydrostatic loading of growth plate chondrocytes. Bone 50:79–84. *

Simon AH (1970) Scale effects in animal joints. Arthrit Rheumatism 13:244–255. *

Smith SM, Cockburn RA, Hemmerich A, Li RM, Wyss UP (2008) Tibiofemoral joint contact forces and knee kinematics during squatting. Gait Posture 27:376–386

Steward AJ, Wagner DR, Kelly DJ (2013) The pericellular environment regulates cytoskeletal development and the differentiation of mesenchymal stem cells and determines their response to hydrostatic pressure. Eur Cell Mater 25:176–178

Tatsamura M, Sakane M, Ochiai N, Mizuno S (2013) Off-loading of cyclic hydrostatic pressure promotes production of extracellular matrix by chondrocytes. Cells Tissues Organs 198:505–413

Torsellini A, Maggi L, Guidi G, Lombardi V (1997) Modifications of platelet function induced by the variations of hydrostatic pressure in an experimental model. Haemostasi Thrombosis:375–383

Toyoda T, Seedholm BB, Kirkham J, Bonass WA (2003) Upregulation of aggrecan and type II collagen mRNA expression in bovine chondrocytes by the application of hydrostatic pressure. Biorheology 40:79–85. *

Urban JPG (1994) The chondrocyte: a cell under pressure. Br J Rheumatol 33:901–908

Whitfield JF (2008) The solitary (primary) cilium–a mechanosensory toggle switch in bone and cartilage cells. Cell Signal 20:1919–1024. *

Wong M, Siegrist M, Goodwin K (2003) Cyclic tensile strain and cyclic hydrostatic pressure differentially regulate expression of hypertrophic markers in primary chondrocytes. Bone 33:685–693. *

Wu Q, Zhangi Y, Chen Q (2001) Indian hedgehog is an essential component of mechanotransduction complex to stimulate chondrocyte proliferation. J Biol Chem 276:35290–35296

# Effects of High Pressure on the Activity of Ordinary Animals, Including Humans, and on the Function of Their Excitable Cells and Ion Channels

**4**

Ordinary animals, meaning shallow water and terrestrial animals which do not normally experience high hydrostatic pressure, are drastically affected by deep-sea pressure. It affects their metabolism and much else besides. The effects of pressure on metabolism and on nerves, and more generally excitable membranes, are interesting for at least two reasons. They reveal some aspects of the molecular structure of individual components and they point the way to understanding adaptation to high pressure.

## 4.1 Early and Modern Observations, Mostly on Invertebrates

Regnard (1891), was the first person to see the effects of high pressure on the activity of animals (Chapters 1 and 13). He saw the immediate effect of pressure (10 MPa) was to stimulate swimming activity in various crustacea, small copepods and amphipods. Higher pressures immobilised the animals, which promptly recovered when the pressure was reduced to normal, i.e., on decompression. These observations may have interested Regnard's fellow biologists but they did not stimulate further work. It was not until the 1930s that another Frenchman, Fontaine, published the results of his high-pressure experiments on a wide variety of cells, isolated tissue and whole animals. He did not observe the swimming activity of animals under pressure but measured their oxygen consumption. The results were uniform; approximately 10 MPa increased the oxygen consumption of fish; plaice (*Pleuronectes platessa*), sand goby (*Gobius minutus*) and the eel (*Anguilla vulgaris*). Their oxygen consumption probably reflected their increased motor activity at pressure. Fontaine also found that isolated muscle, from frogs and eels, contracted under high pressure (60 MPa), a response he compared with the effects of direct electrical stimulation. His monograph, published in the Annals of the Oceanographic Institute in Monaco, provides a great deal of other information, including the names of the forgotten workers who carried out high-pressure experiments at the time.

© Springer Nature Switzerland AG 2021
A. Macdonald, *Life at High Pressure*, https://doi.org/10.1007/978-3-030-67587-5_4

Some years later Ebbecke (1935), a German physiologist, also experimented with pressure and a wider variety of animals. He saw the tentacles of the sea anemone, *Actinia equina*, increase their movement over the pressure range 10–30 MPa but they retracted at 40 MPa. Excised pieces of the subumbrella tissue of the jellyfish *Cyanea* normally show a rhythmic beating and Ebbecke saw this was increased in frequency at pressure. He also observed that pressure increased the activity of a marine worm (*Tomopteris sp.*), a sea snail (Littorina sp., Phylum *Mollusca*, class *Gasteropoda*) shrimps and even the slow-moving sea urchin (*Echinus sp.*, Phylum Echinodermata). Similarly, pressure stimulated the activity of various fish and the Hemichordate, *Branchiostoma lanceolatum*. The implications of the uniformity of these observations were that high pressure affected the animals' nerves and muscles which have in common excitable membranes, that is membranes capable of transmitting a nerve impulse.

A good historical account of the emergence of high-pressure biology is given by Damazeau and Rivalain (2011), but we must press on to the 1960s when biologists in Europe and America resumed and extended these observations.

Menzies et al. (1972), at Florida State University observed the increase in activity caused by pressure on the planktonic Ctenophore *Mnemiopsis*, the worm *Nereis*, the Amphipod *Telorchestia* and the crab *Uca pugilator* and Flugel, Schlieper and colleagues in Germany did similar work, independently. In the UK, in preparation for attempting the isobaric collection of deep-sea animals (Chapter 5), the author studied the Amphipod, *Marinogammarus marinus*, a sea shore animal about a centimetre in length known as a sand hopper. Its segmental structure supports numerous limbs and special pleopods whose beating drives water over its gills, thus facilitating the exchange of dissolved gases, oxygen diffusing into the gills and carbon dioxide diffusing out. The observational high-pressure vessel used (Chapter 13) had a safe working pressure of over 100 MPa and a 2.5 cm diameter window which could be used with a long-range stereo binocular microscope. Regnard and Ebbecke would have been envious.

Subjecting *M. marinus* to pressure steps of 5 MPa at 5 min intervals at 3–6 °C, caused the majority of animals to respond with bursts of swimming and an increase in the pleopod beat frequency.

Then 10 MPa causes spasms (a backwards flexing of the body). This is a good measure of pressure tolerance in deep-sea amphipods, see Chapter 6.

Slower compression, 1 MPa each 6 min, causes spasms at 5 MPa. Pressures higher than 10 MPa immobilise the animals.

Very slow, smooth compression applied with a special apparatus at a rate of 0.03 MPa per minute causes no spasms, but immobilisation occurs at high pressure.

When specimens of *M.marinus* were immobilised (paralysed) by high pressure and then decompressed rapidly they resumed normal activity, typically within minutes. This was the case even when pressure—paralysed animals were held at 20 MPa for 10 h, demonstrating a notable tolerance to high pressure but also to the absence of gill ventilation and presumably diminished respiration (Macdonald 1972, 1976a, b). The Asian equivalent of *M. marinus* is *Eogammarus possjeticus* and it too survives 20 MPa for many hours, but the vessel in which it was pressurised lacked windows so we do not know how its motor activity was affected (Chen et al. 2019).

We need to be sure of our interpretation of the motor activity, namely the animals are responding to the direct effects of pressure on their physiology and are not, for example, displaying a reflex response to a sensory stimulus. The pattern of initial hyperexcitability at moderate pressure followed at higher pressures by paralysis is universal which makes a purely behavioural response very unlikely, although in some cases increased pressure may stimulate upward swimming. Experiments with isolated tissues demonstrate that pressure can act directly on nerves and muscles to elicit such responses, but there is, quite separately, an interesting sensory response to very small increases in hydrostatic pressure in some animals, see Box 4.1.

*Crangon crangon*, a shrimp that lives in soft sediments in shallow water possesses a "tail flip" escape reflex which is activated by a step compression to one or more MPa. Compression to 6 MPa elicits a burst of convulsions based on the tail flip. If held at that pressure for several hours the quiescent animal manifests single tail flip convulsions at widely spaced intervals, apparently spontaneously (Wilcock et al. 1978). After these observations were made, neurophysiological work on the same escape reflex in the crayfish showed that fast conducting abdominal nerves could cause command neurones to activate the reflex, Arnott et al. (1998). Neuronal circuits for the escape reflex have been proposed. Transection of Crangon's abdominal nerves trebled the pressure which elicited convulsions, so presumably pressure acts on nerves "upstream" of the transection to elicit the convulsions, but in the treated animal a higher pressure can act "downstream" (Chapter 13, Fig. 13.2). The lobster *Homarus gammarus*, a giant version of Crangon with a different lifestyle, also undergoes tail flip convulsions when pressurised.

Other Crustacea, the brackish water shrimp *Palaemonetes varians* and the spider crab *Maja brachydactyla*, showed the same pattern (Oliphant et al. 2011; Thatje et al. 2010). To digress into deeper water, the hermit crab *Pagurus cuanensis* (maximum depth 250 m, 2.5 MPa) and the shrimp-like bathypelagic *Systelapsis, Acanthophyra* and *Euphausia*, all show the general pattern; increased activity on compression followed by suppressed activity at around 20 MPa. Other oceanic animals that occupy the shallow water above great depths, for example the planktonic *Anomalocera patersoni* (a Copepod, a Sub Class of Crustacea) and the Amphipod *Parathemisto*, are stimulated by moderate pressures and *Parathemisto* exhibits whole body spasms just like *M. marinus*. The author has used observational pressure equipment at sea to make some of these observations and on one occasion attempted to test the pressure tolerance of a small planktonic squid. It responded to the initial compression with its escape reflex, a discharge of black ink, which terminated the experiment! Other more successful observations on oceanic animals showed some were not particularly responsive to pressure tests. An Amphipod, *Lanceola sayana*, unlike the shrimp-like crustacea mentioned above, failed to respond to compression until 35 MPa was reached, when immobilisation set in. The absence of spasms was very clear, presumably reflecting a significant pressure tolerance. The planktonic giant Ostracod *Gigantocypris mulleri* (fully 1 cm in diameter), also showed no excitation on compression and its normal swimming gradually slowed at pressures greater than 20 MPa but persisted at 50 MPa. It lives at depths of 600–1500 m, 6–15 MPa. Shallow water oceanic ostracods, *Conchoecia*, were more susceptible to high pressure (Macdonald et al. 1972; Macdonald and Teal 1975; Macdonald 1972).

Relating pressure tolerance to the depth at which animals normally live is more easily achieved with animals that live on the ocean floor, see Chapter 5.

---

**Box 4.1: The Problem of Very Small Hydrostatic Pressures**

Some aquatic animals respond to very small changes in hydrostatic pressure. Many fish are buoyant because they possess a gas-filled swim bladder. It is a compressible organ, intrinsically capable of responding to small changes in pressure by decreasing in volume and thus affecting the degree of stretch in the bladder wall. Animals that do not contain a gas phase and yet detect small changes in hydrostatic pressure present a puzzle. Some crustacea, for example intertidal crabs and shrimp-like animals, respond to very small changes in hydrostatic pressure, 10 kPa or less, and they certainly do not contain a gas phase. Quite separately, some cells also appear to be affected by very small or micro-hydrostatic pressures, as discussed in Chapter 3.

---

## 4.2    Effects of Pressure on the Activity of Fish

Fish respond to an increase in hydrostatic pressure broadly in the same way as invertebrates, as mentioned earlier (Barthelemy et al. 1981). Regnard's observations of plaice (*Pleuronectes platessa*) and goldfish (*Carassius auratus*) have been confirmed in modern studies. The eel *(Anguilla)*, for example, becomes agitated, and abnormally active at moderate pressures and convulses at 10 MPa, with powerful writhing movements that promptly subside on decompression. Moderate pressure sometimes stimulates upward swimming (Sebert and Macdonald 1993; Chapter 12). Juvenile Sea bass (*Dicentrarchus labrax*) become excited and swim faster at 5 MPa and convulses at 8 MPa. Eventually high pressure immobilises the animals, reversibly if they are not kept at high pressure too long (Dussauze et al. 2017). The gas filled-buoyancy organ present in many fish, but excluding sharks and their relatives, is compressed when the ambient pressure is increased so it might be thought that this would complicate the interpretation of their response to pressure. However, this seems not to be the case. Two species, *Symphurus palguisa* and *Pleuronectes platessa*, the flounder and plaice, which have no swim bladder, respond to compression in much the same way as species with a bladder (Brauer et al. 1974). The response to high pressure in the Rainbow trout (*Salmo gairdneri*), whose bladder can be surgically removed or catheterised to allow it to equilibrate with the ambient water pressure, is unchanged by the procedure.

These observations on fish are interpreted in the same way as those on invertebrates; the abnormal activity is caused by pressure acting directly, pathologically, on the neuromuscular system.

## 4.3    Effects of Pressure on the Activity of Air-Breathing Vertebrates

How does the activity of air-breathing animals, particularly vertebrates, respond to high pressure? The answer is, much in the same way as aquatic vertebrates. But there is a problem, how can air-breathing animals be pressurised in a way comparable to the way aquatic animals can be exposed to water pressure?

If an air-breathing animal is pressurised in air a number of drastic changes take place which obscure any effect which the increased hydrostatic pressure might have. The pioneer studies of human divers, breathing compressed air, revealed these complications, which were subsequently studied using small mammals, compressed in air-filled hyperbaric chambers. (The term hyperbaric is traditionally used in diving physiology). Air is approximately 80% nitrogen and 20% oxygen, and at normal atmospheric pressure (100 kPa) the oxygen is said to contribute a partial pressure (PO2) of 20 kPa. If the air is compressed to 500 kPa, equivalent to a water depth of 50 m, the partial pressure of oxygen is increased to 100 kPa. Oxygen is toxic at such a PO2, and it will damage the pulmonary tissue. At a partial pressure of 300 kPa or more it causes muscle twitching and eventually seizures, by acting on neurotransmitters in the brain. Although oxygen is popularly thought of as life giving, it is in fact, a very toxic substance at partial pressures slightly higher than normal, which is well recognised in clinical medicine. An increased partial pressure of nitrogen (PN2) can cause euphoria and behaviour similar to alcohol (ethanol) intoxication. This is called inert gas narcosis, and is closely related to general anaesthesia (discussed in Chapter 11).

In human diving these problems of compression in air were solved by using a mixture of the inert gas helium and a small concentration of oxygen (heliox), which at the intended hydrostatic pressure, would exert a safe PO2. During experiments in the 1960s and 1970s, exploring the safe depth range of divers breathing a helium–oxygen gas mixture, it was noted that the divers (i.e., human subjects in a hyperbaric chamber) showed tremors, particularly in their fingers. These "helium tremors" were re-named the high-pressure neurological syndrome, HPNS, when it was later learned that they were the early symptoms of a profound disturbance to the nervous system caused by increased hydrostatic pressure. Thus, high-pressure experiments can be carried out with air-breathing animals using helium to transmit pressure instead of water, and controlling the PO2, and other factors, carefully. At much higher pressures, helium can exert a weak, inert gas-like effect, which is explained in Chapter 11. Furthermore helium, at twice atmospheric pressure, or less, has apparent neuroprotective effects which are being actively studied for clinical reasons, although how helium actually interacts biochemically is a complete mystery (Winkler et al. 2016; Oei et al. 2010). Nevertheless, in diving conditions hyperbaric helium is biochemically inert, transmitting hydrostatic pressure through a gas mixture which is mechanically good to breathe with a safe partial pressure of oxygen.

Heliox compression can be used with air-breathing animals but certain animals, which use their lungs to breathe air and their gills to breathe underwater, can be safely pressurised hydrostatically in either heliox or water. Newts have been used in this way, see below. A much more complicated procedure is to use animals, particularly mammals, whose lungs are filled with special fluids that enable oxygen and carbon dioxide to be exchanged across the pulmonary tissues. Such liquid breathing techniques, using perfluorochemical liquids, are used in clinical studies of several conditions, particularly in premature babies (Cox et al. 2003; Wolfson and Shaffer 2005; Kohlhauer et al. 2015). Oxygen and carbon dioxide are highly soluble in the liquids, which can provide the required gas exchange, with supplementary procedures that need not concern us here. In the development of the liquid breathing technique, small mammals were used and, as liquid breathing mammals, they have been separately used in hydrostatic pressure tests (more soon).

Here are some experiments based on these methods.

The newt *Triturus cristatus carnifex* (Class Amphibia, Sub Class Urodela), has both external gills, retained from the larval stage, and lungs, and can respire adequately in both water and in air. It is agile and has a "righting" reflex, which means if it is turned upside down and dropped it will attempt to land on its feet. Mice have a similar reflex, which provides the basis for comparing the two at high pressure. When pressurised in heliox at the rate of 4 MPa per hour, mice typically show tremors then convulsions at around 8–10 MPa and if pressure is further increased, more severe convulsions follow, culminating in a rigid extension of the limbs. The first convulsion differs in several ways from the latter (see below), but recovery from both is rapid and complete when pressure is reduced. The important point however, is the rigid extension of the limbs prevents the mouse from performing its righting reflex. Using an observational pressure vessel fitted with a rotating inner lining, Miller, Lever and colleagues at Oxford University, UK, scored the ability of the mouse to right itself during helium compression. The reflex score is 100% during the early stage of compression but declined rapidly when severe convulsions and rigidity sets in. Experiments with newts pressurised in heliox showed their righting reflex score was also 100% in the early stages of compression but it abruptly reduced to 50% at about 20 MPa. When newts were pressurised in water the reflex was similarly reduced to 50% at 20 MPa. The animals' activity during the early stages of compression in heliox and in water were also much the same. Thus, compression in heliox is equivalent to compression in water; both subject the animal to hydrostatic pressure.

The laboratory mouse has been subjected to high pressure using the liquid breathing technique, used extensively by Kylstra. A high partial pressure of oxygen and a low body temperature were required to ensure the health of the animals. Compression at 50 kPa per minute caused convulsions at 8.5 MPa depending on the body temperature, a result confirmed in similar experiments carried out in another laboratory. These results are similar to the convulsions seen in heliox compression. In other experiments, liquid breathing dogs were pressurised and compared with control animals compressed in heliox. The criterion used to measure the pressure effect was the time course of evoked electrical potentials, recorded at pressures of up to 10 MPa. There was no difference between the results obtained with the two methods of compression.

The conclusion is, provided other factors are kept within healthy limits, heliox compression of air-breathing animals achieves hydrostatic compression, hence the effects seen are primarily caused by the thermodynamic mechanisms outlined in Chapter 2. The generally held view is that pressure exerts an immediate and direct effect on some molecular "target" and may well trigger a secondary compensatory process which is time dependent. The abnormal activity seen in mammals in particular, the HPNS, is an important issue in human diving. (The term HPNS is best confined to mammals but, with caution, it can be useful with vertebrates but is best avoided with invertebrates.) In humans undergoing "dry" pressure chamber dives, the main symptoms are tremors, starting at around 1.5 MPa with jerky muscle movements appearing at 4 MPa. A decrement in manual dexterity is apparent at around the same pressure. At the higher pressures, episodes of microsleep occur, in which the subjects slip and out of a conscious state and memory can also be impaired. There is abundant evidence from electroencephalograms that all is not normal. In Primates severe muscle spasms and convulsions occur at pressure a little higher than those which cause symptoms in humans. It is thought by some that HPNS resembles, and perhaps illuminates, medical conditions such as epilepsy.

### 4.3.1   The Pressure Tolerance of Aquatic and Air-Breathing Vertebrates

At this juncture we have two ways to go; one is to enquire into the detailed mechanisms underlying the HPNS and the other is to extend the range of vertebrates studied. Brauer, working in North Carolina, has contributed to both these ends and we shall start with the data he and his team have published on the HPNS observed in a wide variety of vertebrates (Brauer et al. 1974). Considerable care was taken to ensure the experimental conditions and state of the animals justify comparing the results. All the animals pass through the same sequence of motor disturbances during the compression, which was carried out at a rate of 2.4 MPa per hour. Twenty-two species were studied and in the following examples the mean threshold pressure, in MPa, at which tremors began is followed by the equivalent for convulsions. Two species of Primate, the Rhesus monkey (Macaca mulatta) and the Squirrel monkey (Saimiri sciureus) scored 3.1/8.3 and 3.4/6.7 respectively. The Carnivore *Procyon lotor*, the racoon, 4.2/8.8; Rodents, Mus musculus (house mouse) 4.4/9.7 and *Epimys rattus*, the rat, 6.6/8.9; The Marsupial *Didelphys virginiana* (Opossum), 3.8/7.9; Aves, i.e., birds, Gallus domesticus (chicken) 5.3/10.2; the Reptile *Anolis carolinensis* (lizard) 7.0/10.6 and the fish *Anguilla rostrate* (eel) 5.6/10.7. Generally, mammals are more susceptible to pressure than other less advanced vertebrates. Data for humans, acquired in separate experiments to extend the safe depth range of human divers, is limited to finger tremors. Brauer quotes the onset pressures of

2.5–2.7 MPa obtained by French scientists with whom he worked. From Brauer's data a convulsion pressure for humans of approximately 6.5 MPa can be predicted. The question of the susceptibility of humans to hydrostatic pressure is important not only in human diving but in submarine escape procedures.

The implication of Brauer's data for diving mammals is also of interest. The Carnivore, the racoon, is probably the appropriate species to use for comparison, and it convulsed at a pressure of 8.8 MPa, during a slow compression. This is the pressure at a water depth of 880 m which is within the depth range of a number of specialist diving mammals. Generally, this strengthens the idea that their neuromuscular and central nervous systems must be adapted to tolerate rapid increases in pressures. So far, hyperbaric experiments with diving mammals have focused on respiratory physiology, with a limited pressure range (Chapter 11).

### 4.3.2   HPNS

In the 1960s the hyperexcitable symptoms in humans were seen in simulated dives in Russia, the UK, the USA and France. These dry dives were carried out by Navies and by commercial diving companies who shared the objective of extending the depth range of free divers. Hyperbaric physiologists, particularly Brauer in America, who knew of the pressure experiments Regnard carried out in nineteenth century Paris, were at an advantage in interpreting the novel symptoms, at first called "helium tremors." Soon they were found to be rather more than tremors and called the High Pressure Neurological Syndrome by Brauer and his colleagues. According to Brauer, the Rhesus monkey was the first mammal in which HPNS was demonstrated. According to Naquet simulated dives with human subjects carried out by the French diving company Comex demonstrated the "principal symptoms" of HPNS in 1968. The HPNS presented a serious problem in extending the depth range of human diving beyond 100 m or so and in the evolution of deep diving animals it must have been a very significant factor.

The HPNS is studied in experiments with laboratory animals, often using invasive neurophysiological, biochemical and pharmacological techniques. The mouse is commonly used, and less often larger mammals, including primates. In the mouse, two types of convulsions or seizures are clearly discernible. The first to appear during compression, the Type I seizure described by Brauer, is much more susceptible to the rate of compression than the Type II seizure. They also differ in their susceptibility to drugs and in the electrical waves recorded from the brain. The Type II seizure is associated with major electrical disturbances, a slowing of the heart rate and often with death. Breeding experiments show that the onset pressure for Type I seizures is largely genetically determined (Chapter 12).

The HPNS is also studied in human subjects (Todnem et al. 1989; Rostain and Lavoute 2010). Recordings of the gross electrical activity of specific regions of the brain (electroencephalograms) and from muscle groups (electromyograms) are made with human volunteers, in much the same way that patients with brain or spinal disorders are monitored. However, the volunteers for HPNS experiments are screened not just for their medical fitness but also to ensure their motivation to be experimental subjects is not too closely aligned to advancing their career. The experiments are demanding. Subjects are confined in a hyperbaric chamber and typically a "dry dive" takes many days, with the decompression stage being particularly critical. The safe management of the subjects requires clinical support, highly trained technicians and a well thought out scientific protocol. The cost of the hyperbaric chamber, its instrumentation, the gas used (helium is expensive) and the catering for everyone involved, is considerable. Accordingly, such work is confined to a limited number of laboratories. Some are run by military organisations, others by commercial diving companies and a few by medical and academic institutions. Often the chambers also provide clinical, emergency decompression treatment for sport and professional divers.

Countries vary but generally the scientific protocols for these types of experiment have to comply with State laws and also require approval by an institution's ethical committee. In some countries, these controls also apply to simple non-invasive experiments with fish, and generally experiments with live invertebrates do not require special approval. Experiments with isolated tissues or brain slices are also an important approach, and these depend on a supply of animals properly reared in compliance with State regulations. The publication of the results of such experiments, using humans or animals, in reputable Scientific journals usually entails a declaration that the work has been carried out in a manner consistent with due regulations.

A great deal of information about the HPNS was gathered in a few years, often during dry dives which combined more than one objective. Combining varied compression profiles and the composition of gases to mitigate HPNS with physiological measurements probably make good use of expensive facilities and valued human subjects, but does not necessarily produce good, repeatable experiments or data which can be closely compared with anything else. There was much enthusiasm, particularly in France, for "divers to conquer the depths"! The area of seafloor likely to be conquered by free divers is small compared to that of the deep ocean, and even then, at that exciting time, some scientists saw the future of working in deep water was with manned and unmanned submersibles and robots. Companies, motivated by cost-benefit and ethical considerations, began to adopt a mixed approach, combining a "diver replacement" policy with traditional free diving. HPNS was not the only risk in ultra-deep diving, as new decompression problems emerged. Oil reserves are now viewed differently. The development of various types of submersible vehicles and the obscure effects of ultra-deep diving on the diver's brain and the novel decompression complications have changed everything. Scientific research into the causes and control of the HPNS in humans has lost both its practical and ethical justification, with the exception of submarine escape.

### 4.3.3  Free Escape from Submarines

The use of submarines is generally accepted as an ethical means of defence and the welfare of the crew is as important as that of other military personnel. The operational depth of submarines is classified information and so is their depth or rupture limit. However, two emergency scenarios are envisaged. One is a damaged submarine in which the crew is trapped in a compartment and subjected to increased air pressure, and therefore in need of both rescue and safe decompression. This situation has arisen several times and weighs heavily in the thinking about safety. NATO submarines are supported by a rescue submersible vehicle which is transported by a mother ship to the rescue site, where it is launched. It can lock on to the stricken submarine and convey the crew, at pressure, back to a hyperbaric chamber on the mother ship (Blatteau et al. 2013). The rescued submariners are then treated to decompression procedures which have been developed for commercial diving. These need refinements to cater for submariners.

The other scenario is that, for whatever reason, the simplest possible procedure, free escape, is the only option. Free escape means just that; the submariner is abruptly subjected to the external environment. The depth limit from which free escape might be a practical possibility is uncertain but it is not difficult to imagine circumstances in which the risk of free escape from a submarine trapped at any depth might be preferable to a slow death. Free escape is part of a submariner's training. The basic procedure involves the trainee, who is breathing air at normal atmospheric pressure, entering a "lockout" chamber wearing an all enclosing survival suit and breathing apparatus. The chamber is then flooded, increasing the pressurise to match that outside the submarine. Assume that is 2 MPa, at a depth of 200 m the edge of the continental slope. This is within the depth range of highly specialised professional divers, whose depth range is a little in excess of 500 m. Flooding might take 10–20 s, giving a rate of compression of 12–6 MPa per minute. When flooding is complete the subjects locks out to enter the sea and ascends rapidly in a vertical posture, allowing the enormous expansion of gas to escape from the lungs. Hydrostatic decompression would be rapid, say 1.5 m per second. Training from shallower depths is routine, with the main risk being a failure to vent expanding gas from the lungs. A paper published over 50 years ago reported successful free escape trials from a submarine at a depth of 183 m (Barnard 1971). More recent reports describe a "lockout" tower or tunnel which can accommodate two people, one of whom enters the water at a time. Speed is essential to minimise air equilibrating with the diver's tissues at pressure, as decompression sickness is a serious risk. Free escape from 300 m depths is judged possible in dire circumstances. (Video recordings of free escape procedures are available on Websites.)

The rate of compression envisaged in free escape is very fast compared to the pressure profiles used with either humans or primate subjects in a hyperbaric chamber, and generally rapid compression exacerbates the symptoms of the HPNS. There seems to be an ethical case for experimenting with primates, and perhaps humans, to mitigate the HPNS and other pathological consequences of free escape. There may be as yet unexplored pharmacological ways of minimising the risks (Gennser and Blogg 2008; Bhutani 2016). Both technically and ethically this is a difficult subject with, perhaps, something in common with problems of exploring space using manned vehicles.

## 4.4 Reflexes

There is a considerable body of knowledge about the onset of the HPNS but our understanding of the mechanisms involved is still incomplete. As with other brain disorders there is a profound conceptual difficulty in understanding the disruption to a complex organisation. For example drugs that affect the activity of certain parts of the brain are a focus of interest, but it is not clear if they shed any light on our main interest here, animals adapted to high-pressure environments. A thorough understanding of the relatively simple responses to pressure, such as the tail flip convulsions in Crangon and the disturbed spinal reflexes in humans at high pressure, is an attractive approach. Bennett, a major player in this field who has organised many multi-day dry chamber dives to explore the HPNS, has also collaborated with Harris to carry out relatively simple reflex experiments on human subjects. One involved the knee jerk reflex, sometimes known as the patellar tendon reflex. Readers may well be familiar with it. A firm tap on the tendon just below the knee causes the quadriceps muscle to contract. It is used medically because a normal response shows the circuitry to be in good order. That circuitry consists of the following. Within the muscle there are small tension-sensing structures called spindles. The tap stretches the tendon, thence the muscle and is sensed by the spindle. It sends impulses to the spinal cord in which alpha motor neurons are stimulated to send impulses to the muscle, causing it to contract. In the pressure chamber subjects were fitted with an apparatus that recorded the muscle response, and much else besides. It took 24 h to pressurise the subjects to 3 MPa, at which pressure they were kept for a week. At 3 MPa the time delay for the reflex to work was unaffected, showing that the nerve conduction velocity and synaptic transmission were unaffected. The reflex jerk of the leg was restrained so the force of muscle contraction could be measured at constant length, and this was found to be considerably increased by pressure. The main cause of this was thought to be the increased state of excitability of the alpha motor neurons, which is determined by higher centres, perhaps by reducing inhibition. So, even the simplest of nervous pathways is hyper responsive at increased pressure. As part of the experimental design the Jendrassic manoeuvre was incorporated. This is a way of reinforcing the reflex by the subject pulling hard with his hands joined. How this actually works is not entirely clear, but it is interesting that the reinforcement achieved in this way was reduced at pressure. So even modifying the simplest of nervous pathways produces complications and uncertainty.

What will be instructive here is an outline of the effects of pressure on ordinary nerve cells and synapses. This will give us an idea of the extent of the adaptations required to evolve animals with a nervous system that can cope with high, and perhaps variable, pressures. It will also link with the very limited information that we already have on the pressure tolerance (adaptation) of nerves and muscles in deep-sea animals (Chapter 5).

## 4.5     Excitable Membranes

Nerve and muscle cells transmit electrical impulses along their membranes at high speed. This is excitability and is the basis of the motor activity described previously, and of complex processes in the brain, which involves many nerve cell impulses interacting at junctions, called synapses. The simplest, somewhat hypothetical, case, is an impulse travelling along the membrane of a nerve cell (neuron) and terminating as a synapse at its junction with a muscle cell. This neuromuscular junction transmits the impulse to the muscle membrane (in a surprisingly elaborate way), where it continues along the muscle cell membrane, penetrating the muscle interior by means of finer membranes, where it initiates contraction. Much of this excitability consists of processes that are not chemical reactions but physical processes, beautifully organised.

### 4.5.1     Effects of Pressure on Isolated Neurons

Nerves, at the anatomical level, are bundles of nerve cells or neurons and are readily seen in various parts of an animal's body. In many animals, there is a conspicuous bundle of nerves running along the length of the body, the spinal cord of vertebrates and the ventral nerve cord of invertebrates such as crustacea, insects and worms. These structures comprise neurons packed together with supporting cells, and they send branches to specific regions to either command an action or receive a sensory signal and transmit it back to a centre. Nerve impulses, arising from individual cell membranes, can be detected in nerves as electrical waves, using methods devised by the pioneer physiologists. The approach is still a useful method today, especially in clinical neurology, producing electroencephalograms and electromyograms mentioned earlier. However, a nerve impulse cannot occur without the organised structure of a cell and so the individual cell is the basic unit to study.

The membrane of a typical eukaryote cell is electrically charged, positive on the outside surface and negative on the inside. This electrical potential, detected by electrodes positioned on either side of the membrane, is usually about 70 mV, which may not seem much but it is quite a significant force across the microscale of the membrane. The molecules therein feel it. How this positive charge is created is explained in Box 4.2, and we can simply say that a nerve impulse is the transient collapse of the charge and the propagation of the collapsed state along the membrane. The velocity of travel depends on many factors which are open to evolutionary and acclimatory changes (Chapter 12), including temperature and pressure (Harper et al. 1990; Macdonald 1994).

**Box 4.2: The Resting Potential**

The electrical charge across the membrane arises from the active transport of $Na^+$ ions out of the cell and the coupled inward transport of $K^+$ ions into the cell. They are normally loosely accompanied by various negative ions. This transport is a biochemical reaction and it requires energy, but it only generates a small part of the electrical potential across the cell membrane, the cell's resting potential. Most of this potential arises from the subsequent outward diffusion of the $K^+$ ions, down their concentration gradient. The intracellular concentration of $K^+$ is high. Diffusion takes place through specific ion channels in the membrane, which are very large proteins, containing a water filled central hole. The hole, the channel pore, caters for the passage of a specific ion and is opened and closed by special mechanisms. The permeability of the cell membrane is determined by its population of open ion channels. Typically, a large number of $K^+$ ions diffuse out of the cell and they create an excess of positive charges on the outer membrane surface, which restrains the process (like charges tend to repel). An equilibrium is reached when the concentration of $K^+$ ions diffusing out is balanced by the restraining influence of the positive charge they create and the negative charge they leave behind. This is the K equilibrium potential. It dominates the cell's resting potential and is the basis of the nerve impulse.

The collapse of the resting potential is self-propagating and for historical reasons, it is called an action potential. The collapse, also called a depolarisation, can be initiated in several ways. Let us assume here that the nerve is stimulated "upstream" by a synapse. The stimulus to collapse the resting potential works in an all or nothing way. A stimulus that fails merely sees the resting potential restore itself. One which succeeds by exceeding a threshold, creates a standard impulse that collapses—depolarises—the resting potential in the adjacent patch of membrane. Impulses are all or nothing, of standard size, and their speed of propagation is determined by the size and other properties of the cell. The "information content" in action potentials lies in their frequency.

The initial depolarisation of the resting potential activates (opens) the population of ion channels which conduct $Na^+$ ions, but only fleetingly. They soon inactivate, reducing the inward diffusion of Na ions but by then the membrane potential has been abolished. The opening of each individual Na channel is transient and their aggregate effect, the depolarisation of the membrane potential, is sensed by the ion channels which are permeable to K ions. As they activate (open) K ions diffuse out of the cell, and the resting potential is restored in the wake of the departing impulse. These K channels do not inactivate. The quantity of $Na^+$ and $K^+$ ions involved in this nicely phased process is very small in relation to the contents of the cell, and the active transport of the ions (Na out and K in) continues unabated and sustains the required concentration gradient.

### 4.5.2   Measuring Ionic Currents by Different Methods

There are several ways in which cell resting potentials, and action potentials with their associated ion currents, can be measured and recorded. For example they can be measured in an individual neuron. Glass microelectrodes can be inserted into the cell without causing any damage to the membrane, which would create short circuits, a condition which high pressure fortunately does not affect (Fig. 4.1).

A biphasic recording of an action potential is a macroscopic method that records the electrical wave travelling along a bundle of neurons, anatomically a nerve. A pair of fine wire electrodes, about 2 mm apart, are placed on the nerve. They detect the change in electrical potential between them as the wave of depolarisation passes along the nerve, the potential being the sum of many individual axons making up the nerve bundle, hence the term compound action potential. One is shown in idealised form in the inset in Fig. 4.2.

A third way of recording ionic currents is the patch clamp method, which is very different from the classical methods just described (Hamill et al. 1981; Ogden 1994).

In the history of chemistry and physics the idea of dealing with single molecules, one at a time so to speak, was something of a Holy Grail. It can be argued that physiologists attained that goal first, by way of the patch clamp technique. Several brilliant electrophysiologists contributed to this achievement, but the Nobel Prize in Physiology or Medicine was awarded to the German scientists, Neher and Sackmann in 1991, for their work. Experts in the subject described the award as particularly uncontroversial. In 1976 Neher and Sackmann reported the first recording of the current passing through a single channel (molecule). Instead of using a traditional intracellular microelectrode made of a glass tube filled with saline and drawn down to a microscopic tip that impales the cell membrane, they used a special glass micropipette. Its blunt, smooth tip (about 1–2 μm in diameter), pressed onto the surface of, for example a frog muscle cell, creates a high resistance seal (a mega-ohm seal) by a mysterious interaction between the cell surface and the glass.

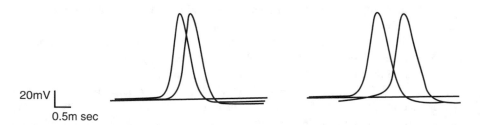

20mV ⌊

0.5m sec

**Fig. 4.1** The action potential of the giant axon of the squid *Loligo forbesii*, at high pressure. The potential is measured with a glass intracellular microelectrode. Left; recording at normal atmospheric pressure. Threshold stimulation elicited the first curve and a twice threshold stimulation the second curve. Right; Similar stimulation at 20 MPa. The axon was immersed in saline and pressure transmitted by liquid, medicinal, paraffin. Note the broadening of the action potential, caused by a slowing of the kinetics of the conformational changes in the ion channels involved. Reproduced with permission from Wann et al. (1979)

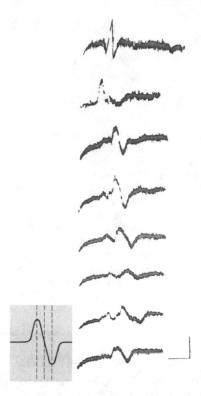

**Fig. 4.2** A compound action potential recorded at high pressure. The sequence of recordings shown is of the evoked compound action potential in the abdominal ventral nerve of the lobster Homarus gammarus at 8 °C, as in Fig. 13.2 in Chapter 13. From top to bottom: at atmospheric pressure. Then 150 min later still at atmospheric pressure. The third recording was at 2.7 MPa, then 5.1, 8.2, 10.3 MPa. The last two recordings were; at 5 min after decompression to normal pressure and then after 70 min at normal pressure. Horizontal bar, 2 ms. Vertical bar 40 µV. Note the compound action potential shows a broadening and a decrease in amplitude at high pressure, which is partially reversed on decompression. Reproduced with permission from Wilcock (1972)

Current flowing from the cell interior, through an ion channel in the membrane beneath the tip of the pipette and into the electrode is accurately measured. Voltage clamp circuitry (more soon) was employed to do this, hence the term patch clamp recording.

Mega-ohm seals form a loose patch, so-called because the electrode could be repositioned on the membrane. It has been used at high pressure (Chapters 11 and 13). Later, the giga-ohm seal was developed and became the definitive technique. It had the great advantage that its high resistance seal greatly reduced the background electronic noise, enabling small channel currents to be resolved (Fig. 4.3a). This type of seal is irreversible. Typically, ion channels open and close at high frequency, at random, with a probability determined by the recording conditions, so patch clamp, single channel recordings have the appearance of a series of current pulses

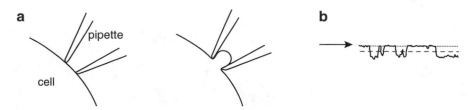

**Fig. 4.3** (a) Patch electrode sealed to the surface of a cell and (b), single channel currents. (a) The giga-ohm seal between the electrode and the membrane means that the currents flowing through open channels in the patch membrane far exceed the background noise. The example shown is of a cell attached patch. If the electrode is pulled away from the cell an inside-out patch is formed, with the solution in the electrode in contact with the outside face of the membrane. The patches are remarkably stable at high pressure, of which more below. (b) A patch with apparently one active channel shows brief random pulses of current whose duration can be measured using a computer program. In the example, the openings, downward deflections, have a duration of about 2 ms. The program sets a threshold of 50% of the current (in this case 10 pA), and successively measures the open and closed times. The arrow shows zero current flowing when all channels are closed. The probability of a channel being open is calculated by dividing the total open time by the total recording time, typically 10–30 s, much longer than the short recording shown. Many thousands of events go into the analysis (From author's unpublished material and common knowledge)

(Fig. 4.3b). Their analysis provides fundamental information about how channel proteins function.

Other techniques for observing the electrical activity of neurons use slices of brain tissue with microelectrodes or separately, special dyes (Wlodarczyk et al. 2008). The latter undergo a change in fluorescence as the membranes in which they are dissolved experience a change in voltage. These techniques can be used at high pressure, with an optical pressure chamber in the case of the fluorescent method (Chapter 13).

Our understanding of ion channels has spread across biology, well beyond excitability. Ion channels exist in the membranes of all forms of life. Bacteria, for example possess channels that open when the cell is swollen by an osmotic influx of water. The membrane stretches, opens the channels which release the fluid pressure, and prevent the cells rupturing. Such mechano-activated channels are ubiquitous and can be studied with the patch clamp technique. Voltage activated channels, already mentioned, probably evolved much later, along with special channels involved in synaptic transmission. They are opened when special compounds, neurotransmitters, bind to them. They are called ligand gated channels; ligands being, historically, a term for specific compounds that bind to enzymes and gating is a term borrowed from physics and used in channel kinetics to mean open/close.

### 4.5.3    How High Pressure Affects Excitable Membranes

The following examples do not provide an explanation for the motor symptoms seen in animals at high pressure, but they do provide food for thought. Deep-sea animals have presumably evolved excitable membranes that function normally at high pressure.

The first experiments in this field were carried out in the 1930s on isolated frog nerves by Grundfest and by Cattell (1936), in America and Ebbecke and Schaeffer (1953), in Germany. Many of their results have been confirmed but there is some doubt about the significance of temperature changes arising from the heat of compression of the inert oil, liquid paraffin, which was used to transmit pressure and insulate the electrodes. Spyropoulos' 1957 experiments with the squid giant axon were a significant step forward. Extracellular recordings from various nerves were also carried out, revealing a compound action potential (as in Fig. 4.2). They have more recently been carried out by Harper and colleagues for special reasons and will serve here to illustrate some basic effects of pressure. The experiments used the visceral branch of the vagus nerve of the cod, *Gadus morhua*, to provide a comparison with the same nerve isolated from deep-sea fish (Chapter 5). The classical, and simple extracellular recording technique was used for several reasons, chief of which was the experiments with nerves from deep-sea fish had to be carried out at sea. In such conditions, vibration can jeopardise more precise modern methods of recording. However, Campenot has successfully achieved intracellular recordings from a crab axon, at sea.

The visceral nerve, immersed in an appropriate solution, was mounted in a plastic holder. The assembly fitted in a cylindrical pressure vessel which was filled with liquid paraffin, and equipped with electrical connections, temperature control and connected to high-pressure plumbing in the usual way. Pressure was increased in steps and the resultant heat of compression was fully dissipated within 1 h, after which recordings started. The nerve was stimulated by applying 20 V square wave pulses. The resultant compound action potential was reduced in amplitude, almost disappearing at 3.1 MPa, but decompression restored it to near normal. The reduction was partly due to a reduced number of axons responding to the stimulus, but we know from other experiments that the pressure slows the response of the ion channels involved. Similar effects occur in most nerves, but there are exceptions. The speed with which the action potential propagates itself along the nerve at high pressure was reduced at pressure and associated with a broader action potential. Thus, generally the responsiveness of the fish nerve to stimulation was reduced by pressure, meaning the critical depolarisation required to fire an all or nothing action potential was increased. The refractory period, that is the time taken for the nerve to recover and be capable of generating a subsequent action potential, was also increased two or threefold at 3.1 MPa. These experiments served as controls for experiments on deep-sea fish nerves, described in Chapter 5.

The intracellular method of recording action potentials from a single neuron provides more precise information (Fig. 4.1). Historically the use of the giant axon from the squid played a key role in this field. The axon is sufficiently big for a macroscopic electrode to be inserted along its length, providing electrical recordings of events across an undamaged membrane. Experiments, mainly in Marine Laboratories such as those at Plymouth, UK, and Woods Hole, USA, using freshly collected squid, led to the detailed analysis of the action potential, summarised above. It comprises a fleeting inflow of Na ions through Na channels which depolarise the membrane, causing the channels to inactivate. The depolarisation triggers an outward K current (i.e., opens K channels), which restores the resting potential. Intracellular recordings from normal neurons required the development of glass microelectrodes which penetrate the membrane in a self-sealing way, without causing short circuits, already mentioned (Fig. 4.1a). A full account of these techniques is clearly well beyond the scope of this book, but classic texts by, for example Junge (introductory) and by Levitan and Kaczmarek and by Hille (advanced), are well written and very rewarding. The comprehensive theory of nerve excitation was developed by two British workers and is referred to as the Hodgkin Huxley Theory. It takes the form of a mathematical model describing the time course of the phased, self-propagating currents, the famous H-H equations. And, at a technical level, central to the experiments with squid axons, was the development of the "voltage clamp" technique, which allowed the separate $Na^+$ and $K^+$ currents to be recorded. (Junge's account of this technique is a good introduction.) The work was started before the second World War and the main paper was published in 1952. The impact of the H-H Theory was immense. Hille pointed out that the work of Hodgkin and Huxley was so new and technical that electrophysiologists in 1952 were unprepared to develop it. In 1963 Hodgkin and Huxley were awarded, with Eccles, the Nobel Prize in Physiology or Medicine. In high-pressure electrophysiology, meaning among a small number of isolated and rather bright individuals, there was much interest in extending these voltage clamp experiments with squid axons to high pressure. The first to achieve this were Henderson and Gilbert in 1975 and more soon followed, some studying other cells (Harper et al. 1981; Shrivastav et al. 1981 (*Loligo pealei*); Kendig 1984; Conti et al. 1982a, b). The work of Conti and his colleagues on the squid axon produced definitive data.

The all or nothing nature of an action potential involves the way voltage is "sensed" by ion channels, that is by specific proteins embedded in the cell membrane. At the time when the H-H model was being developed, currents carried by specific ions were real events but ion channels were a hypothetical necessity and were not yet identified as proteins. Their sensitivity to a voltage gradient across the membrane was imagined to reside in the movement of an exceedingly small electric current, called a gating current, which was predicted to precede the flow of the main ionic current carried by $Na^+$ ions, passing through the membrane and depolarising the resting potential. Conti et al. (1984), using the giant axon of the squid *Loligo vulgaris*, were able to show that high pressure, up to 60 MPa, slowed the Na gating current, roughly halving it at 50 MPa. Note that this current is the movement of charged components of macromolecules, not of freely moving ions. A much greater effect is exerted by the same pressure on the activation of the Na current itself, in the same squid axon. In separate experiments, the K current was also shown to be similarly slowed by pressure. They also showed that the positive activation* volumes (Chapter 2) in each of these processes probably arose from intrinsic conformational changes in the channel protein and were unlikely to result from pressure affecting the fluidity of the lipid bilayer surrounding them. In quite different cells, neurons in the snail *Helix pomatia* (studied by Harper and colleagues), and axons in the lobster *Panulirus interruptus* (Grossman and Kendig), broadly similar effects of pressure were found.

Single channel recording is a fundamentally different method of recording current flowing across a membrane, as explained in Fig. 4.3. As already mentioned, although it grew out of electrophysiology it has proved valuable in other areas of cell physiology, including bacteria and other non-excitable cells, of which more soon. As with voltage clamp experiments using the squid giant axon, extending the technique of single channel, patch clamp recording to high pressure held great promise, and this was achieved by Heinemann, working at the Max Planck Institute for Physical Biochemistry in Göttingen, Germany. Heinemann amazed himself and everyone else when he demonstrated a way of transferring a patch of membrane securely attached to the micro-pipette to a pressure vessel (Chapter 13; Fig. 13.3). Furthermore, the patch survived being subjected to high hydrostatic pressure and decompression. The first single channel recordings made at high pressure were of channels gated, not by voltage but by acetylcholine, the neurotransmitter responsible for transferring the action potential from a motor nerve to a muscle membrane, in this case embryonic rat muscle. The acetylcholine binds to the ion channel, induces a structural change, which allows ions to pass through. The binding is transient so the channel closes when the acetylcholine detaches itself. This is gating kinetics and although more than one channel was present in the patches, the kinetics can be quantified by measuring the mean open and mean closed time of the population of channels. Both were increased by high pressure, implying positive activation volumes occurs in both the opening and closing transitions, in other words, by subtle changes in protein structure. The electrical conductance of the channels, the rate at which ions diffuse through the pore, was unaffected by pressures, up to 60 MPa. This means that pressure does not affect the structure of the channel pore. This first demonstration of single channel recording at high pressure, showing that high pressure affects the gating kinetics of channels but not their conductance, has also been found in many other neurons and channels.

Thus, an action potential, originally recorded as a wave of electrical depolarisation in a nerve, is now understood to be the result of a population of ion channels functioning in harmony. A fundamental property of the channels, which H-H analysis discovered, is their activation and inactivation. Single channel recording has enabled these processes to be studied at high pressure.

### 4.5.3.1 Single Channel Recording at High Pressure

Consider the K channel known as Shaker B and its mutants, derived from the fruit fly Drosophila melanogaster, and expressed (grown) in the oocytes of *Xenopus*. This is a large cell, good at expressing foreign genes and providing channels in a membrane well suited for patch clamp recordings. It does not matter that the channel is alien to the oocyte. Inactivation of the Shaker B channel takes place by one of two mechanisms. The best understood is the "tethered" diffusion of a "ball and chain." The N terminal section of the channel protein is shaped like a ball and is tethered by a polypeptide chain to the body of the channel. After channel opening, the ball can hydrophobically bind to the mouth of the channel pore, on the cytoplasmic end, blocking the entry of K ions. Over time the ball diffuses off, but remains tethered, leaving the channel in a resting state and ready to be opened by the appropriate voltage. Mutants lacking the ball and chain inactivate by a less understood, but certainly quite different mechanism, called C-type inactivation, which appears to act at the other end of the pore, on the outside of the cell. Meyer and Heinemann (1997) and Schmalwasser et al. (1998) showed that the effect of pressure on these two mechanisms differs. Pressure slows the rate of the "ball and chain" inactivation. This implies a positive activation volume $(\Delta V^*)$ occurs in the channel inactivation process, which is consistent with the hydrophobic binding of the ball, and with other thermodynamic parameters. In contrast, the C-type inactivation was accelerated by pressure.

Another example of the effects of high pressure on single channel kinetics is the case of the BK channel, a near ubiquitous K channel, in which the B stands for big and K for $K^+$ ions. It is attractive to experimenters as it has a high conductance whose function is to restore the resting potential in various cells. In patch clamp experiments the BK channel in the chromaffin cells of the cow were used in the inside-out mode (Fig. 4.1c, d). This means the inner face of the detached membrane faces away from the patch pipette and is bathed in the experimental solution, in this case, one with a negligible concentration of calcium ions. Pressures of up to 90 MPa were applied. The unexpected effect of pressure was a 30-fold increase in the probability of channels opening, an effect which was reversed on decompression. The effect of high pressure is to increase the number of active channels in a patch by up to ten or more. It had no effect on channel conductance. The kinetics of this channel is complicated, but it appears that pressure "wakes up" sleepy channels and encourages others, already awake, to open more frequently. There is some evidence that a naturally occurring blocker is bound to the channel, and pressure might cause its dissociation. However, the main point is that here is an example of high pressure stimulating ion channel activity and not impairing it.

The patch clamp technique has proved very versatile and it can be used to record the activity of bacterial channels. This may seem surprising, considering the relative sizes of bacteria (about 1 μm across), and the opening at the tip of a patch pipette, also about 1 μm. However, some bacteria can be converted to giant spheroplasts of up to 6 μm in diameter, which are living cells lacking their cell wall, and these can be "patched." Using this approach Martinac and colleagues found that voltage, combined with gentle suction or pressure applied to the patch pipette, elicited channel gating. They had discovered mechano-activated MscS channels in *E. coli* spheroplasts, which were subsequently found to be present in the inner membrane of the cell, quite distinct from porin channels in the outer membrane. The porins can be reconstituted in liposomes, uniformly orientated, and can also be "patched" (Delcour et al. 1989), and such patches are stable under high pressure. In this way, it was found that the open state of Omp C from *E. coli* is favoured by high pressure (Macdonald and Martinac 1999).

Mechano-activated channels also occur in Eukaryotes of all grades. Some are well studied, such as those in hair cells in the mammalian inner ear, the Pacinian corpuscles and muscle spindle organs. Some are polymodal, meaning they are responsive to stimuli additional to a mechanical force. Vertebrate K channels of a type known as "two pore domain" have been reconstituted in liposomes and shown to respond to stretch mediated by the bilayer, just like the bacterial channel MscS. The bacterial channels are particularly interesting (Booth 2014), as similar versions are present in Archaea, including the deep-sea *Methanococcus jannaschii*. It was with this in mind that the author collaborated with Martinac to study the MscS in *E. coli* spheroplasts subjected to high pressure, using methods illustrated in Chapter 13 (Fig. 13.4). The experiments were carried out at room temperature. Pressures of up to 90 MPa reduced the gating activity, which was driven by a negative pipette voltage (Macdonald and Martinac 2005). (The problem of devising a means of applying gentle suction pressure to the patch pipette inside a pressure vessel was avoided.) The probability of the channels being open was reduced by pressure, which also reduced the channel open time. Treating the open–closed transition as an equilibrium allowed the volume change involved in closing to be calculated as $\Delta V$-147 cubic A per channel molecule. The way pressure acts is not by ordering the bilayer which supports the channel, as a similar ordering by a reduction in temperature of 10 °C had little effect on channel gating.

Experiments have shown that normally the channel is opened by tension transmitted by the bilayer to the channel protein. Its structure has been established at atomic resolution by X-ray methods and its gating movements studied in detail. Returning to the idea that the channels in *M. jannaschii* (designated MscMJ), are similar to MscS, Kloda and Martinac, showed that the pore diameter of MscS is half that of MscMl. However, the latter is similar to MscS in being activated by voltage and comparable stretch. This is apparently so, despite two potential complications. First, MscMJ was reconstituted in liposomes made of lipids quite different to those of which the Archaeal membrane is composed. Second, MscMJ has to function at high hydrostatic pressure, which we have seen has a significant effect on MscS.

MscL, a second mechano-activated bacterial channel mentioned earlier, is the channel of large conductance. It is actually a pentameric structure, smaller than MscS, which is made of seven subunits. The diameter of the open MscL is about 40 A, perhaps more, depending on how it is measured, and it is probably the largest pore of any gated channel. The diameter of the closed state is 18 A at the outer end and 2 A at the cytoplasmic end. It is normally stretch activated and the molecular, indeed atomic, structure of the gating mechanism is being intensively studied. A group of amino acids forming a "hydrophobic lock" seems to act as a gate and mutations changing the amino acids alter the gating kinetics. Increasing the hydrophobicity of the amino acids makes the channel harder to open and, conversely, more hydrophilic amino acids enables it to open more readily.

In a notable experiment, Petrov et al. (2011), subjected MscL to patch clamp recording at high pressure. They used a particular mutant of *E. coli* which spontaneously gated as a result of glutamate substituting for the less hydrophilic glycine, technically known as a gain of function mutant G22E. The channel was studied in both spheroplasts and reconstituted in liposomes using the set up in Fig. 13.4 (Chapter 13) Pressures of up to 90 MPa increased the probability of the channels being open, and increased the length of the open times, in either the normal fully open state, or in a half open (sub-conductance state). Pressure probably acts by favouring the hydration of the "hydrophobic lock." Hydrophobic gating is a feature of many channels (Birkner et al. 2012; Aryal et al. 2015). It relates to the behaviour of water molecules confined in a hydrophobic tube, the "dewetting" which ensues, blocking the flow of ions. The effect of high pressure on this phenomenon has yet to be investigated, but hydrophobic gated channels, adapted to high pressure, are likely to be rewarding targets for research.

An obvious point that should not be overlooked is that although the effect of pressure on the MscS channel is the opposite to its effect on MscL, in both it had no effect on the conductance of the fully open channel.

Finally, the development of the patch clamp technique has "co-evolved" with our knowledge of the structure of channels. Although a channel is regarded as a single protein it is usually a multimeric one. Channels can be classified according to their number of subunits, which are organised in a circle, forming a pore that spans the membrane. Four subunits are characteristic of voltage gated Na and K channels involved in action potentials. Five subunits are common in ligand gated channels, such as the acetylcholine gated channels and MscL. Gap junctions, which are permanently open pores connecting adjacent cells, are formed of six subunits, and as we have just seen MscS is a seven-unit multimer. In all cases the conducting path in the pore is filled with an aqueous solution through which ions, or other solutes, diffuse. The structure of the pore wall selects the solutes which pass through. The voltage gated K channel involved in action potentials has one of the most highly selective pores and its conductance is such that many millions of K ions per second can pass through it, the K current. The unit of electric current is the Amp, but the pico amp is the practical unit for channel currents, that is 10 to the power of $-12$, of an Amp. The picosiemens (pS) is the equivalent unit of conductance.

## 4.6    Effects of Pressure on Synapses

To repeat the point made at the start of this Section 4.4, action potentials, reaching the end of the nerve, are transmitted onwards to another nerve, muscle or gland by means of a synapse (Box 4.3). Synaptic transmission involves the release of a specific chemical compound (a neurotransmitter) but a small number of synapses transmit by a purely electrical mechanism, which we will not concern us here. The properties of synapses are crucial to an understanding of how pressure affects the activity of animals. Synapses are either excitatory or inhibitory. That is synaptic transmission can excite the downstream (postsynaptic) cell or it can render it less responsive to excitation. And when an inhibitory synapse is itself inhibited, perhaps by high pressure, then excitation follows (Box 4.3).

**Box 4.3: The Working Synapse**

The cellular organisation consists of the neuronal terminal almost in contact with the postsynaptic cell membrane of another nerve, muscle or gland. The gap between them, the synaptic cleft, is a fluid-filled zone across which the neurotransmitter diffuses and is thus the site of the slight delay which synaptic transmission incurs. The depolarisation of the presynaptic region of the axon causes special voltage sensitive channels to open. These are permeable to calcium ions which are present in high concentration in the extracellular fluid. Ca ions thus diffuse through the channel pores into the nerve terminal, causing vesicles containing neurotransmitter to discharge their contents to the outside. If calcium ions are lacking, transmission fails.

The neurotransmitter is contained in vesicles which, by exocytosis, discharge their contents into the synaptic cleft. The spontaneous occurrence of this process, causing miniature postsynaptic potentials is highly sensitive to pressure, described in Chapter 2. Transmitter release triggered by an action potential is quantal (i.e., in discrete packets) and synapses vary in the number of quanta normally released per action potential. The neuromuscular junction, of the frog or crustacea for example, typically involves a large number, conferring a safety margin that transmission will occur. Inhibitory synapses also show quantal release.

The released neurotransmitter molecules diffuse across the cleft, bind to specific receptors connected to channels in the postsynaptic membrane, and inducing a conformational change that opens the channel. The effect of the subsequent diffusion of ions, typically Na and other ions, is the depolarisation of the postsynaptic resting potential. In the case of an excitatory synapse this generally it creates an excitatory postsynaptic potential, EPSP. The aggregate effect of EPSPs is excitatory. In contrast, an inhibitory synapse works by increasing the resting potential of the postsynaptic membrane, by opening K channels. The ensuing flux of K ions takes the resting potential further from the threshold which must be crossed to trigger an action potential.

It was earlier stated that the information content of action potentials is in their frequency, and we see examples of that in synaptic transmission. Even simple excitatory synapses can respond to stimulation in a variety of ways, Facilitation is the term used to describe the enhanced release of transmitter during a burst of high-frequency stimulation. It is caused by an accumulation of Ca ions and lasts for a second or so. It is demonstrated by the twin pulse method, in which the response to the second stimulus (action potential) is greater than that of the first. After a burst of high-frequency stimulation, the response of the synapse can be reduced by the depletion of transmitter vesicles available for discharge. This is depression and not inhibition. Also, after a burst of high-frequency stimulation, a single stimulus can elicit an enhanced release of transmitter, a process called potentiation, which occurs on a time

(continued)

**Box 4.3** (continued)

scale of seconds and minutes. The underlying mechanism also involves accumulated Ca ions, but in this case the Ca ions come from mitochondria in the same cell. During high-frequency stimulation the increased concentration of Ca ions, in addition to triggering transmitter release, is also accumulated by mitochondria. When the stimulation stops and Ca no longer enters the synapse from the external fluid, the Ca diffuses from the mitochondria into the cell, and sustains the release of the transmitter. A different form of potentiation which lasts for hours or days arises from a quite different mechanism and is appropriately called long-term potentiation.

We are not finished yet. The neurotransmitter which has been released, bound to postsynaptic receptors and then dissociated, is recycled by an enzyme and the membrane debris, the result of exocytosis which released the transmitter, is returned to the presynaptic cell by the process of endocytosis. This compensatory endocytosis, or membrane retrieval, can be measured by a patch clamp method. The steady state performance of synapses requires synthesis and metabolic energy. Synapses offer multiple targets for high pressure, and in the long term, natural selection, to act on. Finally, synapses, are intrinsically interacting units, so although we need to be aware of their individual properties, which vary a lot, their interaction in simple and complex circuits underlies the motor activity and behaviour of an animal. In the present context, the impaired coordination, hyperexcitability and immobilisation seen in animals subjected to progressively high pressures, looks a nice simple progression but we are some way from understanding it.

How does pressure affect simple, isolated synapses? As with neurons, the examples answering the question illustrate the possibilities of what may occur in an animal subjected to high pressure.

The stellate ganglion in the squid, which connects to the celebrated giant axon, was important in the early research into the synaptic transmission. It enabled workers to show, using both electrical and optical methods, that the arrival of an action potential triggered the flux of Ca ions into the presynaptic region, causing the release of the transmitter, which in turn caused the depolarisation, the excitatory postsynaptic potential, EPSP. So, it was appropriate that the same ganglion was used to show that high pressure slowed the EPSP and increased fatigue, the latter being a diminished response to a train of excitatory impulses (Henderson et al. 1977). The highest pressure used was 20 MPa, but the lowest, 3.5 MPa, also had an effect. Experiments with crustacean nerves and muscle proved more convenient and they confirmed that pressure reduces the influx of Ca ions and the amplitude of the EPSP, triggered in this case by the release of the amino acid transmitter, glutamate (Parmentier et al. 1981; Grossman and Kendig 1990; Golan and Grossman 1992). Similarly, the frog neuromuscular junction, a classic preparation subjected to a thorough investigation by Aviner et al. (2013), responds in the same way. Thus, the voltage gated Ca channels in the presynaptic membrane seem to be pressure sensitive targets.

Similar effects of pressure are seen in the Ca channels in cells in the mammalian brain. One way to study them is to use a slice of a specific part of the isolated brain, which exposes cells that can be impaled by microelectrodes. Grossman and his colleagues, who have contributed much to this subject, found that transmitter release from synapses in the rat corticohippocampal region was reduced by 5 and 10 MPa, an effect which could be largely mimicked by reducing the external concentration of Ca ions. A useful generalisation is that in these and many other synapses, high pressure inhibits transmission in a graded fashion by acting "upstream," reducing the quantal release of transmitter, by reducing the influx of Ca ions.

Glutamate gated channels are widespread in excitatory synapses in the vertebrate central nervous system, and their sensitivity to high pressure has been found to depend on the specific subunits which make up their elaborate tetrameric structure, Box 4.4.

**Box 4.4**

Glutamate gated channels have been resolved into two prime groups, defined by the transmitter which best opens them. They are AMPA receptor- and NMDA receptor-channels. (AMPA is a-amino-3-hydroxy-5-methyl-4-isoxazole proprionic acid; NMDA is N-methyl-D-aspartic acid.) They occur together in the postsynaptic membrane and share common transmitters which can open them experimentally, such as L-glutamate. NMDA receptor channels are normally activated by two transmitters, NMDA and glycine, two molecules of which bind to receptors which are part of the protein structure of the channel. (Glycine is called a co-agonist or co-ligand.) The properties of these two differ markedly. AMPA channels provide brief openings and NMDA channels have much slower kinetics and pass both Na and Ca ions. The molecular structure of these channels has been resolved in detail, and ingenious techniques allow experimenters to synthesise chimaeras, that is functional channel made up of different subunits. They can be expressed in oocytes (which we have previously met) and reveal their properties to well established, but complicated, electrophysiological methods. Such experiments have shown that 10 MPa enhances the post synaptic current generated by certain combinations of subunits, but others are not affected (Mhor et al. 2012). An increased postsynaptic excitation of synapses in the brain by high pressure probably has some significance in the HPNS. With deep-sea fish and diving mammals in mind, the multiplicity of subunits comprising these channels and their varied sensitivity to high pressure suggests that evolving a pressure stable CNS may not be so improbable after all, but investigating it will require molecular-level details.

Inhibitory synapses in the vertebrate nervous system are of obvious interest. In the early HPNS studies with small mammals, the possibility of glycine mediated pathways in the spine playing a role in convulsions became attractive. This idea has been strengthened by the demonstration, using electrophysiological methods, that the affinity (binding) of glycine to the human glycine receptor channel is significantly reduced by 5 MPa or more (Roberts et al. 1996). There is also evidence that the presynaptic release of glycine is inhibited. In a simplistic way, we can say that inhibited inhibition implies excitation.

The activity of pressurised animals and humans is driven by the central nervous system, but it is finally expressed by muscular activity. So, how susceptible are muscles to pressure? Recall that the nineteenth century physiologist Regnard demonstrated that isolated muscles respond to compression. This chapter will close with a brief example of an experiment with humans that Harris and Bennett carried out during a dry dive to a simulated depth to 686 m, at Duke University, the USA. This was a major experimental hyperbaric programme. Heliox was supplemented with a small partial pressure of nitrogen to reduce the HPNS (explained in Chapter 11), and the experimental setup was similar to that previously described to measure the reflexes of subjects in a pressure chamber. The prime target muscle was the soleus muscle, which has a high ratio of slow: fast contracting fibres. It is a calf muscle, but largely covered by the conspicuous gastrocnemius. The subject, seated with the foot resting on a force plate, was fitted with stimulating and recording electrodes. The soleus muscle could be directly maximally stimulated by applying electrical pulses to the tibial nerve, behind the knee. It was found that the time delay and the duration of the recorded electromyogram, EMG, was unaffected at a pressure of 6.8 MPa, but the maximum peak force of a single twitch was markedly increased. Because of the direct electrical stimulation of the motor nerve connected to the muscle, any central (brain) influences can be ruled out. Probably pressure acts inside the muscle cells, on the process known as excitation-contraction and, or, on the contractile process itself. This demonstration of the effects of pressure on human muscle may never be repeated, if the ethical basis of such high-pressure dry dives fails to justify such experiments. But we can look forward to pursuing the matter with deep diving mammals and other deep-sea animals (Chapter 11).

## 4.7    Post Script

The study of how high pressure affects muscular contraction has a long history, rich in elegant experiments. The subject is lightly dealt with in Chapter 5 but it has been largely avoided here to avoid overload. Interested readers can refer to Friedrich (2010), and other references in Chapters 5 and 13.

## 4.8     Conclusion

The general conclusion from the preceding chapters is daunting. From Chapter 2 we must conclude that deep-sea pressures are sufficient to affect most, if not all, chemical equilibria and reaction rates in living organisms. This is confirmed in the review of the pressure effects on some of the cellular processes in ordinary organisms. We therefore conclude that life at high pressure has had to evolve adapted versions of all these processes. The pulses of high pressure in the microenvironment of load-bearing joints seem to have been put to good use by natural selection by means yet to be fully understood.

Subsequent chapters are largely concerned with organisms which normally experience high pressure, either because they permanently occupy, or normally visit, a high pressure environment. The aim is to explain how life proceeds in these extraordinary organisms which occupy the two global high pressure environments, the Deep Biosphere and the Deep Sea. We shall see how much progress has been made.

## References[1]

Arnott SA, Neil DM, Ansell AD (1998) Tail flip mechanism and size dependent kinematics of escape swimming in the brown shrimp Crangon crangon. J Exp Biol 201:1771–1784

Aryal P, Sansom MSP, Tucker SJ (2015) Hydrophobic gating in ion channels. J Mol Biol 427:121–130

Aviner B, Gradwohl G, Moore HJ, Grossman Y (2013) Modulation of presynaptic Ca++ currents in frog motor nerve terminals by high pressure. Eur J Neurosci 38:2716–2729

Aviner B, Gradwohl G, Aviner MM, Levy S, Grossman Y (2014) Selective modulation of cellular voltage -dependent calcium channels by hyperbaric pressure–a suggested HPNS partial mechanism. Front Cellular Nuerosci 8:136, 1–18 *

Barnard EEP (1971) Submarine escape from 600 feet (193meteres). Proc Roy Soc Med 64:1271–1273

Barthelemy L, Belaud A, Saliou A (1981) A study of the specific action of "per se" hydrostatic pressure on fish considered as a physiological model. In: Bachrach AJ, Matzen MM (eds) Underwater physiology VII, Proceedings of the symposium on underwater physiology. Academic Press, Bethesda

Bhutani S (2016) Thermal considerations in escape versus rescue from a disabled submarine. J Marine Med Soc 18:75–78

Birkner JP, Poolman B, Kocer A (2012) Hydrophobic gating of mechanosensitive channel of large conductance evidenced by single-subunit resolution. Proc Natl Acad Sci U S A 109:12944–12949

Blatteau J-E et al (2013) Submarine rescue decompression procedure from hyperbaric exposure up to 6 bar of absolute pressure in man: effects on bubble formation and pulmonary function. PLoS One 8:e6781

Bliznyuk A, Aviner B, Golan H, Hollmann M, Grossman Y (2015) The N-methyl-D-aspartate receptor's neglected subunit GluN1 matters under normal and hyperbaric conditions. Eur J Neurosci 42:2577–2584. *

---

[1]Those marked * are further reading.

Booth IR (2014) Bacterial mechanosensitive channels: progress towards an understanding of their roles in cell physiology. Curr Opin Microbiol 18:16–22

Brauer RW (1972) Barobiology and the experimental biology of the deep sea. North Carolina sea grant program. University of North Carolina, North Carolina. *

Brauer RW, Beaver RW, Hogue CD, Ford B, Goldman SM, Venters RT (1974) Intra- and inter species variability of vertebrate high pressure neurological syndrome. J Appl Physiol 37:844–851

Brauer RW, Mansfield WM, Beaver RW, Gillen HW (1979) Stages in development of high pressure neurological syndrome in the mouse. J Appl Physiol 46:756–765. *

Campenot RB (1975) The effect of high hydrostatic pressure on the transmission at the crustacean neuromuscular junction. Comp Biochem Physiol B 52:133–140. *

Cattell M (1936) The physiological effects of pressure. Biol Rev 11:441–475

Chen J, Liu H, Cai S, Zheng H (2019) Comparative transcriptome analysis of Eogammarus possjeticus at different hydrostatic pressure and temperature exposures. Sci Rep 9:3456

Clark LC, Gollan F (1966) Survival of mammals breathing organic liquids equilibrated with oxygen at atmospheric pressure. Science 152:1755–1756. *

Connor CW, Ferrigno M (2009) Estimates of N2 narcosis and O2 toxicity during submarine escapes from 600 to 1000 fsw. Undersea Hyperbar Med 36:237–240. *

Conti F, Fioravanti R, Segal JR, Stuhmer W (1982a) Pressure dependence of the sodium currents of squid giant axon. J Membr Biol 69:23–34

Conti F, Fioravanti R, Segal JR, Stuhmer W (1982b) Pressure dependence of the potassium currents of squid giant axon. J Membr Biol 69:35–40

Conti F, Inoue I, Kukita F, Stuhmer W (1984) Pressure dependence of sodium gating currents in the squid giant axon. Eur Biophys J 11:137–147

Corry B et al (2019) An improved open-channel structure of MscL determined from FRET confocal microscopy and simulation. J Gen Physiol 136:483–494. *

Cox C, Stavis RL, Wolfson MR, Shaffer TH (2003) Long term tidal liquid ventilation in premature lambs: physiologic, biochemical and histological correlates. Biol Neonate 84:232–242

Cruickshank CC, Minchin RF, Le Dain AC, Martinac B (1997) Estimation of the pore size of the large conductance mechanosensitive ion channel of Escherichia coli. Biophys J 73:1925–1931. *

Cui Y-M et al (2010) Design, synthesis and characterization of BK channel openers based on oximation of abietane diterpene derivatives. Bioorg Med Chem 18:8642–8659. *

Damazeau G, Rivalain N (2011) High hydrostatic pressure and biology: a brief history. Appl Microbiol Biotechnol 89:1305–1314

Delcour A, Martinac B, Adler J, Kung C (1989) Modified reconstitution method used in patch clamp studies of Escherichia coli on channels. Biophys J 56:631–636

Dussauze M et al (2017) Dispersed oil decreases the ability of a mode fish (Dicentrarchus labrax) to cope with hydrostatic pressure. Environ Sci Pollut Res 24:3054–3062

Ebbecke U (1935) Uber die Wirkungen hoher Drucke auf marine Lebewesen. Pflugers Arch Ges Physiol 236:648–657

Ebbecke U, Schaeffer H (1953) Uber den Einfluss hoher Drucke auf den Aktionsstrom von Muskeln und Nerven. Pflugers Arch Ges Physiol 236:679–692

Flugel H, Schlieper C (1970) The effect of pressure on marine invertebrates and fishes. In: Zimmerman AM (ed) High pressure effects on cellular processes. Academic Press, Cambridge. *

Fontaine M (1928) Sur les analogues existant entre les effects d'une tetanisation et ceux d'une compression. CR Hebd Seanc Acad Sci Paris 186:99–101. *

Fontaine M (1930) Recherches experimental;es sur les reactions des etres vivants aux forte pressions. Annals Inst Oceanogr Monaco 8:2–99. *

Friedrich O (2010) Chapter 10: Muscle function and high hydrostatic pressures. In: Sebert P (ed) Comparative High Pressure Biology. Science Publishers, New York

Gennser M, Blogg SL (2008) Oxygen or carbogen breathing before simulated submarine escape. J Appl Physiol 104:50–56

Golan H, Grossman Y (1992) Synaptic transmission at high pressure effects of (Ca ++)o. Comp Biochem Physiol 103A:113–118

Grossman Y, Kendig JJ (1984) Pressure and temperature: time dependent modulation of membrane properties in a bifurcating axon. J Neurophysiol 52:692–708. *

Grossman Y, Kendig JJ (1990) Evidence for reduced presynaptic Ca entry in a lobster neuromuscular junction at high pressure. J Physiol 420:355–364

Grossman Y, Aviner B, Mor A (2010) Pressure effects on mammalian central nervous syndrome. In: Sebert P (ed) Comparative high pressure biology. Science Publishers, New York, pp 161–186. *

Grundfest H (1936) Effects of hydrostatic pressures upon the excitability, the recovery and the potential sequence of frog nerve. Cold Spring Harb Symp Quant Biol 4:179–186. *

Hamill OP, Marty A, Neher E, Sakmann B, Sigworth FJ (1981) Improved patch-clamp techniques for high resolution current recording from cells and cell-free membrane patches. Pflugers Arch 391:85–100

Harper AA, Macdonald AG, Wann KT (1981) The action of high hydrostatic pressure on the membrane currents of Helix neurones. J Physiol 311:325–330

Harper AA, Watt PW, Hancock NA, Macdonald AG (1990) Temperature acclimation effects on carp nerve: a comparison of nerve conduction, membrane fluidity and lipid composition. J Exp Biol 154:305–320

Harris DJ (1979) Observations on the knee-jerk reflex in oxygen-helium at 31 and 43 bars. Undersea Biomed Res 6:55–74. *

Harris DJ, Bennett PB (1983) Force and duration of muscle twitch contractions in humans at pressures up to 70 bar. J Appl Physiol Respirat Environment Physiol 54:1209–1215. *

Harris DJ, Coggin R, Roby G, Turner G, Bennett PB (1982) Slowing of S.E.P. late waves in gas breathing and liquid breathing dogs compressed up to 101 bars. Undersea Biomed Res 9:7; (abstract) *

Heinemann SH, Conti F, Stuhmer W, Neher E (1987a) Effects of hydrostatic pressure on membrane processes. J Gen Physiol 90:765–778. *

Heinemann SH, Stuhmer W, Conti F (1987b) Si9ngle acetylcholine receptor channel currents recorded at high hydrostatic pressures. Proc Natl Acad Sci U S A 84:3229–3233. *

Henderson JV, Gilbert DL (1975) Slowing of ionic currents in the voltage-clamped squid axon by helium pressure. Nature 258:351–352. *

Henderson JV, Lowenhaupt MT, Gilbert DL (1977) Helium pressure alteration of function in squid giant synapse. Undersea Biomed Res 4:19–26

Hille B (2001) Ion channels of excitable membranes. Sinauer Associates Inc., Sunderland. *

Junge D (1976) Nerve and muscle excitation. Sinauer Associates Inc., Sunderland. *

Kendig JJ (1984) Ionic currents in vertebrate myelinated nerve at hyperbaric pressure. Am J Phys 246:C84–C90

Kloda A, Martinac B (2001) Molecular identification of a mechanosensitive channel in Archaea. Biophys J 80:229–240. *

Kloda A et al (2008) Mechanosensitive channel of large conductance. Int J Biochem Cell Biol 40:164–169. *

Kohlhauer M et al (2015) Hypothermic total liquid ventilation is highly protective through cerebral hemodynamic preservation and sepsis-like mitigation after asphyxial cardiac arrest. Crit Care Med 43:e420

Kuffler SW, Nicholls JG, Martin AR (1984) From neuron to brain. Sinauer Associates Inc., Sunderland. *

Kylstra JA (1968) Experiments in water breathing. Sci Am 219:66–74. *

Kylstra JA, Paganelli CV, Lanphier EH (1966) Pulmonary gas exchange in dogs ventilated with hyperbarically oxygenated liquid. J Appl Physiol 21:177–184. *

Kylstra JA, Nantz R, Crowe J, Wagner W, Saltzman HA (1967) Hydraulic compression of mice to 166 atmospheres. Science 158:793–794. *

Lever MJ, Miller KW, Paton WDM, Smith EB (1971) Pressure reversal of anaesthesia. Nature 231:368–371. *

Levitan IB, Kaczmarek LK (2002) The neuron. Cell and molecular biology. Oxford University Press, Oxford. *

Logue PE, Schimdt FA, Rogers HE, strong GB (1986) Cognitive and emotional changes during a simulated 686 m deep dive. Undersea Biomed Res 13:225–2345. *

Lundgren CEG, Ornhagen HC (1976) Hydrostatic pressure tolerance in liquid breathing mice. In: Underwater physiology V, Proc. Fifth symposium on underwater physiology. FASEB, Bethesda, pp 397–404. *

Macdonald AG (1972) The role of high hydrostatic pressure in the physiology of marine animals. Symp Soc Exp Biol 26:209–231

Macdonald AG (1976a) Locomotor activity and oxygen consumption in shallow and deep sea invertebrates exposed to high hydrostatic pressures and low temperature. In: Lambertsen CJ (ed) underwater physiology V, Proceedings of the 5th symposium. FASEB, Maryland, pp 405–419

Macdonald AG (1976b) Hydrostatic pressure physiology. In: Bennett PB, Elliott DH (eds) The physiology and medicine of diving. Bailliere Tindall, London

Macdonald AG (1994) The adaptation of excitable membranes to temperature and pressure: conduction velocity. In: Cossins AR (ed) Temperature adaptation of biological membranes. Portland Press, London

Macdonald AG (1997) Effect of high hydrostatic pressure on the BK channel in Bovine Chromaffin cells. Biophys J 73:1866–1873. *

Macdonald AG, Martinac B (1999) Effect of high hydrostatic pressure on the porin OmpC from Escherichia coli. FEMS Microbiol Lett 173:m327–m334

Macdonald AG, Martinac B (2005) Effect of high hydrostatic pressure on the bacterial mechanosensitive channel MscS. Eur Biophys J 34:434–441

Macdonald AG, Teal J (1975) Tolerance of oceanic and shallow water Crustacea to high pressure. Deep-Sea Res 22:131–144

Macdonald AG, Gilchrist I, Teal J (1972) Some observations on the tolerance of oceanic plankton to high hydrostatic pressure. J Mar Biol Assoc UK 52:213–223

Macdonald AG, Gilchrist I, Wann KT, Wilcock SE (1980) The tolerance of animals to pressure. In: Gilles R (ed) Animals and environmental fitness, vol 1. Pergamon Press, Oxford, pp 385–403. *

Martinac B, Kloda A (2003) Evolutionary origins of mechanosensitive ion channels. Prog Biophys Mol Biol 82:11–24. *

Martinac B, Buechner M, Delcour A, Adler J, Kung C (1987) Pressure-sensitive ion channel in Escherichia coli. Proc Natl Acad Sci U S A 84:2297–2301. *

McCall RD, Frierson D (1981) Evidence that two loci predominantly determine the difference in susceptibility to the High Pressure Neurological Syndrome Type I seizure in mice. Genetics 99:285–307. *

Menzies RJ, George RY, Avent R (1972) Responses of selected aquatic organisms to increased hydrostatic pressure: preliminary results. In: Barobiology and the experimental biology of the deep sea. University of North Carolina Sea Grant Program, Chapel Hill, pp 37–54

Meyer R, Heinemann SH (1997) Temperature and pressure dependence of Shaker K+ channel N- and C- type inactivation. Eur Biophys J 26:433–445

Mhor A, Kuttner YY, Levy S, Mor M, Hollmann M, Grossman Y (2012) Pressure -selected modulation of NMDA receptor subtypes may reflect 3D structural differences. Front Cell Neurosci 6:37, 1–9

Miller KW, Paton WDM, Smith EB, Smith RA (1973) The pressure reversal of anaesthesia and the critical volume hypothesis. Mol Pharm 9:131–143. *

Naquet R (1988) In: Lemaire C, Rostain J-C (eds) The high pressure nervous syndrome and performance. Octares, Marseille, France. *

Neher E, Sackmann B (1976) Single-channel currents recorded from membrane of denervated frog muscle fibres. Nature (London) 260:779–802. *

O'Donnel SW, Horn WG (2014) Initial review of the U.S. Navy's pressurised submarine escape training outcomes. Undersea Hyperbar Med 41:33–40. *

Oei GTML, Hollmann MW, Preckel B (2010) Cellular effects of helium in different organs. Anaesthesiology 112:1503–1510

Ogden D (1994) Microelectrode techniques. The company of biologists. Cambridge University Press, Cambridge

Oliphant A, Thatje S, Brown A, Morini M, Ravaux J, Shillito B (2011) Pressure tolerance of the shallow water caridean shrimp Palaemonetes varians across its thermal tolerance window. J Exp Biol 214:1109–1117

Otter T, Salmon ED (1985) Pressure-induced changes in Ca++ channel excitability in Paramecium. J Exp Biol 117:29–43. *

Parmentier JL, Shrivastav JL, Bennett PB (1981) Hydrostatic pressure reduces synaptic efficiency by inhibiting transmitter release. Undersea Biomed Res 8:175–183

Petrov E, Rhode PR, Martinac B (2011) Flying-patch patch clamp study of G22E-MscL mutant under high hydrostatic pressure. Biophys J 100:1635–1641

Regnard P (1885) Phenomenes objectifs que l'on peut observer sur las animaux soumis aux hautes pressions. CR Seanc Soc Biol 37:510–515. *

Regnard P (1891) Recherches experimentales sur les conditions de la vie dans les eaux. Mason, Paris

Roberts RJ, Shelton CJ, Daniels S, Smith EB (1996) Glycine activation of human homomeric a1glycine receptors is sensitive to pressure in the range of the high pressure nervous syndrome. Neurosci Lett 208:125–128

Rostain JC, Lavoute C (2010) The high pressure nervous syndrome. In: Sebert P (ed) Comparative high pressure biology. Science Publishers, New York, pp 431–464

Rostain JC, Gardette B, Gardette-Chaffour MC, Forni C (1984) HPNS of baboons during helium-nitrogen-oxygen slow exponential compressions. J Appl Physiol 57:341–350. *

Sackmann B, Neher E (1995) Single channel recording. Plenum Press, New York. *

Sakmann B, Neher E (1984) Patch clamp techniques for studying ionic channels in excitable membranes. Annu Rev Physiol 46:455–472. *

Schlieper C (1972) Comparative investigations on the pressure tolerance of marine i9nvertebrates and fish. Symp Soc Exp Biol 26:197–207. *

Schmalwasser H, Neef A, Elliott AA, Heinemann SF (1998) Two electrode voltage clamp of Xenopus oocytes under high hydrostatic pressure. J Neurosci Methods 81:1–7

Sebert P, Macdonald AG (1993) Fish. In: Macdonald AG (ed) Advances in comparative and environmental physiology, vol 17. Springer, Berlin, pp 148–196

Shrivastav BB, Parmentier JL, Bennett PB (1981) A quantitative description of pressure induced alterations in ion channels in the squid giant axon. In: Bachrach AJ, Matzen MM (eds) Proc. of the Seventh Symposium on Underwater Physiology. Undersea Society Inc., North Palm Beach

Spyropoulos CS (1957) Response of single nerve fibres at different hydrostatic pressures. Am J Phys 189:214–218

Talpalar AE, Grossman Y (2003) Modulation of rat corticohippocampal synaptic activity by high pressure and extracellular calcium: single and frequency responses. J Neurophysiol 90:2106–2114. *

Thatje S, Casburn l, Calcagno JA (2010) Behavioural and respiratory response of the shallow water hermit crab Pagurus cuanensis to hydrostatic pressure and temperature. J Exp Mar Biol Ecol 390:22–30

Todnem K, Knudsen G, Riise T, Nyland H, Aarli JA (1989) Nerve conduction velocity in man during deep diving to 360msw. Undersea Biomed Res 16:31–40

Verma R, Mohanty c s, Kondange C (2016) Saturation diving and its role in submarine rescue. J Marine Med Soc 18:72–74. *

Wann KT Macdonald AG, Harper AA, Wilcock SE (1979) Electrophysiological measurements at high hydrostatic pressure: methods for intracellular recording from isolated ganglia and for extracellular recording in vivo. Comp Biochem Physiol 64A:141–147

Westbrook GL (1999) Ligand gated ion channels. In: Sperelakis N (ed) Cell physiology sourcebook. Academic Press, New York, pp 675–687. *

Wilcock SE (1972) PhD Thesis, Aberdeen University

Wilcock SE, Wann KT, Macdonald AG (1978) The motor activity of Crangon crangon subjected to high hydrostatic pressure. Mar Biol 45:1–7

Winkler DA, Thornton A, Farjot G, Katz I (2016) The diverse biological properties of the chemically inert noble gases. Pharmacol Therap 160:44–64

Wlodarczyk A, Macmillan PF, Greenfield SA (2008) Voltage sensitive dye imaging of transient neuronal assemblies in brain slices under hyperbaric conditions. Undersea Hyperbar Med 35:35–40

Wolfson MR, Shaffer TH (2005) Pulmonary applications of perfluorochemical liquids: ventilation and beyond. Paediatric Resp Rev 6:117–127

# The Effects of Decompression and Subsequent Re-compression on the Activity of Deep-Sea Animals and Eukaryote Cells. The Isobaric Collection of Deep-Sea Animals

**5**

In this chapter, we focus on the activity of deep-sea animals and their response to lower, and to higher, than normal pressures. Deep-sea Eukaryote microorganisms are also included.

Hydrothermal vent organisms, Prokaryotes from the Deep Sea and the Deep Biosphere and Deep-Sea divers, both animal and human, are dealt with in subsequent chapters.

Two distinct types of observation are made. One is of organisms collected by simple methods, allowing decompression to be experienced and the other is of organisms collected without decompression, using special isobaric collecting equipment. This then allows observations to be made at their original high pressure followed by a controlled decompression.

## 5.1    Deep-Sea Amphipods

Amphipods, similar to *M. marinus* (Chapter 4), living on the deep seafloor are practical experimental animals. The effects of pressure on the activity of amphipods from a range of depths have been investigated, but unlike the experiments with their shallow water counterparts, deep-sea species experience decompression during collection unless isobaric methods are used. The observations described here were carried out by various workers, including the author, and assumed that amphipods from deep water would be more tolerant to high pressure than the shallow species. We had little idea of how decompression during their collection would affect them. The author collaborating with an engineer, Dr. Ian Gilchrist, planned to test the above assumption about pressure tolerance and to investigate the effects of decompression, by devising an isobaric method of collection. It is described in Chapter 13, Section 13.7.

© Springer Nature Switzerland AG 2021
A. Macdonald, *Life at High Pressure*, https://doi.org/10.1007/978-3-030-67587-5_5

### 5.1.1    Amphipods Collected with Decompression

We begin with the author's observations because they provide meaningful comparisons between animals from different depths. Experiments were carried out in the North Atlantic during several cruises on ships operated by the Natural Environmental Research Council, UK. Simple baited traps were deployed, initially using a deep mooring procedure which was replaced by a conventional free-fall, "pop-up" method designed for the heavy isobaric trap (Chapter 13, Fig. 13.20).

Groups of mixed species of benthic amphipods were studied, all similar in lifestyle and size to *M. marinus*, which were subsequently identified by experts. All the animals behaved in much the same way; no obvious species-specific responses were seen. After collection, the animals were held at a temperature close to that of their environment, in an observational pressure vessel and then subjected to 5 MPa steps at intervals of 5 min.

Animals collected from 1200 to 1300 m depths, with decompression.

In an observational pressure vessel (Chapter 13, Fig. 13.1), they exhibited normal activity at surface pressure, for many hours. Half of the total of 127 individuals collected showed spasms at 30 MPa, but the other half showed no symptoms, which is strange and unexplained. There was no obvious response to the light used. In another experiment the mean pressure for spasms in a group of animals was 27.3 MPa (to be compared to 10 MPa for *M. marinus* Fig. 5.1).

Other animals, which were kept at 13 MPa for 3 h and then pressure tested, showed spasms at a mean of 33.5 MPa, which is significantly higher than that of the untreated group. This suggests that decompression to surface pressure initially sensitises the animals to the subsequent pressure test.

Animals collected from 2000 m depth, with decompression.

Only 2 of 15 individuals were active at surface pressure but at 20 MPa all rapidly resumed normal activity.

Sixty percent underwent spasms at 45 MPa. Thus, decompression from 2000 m depths (20 MPa) to atmospheric pressure caused more extensive inhibition (paralysis) than from 1300 m, but the animals regain normal activity when restored to their normal ambient pressure. They undergo spasms at a higher pressure than shallow water animals, consistent with the hypothesis.

Animals collected from 2700 m depth, with decompression.

Only 3 out of 43 individuals were active at surface pressure and the rest looked dead. After 2 h exposure to 25 MPa 12 were mobile. Thus, decompression from 2700 m (27 MPa) to atmospheric pressure exerts a greater deleterious effect than from 2000 m (20 MPa).

Animals collected from 4000 to 4300 m depths, with decompression.

155 individuals were collected and all were immobile, paralysed, at surface pressure. When restored to 40 MPa 33 individuals resuscitated well and when these were decompressed to surface pressure they became immobilised again. A separate group of resuscitated animals showed some swimming activity when compressed to 65 MPa but at, or above, 70 MPa they became immobile, without showing spasms.

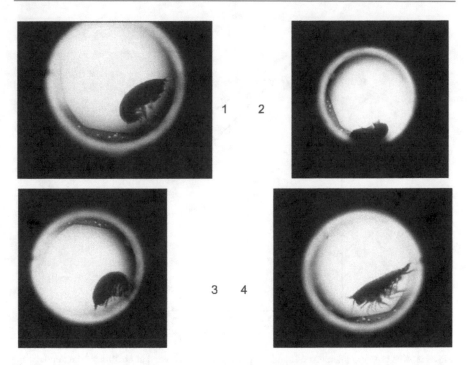

**Fig. 5.1** *Tmetonyx cicada*, collected with decompression from a seafloor depth of 1300 m and then pressure tested in an observational vessel. The photographs record what the experimenter sees. The animal was pressurised at 6 °C in steps of 5 MPa each 5 min. (*1*) Activity was normal up to 15 MPa at which pressure it intermittently curled up and then crawled (*2*). At 25 MPa it vigorously swam with flexing and uncurling. At 30 MPa (*3*) it exhibited violent spasms and uncurling. Swimming became intermittent and jerky. (Shallow water Amphipods would be immobilised at this stage.) At 40 MPa (*4*), the animal straightened out in a sustained spasm and soon the flickering pleopods stopped. It was paralysed. Twenty minutes after swift decompression to atmospheric pressure normal activity was resumed (author's unpublished photographs)

Decompression from 4000 m (40 MPa) immobilises the animals but this condition is reversed in a significant number when they are restored to 40 MPa, after which higher pressures fail to elicit spasms.

The Amphipods used in the experiments were species of the genera *Tmetonyx*, *Eurythenes* and *Paralicella*, all well-known inhabitants of the ocean floor. *Eurythenes gryllus*, for example as defined by a taxonomy based on morphology, has been attributed a depth range of 180–6500 m. However, analysis of its mitochondrial RNA strongly indicates that it is not a single species. The response of animals to compression and decompression is probably best related to the depth and temperature from which the animals are collected; taxonomy is secondary. The data can be summarised in a provisional way (Fig. 5.2), but more data would strengthen the conclusions.

Figure 5.2 indicates there is a graded tolerance to pressure and points three and four show that decompression sensitises the animals to pressure. The former is not

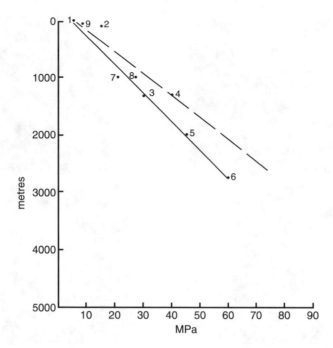

**Fig. 5.2** The pressure tolerance of benthic Amphipods (Macdonald et al. 1980). The depth of the collection is shown on the vertical scale and the convulsion threshold pressure, 50% of group, is on the horizontal axis. Compression 5 MPa steps at 5 min intervals, at 5–8 °C. Animals collected with decompression unless otherwise stated. Lines are drawn for discussion. *1, Marinogammarus marinus*, littoral. *2, Parathemisto sp.* epipelagic. *3, Tmetonyx cicada*, from Rockall Basin. *4*, as *3*, but isobaric collection. *5, Eurythenes sp.* Rockall Basin. *6, Euonyx biscayensis* and *Eurythenes gryllus*, west of Bay of Biscay. *7* and *8*, from Lake Baikal, and pressurised at 1.3 MPa per minute; *Ommatogammarus albinus, Odontogammarus margaritaceus*. *9, Pallascea grubei, Acanthogammarus albus* (see Section 5.1.2). Reproduced with permission from Macdonald et al. (1980)

surprising but the latter was unanticipated. Decompression paralysis was also a surprising new phenomenon, and its reversal clearly demonstrated that the immobilised animals collected from depth are not necessarily dead. The resuscitated animals, from 4000 m, remaining active up to a pressure of 70 MPa, is strong evidence of increased pressure tolerance. These results strengthened the case for isobaric collection. At the time we wondered about the generations of marine biologist who had assumed the immobile animals collected in their deep trawls were dead. Many were, being mechanically damaged, but some were probably showing decompression paralysis and could have been resuscitated. The equipment for doing so has existed since Regnard's day.

Trapping experiments like those described were, and are, a fairly "low tech" business, but a desperate one, heavily dependent on luck with the weather and the catch, with never enough time to acquire all the data ideally needed. So, it is particularly important to relate the results to other observations with deep-sea benthic Amphipods.

Using a time lapse camera fitted to a pop-up trap deployed at a depth of 4850 m, Thurston recorded the activity of a large number of amphipods during the ascent, i.e., decompression. It increased but from a depth of 1600–800 m it declined and at the surface all the animals were immobile. The increased activity is interesting and the immobilisation is consistent with the above. In separate work in the Arabian Sea, a large number of amphipods were recovered in traps which were thermally insulated but not pressure retaining, from depths of 1920–4420 m (Treude et al. 2002). The vast majority of animals were immobilised at the surface and described as dead but that was probably not the case, and many would probably have resuscitated if they had been restored to their normal pressure. In contrast, large numbers of *Stephonyx bicayensis* were collected from depths of 1528–1765 m the temperate North Atlantic and lived happily ever after, well for many weeks, in aquaria at atmospheric pressure, in the UK (Brown and Thatje 2011). Similarly, the specimens of *E. gryllus* which George (1979), collected from a depth of 1850 m in Arctic waters survived well at normal atmospheric pressure and 2 °C. When subjected to increased pressure they showed little response until 52 MPa elicited convulsions. This Arctic version of *E. gryllus* seems to be very tolerant of decompression at 2 °C but is as susceptible to high pressure as those described above. Specimens of nominally the same species of Amphipod, but probably rather separate, were collected with decompression from 3186 m in Antarctic waters and appeared to behave normally at atmospheric pressure. There were no reports of the animals being inactive on exposure to atmospheric pressure, unlike those amphipods described earlier. In the last study the authors (Takeuchi and Watanabe 1998), seemed to be unaware of the significance of hydrostatic pressure! These different degrees of pressure tolerance seen in *E. gryllus* may reflect genetic differences for which several workers have found evidence; (France and Kocher 1996; Eustace et al. 2016; Havermans et al. 2013).

Benthic amphipods have attracted numerous other studies. Feeding experiments with radiolabelled food at the depth of trapping have been carried out by Wirsen and Jannasch (1983). Smith and Baldwin, in a 1982 paper, describe how the submersible Alvin was used to collect and transfer amphipods, *Paralicella caperesca*, to a respirometer. Their respiration was then measured "in situ," and apparently it was stimulated by the odour given off by nearby food.

## 5.1.2 Amphipods from Lake Baikal

Lake Baikal, the deepest freshwater lake in the world, and the largest by volume contains very many endemic species. These include Amphipods with a bottom living, scavenger lifestyle similar to that of their distant marine relatives. The lake was formed over 20 million years ago as a rift valley opened in the earth's crust, and the circulation of water throughout its entire depth range, 1600 m, ensures there are no oxygen-free zones. The second deepest freshwater lake is Lake Tanganyika, with a maximum depth of 1470 m. Time and favourable conditions for life have produced a large number of endemic species in both and research organisations have organised to study them.

Brauer, whose work on the HPNS featured in Chapter 4, made contact with Russian scientists in the late 1970s and organised the investigation of the pressure tolerance of the Lake Baikal Amphipods, funded by American grants (Brauer et al. 1979).

Baited traps were used to collect the animals, which comprised two groups. There were four species of Amphipods from depths of 50 m or less and four species from depths of 935–1400 m. They had all been well studied for reasons other than high-pressure tolerance, and the depth distribution of the populations was known, and surprisingly wide. For example one deep species ranged from a depth of 50–1300 m. All the specimens survived at normal pressure for several days after collection.

Pressure tolerance was studied by observing the activity of the animals in a small observation pressure vessel, kept at a temperature close to that of the animals' habitat. On compression (at 1.3 MPa per minute) the Baikal Amphipods exhibited a sequence of motor responses similar to those seen in marine Amphipods. First there was a hyperexcitable phase, then one of jerky activity with spasms and finally immobilisation set in. This was preceded by weak, abnormal movements of the limbs, with little progression on the floor of the observation chamber. The mean threshold pressure for spasms proved to be a clear feature (Fig. 5.2). This is evidence that the deeper living freshwater Amphipods have adapted to high pressure, a condition which Brauer described as ecologically meaningful, an excellent term. Brauer and his collaborators posed the question, is the adaptation genetically fixed or acquired during the life of the individual, a process called acclimation? Given the wide depth distribution of the freshwater "abyssal" species this is an interesting point that has not arisen in the marine context, but perhaps it should, see Section 5.2.2. Well after this work was published a Russian–German team discovered that the Lake Baikal amphipods contained a compound called trimethylamine oxide (TMAO), with higher concentrations present in the deeper living animals (Zerbst-Boroffika et al. 2005). This does not answer the question about genetic versus acclimated tolerance, but it does provide very significant information about adaptation to high pressure, which will be explained in Chapter 6.

## 5.2     The Isobaric Collection of Deep-Sea Animals: Retrieval Without Decompression

### 5.2.1   The Isobaric Trap

The engineering underlying isobaric collection and the development of an isobaric trap are described in Chapter 13, Section 13.7. The first isobaric collection of benthic amphipods was reported in 1976 (Macdonald and Gilchrist 1972; Gilchrist and Macdonald 1983). Some years earlier, in 1969, essentially the same equipment was used to make an isobaric collection of plankton, mostly small crustacea, from a depth of 530–660 m. It was a technical exercise, as explained in Chapter 13.

This account continues with the author and colleagues pursuing their long-sought series of isobaric collections of amphipods with a cruise in the Rockall Basin. Amphipods from a depth of 1200–1300 m, at a pressure of 13.4 MPa were collected on three occasions. Once on board, the pressure trap was kept cold and the animals within (all *Tmetonyx cicada*) were observed through the trap's pressure window. For 30 min at the trapped pressure they appeared normal, doing all the usual amphipod activities, close to the pressure window. (We had learned to bait the isobaric trap, not with fish bait but mammalian meat, which may seem surprising. Fish bait generates an oily emulsion that adheres to the surface of the pressure window, totally obscuring the view. Mammalian meat, with its higher melting point fat, causes no such problem.)

The number of animals close to the window and therefore observed in detail at high pressure was 20, 7, and 15 in the three recovery experiments. They showed no obvious response to light or to the pressure when it was raised from 13 MPa to 15 MPa. From 15 MPa the pressure was increased in 5 MPa steps at 5 min intervals to see if spasms resulted. In the first experiment the maximum pressure applied was 33 MPa, for technical reasons, and no spasms were seen, but in the other two experiments, the majority of animals underwent spasms at 40 MPa. At that pressure their activity clearly declined so the pressure was smoothly reduced to 13 MPa over a minute, held constant for 40 min and then reduced to surface pressure in 5 MPa steps at 5 min intervals. The animals then recovered normal activity.

An important conclusion is that *Tmetonyx cicada*, collected at their ambient pressure of 13 MPa, underwent spasms at 40 MPa. The same species collected with decompression from the same depth and tested in the same way showed spasms at pressures of 27–33 MPa, depending on treatment, see above. This confirms the earlier finding that decompression sensitises the animals to high pressure, that is, it reduces the pressure required to cause spasms. Furthermore 40 MPa is to be compared with the pressure at which *M. marinus* undergoes spasms, 10 MPa. This is a simple but clear demonstration of increased pressure tolerance in deep-sea animals.

Subsequent cruises deployed the isobaric trap in full ocean depths. The total list of species collected in the trap, but mostly out of sight during the experiments, is given in Table 5.1 but as before the mixture did not influence the results. Those few animals close to the pressure window remained in view for the duration of the observations. They were first observed at their trapped pressure for 20–30 min and the full repertoire of amphipod activity was apparent; short bursts of swimming and crawling, with intermittent pleopod beating in those animals lying on their side in the characteristic ventrally curved posture. If the animals were affected by the light used for the observations it was not obvious. In experiments 3 and 4 the temperature was kept between 2.5 and 3 °C, but in the first two experiments, it increased slightly, but without any obvious effect. Pressure was first increased to 45 MPa (experiment 1) and 40 MPa (experiment 2). It was then raised in 5 MPa steps at 5 min intervals to test the animals' pressure tolerance. Mild excitable activity was apparent at 60 MPa

**Table 5.1** Isobaric collection of Amphipods from abyssal depths (~4000 m)

| Experiment number | Pressure by depth (MPa) | Pressure in trap (MPa) | Number of amphipods | |
|---|---|---|---|---|
| | | | Total trapped | Observed |
| 1. | 43.6 | 44.2 | 71 | 3 |
| 2. | 43.2 | 36 | 14 | 11 |
| 3. | 43.1 | 39.4 | 26 | 7 |
| 4. | 42.0 | 40.8 | 32 | 5 |

Species: Mostly *Paralicella caperesca* but also *Orchomene sp. Valettiopsis, sp. Eurythenes gryllus*, and *Paracalisoma alberti*, Data compiled from Macdonald and Gilchrist (1978, 1980, 1982)

and at 75 MPa activity began to decline. No spasms were seen. At 80–85 MPa all the animals were stationary. In experiment 3 the pressure was reduced from 80 MPa to 40 MPa. The animals recovered normal activity within 30 min and sustained it for 9 h. In contrast, in experiment 4 the pressure was reduced from 90 MPa to 40 MPa and only one individual recovered, after 50 min, during a total of 3 h of observation.

Two results are clear. First, the animals were just as active at their normal ambient pressure as those observed in shallow water, a point subsequently confirmed by others using deep submersibles or deep water cameras. This is consistent with the light used in our observations having no effect on the animals' activity. Second, no spasms were seen in any of the animals isobarically collected from approximately 4000 m. Many showed a mild increase in activity after the pressure was increased well above that at which they had been trapped. It is difficult to avoid the conclusion that the absence of pressure-induced spasms reflects the natural tolerance of the neuromuscular system of these animals to high pressure. Furthermore, the absence of pressure-induced spasms in these animals shows that the animals collected from 4000 m (40 MPa) with decompression, and which were resuscitated at 40 MPa, achieved a fair degree of normality as they too failed to show spasms at higher pressures. We concluded that resuscitation from the decompression experienced during collection deserved more study (Section 5.3). It might, eventually, reveal some of the critical site (sites) at which high pressure acts.

## 5.2.2  Other Isobaric Collections

The isobaric collection of benthic amphipods was also carried out by Yayanos (1978) in the northern Pacific. The entrance to the trap was 5 cm in diameter (Chapter 13), and the amphipods collected were large, centimetres in length, but unidentified. One purpose of the trap was to collect animals for subsequent transport to a land laboratory; but its window enabled periodic observations to be made at high pressure (and low temperature). In the case of Station 654, animals collected from 5782 m at close to their ambient pressure were accidentally subject to a pressure decrease to 17 MPa, which was rapidly rectified. No activity was seen for 2 h but swimming was seen in two animals 8 h after the restoration of normal pressure. In another case, Amphipods from 5741 m were collected at a pressure of 41 MPa or 28% below their ambient pressure. When this was restored to 57 MPa the quiescent animals (five in all) were seen to regain normal activity after 4 h. They were kept alive at sea for a total of 9 days at their normal ambient pressure and temperature. On a subsequent cruise Yayanos (1981), extended the observations. Two specimens of *Paralicella caperesca* were collected from the ocean floor at a depth of 5900 m with their ambient pressure and temperature held constant (60 MPa and 2 °C). They were clearly healthy but when decompressed over a period of 2.5 h the pattern of intermittent pleopod activity was disrupted at around 38 MPa and ceased at 21.5 MPa. A period of 4 min at surface pressure was followed by an increase in pressure to 35 MPa, which was accompanied by the restoration of normal swimming. These observations are consistent with those described above but were extended in 1980 with the isobaric collection of amphipods from a depth of 10,900 m in the Mariana Trench (Yayanos 2009). The unidentified animals were observed within the trap at 102.6 MPa and 2.8 °C for 5 days. Pleopod frequency was measured and the lethal consequence of decompression to atmospheric pressure noted, but the motor activity of the whole animal as a function of pressure was not scored. These animals clearly occupy a significant vertical range (Blankenship et al. 2006; Jamieson et al. 2013), which, in conjunction with their abundance and often large size, makes them attractive prospective experimental material for a variety of experiments.

Shillito et al. (2008), developed an isobaric collection apparatus that requires the support of a submersible (Chapter 13, Figs. 13.25 and 13.26). It produced some interesting observations of shrimps collected from 1700 to 2300 m depths on the Mid Atlantic Ridge hydrothermal vent sites. *Rimicaris exoculata*, for example, collected at a pressure close to that of its depth of origin, 2300 m. were seen to swim normally. When collected with decompression the animals appeared to have lost their balance and were inactive except for the occasional abdominal spasm. Another isobaric system that requires a submersible has been developed by Kooyama et al. (Chapter 13) with which the shrimp *Alvinocaris* was collected from 1157 m depth, but acute pressure tolerance experiments were not carried out (Koyama et al. 2005a).

These observations of the pressure tolerance of abyssal Amphipods and other crustacea set the scene for analytical experiments that the author would have liked to have carried out. However, ship time was, and always will be, limited, and funding for more analytical work was difficult to secure. The required experiments are probably best carried out by using the isobaric vessels to transport the animals to a land laboratory close to deep water. Several exist, for example on Bermuda and Hawaii. Isobaric equipment requiring a submersible is a much more expensive operation than using a surface ship, but given the funding, it should be productive.

One way forward was demonstrated a long time ago (1975) by Campenot's pressure experiments with the neuromuscular junction of the red crab *Geryon quinquedens*. It was acclimated at atmospheric pressure but the species occupies a depth range of 300–1600 m, on the continental slope of the North Atlantic. A pressure of 10 MPa depressed the excitatory junction potential of muscle fibres in the walking leg which Campenot suggested could be offset by increasing the natural frequency of stimulation (Chapter 4). This would be a physiological adaptation to pressure, which is an interesting idea, especially applicable to moderate pressure, but molecular adaptations would probably also be involved. It has no obvious bearing on the pattern of excitation, spasms and immobilisation caused by pressure in shallow water animals and which is modified in their deep-sea counterparts. Crustacean neurobiology is a sophisticated field and is surely capable of tackling problems of adaptation to high pressure.

### 5.2.3    When to Use Isobaric Collecting Equipment?

The need for isobaric collection has to be considered in each individual case. In Chapters 6 and 7 it will be apparent that deep-sea fish collected with decompression can provide enzymes and membranes which can be effectively studied. Can the same be said for other types of investigation? For example Ritchie et al. (2018), collected deep-sea amphipods with decompression and then promptly fixed them in ethanol for subsequent extraction of their DNA, PCR amplification and the isolation of stress proteins In this case of potentially rapidly changing events and pressure labile proteins it would surely be preferable to block metabolism prior to decompression either at depth or after isobaric recovery to enable the stress proteins thought to exist in the animals at depth, to be distinguished from those which might arise from the trauma of hydraulic decompression. The same concern arises with other stress protein experiments using deep-sea crustacea, discussed in Chapter 12. That said, Ritchie's experiments are notable for using hadal amphipods, including *Hirondellea gigas* from a depth of 9316 m. Hadal amphipods are proving valuable for a variety of purposes (see Chapter 6, Section 6.1.1.2). Pictures in Chapter 1 and those available on Websites, make the point that their appearance gives no sign of the special molecular adaptations which they have evolved. Like the deeper water amphipods in Lake Baikal, hadal amphipods are reported to contain high concentrations of the compound TMAO whose significance is explained in Chapter 6, Section 6.2.3.

The study of high-pressure microorganisms poses the same problem of when to use isobaric methods of collection and transfer (Section 5.6; Chapter 8).

### 5.2.4    The Isobaric Collection of Deep-Sea Fish

This is obviously more difficult than collecting small and robust Amphipods from the seafloor. A paper published by Wilson and Smith in 1985 reported attempts to collect *Coryphaenoides acrolepis* in a pressure retaining vessel (Chapter 13, Fig. 13.22). An interesting point to emerge was that a fish collected from a depth of about 1100 m and partially decompressed benefitted from being recompressed. *C. acrolepis* has a gas-filled swim bladder which is not connected to the alimentary canal and thus not readily vented, so decompression causes the gas to expand and rupture tissues. Recompression would obviously reduce the mechanical damage as well as any hydraulic decompression symptoms. Twenty years later Drazen and colleagues achieved the near isobaric collection of the same species from 1300 to 1500 m depths using their trap-respirometer (Chapter 13). One specimen was kept at 94% of the trapped pressure, about 14 MPa, during which time its condition deteriorated and it died after 84 h. Partial decompression during this time upset the animal's buoyancy. No observations of the animal's motor response to changes in pressure were made but its oxygen consumption was measured. The pressurised recovery apparatus described by Shillito and colleagues (Chapter 13, Fig. 13.26), collected a fish, *Pachycara salganhai* from 2300 m depths. It measured 28 cm long and was active at the trapped pressure which was about 20% less than the pressure of capture. When pressure was reduced to atmospheric the animal promptly became motionless. The system designed by Koyama, just mentioned, has isobarically collected a fish, Zoarcidae sp., at a depth of 1170 m and a pressure of 9 MPa, which survived in apparently good health for 3 days (Chapter 13, Fig. 13.23a), No observations of the animal's response to acute changes in pressure were made.

## 5.3   Resuscitation of Animals and Their Tissues from the Effects of Hydraulic Decompression

The paralysis seen in amphipods collected from the depths with decompression presented an unexpected opportunity when it was discovered that resuscitation occurred if the animals were promptly restored to their normal pressure. If the sites of the pathological effects of decompression could be identified, then they would surely be sites of adaptation to high pressure. The slow resuscitation of amphipods is in marked contrast with the prompt immobilisation caused by pressure, but the repair is, intuitively, likely to be slower than injury. Are amphipods the best material for studying resuscitation? The author decided to study decompression paralysis and resuscitation in fish as they and their various tissues would be much better to experiment with after resuscitation. So, the author assembled a team of enthusiastic scientists to conduct specific experiments on resuscitating deep-sea fish tissue at sea. Fortunately, both the funds and the ship time were awarded to do this. The account continues in a personal style to illustrate some of the practicalities of this sort of work.

### 5.3.1   The Resuscitation of Deep-Sea Fish by High Pressure

The traditional methods of collecting fish from deep water inflict damage, usually external abrasion and internal rupture, the latter due to the expansion of a gas-filled swim bladder. Fish lacking a swim bladder can be collected from significant depths by conventional means with a good chance of being in a condition adequate for the experiments we had in mind. The work was carried out near the Porcupine Sea Bight, southwest of Ireland, where the seafloor provided depths down to 4000 m suitable for bottom trawling. Although this is a routine sort of operation, deep-sea trawling should not be underrated. The RRS Challenger was well equipped, and manned, for the work, which was directed by a Fishing Skipper. Bottom trawling at a depth of 4000 m involves paying out 12 km of wire from a special winch. If the procedure goes wrong, perhaps the trawl becoming caught on the seafloor or the wire becoming entangled during its ascent, then the skills of the Fishing skipper would be essential. (A trawl wire can become entangled because of the loss of tension and the growth of a monstrous ball of entangled wire ensues. It takes hours to sort out.)

Once we were on station the ship was trawling, hauling in or paying out, 24 h our per day and the scientific team worked 12 h watches. The actual depth at which the trawling took place was measured by the ship's acoustic equipment. When the trawl was hauled up through the A-frame at the stern of the ship, its contents, released by a tug on the special knot securing the cod end, poured onto the deck. Usually it comprised a mass of flapping and moribund fish, benthic creatures such as echinoderms, a few Crustacea and debris such as coke from old coal-burning ships, tin cans, plastic containers and mud. The scientists would seize the fish they required and hurry off to begin their experiment. The deck would then be swept clean, the trawl checked and promptly returned to the depths.

In this way two species of benthic fish from 900 m depths were collected for resuscitation, *Trachyscorpia cristulata echinata* and *Synaphobranchus kaupi* (Fig. 5.3a). These are common fish in the area and have a depth range that extends from a few 100 to 1000 m and 3200 m, respectively. It is very probable that they were trawled from 900 m depths and not caught by the ascending trawl. Decompression from 9 MPa in the trawl took an hour and at the start of their resuscitation treatment they were promptly and individually restored to 10 MPa, fresh from the trawl. The pressure vessel used was a vertically mounted cylinder with an internal diameter of 26 cm and length 80 cm. A window with a 10 cm viewing diameter and light guides provided adequate visibility within. Appropriate plumbing ensured good pressure control and clean seawater and the vessel was controlled at 5 °C. It was quite a large and heavy apparatus so it was welded to the aft working deck of the ship, in the open air but sheltered (Macdonald et al. 1987). Three specimens of *Trachyscorpia* were used and they all recovered from their moribund state at 10 MPa, showing normal posture, opercular pumping and movement of the pectoral fins. This was consistent with the observations of Wilson and Smith, mentioned earlier. Compression to 15 MPa caused the body to shudder and perform almost normal swimming movements. Decompression to 10 MPa reduced activity. Three *Synaphobranchus*, deep-sea eels, were kept in the cold at normal pressure for 3 h prior to re-compression. They were immobile but started to swim at 10 MPa with normal fluid S-shaped movements, typical of eels. Compression to 19 MPa elicited violent body convulsions. Clearly both these species benefitted from being restored to their normal pressure and the onset pressure for convulsions, 15 MPa and 19 MPa respectively, is higher than seen in shallow water fish. In the case of *S. kaupi* the convulsion pressure was within the depth range of the species so it would be interesting to compare the pressure tolerance of individuals collected from different depths.

**Fig. 5.3** Deep-sea fish photographed by a camera fitted to the ROBIO Lander (Fig. 13.36a). Courtesy of Alan Jamieson, University of Newcastle, UK. (**a**) *Synaphobranchus kaupi* at 1000 m depth, N, Atlantic. The specimens mentioned in the text were 20–25 cm long but these are bigger. (**b**) *Coryphaenoides armatus* at 4500 m depth in the Kermadec Trench. These fishes are about a metre long

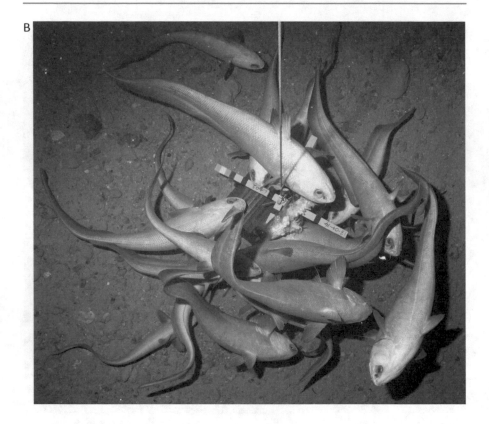

**Fig. 5.3** (continued)

### 5.3.1.1 Fish from Lake Baikal

Lake Baikal, whose greatest depth is 1600 m, supports a population of fish all of which lack a swim bladder. Brauer and his colleagues (1984), found that the deep water family of *Abyssocottids*, most of which live deeper than 600 m, were adversely affected when collected with decompression but improved when recompressed. They manifest convulsions over the range of 12–27 MPa. The shallow water family of Cottids, from less than 75 m depths, convulsed at an average of 8.8 MPa. The pressure which immobilised the fish was similarly related to their living depth; 24 MPa and 18 MPa respectively. The distinction between genetically determined pressure tolerance and acclimated tolerance in these species has been considered by Brauer with some care, with the conclusion that in certain species acclimation is unlikely. The obvious experimental test of the ability of species to acclimate to higher than normal pressures has not been carried out. It would not be difficult to hold fish at selected depths in cages, as has been done with eels in the Mediterranean Sea (Chapter 12). Such experiments would amount to long term pressure tolerance tests. The neurobiology of these highly accessible fish is surely another subject for investigation.

## 5.3.2   The Resuscitation of Isolated Fish Tissue by High Pressure

The resuscitation of the muscle and neurons of deep-sea fish was undertaken during the Challenger cruise already mentioned. Two kinds of muscle were selected, swimming and heart muscle.

### 5.3.2.1 Fish Swimming Muscle

The resuscitation of the lateral swimming muscle of deep-sea fish after decompression from 4000 m depths was carried out by Wardle, of The Marine Laboratory, Aberdeen, and his colleagues (1987). Fish swimming muscle has a distinctive gross structure. Unlike the familiar muscles of terrestrial vertebrates, which are generally linear structures, fish swimming muscle is organised in blocks. These are myotomes, the unit of motor control that contracts as an entity. They are segmentally arranged along the body, usually W shaped and overlapping. Commonly eaten fish such as haddock, cod, and mackerel display this fundamental organisation on our plates and cooks refer to the myotomes as flakes. When myotomes contract serially they bend the fish into a wave that travels the length of the body. The waves push against the water and drive the fish forward. Within the myotome, the muscle's microscopic structure is the same as that in other vertebrate muscle, comprising bundles of protein fibres which, following electrical stimulation, slide past each other to shorten and generate force.

Typically, fish cruise slowly most of the time and only indulge in short sprints. Cruising is driven by aerobic red muscle, which contracts slowly, and the short sprints are powered by fast contracting, anaerobic white muscle. Mackerel exhibit these two muscle types clearly, on the plate, with the cooked aerobic red muscle appearing dark grey or brown. A fish's maximum swimming speed is determined by the maximum tail beat frequency. Serial contraction of the myotomes comes in two styles. Anguilliform swimming, as in the eel, involves the wave travelling along an extensive length, and can be highly energy efficient. Shorter bodied fish use carangiform swimming in which tail flapping is important and this can produce high speeds The properties of fish swimming muscle, and in particular its twitch contraction time, determine the swimming speed.

Most higher vertebrate muscles provide the experimenter with attachment points for measuring. Fish myotomes inconveniently do not, a problem which Wardle has solved by using a rectangular block of muscle ($2 \times 1 \times 1$ cm) cut from the fish and laid on a matching rectangle of plastic 1 mm thick. Four hypodermic needles, 25 mm long, project from the corners, like the legs of a table, piercing the rectangle of muscle. A foil strain gauge is glued to the tabletop and fine wires connect both the foil and the needles to the appropriate circuitry. The needles piercing the muscle deliver an electrical voltage pulse which stimulates contraction, a twitch, and the ensuing flexing of the plastic tabletop is detected by the strain gauge. Tests with plaice muscle showed that the device worked well at pressures up to 60 MPa.

Fish muscle is normally stimulated to contract by fine branches of motor nerves that terminate in neuromuscular junctions that release acetylcholine. This causes a nerve impulse to spread onto the muscle membrane, a process that is blocked by the drug curare. The effect of high pressure on plaice muscle in the presence and in the absence of curare showed no significant difference, so acetylcholine played no part in the stimulation. The muscle membrane was directly electrically stimulated. However, to be sure that at high pressure the direct electrical stimulation of the muscle was not complicated by a contribution through neuromuscular junctions, curare was used. The excitation normally spread across the muscle membrane, penetrating deep into the muscle cells where it causes Ca ions to enter the cell. It is these ions that actually initiate the contractile mechanism.

A benchtop pressure vessel of 6 cm internal diameter and 20 cm length, fitted with electrical connections, was sufficient to accommodate the muscle block, mounted in 50 mL of the appropriate solution, the whole being submerged in liquid paraffin which provided electrical insulation and transmitted pressure. The temperature was measured by an internal probe and controlled by thermostated water circulating through a copper coil surrounding the vessel. Lots of practice ensured the experiments were not too difficult to carry out at sea. We were fortunate in collecting deep-sea eels, *Histiobranchus bathybius*, of a suitable size on seven successive trawls from a bottom depth of 4000 m, something of a luxury. Those used ranged in length from 61 to 71 cm. They had experienced a decompression of 9 MPa per hour in the trawl, from a temperature of 2.5 °C to the surface waters at 14 °C. On deck the eels were unresponsive but were promptly transferred to 5 °C for dissection. Initially, at normal pressure and 6 °C, the muscle responded only weakly to the stimulation but on compression the force of contraction increased, reversibly, peaking at around 20 MPa or higher in two cases. The force generated by muscle kept at normal atmospheric pressure and low temperature declined steadily to a very low level over 20 min. These control experiments show that the isolated muscle of the deep-sea eel requires high pressure to function, and that function is reversibly lost after a limited exposure to atmospheric pressure. Longer exposure leads to irreversible loss of function.

The contraction time also increased with pressure. One individual, 70 cm in length, provided muscle which achieved a contraction time of 250 ms at 40 MPa, implying a tail beat cycle of 500 ms. The general rule is that a fish advances by 70% of its body length each tail beat cycle, which means this individual would have swum at $0.7 \times 70$ cm per half second or about 1 minute per second, or 1.4 body lengths per second. This is theoretically a vigorous fish, but we should not be carried away by these figures. They merely illustrate that the degree to which deep-sea fish muscle can recover from decompression can be assessed by comparing its twitch properties with actual swimming speeds recorded by deep-sea cameras. *Synaphobranchus kaupi*, mentioned earlier (Fig. 5.3a) has been filmed from a lander at a depth of 1500 m by Priede and his colleagues (1992). Gentle cruising speeds of less than 0.2 minute per second were typical and a peak burst of swimming attained 1 minute per second. In other work they deployed ingestible transmitters to track animals. The vertical distribution of this species is 235–3200 m so it would be interesting to experiment with individuals collected over that depth range. Like the case of Amphipod *Eurythenes gryllus* such a depth range raises questions about genetically fixed or more variable, acclimated, physiological properties. The convulsion threshold pressure of 19 MPa seen in *S. kaupi* from 900 m would not be expected of fish collected from 3000 m. The properties of its muscle over that pressure/depth range would be particularly interesting, especially if individuals were collected by isobaric means.

An attempt by Wegner et al. (2008) to measure the resting potential of skeletal muscle in fish collected from depths of several 100 m clearly describes the problem of vibration experienced on board ship. The measurements were at normal atmospheric pressure and produced unsurprising results, considering the limited depth range of the animals. The use of pressure to resuscitate individual fish prior to the measurements would have probably increased the quality of the data. The workers seem unaware of the work of Campenot (1975) who used intracellular recording methods at sea. He reported the resting potentials of crab muscle, from 300 to 1100 m depths, were similar to those of muscle from a shallow water lobster.

### 5.3.2.2 Fish Heart Muscle

The resuscitation of the heart isolated from deep-sea fish was investigated by Pennec et al. (1988), using a preparation he had developed to study the trout heart. The experiments were complicated and difficult to summarise, but it can be said that various degrees of recovery were observed in hearts isolated from *Mora moro* from 900 m, *Coryphaenoides armatus*, 4000 m, and *H. bathybius* 4000 m (see also Hogan and Besch 1993).

### 5.3.3  Fish Neurons

The resuscitation of the nerves of deep-sea fish, undertaken by Harper, was assessed using the recording technique described earlier (Chapter 4, Section 4.5.2). The visceral branch of the vagus nerve was dissected from freshly trawled, and chilled specimens; Mora moro (900 m) and *Bathysaurus mollis* and *Coryphaenoides armatus* (4000 m) (Fig. 5.3b). The shallow water cod, *Gadus morhua*, served as a control as explained in Chapter 4. Following dissection, the nerve was mounted in the recording apparatus and pressurised as swiftly as possible. The temperature increased by 1.5 °C which took an hour to equilibrate, whilst the nerve simultaneously resuscitated. Voltage pulses were applied to stimulate all the fast conducting fibres.

Consider first the amplitude of the compound action potential. In the shallow water cod, it was decreased to near zero at 30 MPa, but was restored to normal on decompression to atmospheric pressure. The nerve from *Mora* produced a normal action potential at 10 MPa and a reduced amplitude at 30 MPa, well in excess of its normal pressure. At atmospheric pressure, the amplitude returned to normal. The nerves from *Mora* and *Gadus* all responded to stimulation at atmospheric pressure. Those from 4000 m depths, *Bathysaurus* and *Coryphaenoides*, were less responsive. In only three of six cases did they respond at atmospheric pressure and in one other case, a response was elicited at 21 MPa. *Bathysaurus*, for example produced a bigger and more normal action potential after several hours at 40 MPa. After a staged decompression, the action potential was unchanged. The amplitude of the action potential is a good indication of the number of neurons in the nerve bundle which respond to stimulation, so clearly the nerves from 4000 m became more excitable at their normal pressure. They retained excitability when subsequently decompressed, at least in the short term.

A second parameter is conduction velocity, estimated from the time taken for the action potential to reach its peak. In the cod nerve, the velocity was maximal at normal atmospheric pressure and it was reduced by about 60% at 31 MPa. The nerves from *Bathysaurus* had a maximum conduction velocity at 40 MPa which was reduced by 25% at atmospheric pressure.

The general conclusion is that nerves from 4000 m fish are impaired by decompression during collection and recover when restored to their pressure of origin (Harper et al. 1987). The results hint at a nice relationship between normal pressure and function. For example, are *Mora*'s nerves similar to those of the cod and somewhat apart from those from the 4000 m depths? The question can be re-phrased as a hypothesis which these experiments show can be tested.

## 5.4    The Tolerance of Other Deep-Sea Animals to Decompression

There are numerous observations of the activity of deep water animals which have been carried out for diverse reasons, but which provide some information about their pressure tolerance and their recovery from the decompression experienced during collection. Similarly, cells from deep-sea animals and free-living eukaryote microorganisms have been investigated for various reasons, providing further information about recovery from decompression. Hydrothermal vent organisms are considered in Chapter 9.

### 5.4.1    Animals from Great and Moderate Depths

Childress and his colleagues, interested primarily in the metabolic rate of deep water animals, found that the bathypelagic crustacea *Gnathophausia ingens* and the deeper living *G. gracilis* could be collected with decompression and kept at their normal pressure (7.7 MPa and 10 MPa respectively) for many days. However, they also remained apparently healthy when kept at normal atmospheric pressure for similar periods. Increasing the pressure to their normal level had no particular effect on the pleopod activity (Mickel and Childress 1982). These animals tolerate pressures ranging from the surface to their bathypelagic zone. Pleopod activity persisted at up to 30 MPa and it would be interesting to know at what pressure it ceased. Do such tolerant animals exhibit a stress response in these circumstances?

The shrimp *Alvinocaris sp.*, which is an active swimmer that stays close to the seafloor, is similarly tolerant of decompression from 11.5 MPa (Chapter 9). In Japan a commercial aquarium that displays animals collected with decompression from hydrothermal vents provides some particularly interesting examples. Numerous animals from depths of 430–1500 m survive well in the aquarium which provides favourable conditions of temperature, pH, carbon dioxide and a source of hydrogen sulphide, all at normal atmospheric pressure. A good example is the barnacle *Neoverruca*, collected from 1340 m depths (13.4 MPa), which survived so well at atmospheric pressure it produced larvae which grew through the early stages of their development (Miyake et al. 2007).

The respiration of a bathypelagic fish, *Anoplogaster cornuta*, trawled from 600 to 700 m mid-water depths, was investigated by Childress, as experience showed it appeared healthy at normal atmospheric pressure. Its respiration was slightly increased when it was restored to its normal pressure, about 7 MPa, but no observations of its motor activity were made. The respiratory rate, metabolic rate, of mid and deepwater animals has interested many marine biologists and generated a lot of data, much of it obtained at normal atmospheric pressure and hardly any from animals collected by isobaric means. In situ measurements are probably the best (e.g., Smith and Laver 1981; Hughes et al. 2011; Nunnally et al. 2016). However, the general conclusion is that the metabolic rate of deep-sea animals is determined by behavioural needs and lifestyle of the species and not by pressure or other physical factors (Childress 1995; Siebel and Drazen 2007; Siebenaller 2010). It so happens that many deep-sea animals have adapted to their dark, cold and food scarce environment by an energetically economic lifestyle, so that many, but certainly not all, have a low rate of metabolism. Animals that require light and speed to feed tend to have higher metabolic rates than others. The visual interaction hypothesis attempts to explain depth-dependent metabolic rates, particularly of swimming muscles. It is not a simple hypothesis and is discussed in detail by Drazen et al. (2013). The authors attempt to reconcile muscle enzyme activities, measured at standard temperature and normal atmospheric pressure, obtained from 18 species of oceanic fish with different lifestyles, occupying different depths (pressures). The role of pressure seems to be omitted, but this is inconceivable in view of the expertise of the authors, so I am at a loss to comment further.

We should not imagine, as was once the case, that deep-sea animals are struggling against high pressure, like climbers on Everest struggling against the lack of oxygen. They are better adapted than that. However, one consequence of a low rate of metabolism, hence low oxygen consumption, is that the consequential oxidative stress is also low. That is the case with normal animals, but does that apply to deep-sea animals at high hydrostatic pressure? It is generally assumed that it does, but experimental evidence appears to be lacking (see below, Section 5.4.2).

A particularly intriguing example of the effects of decompression on animals from moderate depths has been described by Bailey et al. (1994). Using a submersible, fragile specimens of planktonic animals were collected and carefully transferred to small containers at depths of around 850 m, close to the seafloor. The ctenophore *Bathocyroe fosteri* was one of various species that were subject to the following experiment. Some containers were left at depths of 680–750 m, during which time the oxygen in the sealed containers was consumed and monitored, in situ. Other containers were taken to the surface and to normal atmospheric pressure at which the consumption of oxygen was also measured. The results were striking. All the decompressed samples respired at a much slower rate than those kept at depth at 6.8–7.5 MPa. *Bathyocyroe* was the most extreme case, respiring at depth at least six times faster than at atmospheric pressure. The workers concluded that decompression inhibited motor activity and thus oxygen consumption, a conclusion which would be strengthened by two additional points. The conditions which accompanied decompression, such as light and vibration, are important to know, along with a score of the motor activity. Re-compressing the decompressed samples and measuring their oxygen consumption at 6.8–7.5 MPa should restore the respiration to the in situ rate. These results inversely resemble the early experiments with crustacea and fish, whose rate of oxygen consumption at pressure was found to be high, due to pressure stimulating their motor activity.

The apparently healthy and sustained activity of animals collected from depths of around 1000 m (10 MPa) and kept at normal atmospheric pressure, 0.1 MPa, may arise through a process of acclimation (Chapter 12). In which case they are, by definition, in a physiological state different to that in their normal environment. The survival of the hydrothermal vent mussel *Bathymodiolus azoricus*, discussed in Chapter 12, a species that occupies a significant depth range, is particularly interesting. Specimens collected from 850 m depths with decompression survive at normal atmospheric pressure for a number of weeks and appear to begin to acclimate. Those similarly collected from 3000 m survive no more than a few days. The extent of acclimation in such cases will only be properly understood with more experimental experience.

Carbon dioxide storage schemes, intended to mitigate the global rise in atmospheric $CO_2$, envisage liquid $CO_2$ being stored in the ocean crust and it is inevitable that unintended seepage will locally acidify the seawater. An experimental attempt was made by Pane & Barry to assess the tolerance of the deepwater crab *Chionaecetes tanneri* to such an event, by measuring its ability to regulate the pH of its blood (called haemolymph in crustacea). The animals were collected with decompression from 1000 m depths in the Monterey Canyon by an ROV (remotely operated vehicle), and kept at normal atmospheric pressure in an aquarium, where they appeared healthy for months. Their response to being exposed for 24 h to seawater suitably enriched with dissolved $CO_2$ was compared to the response of a shallow water crab, Cancer magister, the exposure being carried out at normal atmospheric pressure. The regulation of the pH of haemolymph is a complicated process so we will skip the details. The main conclusion was the shallow water crabs coped better than the deep water crabs. $CO_2$ storage schemes will therefore be "physiologically challenging" to deep-sea animals. This last point was well known anyway, but the real conclusion is that the deep water crabs, after a long time at normal atmospheric pressure, perhaps adapted?, and exposed to $CO_2$ rich seawater, at atmospheric pressure, coped less well than the shallow water crabs. We do not know how the deep water crabs, kept at 10 MPa, would have responded to $CO_2$ rich seawater at 10 MPa, surely the point of the experiment. However, the attempt to use a deep water animal for these tests is preferable to using a "surrogate" shallow water animal, as some workers have advocated, for testing the toxicity of crude oil at depth. Although Olsen and colleagues subjected *Eurythenes gryllus* (from 1000 m depths) to a light Arabian oil dispersion, unfortunately the tests were carried out at atmospheric pressure. The results showed that *E. gryllus* was affected in the same way as shallow water amphipods, leaving unanswered the effect at 10 MPa. And returning to the crab experiments, many other aspects of the animals' physiology would have to be tested over prolonged periods before any realistic assessment could be made of the susceptibility of the crab, and other deep-sea animals, to $CO_2$ seepage or other pollutants. Understanding the toxicity of alien substances to life in the deep sea is not a simple. A good start is to measure the effect of pressure on the oil: water partition coefficient of the toxin.

## 5.4.2 The Antioxidative Defences of Deep-Sea Animals

Oxidative damage to macromolecules, generally referred to as oxidative stress, is a fact of life for organisms of all grades of complexity. It is a silent background process that is driven by the presence of highly reactive species of the molecule which arise as a by-product of essential metabolic reactions, especially in mitochondria. Rather less than 5% of the oxygen consumed by mitochondria is converted to reactive oxygen species (ROS), so a high metabolic rate generates more oxidative stress. The reactive molecules have a fleeting existence and their damage, peroxidation of membrane lipids for example, is normally kept within safe limits by intrinsic defence mechanisms. Well oxygenated shallow sunlit seas present a significant oxidative threat because sunlight can generate ROS. In contrast, the dark depths of the ocean, which are variably but moderately well oxygenated, presumably lack this source of ROS (Janssens et al. 2000). However, there are two unknowns lurking in the depths. One is the way hydrostatic pressure increases the activity of the dissolved oxygen, which is well established (Chapter 10), but the possible consequential production of ROS is not at all clear. In fact, the possibility does not seem to be recognised. The second uncertainty is the effect of high pressure on the production and lifetime of ROS. They are charged entities and therefore likely to be favoured by high pressure. A slight increase in their fleeting lifetime would be very significant. If that were the case it could be a contributory reason for the evolution of reduced metabolic rates in deep-sea animals, to reduce endogenous ROS, additional to ecological factors. This speculation is open to experiment tests!

Even if the oxidative stress of the deep sea is not entirely clear the oxidative defences of organisms living there are important and interesting. With this in mind, Janssens et al. (2000), investigated the "antioxidative arsenal" of fish living at depths of less than 2000 m, with the three deepest species occupying 650–1000 m depths. They found evidence of antioxidative capability matching the low metabolic rates, but the full picture is more complicated than that. The main point of interest here is that the activity of the antioxidative enzymes was measured at normal atmospheric pressure, not at the pressure where they operate, which is around 10 MPa. That may not be too important but for deeper animals it might be. *E. gryllus* from 2000 m depths in the Arctic Ocean has been compared with *Gammarus wilkitzkii*, an amphipod which lives in the sunlit surface of the same ocean. Here too the biochemical reactions which defend *E. gryllus* from oxidative damage was measured at normal atmospheric pressure, ignoring the possibility that 20 MPa might be significant (Camus and Gulliksen 2005).

However, in the hydrothermal vent mussel *Bathymodiolus* (Chapters 9 and 12), whose response to toxic elements such as copper and cadmium has been studied, the activity of one antioxidant enzyme, superoxide dismutase (SOD), has been measured at high pressure. Animals from 1700 m depths were kept at 17 MPa and at atmospheric pressure and their SOD activity in both gill and mantle tissue was found to be the same, i.e., independent of pressure. The lipid peroxidation in the same tissues was also independent of pressure.

### 5.4.3   Deep-Sea Fish Cells and Tissues In Vitro

Some investigators have used isolated cells to study the pressure tolerance of deep-sea fish, with mixed results.

One obvious possibility is the fish erythrocyte, the physiological properties of which are well known in the case of shallow water fish and other vertebrates including humans (Thorne et al. 1992). The effect of hydrostatic pressure on the ionic regulation of red cells has been studied by various workers and Shelton showed that those of the plaice, *Pleuronectes platessa*, had specific pressure labile transport mechanisms. Surely, red cells from deep-sea fish would be excellent material in which to study the adaptation to high pressure of ion transport mechanisms? (Shelton and Macdonald 1987). The cells would probably resuscitate well after collecting the fish (with decompression), and in any event, their recovery could be easily monitored and assessed. However, Shelton's measurements of the composition of deep-sea fish plasma (i.e., blood minus the cells) yielded surprisingly high concentrations of magnesium and chloride ions, which might have arisen during trawling and decompression, from pathological changes in the individual's tissues or from an influx of seawater. Erythrocytes from such plasma samples were restored to the donor animals' normal pressure, and the internal concentration of Na ions was measured. In the case of the erythrocytes from *Coryphaenoides armatus* trawled from 4000 m depths, and restored to 40 MPa, little difference could be detected between pressurised cells and control cells at normal atmospheric pressure. The cells were unresponsive, as if poisoned, perhaps by the high magnesium concentration (Shelton et al. 1985). If this is true, then the resuscitation of the muscle and nerves from deep-sea fish described earlier is puzzling. Perhaps such tissues, which are bulky compared to erythrocytes, are less exposed to the magnesium ions during collection. The key to progress probably lies in knowing the precise composition of the plasma from which they are taken.

Among biologists, the technique of tissue culture is regarded as a black art. However, Koyama and Aizawa (2000), have demonstrated its use in deep-sea physiology by first culturing cells from a deep-sea clam, *Calyptogena soyoae*, and then from a deep-sea eel, *Simenchelys parasiticus*, whose vertical range is 366–2630 m (Chapter 13, Section 13.6, Fig. 13.12). The eels were collected from a depth of 1162 m by the submersible Shinkai 2000, using a suction-capture device,

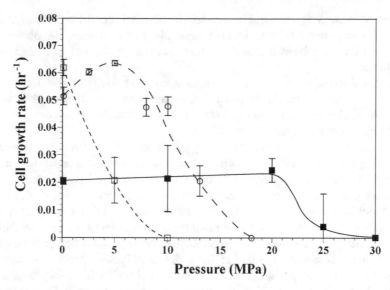

**Fig. 5.4** The rate of cell growth (multiplication) of cultured cells from deep-sea eels, *Simenchelys* sp. (black squares) and shallow water eels (open squares) at different pressures. The open circles refer to cultured mouse cells, but they are not relevant here (Koyama et al. 2005b). Reproduced with permission from Koyama et al. (2005b)

and brought to the surface without pressure loss. In the laboratory decompression over 15 h to atmospheric pressure was carried out, after which the fish were kept alive for several days. Fibroblast-like cells from the pectoral fins were removed and cultured successfully for a year at normal atmospheric pressure, with a doubling time of 77 h. Their tolerance to high pressure was compared with that of similar cells from the shallow water Conger eel, *Conger myriaster*, also grown in flasks pressurised in a steel vessel. The Conger eel cells (open squares, Fig. 5.4) did not grow at pressures greater than 10 MPa and 25 °C, whereas the deep-sea eel cells (black squares, Fig. 5.4) grew well at pressures from atmospheric up to 20 MPa, marginally at 25 MPa and not at all at 30 MPa, at 15 °C.

The cells' resistance to high pressures also differed. The Conger eel cells were all killed by a 20-min exposure to 100 MPa, but to kill all the deep sea cells required 20 min of 200 MPa (Fig. 13.12). Particularly interesting were the experiments on the state of actin and tubulin filaments. A pressure of 40 MPa for 20 min disrupted the filaments in the Conger eel cells but it had little effect on the deep-sea cells. Their filaments were visibly disrupted by 100–130 MPa, the tubulin being more susceptible than the actin filaments. The pressure tolerance of the deep-sea cells may seem surprising as they were derived from eels collected at around 10 MPa, but the vertical distribution of *S. parasiticus* extends to 2630 m. As we shall see in Chapter 6 these same proteins have been extracted from fish collected from abyssal depths. Another interesting observation was that the deep-sea cells were relatively resistant to the effects of the compound G418, which in normal shallow water eukaryote cells binds to 80s ribosomes and interferes with "proofreading," the process which checks the fidelity of protein synthesis. The ribosomes of deep-sea cells differ, reflecting adaptation to pressure (Koyama and Aizawa 2008). Considering the accessibility of these cultured cells it is surprising that more work has not been carried out with them.

An allied technique has been applied to study the toxicity of pollutants, sometimes called xenobiotics, in deep-sea fish, using liver slices incubated in special solutions, at elevated pressure. The liver is the organ that carries out detoxification, involving the enzyme cytochrome P4501A.

This enzyme is thought to be induced by pollutants in *Coryphaenoides armatus* at depth (Stegeman et al. 1986), but Lemaire and a team of workers experimented with the liver of *C. rupestris*, conveniently collected from 530 m in a Norwegian fjord. Slices were prepared at sea, but the subsequent experimental incubations were carried out in a land laboratory. The test compound, 3-methylcholanthrene, binds to a receptor which, in turn, sets in motion the transcription of the enzyme P4501A which carries out the detoxifying reaction. Induction was marked at atmospheric pressure but was reduced at 5 and 15 MPa, the latter being a pressure greater than that from which the specimens were collected, but well within the range of the species. Does this mean the fish are responding weakly to the pollutants in their normal environment and is the induction sufficient to protect the animal? The same procedure was used with liver slices from a shallow water fish, *Dicentrarchus labrax*, and also showed that high pressure depressed the induction, essentially the same result. These are preliminary studies, apparently resuscitating or conserving a certain amount of liver function (LeMaire et al. 2012, 2016; LeMaire 2017). It has limitations but avoids the problem of measuring P4501A levels in deep-sea fish kept healthy in a high-pressure aquarium, which has yet to be achieved (Chapter 13). And when that procedure becomes routine the next question will be, how do you "fix" the experimental animals, and their liver, prior to decompression? Studying sedentary animals like deep-sea mussels by means of Lander cages might be easier, but would isobaric retrieval and fixation be required? Another approach has been to study the effects of pressure on the toxicity of crude oil compounds on larvae of a shallow water animal, *Pomatoceros lamarcki*, Vevers et al. 2010). As mentioned earlier, gaining an understanding of the toxicity of pollutants in deep-sea organisms is not easy.

## 5.5    Other Deep-Sea Animals

A number of observations have been made on the pressure tolerance of deep-sea animals and their larvae. As many of these are from hydrothermal vent waters they are considered in Chapter 9.

# 5.6    Microbial Eukaryotes

## 5.6.1    Protists

Microorganisms are not necessarily Bacteria or Archaea. Within the Eukaryotes there are major groups of free-living cells and microscopic animals which qualify as microorganisms, such as yeasts and what used to be called protozoa, now protists. The latter include Foraminifera (Rhizaria) and Flagellates, which are heterotrophic, as they typically ingest bacteria. In contrast, Fungi (including yeasts) which are also heterotrophs, generally absorb organic compounds. The point is they all inhabit the oceans in surprising numbers and this last section of Chapter 5 deals with their pressure tolerance and their ability to survive and recover from decompression. Prokaryote microorganisms are separately discussed in Chapter 8.

Remember the scale of the environment in which these eukaryote microorganisms live (Scheckenbach et al. 2010). It comprises the bulk of the ocean plus the abyssal floor, i.e., at depths of 3000–6000 m, which is covered by soft sediment into which oxygen penetrates to a depth of, say, 10 cm. This benthic environment covers more than half of the earth's surface. The deeper hadal regions are smaller in the area but often richer in organic material, and only recently have been subject to investigations. The ocean floor accumulates particulate material which descends from the ocean water above. How do the organisms which live off this material cope with the pressures encountered? Do the shallower water species succumb to the increased pressure as they descend? How pressure tolerant are those which live in the deep sediments? Whilst microorganisms are our main concern here, there is a distinct infauna, comprising a wide variety of miniature animal species including familiar molluscs, crustacea and worms and less familiar groups such as nematodes (roundworms) living in the top few centimetres of sediment. The term meiofauna is also used to classify these miniature creatures, defined by being retained by a 44 um, or even 31 μm, mesh sieve (Giere 2009). For some reason, they have escaped the attention of experimenters. However, they are included with microorganisms in the new questions being posed about the tolerance of these abyssal and hadal organisms to pollution and most recently, to the prospect of localised storage of $CO_2$ beneath the seafloor and its consequential seepage.

The state of knowledge is illustrated by a paper published by Jing et al. (2018). They examined water samples from various depths above the Mariana Trench, down to 5900 m. Picoeukaryotic organisms were studied. They were not randomly distributed but occurred in groups or communities. Four major groups were conspicuous. (1) Alveolata, which includes ciliated protists like Tetrahymena and Dinoflagellates. (2) Opisthokonta, which includes yeasts (i.e., fungi) and choanoflagellates. (3) Rhizaria, Foraminifera, diverse Flagellates, Radiolaria and amoebae. (4) Stramenopiles (Heterokonts) which include algae and diatoms. Their numbers are huge and best conveyed by the high proportion of the ocean floor covered by their remains, which fall as marine snow, mentioned earlier. Siliceous oozes are formed from Diatoms and Radiolaria and calcareous oozes from algae called coccolithophores when their rate of deposition exceeds the rate at which their mineral component dissolves. Unfortunately, although the molecular biology techniques produce impressive results, they do not throw any light on the response of these organisms to high pressure.

The first examples of living eukaryote microorganisms in the sediments in the Mariana Trench (10,897 m) have been described by Akimoto and by Takami et al. (2001), namely a Foraminiferan and fungi. Burnett (1977) was among the first to collect deep-sea benthic Eukaryotic microorganisms using a refined but simple coring device, finding Ciliates, Flagellates and amoeboid cells in the top layer of sediments at 5498 m depths in the Pacific. Generally, the cells which inhabit the soft sediments at abyssal pressure can occur in great numbers and provide the experimenter with a specific depth of origin. For example Turley et al. (1988) used a corer to collect the benthic flagellate Bodo sp. from 4500 m depth. It survived decompression and although it failed to grow at 2 °C and normal atmospheric pressure, at 45 MPa it started to grow after a lag period of 16 days. Its doubling time was between 1 and 3 days. Microscopy revealed microtubules forming part of the cytostome, the site where bacteria are ingested. In a similar fashion several species of foraminifera, for example *Tinogullmia sp.*, were collected from similar depths. They were kept at 43 MPa and 5 °C for 36 days during which time they ingested bacterial cells and fluorescent particles and appeared healthy.

In one of the early experiments with eukaryote microorganisms Turley and colleagues collected various flagellates from a depth of 2665 m in the north Atlantic and demonstrated that many grew at pressures up to 20 MPa. Specifically *Paraphysomonas butcheri* (Chrysophyte) had a doubling time of 2–3 days at 5 MPa. In some experiments at 20 MPa this increased to nearly 10 days and in others to infinity, i.e., growth ceased. Another species, *Bodo curvifilus*, had a doubling time of 2 days at pressures up to 30 MPa and *Cercomonas sp.* failed to grow at pressures of 20 MPa or less, but grew well at 30–45 MPa. The first two are barotolerant and *Cercomonas* is barophilic. These cells are surprisingly robust (Turley et al. 1988, 1993; Turley and Carstens 1991).

Two species of flagellates that are widespread in the open ocean have been grown over slightly longer periods at high pressure (Morgan-Smith et al. 2013). They are *Cafeteria roenbergensis*, a Bicosoecid with a global distribution and *Neobodo designis*, a Kinetoplastid, with a similar distribution, but more common in deeper water. The experiments used cells originally isolated from shallow water and consisted of growth experiments at high pressure (50 MPa, i.e., 5000 m depth) and at 0.8 MPa, corresponding to a depth of 80 m. Batch cultures were grown and repeatedly sampled and renewed over several days. The bacteria in suspension, on which the cells fed, became progressively diluted but this was not significant. The initial response of the cells was to decline in number, both at high pressure and in the low-pressure controls, but the decline was greater in the former. However, in some experiments, the residual population of cells grew at 50 MPa. This implies some acclimation or adaptation to high pressure takes place in a few days, at least in vitro, suggesting that in Nature these shallow water cells may not all die as they sink.

Some particularly careful growth experiments were carried out by Atkins et al. (1998) with cells collected by the submersible Alvin from hydrothermal vent waters (2500 m depth and 20 °C) and from colder water on the continental slope (1500 m depth 3.5 °C). Generally the cells grew at the pressure from which they were collected but in the case of *Caecitellus parvulus*, from vent waters, growth at a pressure of 30 MPa, i.e., 5 MPa more than their pressure of origin, led to encystment. On decompression to atmospheric pressure vegetative cells emerged and grew, a remarkable example of pressure tolerance.

Although microbial eukaryotes are to be found on the abyssal plain, for example Ciliates living as epibionts on the surface of copepods which, in turn, live in abyssal sediments (Sedlacek et al. 2013), they also inhabit the sediments of the more restricted deep trenches in significant numbers. Living foraminifera, for example *Lagenamimma difflugiformis*, have been found alive in the Mariana Trench (Akimoto et al. 2001; Todo et al. 2005). Those in the Kuril–Kamchatka Trench are well illustrated in a paper by Lejzerowicz et al. (2015). Chinese workers have also made a substantial study of the microbial Eukaryotes, including the smallest category, Picoeukaryotic cells, inhabiting the waters of the Mariana Trench, from top to bottom (Jing et al. 2018; Xu et al. 2018).

In 2017 Zivaljic and co-workers published a review listing numerous heterotrophic flagellates which, in common with the organisms mentioned above, survive decompression from considerable depths and manifest piezotolerance or piezophilic growth. It would be interesting to know what proportion of the population fails to survive decompression.

Mention must also be made of giant eukaryotic cells. One is a Foraminiferan called Xenophyphores, which are only found on the deep seafloor. They appear as large lumps, or frilly, plant-like structures, as much as 20 cm or in diameter. They are the converse of a microorganism. The visible surface (test) is a "skin" of agglutinated material containing foreign particles. In the case of *Shinkaiya* lindsayi, found in the Japan Trench (Lecroq et al. 2009), it is 8 cm in diameter. Often within the test lies a multinucleate plasmodium, various tubular structures and crystals of barium sulphate and other minerals. They do not exhibit any motor activity. Another example is *Gromia sphaerica*, classified as among amoebae of uncertain affinities. It is a blob in shape, with a diameter of a few mm ranging up to 40 mm, often seen in photographs in small clusters on the seafloor at depths of 1200–1600 m. It too has a test and a quite complicated internal structure. These giant cell/organisms are surely potentially rewarding, albeit very delicate creatures whose careful retrieval by a deep-sea vehicle is long overdue. The Xenophyophores have recently been investigated in the Pacific, in the context of possible mining operations, by Gooday, from the National Oceanographic Centre, UK, and colleagues (2000, 2017).

## 5.6.2  Fungi

Fungi are heterotrophs; they subsist on organic compounds in their environment, unlike autotrophs which use an external energy source such as the sun or reduced inorganic compounds. There are many thousands of species of fungus but only a small proportion are marine. Fungi, with the status of Kingdom, comprise five main

Classes of which two, the Ascomycetes and the Basidiomyctes are "higher" fungi and the Myxomycetes, Oomycetes and Chytriodiomyctes are the "lower" fungi.

The deep oceans are certainly infected with fungi (Nagano and Nagahama 2012), which qualify as eukaryote microorganisms as they exist as free-living cells or, if multicellular, are less than 3 mm in size. An example is the Ascomycete *Oceanitis scutella*, living at 4000 m depths in the Atlantic, which has an ascocarp 2 mm long. Fungi have been isolated from the sediments in the Mariana Trench, from 10,897 and 10,500 m depths. In the late 1990s the unmanned submersible Kaiko, operated by its submersible mothership, collected sediment samples from which cultures were subsequently isolated and grown at normal and high pressure. Two fungi were identified; *Rhodotorula mucilaginosa*, a unicellular yeast and *Penicillium lagena*, a filamentous and apparently terrestrial fungus. From more widespread abyssal waters fungi and fungal spores have also been collected, a procedure which requires precautions to be taken to avoid contamination with surface species. Assuming such precautions work there is a lot of evidence to suggest that many fungi in the deep oceans may well be terrestrial rather than marine, and merely survive their chance presence in the depths. Both spores and hyphae (ie the vegetal growth stage) occur and although fungal spores are amazingly durable in all sorts of environments, the deep-sea hyphal stages are sometimes surprisingly pressure tolerant. The problem is, of course, that although the presence of fungi in sediments can be established microscopically, identification requires growth. The samples obtained by the submersible Kaiko were not isobarically collected so we do not know what cells may have failed to survive decompression.

A careful study by Lorenz and Molitoris (1997), demonstrated that fungi from shallow seas are able to grow at pressures up to 20 MPa. Two species were able to grow at 40 MPa, *Rhodotorula rubra* and *Rhodospirillum sphaerocarpum*, although their depth range in the ocean is not clear. Raghukumar and Damare, at the Indian National Institute of Oceanography, have focussed on fungi in the Central Indian Ocean Basin and the Chagos Trench. From 5000 m depths in the former, numerous fungal species have been collected and grown at 20 MPa. Many grew well at 30 °C from either spores or mycelia inocula, but those which grew better at 5 °C are regarded as deep-sea species, although the familiar concept of species is problematic. From the Chagos Trench comes a remarkable example of fungal survival. Spores were obtained from a core from 370 cm depth of sediment, on the Trench floor at a water depth of 5904 m. A small percentage germinated at pressures up to 50 MPa and at 5 °C, for example *Aspergillus sydowii*, known from terrestrial environments. This implied a deep sea form of the species exists, but of even greater interest is the age of the sediment from which the spores were collected. This was estimated to be from 180,000 to 430,000 years. It appears that deep-sea sediments can preserve spores in a way similar to glacial ice. Incidentally the oldest spores yet to be revived are bacterial symbionts from a bee preserved in amber, with the age of at least 25 million years.

It is interesting to reflect on the status of fungi in the marine world and in the study of the deep sea in particular. Neglected and minor are terms which come to mind, yet in biology generally yeasts, particularly *Saccharomyces cerevisiae* which produces carbon dioxide (which makes bread dough rise) and ethanol (which has always been popular with Homo sapiens) featured in the first biotechnological revolution and much basic research. Japanese scientists have made a particular study of yeast at high pressure. Fungi are also present in the waters associated with hydrothermal vents (Chapter 9). Prokaryotes, a major form of life in high-pressure environments, have been conspicuously absent so far, and they will be discussed in Chapter 8.

## 5.7    Conclusion

The main point is that decompressing deep-sea organisms matters. Often the lethal effects of decompression are obvious but, milder effects can be graded and far from clear. Experimental work on deep-sea organisms has to proceed with this in mind. Isobaric collection from deep-sea depths is required for the study of the integrated whole animal and probably the whole cell, but resuscitation of either may prove effective and adequate for certain purposes. The isobaric collection is probably not required if molecules, such as many proteins and lipids, are to be extracted for study (Chapters 6 and 7). Dynamic structures such as microtubules and other proteins with fast turnover kinetics will probably have to be treated to isobaric collection, rapid fixation/quenching and due control experiments.

## References[1]

Akimoto K, Hatton M, Uematsu K, Kato C (2001) The deepest living foraminifera, challenger deep, Mariana Trench. Mar Micopaleontol 42:95–97

Armstrong JD, Bagley PM, Priede IG (1992) Photographic and acoustic tracking observations of the behaviour of the grenadier Coryphaenoides ( Nematonurus) armatus, the eel Synaphopbranchus bathybius, and other abyssal demersal fish in the North Atlantic Ocean. Mar Biol 112:535–544

Atkins MS, Anderson OR, Wirsen CO (1998) Effect of hydrostatic pressure on the growth rates and encystment of flagellated protozoa isolated from a deep sea hydrothermal vent and a deep shelf region. Mar Ecol Prog Ser 171:85–95. *

Bailey TG, Tortres JJ, Youngbluth MJ, Owen GP (1994) Effect of decompression on mesopelagic gelatinous zooplankton: a comparison of in situ and shipboard measurements of metabolism. Mar Ecol Prog Ser 113:13–27. *

Bailey DM et al (2005) High swimming and metabolic activity in the deep sea eel Synaphobranchus kaupii revealed by integrated in situ and in vitro measurements. Physiol Biochem Zool 78:335–346. *

Barry JP et al (2013) The response of abyssal organisms to low pH conditions during a series of CO2 release experiments simulating deep sea carbon sequestration. Deep-Sea Res II 92:249–260. *

Blankenship LE, Yayanos AA, Cadien DB, Levin LA (2006) Vertical zonation patterns of scavenging amphipods from the Hadal Zone of the Tonga and Kermadec Trenches. Deep Sea Res I 53:48–61

[1]Those marked * are further reading.

Brauer RA, Bekman MY, Keyser JB, Nesbit DL, Sidelelev GN, Wright SL (1979) Adaptation to high hydrostatic pressures of abyssal gammarids from Lake Baikal in Eastern Siberia. Comp Biochem Physiol 65A:109–117

Brauer RW, Sidelyova VG, Dail MB, Galazii GI, Roer RD (1984) Physiological adaptation of Cottoid fishes of Lake Baikal to abyssal depths. Comp Biochem Physiol 77A:699–705

Brown A, Thatje S (2011) Respiratory responses of the deep sea Amphipod Stephonyx biscayensis indicates bathymetric range limitation by temperature and hydrostatic pressure. PLoS One 6 (12):e28888562

Burnett BR (1977) Quantitative sampling of microbiota of the deep sea benthos—1 Sampling techniques and some data from abyssal central. N Pacific Deep Sea Res 24:781–789

Campenot RB (1975) Effects of high hydrostatic pressure on transmission at the crustacean neuromuscular junction. Comp Biochem Physiol 52B:133–140

Camus L, Gulliksen B (2005) Antioxidant defense properties of Artic amphipods: comparison between deep-, sublittoral and surface water species. Mar Biol 146:355–362

Childress JJ (1995) Are there physiological and biochemical adaptations of metabolism in deep sea animals? Trends Ecol Evol 10:30–36

Drazen JC, Bird LE, Barry JP (2005) Development of a hyperbaric trap-respirometer for the capture and maintenance of live deep sea organisms. Limnol Oceanogr Methods 3:488–498. *

Drazen JC, Dugan B, Friedman JR (2013) Red muscle proportions and enzyme activities in deep sea demersal fish. J Fish Biol 83:1592–1612

Eustace RM, Ritchie H, Kilgallen NM, Piertney SB, Jamieson AJ (2016) Morphological and ontogenetic stratification of abyssal and hadal Eurythenes gryllus sensu lato (Amphipoda: Lysianassoidea) from the Peru-Chile trench. Deep-Sea Res 109:91–98

Fontanier C et al (2016) Living (stained) benthic foraminifera from the Mozambique Channel (eastern Africa): exploring ecology of deep sea unicellular meiofauna. Deep Sea Res I 115:159–174. *

France SC, Kocher TD (1996) Geographic and bathymetric patterns of mitochondrial 16S r RNA sequence divergence among deep sea amphipods, Eurythenes gryllus. Mar Biol 126:633–643

George RY (1979) What adaptive strategies promote immigration and speciation in deep sea environment? Sarsia 64:61–65

Giere O (2009) Meiobenthology. The microscopic motile fauna of aquatic sediments. Springer, Berlin, Heidelberg

Gilchrist I, Macdonald AG (1983) Techniques for experiments with deep sea animals. In: Macdonald AG, Priede IG (eds) Experimental biology at sea. Academic Press, London

Gooday AJ, Bowser SS, Bett BJ, Smith CR (2000) A large testate protist, Gromia spherical sp. nov. (Order Filosea), from the bathyal Arabian Sea. Deep-Sea Res I 47:55–73

Gooday AJ et al (2017) Giant protists (xenophyophores, Foraminifera) are exceptionally diverse in parts of the eastern Pacific licensed for polymetallic nodule exploration. Biol Conserv 207:106–116

Griffith RW (1981) Composition of the blood serum of deep sea fishes. Biol Bull 160:250–264. *

Harper AA, Macdonald AG, Wardle CS, Pennec J-P (1987) The pressure tolerance of deep sea fish axons; results of challenger cruise 6B/85. Comp Biochem Physiol 88A:647–653

Havermans C et al (2013) Genetic and morphological divergences in the cosmopolitan deep sea amphipod Eurythenes gryllus reveal a diverse abyss and a bipolar species. PLoS One 8:e74218

Hogan PM, Besch SR (1993) Vertebrate skeletal and cardiac muscle. In: Macdonald AG (ed) Effects of high pressure on biological systems, Comparative and Environmental Physiology, vol 17. Springer-Verlag, Berlin, pp 124–146

Hughes SJM et al (2011) Deep sea echinoderm oxygen consumption rates and an interclass comparison of metabolic rates in Asteroidea, Crinoidea, Echinoidea, Holothuroidea and Ophiuroidea. J Exp Biol 214:2512–2521

Jamieson AJ, Lacey AC, Lorz A-N, Rowden AA, Piertney SB (2013) The supergiant Amphipod Alicella gigantea (Crustacea: Allicellidae) from hadal depths in the Kermadec. S W Pacific Ocean, Trench

Jamieson AJ et al (2019) Microplastics and synthetic particles ingested by deep sea amphipods in six of the deepest marine ecosystems on Earth. R Soc Open Sci 6:180667. *

Janssens BJ, Childress JJ, Baguet F, Rees J-F (2000) Reduced enzymatic antioxidative defense in deep sea fish. J Exp Biol 203:3717–3725

Jing H, Zhang Y, Yingdon L, Zhu W, Liu H (2018) Spatial variability of Picoeukaryotic communities in the Mariana trench. Sci Rep 8:15357

Kohlmeyer J, Kohlmeyer E (1979) Marine mycology, the higher fungi. Academic Press, New York, p 690. *

Koyama S, Aizawa M (2000) Tissue culture of the deep sea bivalve Calyptogena soyoae. Extremophiles 4:385–389

Koyama M, Aizawa M (2008) Piezotolerance of the cytoskeletal structure in cultured deep sea fish cells using DNA transfection and protein introduction techniques. Cyototechnology 56:19–26

Koyama S, Miwa T, Horii M, ishikawa Y, Horikoshi K, Aizawa M (2002) Pressure-stat aquarium system designed for capturing and maintaining deep sea organisms. Deep Sea Res I 49:2095–2102. *

Koyama S, Horii M, Miwa T (2003) Tissue culture of the deep sea eel Simenchelys parasiticus collected at 1,162 m. Extremophiles 7:254–248. *

Koyama S et al (2005a) Survival of deep sea shrimp (Alvinocaris sp.) during decompression and larval hatching at atmospheric pressure. Mar Biotechnol 7:272–2728

Koyama S, Kobayashi H, Inoue A, Miwa T, Aizawa M (2005b) Effects of the piezo-tolerance of cultured deep sea eel cells on survival rates, cell proliferation, and cytoskeletal structures. Extremophiles 9:449–460

Lecroq B, Gooday AJ, Tsuchiya M, Pawlowski J (2009) A new genus of xenophyophores (Foraminifera) from Japan trench; morphological description, molecular phylogeny and elemental analysis. Zool J Linnean Soc 156:455–464

Leduc D et al (2016) Comparison between infaunal communities of the deep sea floor and edge of the Tonga Trench: possible effects of differences in organic matter supply. Deep Sea Res I 116:264–275. *

Lejzerowicz F, Voltski I, Pawlowski J (2015) Foraminifera of the Kuril-Kamchatka area: the prospects of molecular study. Deep-Sea Res II 111:19–25

Lemaire B (2017) Hydrostatic pressure and the experimental toxicology of marine fishes: the elephant in the room. Mar Pollut Bull 124:206–210

Lemaire B et al (2012) Precision-cut liver slices to investigate responsiveness of deep sea fish to contaminants at high pressure. Environ Sci Technol 46:10310–10316

Lemaire B et al (2016) High hydrostatic pressure influences the invitro response to xenobiotics in Dicentrarchus labrax liver. Aquatic Toicol 173:43–52

Lorenz R, Molitoris HP (1997) Cultivation of fungi under simulated deep sea conditions. Mycol Res 101:1355–1365

Ma L-J, Rogers SO, Catranis CM, Starmer WT (2000) Detection and characterization of ancient fungi entrapped in glacial ice. Mycologia 92:286–295. *

Macdonald AG, Gilchrist I (1972) In: Brauer RW (ed) An apparatus for the recovery and study of deep sea plankton at constant temperature and pressure in Barobiology and the Experimental Biology of the Deep Sea. University of North Carolina Press, North Carolina, pp 394–412

Macdonald AG, Gilchrist I (1978) Further studies on the pressure tolerance of deep sea Crustcea with observations using a new high pressure trap. Mar Biol 45:9–21

Macdonald AG, Gilchrist I (1980) Effects of hydraulic decompression and compression on deep sea Amphipods. Comp Biochem Physiol 67A:149–153

Macdonald AG, Gilchrist I (1982) The pressure tolerance of deep sea amphipods collected at their ambient high pressure. Comp Biochem Physiol 71A:149–152

Macdonald AG, Teal JM (1975) Tolerance of oceanic and shallow water crustacea to high hydrostatic pressure. Deep-Sea Res 22:131–144. *

Macdonald AG, Gilchrist I, Teal JM (1972) Some observations on the tolerance of oceanic plankton to high hydrostatic pressure. J Mar Biol Assoc UK 52:213–223. *

Macdonald AG, Gilchrist I, Wann KT, Wicock SE (1980) The tolerance of animals to pressure. In: Gilles R (ed) Animals and environmental fitness, Proc. 1st Conference of the European Society for Comp. Physiol. Biochem. Pergamon Press, Oxford, pp 385–403

Macdonald AG, Gilchrist I, Wardle CS (1987) Effects of hydrostatic pressure on the motor activity of fish from shallow water and 900 m depths; some results of Challenger cruise 6B/85. Comp Biochem Physiol 88A:543–547

Mickel TJ, Childress JJ (1982) Effects of pressure and pressure acclimation on activity and oxygen consumption in the bathypelagic mysid Gnathophausia ingens. Deep-Sea Res 29:1293–1301

Miyake H, Kitada M, Tsuchida S, Okuyama Y, Nakamura K-I (2007) Ecological aspects of hydrothermal vent animals in captivity at atmospheric pressure. Mar Ecol 28:86–92

Morgan-Smith D, Garrison CE, Bochdansky AB (2013) Mortality and survival of cultured surface-ocean flagellates under simulated deep sea conditions. J Exp Mar Biol Ecol 445:13–20

Nagano Y, Nagahama T (2012) Fungal diversity in the deep sea extreme environments. Fungal Ecol 5:463–471

Nunnally CC, Friedman JR, Drazen JC (2016) In situ respiration measurements of megafauna in the Kermadec Trench. Deep Sea Res I 118:30–36

Olsen GH et al (2016) Sensitivity of the deep sea amphipod Eurythenes gryllus to chemically dispersed oil. Environ Sci Pollut Res 23:6497–6505. *

Pane EF, Barry JP (2007) Extracellular acid-base regulation during short term hypercapnia is effective in a shallow water crab but ineffective in a deep sea crab. Mar Ecol Prog Ser 334:1–9. *

Pennec J-P, Wardle CS, Harper AA, Macdonald AG (1988) Effects of hydrostatic pressure on the isolated hearts of shallow water and deep sea fish: results of challenger cruise 6B/85. Comp Biochem Physiol 89A:215–218

Quetin LB, Childress JJ (1980) Observations on the swimming activity of two bathypelagic mysid species maintained at high hydrostatic pressures. Deep-Sea Res 27A:383–391. *

Raghukumar C, Damare S (2008) Deep sea Fungi. In: Michiels C, Bartlett DH, Aertsen A (eds) High pressure microbiology. ASAM Press, Washington, DC. *

Ritchie H, Jamieson AJ, Piertney SB (2018) Heat shock protein adaptation in abyssal and hadal amphipods. Deep Sea Res I 155:61–69

Scheckenbach F, Hausmann K, Wylezich C, Weitere M, Arndt H (2010) Large scale patterns in biodiversity of microbial eukaryotes from the abyssal sea floor. Proc Natl Acad Sci U S A 107:115–120

Sedlacek L, Thistle D, Fernandez-Leborans G, Carman KR (2013) First report of ciliate (Protozoa) epibionts on deep sea harpacticoid copepods. Deep-Sea Res II 92:165–171

Shelton CJ, Macdonald AG (1987) Effect of high hydrostatic pressure on Rb+ influx in the erythrocyte of the plaice, (Pleuronectes platessa). Comp Biochem Physiol 88A:481–485

Shelton C, Macdonald AG, Pequeux A, Gilchrist I (1985) The ionic composition of the plasma and erythrocytes of deep sea fish. J Comp Physiol B 155:629–633

Shillito B et al (2008) Live capture of megafauna from 2300 m depths, using a newly designed pressurised recovery device. Deep Sea Res I 55:881–889

Siebel BA, Drazen JC (2007) The rate of metabolism in marine animals: environmental constraints, ecological demands, and energetic opportunities. Philos Trans R Soc B 362:2061–2078

Siebenaller JF (2010) Effects of deep sea environment on invertebrates. In: Sebert P (ed) Comparative high pressure biology. Science Publishers, New York

Smith KL, Baldwin RJ (1982) Scavenging deep sea Amphipods: Effects of food odor on oxygen consumption and a proposed metabolic strategy. Mar Biol 68:287–298

Smith KL, Laver MB (1981) Respiration of the bathypelagic fish Cyclothone acclinidens. Mar Biol 61:261–266

Stegeman JJ, Kloepper-Sam PJ, Farrington JW (1986) Monooxygenase induction and chlorobiphenyls in the deep sea fish Coryphaenoides armatus. Science 231:1287–1289

Takami H, Inoue A, Fuji F, Horikoshi K (1997) Microbial flora in the deepest sea mud of the Mariana trench. FEMS Microbiol Lett 152:279–285. *

Takeuchi I, Watanabe K (1998) Respiration rate and swimming speed of the necrophagous Eurythenes gryllus from Antarctic deep waters. Mar Ecol Prog Ser 163:285–288

Thorne SD, Hall AC, Lowe AG (1992) Effects of pressure on glucose transport in human erythrocytes. FEBS Lett 301:299–302

Thurston MH (1979) Scavenging abyssal amphipods from the north-east Atlantic Ocean. Mar Biol 51:55–68. *

Todo Y, Kitazato H, Hashimoto J, Gooday AJ (2005) Simple Foraminifera flourish at the ocean's deepest point. Science 307:689

Treude T, Jansen F, Queisser W, Witte U (2002) Metabolism and decompression tolerance of scavenging lysianassoid deep sea amphipods. Deep Sea Res I 49:1281–1289

Turley CM, Carstens M (1991) Pressure tolerance of oceanic flagellates: implications for remineralisation of organic matter. Deep-Sea Res 38:403–413

Turley CM, Mackie PJ (1995) Bacterial and cyanobacterial flux to the deep N E Atlantic on sedimenting particles. Deep Sea Res I 42:1453–1474. *

Turley CM, Lochte K, Patterson DJ (1988) A barophilic flagellate isolated from 4500 m in the mid-North Atlantic. Deep-Sea Res 35:1079–1092

Turley CM, Gooday AJ, Green JC (1993) Maintenance of abyssal benthic foraminifera under high pressure and low temperature: some preliminary results. Deep-Sea Res 40:643–652

Vevers WF, Dixon DR, Dixon LRJ (2010) The role of hydrostatic pressure on developmental stages of Pomatoceros lamarcki, (Polychaeta:Serpulidae) exposed to water accommodated fractions of crude oil and positive genotoxins at simulated depths of 1000–3000 m. Environ Pollut 158:1702–1709

Wardle CS, Tetteh-Lartey N, Macdonald AG, Harper AA, Pennec J-P (1987) The effect of pressure on the lateral swimming muscle of the European eel, Anguilla Anguilla, and the deep sea eel Histiobranchus bathybius: results of challenger cruise 6B/85. Comp Biochem Physiool 88A:595–598

Wegner v F, Koyama S, Miwa T, Freidrich O (2008) Resting membrane potentials recorded on site in intact muscle from deep sea fish, (Sigmops gracile), salvaged from depths of up to 1000m. Mar Biotechnol 10:478–486

Wilson RR, Smith KL (1985) Live capture, maintenance and partial decompression of a deep sea grenadier fish (Coryphaenoides acrolepis) in a hyperbaric trap-aquarium. Deep-Sea Res 33:1571–1582. *

Wirsen CO, Jannasch HW (1983) In situ studies on deep sea amphipods and their intestinal flora. Mar Biol 78:69–73

Xu Z et al (2018) Vertical distribution of microbial Eukaryotes from surface to the hadal zone of the Mariana Trench. Front Microbiol 9:2023

Yayanos AA (1978) Recovery and maintenance of live amphipods at a pressure of 580 bars from an ocean depth of 5700 metres. Science 200:1056–1059

Yayanos AA (1981) Reversible inactivation of deep sea Amphipods (Paralicella capresca) by a decompression from 601 bars to atmospheric pressure. Comp Biochem Physiol 69A:563–565

Yayanos AA (2009) Recovery of live Amphipods at over 102 MPa from the Challenger Deep. Mar Technol Soc J 43:132–136

Zerbst-Boroffika I, Kamaltynow RM, Harjes S, Kinne-Saffran E, Gross J (2005) TMAO and other organic osmolytes in the muscles of amphipods (Crustacea) from shallow and deep water of Lake Baikal. Comp Biochem Physiol A 142:58–64

Zhao F, Xu K (2016) Molecular diversity and distribution pattern of ciliates in sediments from deep sea hydrothermal vents in the Okinawa Trough and adjacent seas. Deep Sea Res I 116:22–32. *

Zivaljic S et al (2017) Survival of marine heterotrophic flagellates from the surface and the deep sea at high hydrostatic pressure: literature review and own experiments. Deep-Sea Res 148:251–259. https://doi.org/10.1016/j.dsr2.2017.04.022. *

# Molecular Adaptation to High Pressure: Proteins in Deep-Sea Animals

**6**

In the latter half of the twentieth century a group of North American biochemists embarked on studies of the enzymes of deep-sea fish, using the established methods of enzyme biochemistry mentioned earlier, in Chapter 2. Their objective was to understand the molecular adaptations required of deep-sea animals in general and of fish in particular, to high pressure. They were led by the young but soon to be distinguished Canadian biochemist Peter Hochachka, who organised the highly significant 1971 cruise of the R V Alpha Helix, a ship designed for experimental work at sea, operated by the Scripps Institution of Oceanography, (Somero and Suarez 2005). The Alpha Helix was based in the deep water close to, and sheltered by, the Galapagos Islands. From this and subsequent cruises, Hochachka, together with George Somero and their numerous colleagues from the USA, Canada and elsewhere, developed an understanding of biochemical adaptations to high pressure, (Somero and Suarez 2005). Many other cruises followed and some of the experiments arising from them are described below. Over time the North American biochemical work became a major contribution to deep-sea Biology, supplemented by scientists from Europe and particularly from Japan.

At that time the expectation of most biologists, be they biochemist, physiologist, microbiologist or ecologist, was that the adaptation of organisms to high pressure would involve a multiplicity of small changes arising from chance mutations. The adaptation of proteins and of enzymes in particular was an obvious focus of research, hence the importance of biochemistry. Sometime later the investigation of the adaptation of cell membranes to high pressure emerged as an offshoot of the study of their adaptation to high and low temperatures. It involves both biochemical and biophysical techniques and is dealt with in Chapter 7.

The account which follows is based on the consensus that adaptation of proteins to high pressure is best understood through the "fitness" of the protein in its physiological role, and in the case of enzymes, that essentially means regulated function. Before getting immersed in the account it is worth noting that at least one other approach to understanding protein adaptation to high pressure has been proposed. Its rationale is as follows. All forms of life have access to twenty amino

© Springer Nature Switzerland AG 2021
A. Macdonald, *Life at High Pressure*, https://doi.org/10.1007/978-3-030-67587-5_6

acids from which to assemble their proteins. Each amino acid has its distinctive combination of physical properties, size, polarity and so on. Attempts to correlate amino acid composition and sequences with adaptation to high pressure have been made by Nath and Subbiah (2016) and by Giulio (2005). The latter compared the amino acids in the proteins of the shallow water archaeon *Pyrococcus furiosus* with those of the barophilic (i.e. pressure tolerant) *P. abyssi*, from 2000 m. A total of 141 proteins were selected and the amino acid substitutions seen in the barophilic proteins were noted and statistically evaluated by a complicated procedure we can skip. It scored the extent to which amino acids occurred in the barophilic protein, producing a league table topped by the most frequent substitution, arginine. Thus, pressure-adapted proteins are likely to have acquired arginine in place of some other amino acid. The league table, with arginine at the top and tyrosine at the bottom, correlated with two physical properties; polarity and smallness. Tyrosine, at the bottom of the list was the least polar and large. Whilst this is interesting it is not clear how this sort of data can actually help us understand pressure-adapted proteins and how they have been naturally selected. It is unfortunate that an organism from such a moderate depth, 2000 m, was selected, but it would be interesting to see if a similar exercise, carried out with organisms from the full ocean depths confirms Giulio' result. If such a confirmation is found it will certainly stimulate a deeper analysis. For the present, we will stay with the consensus.

## 6.1    Protein Adaptation

To study the proteins from deep-sea animals, here mainly fish, the animals first have to be collected in such a way that their proteins are in a healthy state. The problem of deciding if an animal is dead, or alive (and healthy) or somewhere in between, which concerned us in the previous chapter also applies to proteins. Reproducibility of experimental results is generally regarded as consistent with the material being in a healthy state, but supplementary evidence is always welcome. It is generally found that proteins extracted from a recently collected dead animal and re-pressurised for experimental purposes functions reproducibly and, it is thought, in a way similar to its performance in vivo, given the appropriate reaction conditions.

The adaptation of proteins to function at high pressure is generally assumed to arise from chance mutations, a view strengthened by the case of sickle cell haemoglobin in which a single amino acid substitution drastically affects the molecule's pressure stability, (Chapter 2). To that example, we can add the case of the yellow fluorescent protein in which glycine insertions increase the intensity of the light emitted under high pressure, (Watanabe et al. 2013).

The obvious experimental approach is to study the susceptibility of structural proteins and of enzymes from deep-sea fish, to high pressure, i.e. their pressure tolerance. The present state of knowledge provides us with a convenient basis for organising this chapter into Intrinsic and Extrinsic mechanisms of adaptation to high pressure, a distinction which also applies to proteins in other deep-sea organisms.

## 6.1.1 Intrinsic Adaption of Proteins to High Pressure

### 6.1.1.1 Structural Proteins

We know that the ubiquitous structural proteins in ordinary cells can be disaggregated by deep-sea pressures, so it is natural to ask if, in deep-sea organisms, these proteins are adapted to retain their polymerised state. Actin is one such protein and it is the form that occurs in deep-sea fish muscle which has been investigated in some detail. Although actin is a structural protein it is not an inert one, and it is active in various binding and biochemical reactions.

Swezey and Somero (1982), compared the properties of actin extracted from the muscles of a wide range of animals. They included fish from different depths, which were frozen on collection for analysis in the home laboratory. There is no reason to suppose the muscle proteins are drastically affected by the temperature rise and the decompression experienced during collection. During the first Alpha Helix cruise it was established that the myofibrillar structure of the muscle of deep-sea fishes was normal after collection, (Herman and Dreizen 1971). The resuscitation experiments described in Chapter 5 are consistent with the muscles being reversibly and not drastically affected by decompression.

The investigation of actin was a sophisticated one but some of the main conclusions are within our reach here. A now familiar parameter (Chapter 2), the $\Delta V$ of the polymerisation of actin, was measured in physiological reaction conditions (unlike the pioneer Japanese work), using material from the muscles of deep-sea fish. Remember, however, it is the actin of cleaving cells that is depolymerised in vivo; actin in muscle is only depolymerised once it is extracted. It is not depolymerised by pressure in vivo, and we need to know how it is stabilised. Remember also that the polymer–monomer equilibrium with which we are concerned is not a simple physical equilibrium as first imagined. Polymerisation and depolymerisation are the result of chemical reactions whose rates differ at each end of the actin molecule.

Two species, *Coryphaenoides acrolepis* and *C. armatus* were used. They live at much the same temperature (2–3 C), but *C. acrolepis* inhabits a depth range of 250–2100 m and *C. armatus* lives much deeper, at 1900–4800 m. The $\Delta V$ of the gross polymerisation, (self-assembly) of the actin from *C. armatus* is much smaller than that of *C. acrolepis*. It is, in fact, one seventh smaller. The $\Delta V$ for the polymerisation of the actin from *C. acrolepis* was much the same as that from the shallow water *Sebastolobus alascanus* (from depths of 180–440 m), which in turn was similar to that of *Halosauropsis macrochir* (1500–5200 m). This latter comparison is puzzling and without an explanation. However, the undoubted small $\Delta V$ in the case of the actin from *C. armatus* was interpreted as adaptive, enabling actin self-assembly to proceed not just at great depths, but at shallower depths too. We can imagine that the adaptive feature comes into play when the actin is freshly synthesised and organised within the muscle cell, where it lies parallel to myosin filaments, with which it interacts to create tension. The difference between the $\Delta V$s of the actins from the two species of *Coryphaenoides* suggests that the critical depth/ pressure for natural selection to favour the adaptation of actin was around 2000 m per 20 MPa, which is a higher pressure than separates the adaptation of certain deep-sea enzymes, of which more below.

The thermodynamic parameters ($\Delta$ H, (i.e. $\Delta$ U + P $\times$ $\Delta$V) and $\Delta$S), for the polymerisation of the actins studied, indicates the type of intermolecular (inter-monomer) bonds involved in the process. Hydrophobic interactions are expected to predominate where large, positive $\Delta$ H and S values are found, as in the actin from *C. acrolepis* but the low values found in the actin from *C. armatus* strongly suggest such interactions are overshadowed by ionic bonds. This idea is consistent with other properties of the actin which are beyond the scope of the present account. Additionally, a particularly intriguing feature of the actin from *C. armatus* is its marked resistance to thermal denaturation. This is not interpreted as an adaptation to a wildly fluctuating environmental temperature, (the temperature of its habitat is, of course, notably cold and constant), but a consequence of the protein's increased structural stability arising from the adaptive changes.

Some twenty years after Swezey and Somero's work a more detailed examination of the actin from deep-sea fish was undertaken by Morita (2003, 2010). He studied muscle actin from *C. acrolepis* (locally 180–2000 m) and compared it with actin from the abyssal species *C. armatus* (2700–5000 m) and *C. yaquinae* (4000–6400 m). The last two occur in the North Atlantic and the North Pacific, respectively. He confirmed that the $\Delta$V of the gross polymerisation of actin from the two abyssal species was much smaller than that of actin from *C. acrolepis*, (one fifth at atmospheric pressure and one third at 60 MPa). Additionally, the pressure sensitivity of the binding of calcium ions and of ATP to actin was measured. These play a role in the general stability of F actin and, as ionic bonds, their formation is expected to involve a positive activation volume. This indeed proved the case. Significantly the $\Delta$V* for both was, in the case of the abyssal actins, much smaller than that of the actin from *C. acrolepis*. This implies that the abyssal actins hold on to their calcium and ATP at great depth more effectively than the actin from *C. acrolepis* would. The next stage in the investigation used molecular biology techniques which provided the amino acid sequences of the different forms of actin. Remember that the actin monomer comprises a string of 375 amino acids. Wakai and colleagues (2014), including Morita, were able to correlate the differences in amino acid sequences with the effects of high pressure mentioned above. Some of their findings are summarised in Table 6.1.

**Table 6.1**   Amino acid substitutions in the muscle actins of deep-sea fish

| Position in sequence | 54 | 67 | 137 | 155 |
|---|---|---|---|---|
| Actin-carp (shallow) | Valine | Leucine | Glutamine | Alanine |
| Actin-*C. acrolepis* (180–2000 m) | Valine | Leucine | Glutamine | Serine |
| Actin-*C. armatus* (2700–5000 m) | Alanine | Leucine | Lysine | Serine |
| Actin-*C. yaquinae* (4000–6400 m) | Valine | Proline | Lysine | Serine |

Based on Morita (2003)

First consider positions 54 and 67. These lie in a stretch of the molecule which, when folded, is involved in actin–actin binding. It was argued that the substitutions contributed to the small $\Delta V$ of assembly. Although the details are not clear it is plausible that the changes in positions 54 and 67 are related to high-pressure adaptation of the protein. It is interesting that the armatus and yaquinae actins differ in this respect. The positions 137 and 155 show the same substitutions in the two abyssal actins. The amino acids in positions 137 and 155 are involved in the binding of calcium ions and ATP. They contribute to the structure of actin and in the abyssal species their binding is less affected by pressure than in shallow species. Wakai and collaborators used molecular dynamics simulations to investigate these structural features in greater detail. They seem to confirm the earlier work and more emphasis is given to the existence of salt bridges in the actins of *C. armatus* and *C. yaquinae* in stabilising ATP binding and the assembled subunits. These deep-sea actins are stabilised at, and thus adapted to, high pressure, essentially by two amino acid substitutions. Individuals carrying the original mutations presumably acquired a selective advantage, although we do know exactly how that might have been manifested. In conjunction with changes in the animals' physiology arising from numerous, seemingly micro changes in other proteins, deep-sea species such as *C. armatus* and *C. yaquinae* have evolved. This is what the sickle cell haemoglobin story suggested might happen in the process of high-pressure adaptation. However, pressure adaptations arising from quite a different route are also important, and these will be explained in a subsequent section.

The evolutionary history (phylogeny) of the genus *Coryphaenoides* has been the subject of several studies, (Morita 1999; Gaither et al. 2016). They show that *Coryphaenoides*, familiar to marine biologists as a rattail or a grenadier, belongs in the family Macrouridae, which is in the same Order, the Gadiformes, which includes *Gadus morhua*, (family Moridae), the cod, (familiar to many of us as the fish in fish and chips). The most recent study used mitochondrial DNA analysis and concluded that the two abyssal species *C. armatus*, (Atlantic) and *C. yaquinae*, (Pacific), are closely related and diverged from the common ancestor of the other species 5.2 million years ago, i.e. at the junction of the Miocene and the Pliocene. Subsequently their common ancestor gave rise to a succession of species; *C. nasutus, C. cinereus, C. longifilis, C. acrolepis* and *C. pectoralis*, none of which occupy abyssal depths.

### 6.1.1.2 Enzymes

The pressure tolerance of enzyme kinetics can, in many cases, be studied in a simple way. Often reactions can be measured and recorded at high pressure using an optical method, (Chapter 13). A cuvette holds a reaction mixture and the addition of the final ingredient starts the reaction. The cuvette is sealed in the high-pressure optical cell and pressurised in a matter of seconds. Hochachka's group described an example in which this procedure took 10 s, equal to 10% of the time of interest in the reaction, (Mustafa et al. 1971). The condition of the enzyme is obviously important. Lactic dehydrogenase, LDH, isolated from the white muscle of deep-sea fish, is an example. It proved to be stable at surface pressures even after the warming and decompression experienced during collection by a free fall "pop up" method, (as in Fig. 13.20, Chapter 13). The catalytic function of LDH is to convert pyruvate to lactic acid, and it does this in conjunction with a cofactor, or co-substrate, nicotinamide adenine dinucleotide, (NADH). Remember the white muscle functions to generate short, fast bursts of swimming. It can contract vigorously for a while using ATP generated anaerobically. In the absence of oxygen, LDH converts pyruvic acid to lactic acid which acidifies the muscle and contributes to limiting its function. Much of the lactic acid is transported by the blood to the liver where it is converted first to pyruvic acid by LDH, and thence to glucose.

$$\text{Pyruvic acid} + \text{NADH} + \text{H}^+ = \text{Lactic acid} + \text{NA}$$

Experiments carried out by Siebenaller and Somero (1978) to see how high pressures affected the enzyme used the criterion of the Km value as a measure of the binding affinity of NADH, and of pyruvate, to LDH (Chapter 2, Section 2.3). Enzymes have to bind to their substrates, which are normally at a low concentration, and to co-factors, to function, and their effectiveness in doing so is a guide to their adaptation. Reproducibility of the results proved satisfactory.

Two closely related species that experience similar temperatures at different depths were studied; *Sebastolobus alascanus* (180–440 m) and *S. altivelis* (550–1300 m). The muscle LDH from shallow water fish has Km values for NADH and for pyruvate which are increased by high pressure, meaning pressure decreases their binding to the enzyme. In contrast, the LDH from the deeper living fish has corresponding Km values which are relatively unaffected by pressure, as high as 47 MPa. The increased Km values in the case of the shallow water LDH is "highly maladaptive," in the words of Somero, for life in the deep sea. Subsequent work has shown the prime adaptive feature of the *S. altivelis* LDH is a single amino acid substitution in position 115 of the primary sequence, (Siebenaller 1984).

Other studies have shown similar adaptations reflected in the Km values of other fish enzymes, such as malate dehydrogenase (MDH) and glyceraldehyde-3-phosphate dehydrogenase (GAPDH), whose normal catalytic role need not concern us here. The modest pressure difference experienced by *S. alascanus* and *S. altivelis*, around 5–10 MPa, appears sufficient for natural selection to work on, much smaller than in the case of actin. Adapting enzymes to high pressure can be a subtle process.

Molecular biology techniques have opened up numerous possibilities for studying the details of enzymes. Brindley and her colleagues (2008), synthesised LDH starting with RNA extracted from the livers of *Coryphaenoides armatus* and the cod, *Gadus morhua*. From the resultant synthetic enzymes, they concluded that their primary structures only differed by 21 amino acids. The catalytic activity of the LDH from *C. armatus* was found to be less affected by high pressure than the cod LDH and, quite separately, optical measurements that measure the unfolding of the enzymes were carried out. They showed the cod LDH unfolded at pressures much lower than those which affected the LDH from *C. armatus*. Relating these differences in pressure tolerance to the 21 amino acids mentioned above is a demanding exercise, requiring detailed three-dimensional structures. Some tentative, and plausible, conclusions were put forward. The general conclusion was that pressure tolerance in the *C. armatus* LDH seemed to arise from distinct amino acids in the N-terminal region of the primary structure. The pressure tolerance of fish LDH, including the synthetic version from *C. armatus*, is also enhanced by a separate mechanism which will be explained in Section 6.2.2.

Km values have also been used to assess the adaptation of invertebrate enzymes to high pressure, for example malate dehydrogenase, MDH. In particular, the Km of NADH for the version of MDH which occurs in the cytoplasm of muscles has been studied. (MDH is usually thought of as a mitochondrial enzyme, part of the Citric acid cycle). The enzyme from the clam *Calyptogena elongata*, from 700 m depths was compared with that from. *C. phaseolithoformes*, from depths of more than 3000 m. A pressure as high as 50 MPa had little effect on the latter but the former was nearly doubled by 20 MPa, (Dahlhoff and Somero 1991). This type of adaptation, the conservation of km in natural conditions, is seen in the adaptation of many enzymes to other physical conditions, and not just to high pressure.

Another approach to detecting differences between the adaptation of shallow water and deep-sea enzymes may have occurred to the reader, namely the stability of multimeric enzymes. Are the deep-sea versions more stable at high pressure than their shallow water counterparts? Hennessey and Siebenaller have determined the stability of the tetrameric LDH, extracted from a variety of fishes (Table 6.2). Incubating the enzymes for an hour at different pressures and subsequently measuring their catalytic activity provides a measure of the depolymerisation which has taken place.

The pressure required to reduce the activity of the enzyme to one half generally increased with the depth of the origin of the enzyme, indicating an increase in pressure stability with depth. The top three in the list are clearly different from the lower three. The way this stability is achieved is not known.

It is interesting that adaptation to pressure has been revealed in hagfish, formally Cyclostomes, whose anatomical structure indicates they derive from creatures much more ancient than proper fish. The activity of the muscle LDH from *Eptatretus okinoseanus*, a hagfish living at 1000 m depths, proved much more pressure resistant than that from *E. burgeri*, living at less than 60 m. Its amino acid sequence differs from that of E. burger in three positions, 10, 20 and 269. The first two are probably involved in forming the tetramer and position 269 is in the active site of the enzyme

**Table 6.2** The pressure which half inactivates (depolymerises) LDH from fish which occupy different depths

| Species | Depth range in the region sampled | P ½ MPa |
|---|---|---|
| *Nezumia bairdii* | 260–1965 m | 84 |
| *Coryphaenoides rupestris* | 550–1960 | 56 |
| *C. acrolepis* | 475–2825 | 77 |
| *C. carapinus* | 1250–2740 | 126 |
| *C. armatus* | 1885–4815 | 171 |
| *C. leptolepis* | 2288–4639 | 157 |

Based on Hennessey and Siebenaller (1985)

and is probably involved in catalysis, (Nishiguchi et al. 2008). This is another example of 10 MPa apparently influencing the evolution of a protein. Anyone who can work with hagfish, which are ugly ectoparasites, has the admiration of the author. They resemble large eels but lack a proper mouth and eyes, which gives them a revolting appearance. They are commonly collected in deep-sea traps intended for amphipods. When retrieved from several thousand metres depths and put in a large circular tank of seawater on deck, they swim round and round for days, seemingly unaffected by decompression. Their lack of a proper brain probably plays a part in their tolerance to decompression. Their capacity to secrete copious quantities of mucus reinforces their unappealing presence.

As part of a programme to produce enzymes for the food industry and medical technology, Japanese workers, Kobayashi and colleagues (2012, 2018), have extracted enzymes from hadal amphipods. Three trench sites provided the material; The Challenger Deep, (in the Mariana Trench), The Japan Trench and the Izi-Ogasawara trench. Huge numbers of *Hirondellea gigas*, Chapter 1, Fig. 1. 2 (Lan et al. 2016), the only Amphipod species collected, with decompression, provided material from which polysaccharide hydrolases were extracted and purified. The activity of one, named HGcel, a cellulase which hydrolyses cellulose to glucose, was significant at 35 C, atmospheric pressure but was doubled at 2 C and 100 MPa. The conversion of sawdust to glucose is an intriguing commercial prospect. As part of the same project the evolutionary relationships of the enzymes was studied and the work of Ritchie and colleagues on the relationship of the species was confirmed. All three trenches, which came into existence 17–50 million years ago and are connected by depths of 6000 m or more, which renders the population of hadal Amphipods genetically connected.

### 6.1.1.3   Visual Pigments

Another example of the intrinsic adaptation of proteins to high pressure has been demonstrated in the visual pigments of deep-sea squid (Cephalopods) and of bony fish (Teleosts). The evidence lies in the physical properties of the proteins, somewhat apart from their physiological function. The vision of these two major groups of deep-sea animals has attracted a great deal of interest. Their eyes are remarkably similar. Many of us will have learned about the vertebrate eye at school, perhaps by dissecting one obtained from the local abattoir. A drawing of the round lens suspended by small muscles in front of a semi-circular eyeball, with the light-sensitive retina lining the back, shows the essential features of an image forming eye, a camera eye. The same drawing would also illustrate the eye of a squid. This is a remarkable case of convergent evolution, because the evolution of each of these structures has been totally independent, and the embryology of each eye clearly shows that the different parts are assembled in their own way. The light-sensitive retina cells of the vertebrate eye derive from cilia bearing cells and those in the squid arise from micro villi bearing cells, rhabdomers. However, both contain stacks of discs, each formed from flattened membranous vesicle, and it is in the lipid bilayers of these membranes that the molecules of the visual pigments are positioned. They are there to capture photons, packets of light energy, as efficiently as possible. The extraordinary structural similarity of these eyes demonstrates the power of natural selection, working in harmony with optical principles, (Warrant and Locket 2004) (Box 6.1).

---

**Box 6.1 Opsins**

A molecule of visual pigment comprises two parts, a light absorbing chromophore covalently bound to a protein called opsin, and it is the latter that is our main concern here. The opsin protein is long and inserted in the membrane bilayer across which it passes seven times. The role of opsins in cells is to convey external signals to the metabolism proceeding in the interior. The system involves proteins called G-Proteins, coupled to a molecule (the opsin) in the membrane which presents a receptor surface to the exterior. The sensor responds to specific chemicals or to light, as here, and it initiates a sequence of molecular reactions, called a cascade, which leads to the effector carrying out the intended response. The term G-protein has arisen as an abbreviation for guanine-nucleotide binding protein. G-protein coupled signal transduction is a major aspect of cell biochemistry. Its extraordinary complexity has been conserved throughout evolution, which means it must work well. It is an important aspect of biochemical adaptation which Siebenaller in particular has studied in deep-sea animals, demonstrating a number of adaptations to high pressure. The case of opsins in visual pigments provides an example of high-pressure adaptation in the vision of deep-sea animals and of convergent evolution.

---

A recent study by Porter and her colleagues (2016), focussed on the amino acid sequence of the opsins in the visual pigments of Cephalopods and Teleosts which live deeper than 500 m, extending to depths of nearly 4000 m. There are two major classes of opsin molecule; the r-(rhabdomer photoreceptor) type found in Arthropods and Molluscs (including squid) and the c-(ciliary photoreceptor) type in Vertebrates. Each is bound to a chromophore which, on capturing a photon, undergoes an instant change to a mirror image structure. The process is called photoisomerisation, and it activates the opsin to bind to a G-protein, initiating the cascade reaction which ends with a change in the electrical potential across the photoreceptor cell membrane. The details are quite different in the two groups of animals, and so are the micro anatomy and subsequent neural processing to create an image.

However, returning to the opsins, the study was primarily concerned with their relationship to the pressure at which they functioned. The molecular property chosen was not enzymic but, surprisingly, the elementary property of compressibility. One reason for this was that proteins embedded in a lipid bilayer are difficult to extract for closer study without altering them, (just like deep-sea animals), and compressibility can be estimated theoretically. In turn compressibility is likely to influence function, so the hypothesis tested was that high-pressure-adapted animals will have less compressible opsins than those in animals living at lower pressures.

The compressibility of a protein can be predicted from its sequence of amino acids and twelve of their physical properties, such as molecular weight, partial specific volume, hydrophobicity and so on, (Gromiha and Ponnuswamy 1993). The compressibility of the opsins was thus obtained from their amino acid sequences. This was done for 128 species of teleost and plotted against the depth of capture. The decrease in the compressibility of the opsins with the increase in depth was clear and statistically significant. Similarly, the compressibility of the opsins from 65 species of Cephalopods decreased with depth, consistent with the hypothesis. Separately the different sequences of amino acids were subject to phylogenetic analysis, producing the conclusion that selection for amino acids that confer reduced compressibility on the opsin was indeed significant. I am curious that compressibility in an aqueous environment can apply to the state of the opsin in a lipid bilayer. However, it is difficult to avoid the conclusion that the opsins in these two groups of animals have independently evolved to be less compressible the greater the pressure in which they function. It is equally difficult to avoid concluding that this is in some way related to their function.

The visual pigments of deep-sea animals absorb light particularly well in the wavelengths corresponding to the colour of the light which penetrates the depths, i.e. wavelengths of 460–480 nm, or to the colour of bioluminescent light, which is variable but in many cases also blue. This adaptation has been known for many years but in a paper published in 2006 Partridge and colleagues (2006), pointed out the absorption measurements had all been made at atmospheric pressure. They extracted a number of visual pigments from deep-sea fish, including *C. armatus*, and found that pressures up to 45 MPa had very little effect on the wavelength of maximum absorbance. So the match between the wavelength of the ambient light and the visual pigments' absorption is now better established. They also pointed out that high pressure might also have other effects on the molecular processes involved in vision, and we should add that the neurological components involved in processing vision, like other aspects of brain function, must also have undergone adaptation to high pressure.

## 6.2     Extrinsic Adaptation of Proteins to High Pressure

The difference between the intrinsic adaptation of proteins to high pressure and extrinsic adaptation is fundamental. As we have seen, the former involves the primary sequence of amino acids and is thus coded in the DNA whereas the latter secondarily adapts or protects the protein. The two mechanisms can work together. An enzyme might be intrinsically but partially adapted and extrinsically further adapted to function at the required pressure.

In extrinsic adaptation, proteins are protected from the effects of high pressure by changes in the structure of water in their immediate vicinity. The changes are brought about by the presence of inert molecules called compatible solutes which impose what chemists call a solvent effect. Its specific role in adapting the proteins of deep-sea organisms to pressure has been investigated by biochemists, notably by Yancey, Somero and their collaborators.

## 6.2.1   Pressure Protection

The idea that proteins can be protected from high pressure by the solvent immediately surrounding them seems pretty extraordinary. However, in the historical development of high-pressure biology we can, with hindsight, see evidence of "pressure protection" which might have provided the hint of a model for pressure adaptation.

One example is the way heavy water was shown to protect cytoplasmic structures from the effects of high pressure. Heavy water looks like water but its hydrogen is replaced by its isotope, deuterium, creating the molecule $D_2O$. It has a number of distinctive properties; a freezing point of 3.82 °C and a boiling point of 101.42 °C, both of which are consistent with the D bonds being stronger than hydrogen bonds. Cells equilibrated with $D_2O$ have microtubular structures such as the mitotic apparatus or the axopods of Actinosphaerium, which are much more resistant to high pressure than the normal structures. Cell cleavage is also more resistant to pressure in the presence of $D_2O$, (Marsland 1965). The strong D bonds diminish the depolymerising effect of pressure and perhaps, being less electrostrictive than $H_2O$, smaller volume changes occur in polymer–monomer equilibria. More recently $D_2O$ was shown to protect yeast cells from the damage inflicted by a one hour exposure to 150 MPa, (Komatsu et al. 1991). The polyhydric alcohol glycerol also counteracts the effect of high pressure on the mitotic apparatus.

Another example of how macromolecules can be made less susceptible to high pressure involves salt concentration. Recall the pressure depolymerisation of poly Valyl RNase, (Chapter 2). The hydrophobic interactions between the valyl groups cause a gain in solvent entropy and a volume change as aggregation proceeds. The rate of aggregation is reduced by a moderate pressure, but the salt concentration present in the reaction exerts an effect too. A low salt concentration enhances the pressure effect and a high concentration diminishes it. The presence of salts reduces the electrostriction of the water molecules around the valyl groups such that the $\Delta V$ of valyl-valyl interactions is reduced. Hence the effect of pressure is also reduced. In other words, a high salt concentration partially protects poly valyl RNase from pressure dissociation.

Yet another case of "pressure protection" is seen in some experiments carried out with high pressures of inert gases. These experiments are explained in Chapter 11, so here a brief introductory comment will suffice. The inert gases are notable in chemistry as a group in the Periodic Table which, in a biological context, are chemically unreactive. The least reactive of all is helium, followed by hydrogen (surprisingly), neon, argon, krypton, xenon and nitrogen. They all have weak anaesthetic properties, (Chapter 11). That is, they require a significant pressure (called a partial pressure) to drive a sufficient number of molecules into solution where they can perturb physiological target molecules, which in turn affect neurons and ultimately consciousness. No chemical reactions are involved. Familiar clinical gaseous anaesthetics are equally inert chemically but vastly more potent, requiring just a whiff to do their job.

Helium was suspected of having its weak "anaesthetic" effect masked by the hydrostatic pressure component of its partial pressure. So, the experiments relevant here consisted of comparing the effects of a high pressure of helium with those brought about by the same hydrostatic pressure. The experiment used cells whose susceptibility to hydrostatic pressure was well known, the cell division of *Tetrahymena*. The experiments were carried out by the author as a visitor in Brauer's laboratory in North Carolina. A microscope pressure chamber allowed the division of a small number of cells to be observed under either purely hydrostatic pressure (applied to the cells in their culture medium by liquid, medicinal paraffin) or helium pressure (applied as a gas with the cells retained in their tiny drop of culture medium adhering to the pressure window). Contrary to expectations the optics worked out well and it was possible to observe the dividing cells whilst changing smoothly from high hydrostatic pressure to the same helium pressure, and vice versa. The effect of a high pressure of hydrogen was also measured, with an initial 1 MPa of helium present to avoid creating an explosive mixture. (The risk was minimal because of the small volumes involved). The results were clear. 25 MPa hydrostatic pressure completely blocked cell division whereas the same pressure of helium permitted a low rate of division. Hydrogen (24 MPa plus 1 MPa helium) enabled division to proceed more rapidly than the same pressure of helium, (Macdonald 1975). These inert gases thus counteract the effects of hydrostatic pressure in dividing cells, (as they do in anaesthetised animals, see Chapter 11). At the time the preferred interpretation was that the gases weakly interacted (bound) at molecular sites in such a way as to counteract the pressure effect, but it is conceivable that their presence, as inert, apolar molecules, affects the structure of water close to the sites primarily affected by hydrostatic pressure—a solvent effect, of which more soon.

Should these examples of "pressure protection" have alerted us to the possibility that Nature might do something similar in adapting organisms to deep-sea pressures? Perhaps somewhere in the scientific literature that very point was made; it would be an interesting piece of historical research to establish the antecedents of pressure protection.

## 6.2.2   Compatible Solutes

To explain how compatible solutes achieve their solvent effect and extrinsically adapt proteins to high pressure, we need to begin with the seemingly unrelated topic of osmoregulation in marine organisms.

Seawater is a concentrated solution of salts, mainly sodium chloride, which generates a significant osmotic pressure. Its concentration, expressed in terms related to its osmotic pressure, is expressed as osmolarity, using units called Osmoles. (The osmolarity of seawater is 1100 Osmol per Kg). The body fluids of marine animals contain salts and organic compounds, called osmolytes, as they generate an osmotic pressure. The interesting point is how the two, seawater and body fluids, match. The hagfish, mentioned earlier, which, in evolutionary terms, is a much more primitive than fish, has a body fluid whose osmotic pressure is similar to that of seawater. Hagfish are osmoconformers, as are many marine invertebrates. The osmotic pressure of their body fluids matches that of seawater so they have no osmoregulation to worry about. However, many marine fish have body fluids which are a more dilute salt solution than seawater, a fascinating evolutionary legacy, which we cannot pursue here. The consequence is that there is a net osmotic flux across the semi-permeable membranes which separate the two solutions. Water moves from the fish' body fluids to the seawater, thus tending to dehydrate the animal. The cells bathed in the body fluids of the fish also respond osmotically to their aqueous environment. To minimise the osmotic difference between the external seawater and the animals' body fluids, and between the body fluids and the cells, many animals have evolved body and cellular fluids which contain special solutes, called compatible solutes, which contribute osmotic pressure without affecting anything else, (although they may have additional beneficial effects, as antioxidants and contributing to buoyancy). These inert osmolytes simply contribute to the osmolarity of the solution. Typically, they are organic molecules with no electrical charge such as glycine (an amino acid), glycerol (a polyhydric alcohol), inositol (a cyclic polyhydric alcohol), trehalose (a sugar) and betaine (a methylamine). Crucially these osmolytes do not affect the proteins with which they come in contact. The fish and other organisms with body or cellular fluids which, although containing some osmolytes, are still not in osmotic equilibrium with seawater, osmotically regulate to avoid dehydration. Typical bony fish, for example offset their osmotic loss of water by osmosis by drinking seawater and excreting salts.

In this context, one particular group of marine fish presented a puzzle because one of their osmolytes was known to be anything but an inert molecule. The anomalous osmolyte is urea, a substance most animals excrete, as we do in our urine. Urea denatures proteins by upsetting the hydration of the protein and hence its folded structure—a drastic example of an adverse solvent effect. The fish which retain urea include sharks, i.e. Elasmobranchs in the Class Chondrichthyes, the Cyclostomes (e.g. hagfish) and the Class Choanichthyes (*Crossopterygii*), e.g. the "living fossil" *Coelacanth*. They all have a significant concentration of urea in their body and cellular fluids. Generations of zoology students have learned this fact and have certainly been puzzled by it (as I was) and moved on, or perhaps converted to biochemistry to study the puzzle more closely.

In 1979 and 1980 Yancey and Somero published seminal papers, pointing out that the high urea concentration was accompanied by other osmolytes which counteracted urea's tendency to destabilise the structure (denature) proteins. The most effective of these is trimethylamine N-oxide, abbreviated to TMAO, but there are other, less effective stabilisers, such as betaine.

TMAO

Betaine

Numerous experiments demonstrate this "protective" effect, (Bennion and Daggett 2004). For example the enzyme lactate dehydrogenase from the white shark *Carcharodon carcharias* was shown to be denatured by urea and then re-natured by TMAO. The animal's urea rich body fluids are normally accompanied by a concentration of TMAO in a ratio of 2 (Urea): 1 (TMAO), which experiments show is the ratio that nicely balances the denaturing and stabilising effects, and which, of course, contributes to the osmolarity to match that of seawater. This extraordinary osmoregulatory arrangement presumably evolved hundreds of million years ago, and it clearly works well.

Hatori and colleagues (2014), demonstrated a nice example of TMAO counteracting the denaturing effect of urea in the polymerisation of actin, extracted not from an elasmobranch but from a rabbit. This makes the point that we are here dealing with a general chemical phenomenon and not just a specialised case of adaptive biochemistry. Quite simply, urea slows the rate of actin polymerisation whilst TMAO restores it. The ratio of 2 (urea): 1 (TMAO) did so most effectively, by balancing their separate effects.

Animals like us, with kidneys that carry out the excretion of urea in solution, (urine), have the interesting problem of protecting certain kidney cells that are exposed to a significant concentration of urea. Mammalian kidney cells in culture grow poorly in the presence of urea but best in the presence of the compatible solute betaine, which counteracts the adverse effect of urea.

Further research revealed that TMAO is present in marine fish in general, but in deep-sea fish it is present at high concentration, and one which is related to the depth at which the fish live, (Samerotte et al. 2007; Bockus and Seibel 2016). Experiments using enzymes extracted from deep-sea fish show that TMAO also counteracts the adverse effects of high pressure. One of the early examples of this, from Yancey's laboratory, featured the LDH from the white muscle of the abyssal *Coryphaenoides leptolepis*. The Km for the enzyme's cofactor, NADH, was determined over a range of pressures and found to be increased by 30% at 30 MPa, clearly indicating its binding to the enzyme was weakened by pressure. In the presence of 250 mM TMAO the same pressure had no effect on the Km, clearly an adaptation to pressure, (Gillett et al. 1997). The term piezolyte is given to osmolytes with this property. The mechanism for this pressure protection will be discussed later, but for the present, the account concentrates on the occurrence of TMAO in deep-sea organisms.

To digress briefly, recall the disappointing experiments with blood cells taken from deep-sea fish and subjected to high pressure, (Chapter 5, Section 5.4.3). At the time the solution in which the cells were suspended was thought to be unsuitable in some way. Perhaps it should have contained TMAO.

## 6.2.3   TMAO in Deep-Sea Animals

The occurrence of TMAO in the cellular and body fluids of bony fish presumably initially evolved simply as an osmolyte, but it was available for selection as a protein stabiliser, enabling certain species to evolve into greater depths. The concentration of TMAO in the muscle fluids of many teleosts increases with the pressure at which they live. The most extreme case is the snail fish (*Notoliparis kermadecensis*), retrieved from a depth of 7000 m in the Kermadec Trench, whose muscles proved to have a TMAO concentration of 981 Osmol per Kg. This is close to the osmolarity of seawater, 1100 Osmol per Kg. The collection of the 5 specimens used to establish this point was a triumph of perseverance and well-deserved luck for Yancey and his colleagues (2014). The data imply that at a depth of 8300 m +/− 100 m the extrapolated concentration of TMAO would be isosmotic with seawater. This indicates that there is a depth (pressure) limit to the stabilisation which TMAO can provide without creating new osmoregulatory problems. Significantly teleosts have been found no deeper than 8370 m and yet greater depths exist for them to occupy, (Priede and Froese 2013).

Generally shallow water Elasmobranchs and similar osmoconformers have a significant concentration of urea matched with TMAO in the ratio of 2:1 but this changes in deep water species. They have an increased concentration of TMAO which is accompanied by a reduced concentration of urea. The depth range of Elasmobranchs is surprisingly limited. They are scarce at 3000 m depths and probably absent from depths of 4000 m. Treberg and Speers-Roesch (2016) have set out the possible reasons for this without reaching a firm conclusion but Musick and Cotton (2015), argue the case for energy (food) playing an important role. However, there is the possibility that the urea: TMAO mix might create other stresses at significant pressures, (Withers et al. 1994). Similarly, Yancey and colleagues had previously suggested that decreasing the concentration of urea and increasing that of TMAO may be limiting in some way.

The vertical distribution of teleosts is much greater than that of Elasmobranchs, 8000 m as we have seen. And it has been pointed out that the concentration of TMAO expected to adapt proteins to pressure at a depth of around 8000 m would create novel osmoregulatory problems. Therefore, this might set a depth limit for fish whose proteins are extrinsically adapted to pressure by TMAO. Which prompts the question, why have fish, including the Elasmobranchs and others, not (yet) evolved intrinsically adapted proteins to extend their respective depth ranges? After all, a considerable volume of the ocean is available below 8000 m, which is occupied by invertebrates. Numerous other interacting factors may limit the depth range of teleosts, (Yancey et al. 2014). They include metabolism and the energy costs of swimming to feed and avoid predators, growth and fecundity in a food scarce environment, all discussed by Laxon and colleagues and by Priede and Froese (2013).

Invertebrates that live in the depths have also evolved extrinsic adaptations to counteract the effects of pressure on their proteins, (Kelly and Yancey 1999; Siebenaller 2010). Examples which have special osmolytes include shrimps living at 2900 m in which TMAO is the dominant osmolyte whereas their shallow water counterparts have only the osmolyte glycine. Abyssal crabs, squids, clams and an anemone have been shown to have concentrations of TMAO which is much higher than that of their equivalents in shallow water. Some deep-sea echinoderms, molluscs and polychaetes appear to lack TMAO but instead have other osmolytes that are known to be compatible solutes.

Recall the amphipods living in Lake Baikal which contained a significant concentration of TMAO, (Chapter 5, Section 5.1.2). It occurs in the muscle tissue and those amphipods living at 1500 m depths had a much higher concentration than the shallow water amphipods, (Zerbst–Boroffika et al. 2005). We know that 10 MPa is sufficient to elicit pressure adaptation of proteins in marine fish and 15 MPa is sufficient to drive the natural selection of TMAO in the Lake Baikal amphipods. At the other end of the depth-pressure range, hadal amphipods, *Hirondellea sp.*, *Lysianassidae sp.* and *Alicella sp.*, collected from the Kermadec, Mariana and other Pacific trenches, were found to contain TMAO and other putative compatible solutes (Wallace et al. 2014). Their details have yet to be published.

The proteins in the hydrothermal vent fauna and flora have similar high-pressure adaptations, (Chapter 9).

### 6.2.4   Some Implications of the Extrinsic Adaptation of Proteins to High Pressure

Readers may be wondering about the origin of the diverse osmolytes which are pressure protective. They were, and are, involved in various metabolic pathways and were thus available for opportunist selection.

Presumably the intracellular concentration of TMAO and other protective osmolytes has to be regulated to achieve the required pressure protection. Does the protection extend to many proteins? If so, it presumably greatly simplifies the process of co-evolving other proteins and macromolecules to high pressure. How specific might the protection be? Does it lead to a narrow band of pressure in which a particular protein functions well, or does it protect over a wide range of pressure? Can TMAO be "switched on" during the life of an animal, for example during the migration or in laboratory acclimation? And in the long term does extrinsic adaptation preclude natural selection for intrinsic adaptation? (Somero 2003). And finally, as the phenomenon of extrinsic adaptation has been described as an aspect of the solvent effect, how universal is it? This last point will be taken up after considering the mechanism of protection.

### 6.3   How Does TMAO Counteract the Effects of High Pressure on Proteins?

The pressure protection of proteins by TMAO and other osmolytes is a special case of a general and fundamental property of aqueous solutions, (Lin and Tmasheff 1994). The "salting out" of proteins, which has occupied chemists for a hundred years, was the first of several phenomena to be discovered which demonstrated that even a simple aqueous solution has unexpected properties. Salting out described the precipitation of proteins following the addition of certain salts to a solution. We can imagine the hydration of the salts denies water to dissolve the proteins. This excluded volume effect is apparent in the cellular environment, which can be very crowded with proteins, smaller molecules and hydrated membranous surfaces, all reducing the availability of water to act as a solvent. A high proportion of the water nominally present is unavailable as bulk water and in recent years biochemists have become concerned about the significance of this. Generally, the crowding effect favours the compact state in a folded–unfolded protein equilibrium, (Gao et al. 2017; Golub et al. 2018). Experiments in the proverbial test tube, weakly mimicking the cell interior, have to be interpreted with caution. A value for a $\Delta V$ of for protein unfolding, for example may be accurate in the experimental conditions, but quantitatively different in the cell.

So how should we regard the way TMAO counteracts the unfolding induced by urea and how is that property related to the pressure protection it confers on proteins? Urea causes proteins to unfold by acting rather like a super solvent. It interacts with the peptide chain and amino acid side chains intimately, maximising their "solvent accessible surface area." It has little effect on water structure per se. The way TMAO opposes this is somewhat mysterious, a work in progress, (Sarma and Paul 2013; Ma et al. 2014). TMAO is amphipathic, with hydrophobic methyl groups and a highly polar N-O group, which exert a marked effect on solvent water. Its pressure protection has been described as "nanocrowding." Recall the optimum ratio of Urea: TMAO for neutralising the unfolding effect of urea is 2:1. TMAO forms strong hydrogen bonds with water, decreasing the hydrogen bonding between urea and protein. The effect of urea is thus effectively removed. This is interesting and important to know because it explains how TMAO evolved in Chondrichthyes and was thus available for selection to adapt their proteins to high pressure. Independently TMAO was selected in Teleosts for the same purpose as it was available, part of the biochemical inventory. All organisms seem to have osmolytes available from their metabolic resources.

There is a consensus that TMAO does not bind directly with proteins in any way which might obviously stabilise them, but the possibility of some form of interaction cannot be excluded, (Canchi et al. 2012). Experiments and molecular simulations have produced a confusing plethora of ideas. The following seem to have a consistent thread. The strong hydrogen bonds which TMAO forms with water, mentioned above, are outside the protein's hydration layer and seem to be reinforced by pressure. This general increase in the hydrogen bonding of water structure opposes the penetration of water into voids within the protein structure, a major source of the volume decrease which occurs during unfolding at high pressure. TMAO thus opposes unfolding.

It is not at all clear to what extent these ideas of how TMAO opposes protein unfolding apply to other protective osmolytes.

The general, some might say universal, nature of the pressure protection conferred by TMAO and other osmolytes is illustrated by the following examples.

The enzyme trypsin cleaves the amino acid subunits from proteins and thus inactivates enzymes. For example LDH extracted from *Sebastolobus altivelis*, subjected to a standard exposure to trypsin, undergoes a reduction in the activity of 13%. The presence of TMAO reduced this to only 7%. A standard exposure to 100 MPa reduced the enzyme's activity by 9% and when combined with trypsin, by 32%. Pressure opens the enzyme's structure, exposing more sites where trypsin can act. This latter effect is halved in the presence of TMAO. Similar experiments using LDH extracted from the cow, Bos taurus, demonstrated the same phenomenon, (Yancey and Siebenaller 1999).

In food technology, high pressure is used to inactivate microorganisms. Smiddy et al. (2004), showed that the survival of *Listeria innocua* after exposure to 400 MPa is increased when pressure protecting compatible solutes such as betaine are present.

TMAO affects at least one type of ion channel (Chapter 4). Petrov and colleagues showed that pressures of up to 80 MPa reduced the opening probability of the bacterial mechanosensitive channel. This effect was counteracted, up to 40 MPa, by TMAO present on the cytoplasmic side of the channel. At atmospheric pressure TMAO was without effect. Curiously TMAO was also without effect at pressures above 40 MPa. The channel protein protrudes well clear of the supporting membrane through which it passes, so it is perhaps not surprising that TMAO exerts its characteristic effect, but this result has important implications. Are channels in general adapted to high pressure by TMAO and other compatible solutes? Can a nervous system be protected from disruption by TMAO?

Another membrane-bound protein, the mitochondrial enzyme FoF1-ATPase, (Chapter 2, Section 2.6.3), extracted from the heart muscle of the cow, is also susceptible to high pressure, (Saad-Nehme et al. 2001). When the enzyme, present in the form of sub-mitochondrial particles, was incubated at pressures of between 50 and 200 MPa for a short time its activity was found to be decreased after decompression. When pressurised in the presence of the compatible solutes sucrose, sorbitol, methylamine, betaine or glycerol, the loss of activity was much reduced. It would appear that pressure reduced the activity of the enzyme by dissociating it to subunits and that the compatible solutes prevented the dissociation.

Another example comes from the work of Winter and colleagues. The tertiary structure of a small monomeric protein, abbreviated SNase, from the bacterium *Staphylococcus aureus*, was examined with small angle X-ray scattering at very high pressures, up to several hundred MPa. The effects of pressure, glycerol, TMAO, other osmolytes and of urea on the enzyme's structure were clear. TMAO and glycerol each counteracted the effect of pressure, which, in their absence, unfolded the protein, increasing its radius of gyration. There was also evidence that TMAO counteracted the denaturation of the SNase caused by urea, (Krywka et al. 2008).

The universality of the solvent effect, if it was in doubt, is dramatically illustrated in the treatment being developed for the immensely important phenomenon of the misfolding of freshly synthesised proteins in human cells, (Jaworek et al. 2018). Some misfolded proteins escape the mechanism meant to capture them and so survive to create serious diseases. These diseases are not caused by infectious agents such as viruses but by errant proteins, which may initiate pathological processes, as in Alzheimer's disease and glaucoma, or they can cause a disease by blocking essential reactions, as in cystic fibrosis. In many cases, the misfolding of the protein arises from a mutation. There are established lines of research attempting to cure some of these conditions by treating the misfolded proteins. An example is the use of TMAO in isolated cells to "cure" the misfolded proteins responsible for a particular form of glaucoma, (Lia et al. 2009). The TMAO solution provides the microenvironment favourable for the protein to re-acquire its native, correctly folded, state. These experiments are complicated and clearly well outside the scope of this book, but they illustrate rather well the unity of Nature. Who would have thought that the evolution of osmolytes in osmoregulation, the adaptation of proteins to high pressure in deep-sea animals and the possible cure of some rare diseases in humans, had anything in common?

## 6.4 Conclusions

Some structural proteins and enzymes extracted from deep-sea animals have properties consistent with their functioning at a high pressure, similar to that from which they came. These molecules are described as adapted, intrinsically or extrinsically. This really means their properties are consistent with the conditions in which they function and different to those of homologous proteins that function in other conditions.

## References[1]

Barton KN, Buhr MM, Ballantyne JS (1999) Effects of urea and trimethylamine – N-oxide on fluidity of liposomes and membranes of an elasmobranch. Am J Phys 276:R397–R406. *

Bennion BJ, Daggett V (2004) Counteraction of urea-induced protein denaturation by trimethylamine N oxide: a chemical chaperone at atomic resolution. Proc Natl Acad Sci USA 101:6433–6438

Bennion BJ, DeMarco ML, Daggett V (2004) Preventing misfolding of the prion protein by trimethylamine N-oxide. Biochemist 43:12955–12963. *

Bockus A, Seibel BA (2016) Trimethylamine oxide accumulation as a function of depth in Hawaiian mid-water fishes. Deep Sea Res I 112:37–44

Brindley AA et al (2008) Enzyme sequence and its relationship to hyperbaric stability of artificial and natural fish lactate dehydrogenases. PLoS One 3:e2042

Canchi DR et al (2012) Molecular mechanism for the preferential exclusion of TMAO from protein surfaces. J Phys Chem B 116:12095–12104

Dahlhoff E, Somero GN (1991) Pressure and temperature adaptation of cytosolic malate dehydrogenase of shallow and deep-living marine invertebrates: evidence for high body temperatures in hydrothermal vent animals. J Exp Biol 159:473–487

Drazen JC, Dugan B, Friedman JR (2013) Red muscle proportions and enzyme activities in deep sea demersal fishes. J Fish Biol 83:1592–1612. *

Gaither MR et al (2016) Depth as a driver of evolution in the deep sea: insights from grenadiers (Gadiformes:Macrouridae) of the genus Coryphaenoides. Molec. Phylogenetics Evolut 104:73–81

---

[1]Those marked * are further reading.

Gao M et al (2017) Crowders and cosolvents – major contributors to the cellular milieu and efficient means to counteract environmental stress. ChemPhysChem 18:2951–2972

Gerringer ME, Drazen JC, Yancy PH (2017) Metabolic enzyme activities of abyssal and hadal fishes: pressure effects and a re-evaluation of depth related changes. Deep Sea Res I. https://doi.org/10.1016/jdsr.2017.05.010. *

Gillett MB, Suko JR, Santoso FO, Yancey PH (1997) Elevated levels of trimethylamine oxide in muscles of deep sea gadiform teleosts: a high pressure adaptation? J Exp Zool 279:386–391

Giulio M (2005) A comparison of proteins from Pyrococcus furiosus and Pyrococcus abyssi: barophily in the physicochemical properties of amino acids and in the genetic code. Gene 346:1–6

Golub M et al (2018) The effect of crowding on protein stability, rigidity, and high pressure sensitivity in whole cells. Langmuir 34:10419–10425

Gromiha M, Ponnuswamy P (1993) Relationship between amino acid properties and protein compressibility. J Theoret Biol 165:87–100

Hatori K, Iwasaki T, Wada R (2014) Effects of urea and trimethylamine N-oxide on the binding between actin molecules. Biophys Chem 193–194:20–26

Hennessey JP, Siebenaller JF (1985) Pressure inactivation of tetrameric lactate dehydrogenase homologues of confamilial deep living fishes. J Comp Physiol B 155:647–652

Herman L, Dreizen P (1971) Electron microscopic studies of skeletal and cardiac muscle of a benthic fish. 1. Myofibrillar structure in resting and contracted muscle. Am Zool 11:545–557

Hochachka PW (1971) The 1971 Alpha helix expedition to the Galapagos Archipelago: introduction and acknowledgments. Am Zool 11:401–403. *

Hochachka PN, Somero GN (2002) Biochemical adaptation. Mechanism and process in physiological evolution. Oxford University Press, New York. *

Jaworek MW, Schuabb V, Winter R (2018) Pressure and cosolvent modulation of the catalytic activity of amyloid fibrils. Chem Commun 54:5696–5699

Kelly RH, Yancey PH (1999) High contents of trimethylamine oxide correlating with depth in deep sea teleost fishes, skates, and Decapod crustcaeans. Biol Bull 196:18–25

Klink BU, Winter R, Engelhard M, Chizhov I (2002) Pressure dependence of the photocycle kinetics of bacteriorhodopsin. Biophys J 83:3490–3498. *

Kobayashi H et al (2018) Polysaccharide hydrolase of the hadal zone amphipods Hirondellea gigas. Biosci Biotechnol Biochem 82:1123–1133

Kobayashi H, Hatada Y, Tsubouchi T, Nagashama T, Takami H (2012) The hadal amphipod Hirondellea gigas, possessing a unique cellulase for digesting wooden debris buried in the deepest seafloor. PLoS One 7:e42727

Komatsu Y et al (1991) Deuterium oxide, dimethylsulfoxide and heat shock confer protection against hydrostatic pressure damage in yeast. Biochem Biophys Res Commun 174:1141–1147

Krywka C, Sternemann C, Paulus M, Tolan M, Royer C, Winter R (2008) Effect of osmolytes on pressure-induced unfolding of proteins: a high pressure SAXS study. ChemPhysChem 9:2809–2815

Lan Y et al (2016) The deepest mitochondrial genome sequenced from Mariana Trench Hirondellea gigas (Amphipoda). Mitochondrial DNA B 1:802–803

Laxson CJ, Condon NE, Drazen JC, Yancey PH (2011) Decreasing urea: Trimethylamine N-Oxide ratios with depth in Chondrichthyes: a physiological depth limit? Physiol Biochem Zool 84:494–505. *

Lia L-Y et al (2009) Correction of the disease phenotype of myocilin-causing glaucoma by a natural osmolyte. Invest, Opthalm & Visual Sci 50:3743–3749

Lin T-Y, Timasheff SN (1994) Why do some organisms use urea-methylamine mixture as osmolyte? Thermodynamic compensation of urea and trimethylamine N-oxide interactions with protein. Biochemist 33:12695–12701

Ma J, Pazos IP, Gai F (2014) Microscopic insightsinto the protein stabilizing effect of trimethylamine N oxide (TMAO). Proc Natl Acad Sci U S A 111:8476–8481

Macdonald AG (1975) The effects of helium and of hydrogen at high pressure on the cell division of Tetrahymena pyriformis. J Cell Physiol 85:511–528

Marsland D (1965) Partial reversal of the anti-mitotic effects of heavy water by hydrostatic pressure. Exp Cell Res 38:592–603

Morita T (1999) Molecular phylogenetic relationships of the deep sea fish genus Corpheanoides (Gadiformes: Macrouridae) based on mitochondrial DNA. Mol Phylogenet Evol 13:447–454

Morita T (2003) Structure based analysis of high pressure adaptation of a-Actin. J Biol Chem 278:28060–28066

Morita T (2010) High pressure adaptation of muscle proteins from deep sea fishes, *Coryphaenoides yaquinae* and *C. armatus*. Annal NY Acad Sci 1189:91–94

Musick JA, Cotton CF (2015) Bathymetric limits of chondrichthyans in the deep sea: a re-evaluation. Deep Sea Res II 115:73–80

Mustafa T, Moon TW, Hochachka PW (1971) Effects of pressure and temperature on the catalytic and regulatory properties of muscle pyruvate kinase from an off-shore benthic fish. Am Zool 11:451–466

Nath A, Subbiah K (2016) Insights into the molecular basis of piezophilic adaptation: extraction of piezophilic signature. J Theoretical Biol 390:117–126

Nishiguchi Y, Miwa T, Abe F (2008) Pressure adaptive differences in lactate dehydrogenases of three hagfishes: Eptatretus burger, Paramyxine atami and Eptatretus okinaseanus. Extremophiles 12:477–480

Papini CM, Pandharipande PP, Royer CA, Makhatadze GI (2017) Putting the piezolyte hypothesis under pressure. Biophys J 113:974–977. *

Partridge JC, White EM, Douglas RH (2006) The effect of elevated hydrostatic pressure on the spectral absorption of deep sea fish visual pigments. J Exp Biol 209:314–319

Petrov E, Rhode PR, Cornell B, Martinac B (2012) The protective effect of osmoprotectant TMAO on bacterial mechanosensitive channels of small conductance MscS/MscK under high hydrostatic pressure. Channels 6:262–271. *

Porter ML, Roberts NW, Partridge JC (2016) Evolution under pressure and the adaptation of visual pigment compressibility in deep sea environments. Mol Phylogenet Evol 105:160–165

Priede IG, Froese R (2013) Colonization of the deep sea by fishes. J Fish Biol 83:1528–1550

Ritchie H, Jamieson AJ, Piertney SB (2015) Phylogenetic relationships among hadal amphipods of the superfamily Lysianassoidea: implications for taxonomy. Deep Sea Res I 105:119–131. *

Saad-Nehme J, Silva JL, Meyer-Fernandes JR (2001) Osmolytes protect mitochondrial FoF1 – ATPase complex against pressure inactivation. Biochim Biophys Acta 1546:164–170

Samerotte AL, Drazen JC, Brand GL, Seibel BA, Yancey PH (2007) Correlation of trimethylamine oxide and habitat depth within and among species of teleost fish: an analysis of causation. Physiol Zool 80:197–208

Sarma R, Paul S (2013) Crucial importance of water structure modification on trimethylamine N-oxide counteracting effect at high pressure. J Phys Chem B 117:677–689

Siebenaller JF (1984) Structural comparison of lactate dehydrogenase homologs differing in sensitivity to hydrostatic pressure. Biochim. Biophys Acta 786:161–169

Siebenaller JF (2010) Effects of the deep sea environment on invertebrates. In: Sebert P (ed) Comparative high pressure biology. Science Publishers

Siebenaller J, Somero GN (1978) Pressure adaptive differences in lactate dehydrogenases of congeneric fishes living at different depths. Science 201:255–257

Smiddy M, Sleator RD, Patterson MF, Hill C, Kelly AL (2004) Role for compatible solutes glycine betaine and L-carnitine in Listerial barotolerance. Appl Environ Microbiol 70:7555–7557

Somero GN (2003) Protein adaptation to temperature and pressure: complementary roles of adaptive changes in amino acid sequence and internal milieu. Comp Biochem Physiol B 136:577–591

Somero GN, Suarez RK (2005) Peter Hochachka: adventures in biochemical adaptation. Annu Rev Physiol 67:25–37

Swezey RR, Somero GN (1982) Polymerization thermodynamics and structural stabilities of skeletal muscle actins from vertebrates adapted to different temperatures and hydrostatic pressures. Biochemistry 21:4496–4503

Swezey RR, Somero GN (1985) Pressure effects on actin self assembly: interspecific differences in the equilibrium and kinetics of the G to F transformation. Biochemistry 24:852–860. *

Treberg JR, Speers-Roesch B (2016) Does the physiology of chondrichthyan fishes constrain their distribution in the deep sea? J Exp Biol 219:615–625

Wakai N, Takemura K, Morita T, Kiao A (2014) Mechanism of deep sea fish a-Actin pressure tolerance investigated by molecular dynamics simulation. PLoS One 9:e85852

Wallace GT, Jamieson AJ, Bartlett DH, Cameron J, Gotz MT, Yancey PH (2014) Depth adaptation in hadal crustaceans: potential piezolytes increase with depth in the tissues of marine amphipods. Integ, Comp Biology 54:P3.101 abstract

Warrant EJ, Locket NA (2004) Vision in the deep sea. Biol Rev 79:671–712

Watanabe TM et al (2013) Glycine insertion makes yellow fluorescent protein sensitive to hydrostatic pressure. PLoS One 8:e73212

Withers PC, Morrison G, Hefter GT, Pang T-S (1994) Role of urea and methylamines in buoyancy of elasmobranchs. J Exp Biol 188:175–189

Yancey PH (2005) Organic osmolytes as compatible, metabolic and counteracting cytoprotectants in high osmolarity and other stresses. J Exp Biol 208:2819–2830. *

Yancey PH, Siebenaller JF (1999) Trimethylamine oxide stabilizes teleost and mammalian lactate dehydrogenases against inactivation by hydrostatic pressure and trypsinolysis. J Exp Biol 202:3597–3603

Yancey PH, Siebenaller JF (2015) Co-evolution of proteins and solutions: protein adaptation versus cytoprotective micromolecules and their roles in marine organisms. J Exp Biol 218:1880–1896. *

Yancey PH, Somero GN (1979) Counteraction of urea destabilization of protein structures by methylamine osmoregulatory compounds of elasmobranch fishes. Biochem J 183:317–323. *

Yancey PH, Somero FN (1980) Methylamine osmoregulatory solutes of elasmobranch fishes counteract urea inhibition of enzymes. J Exp Zool 212:205–213. *

Yancey PH et al (1982) Living with water stress: evolution of osmolyte systems. Science 217:1214–1222. *

Yancey PH, Blake WR, Conley J (2002) Unusual organic osmolytes in deep sea animals: adaptations to hydrostatic pressure and other perturbants. Comp Biochem Physiol A 133:667–676. *

Yancey PH et al (2014) Marine fish may be biochemically constrained from inhabiting the deepest ocean depths. Proc Natl Acad Sci U S A 111:4461–4465

Zerbst-Boroffika I et al (2005) TMAO and other organic osmolytes in the muscles of amphipods (Crustacea) from shallow and deep water of Lake Baikal. Comp Biochem Physiol A 142:58–64

# Molecular Adaptation to High Pressure: Membranes

**7**

Membranes, as we have seen, are multimolecular structures composed of lipids and proteins (Chapter 2). It is the lipids that contribute to the main characteristics of membranes, be they a membrane enclosing a cell or an intracellular membrane enclosing a special compartment. Both provide a selective permeability barrier and support and orientate incorporated proteins. In Chapter 2, the basic structure of Bacterial and Eukaryote membranes was considered, and the effect of high pressure on the fluidity of their type of lipid bilayer was explained. Pressure reduces their fluidity, 50 MPa having an effect equivalent to a 10 °C reduction in temperature. However, there is more to membranes under pressure than this; see Winter and Jeworrek (2009) and Ding et al. (2017).

We have now to consider membranes in all three domains of life, which means we must include the rather different Archaeal membranes. They are built of phospholipids which are based on glycerol 1-phosphate linked to isoprenoid chains. In contrast, bacterial and eukaryote membranes are made of phospholipids based on glycerol 3-phosphate (Chapter 2, Fig. 2.1). This distinction in the composition, and hence structure, of membranes is called the Lipid Divide, which fits somewhat uneasily with the trio of life domains. In the early evolution of life, the bacteria and the archaea are thought to have preceded the eukaryotes, which probably arose as a sort of chimaera of the two. Yet the Lipid Divide sets the Archaea apart from the Bacteria and the Eukaryotes. The Lipid Divide means the lipids and their synthesis differ between the two groups and the resultant membrane structures are different. Does this Lipid Divide affect the way in which the membranes have adapted to high pressure? We know that all three domains of life have adapted to high pressure, so their membranes must also be pressure-adapted. The question is, how?

© Springer Nature Switzerland AG 2021
A. Macdonald, *Life at High Pressure*, https://doi.org/10.1007/978-3-030-67587-5_7

## 7.1    Homeoviscous Adaptation

In Chapter 2, it was stated that the composition of the membrane bilayers in all organisms was regulated to provide the "correct" fluidity for proper function. This sweeping statement now needs some explanation and qualification, best provided by first discussing the membranes of animals and their adaptation, followed by bacterial membranes. Archaeal membranes will be dealt with in the context of hydrothermal vents, in Chapter 9.

It was shown by Sinensky in 1974 that *E. coli* grown in the range of 15–43 °C contained lipids whose fatty acids varied with the growth temperature. High temperatures resulted in more saturated and longer chain fatty acids. Correspondingly, cells grown at low temperature contained lipids whose fatty acids had carbon chains which were shorter and more unsaturated, the latter implying fatty acid chains with cis double bonds (Chapter 2, Fig. 2.2). This was not an entirely new observation, but the follow-up was. The extracted lipids, which can only come from the bacterial membrane, were subjected to a spectroscopic technique called Electron Spin Resonance which provides a measure of viscosity. It was found that the various lipid extracts, when measured at the temperature at which they were synthesised by the growing cells, were of the same viscosity. This is homeoviscous adaptation. Other workers, using a fluorescence technique, showed the lipid bilayers of membranes in a wide variety of organisms manifest homeoviscous adaptation. For example, Cossins (1994) and colleagues have measured the fluidity of brain myelin membranes prepared from an Antarctic fish, the common trout, the rat, a bird and the pigeon. They have very different fluidities at a common temperature but at the temperature at which membranes normally function (0, 6, 37 and 42 °C, respectively), the fluidities are much more similar. The measurements Sinensky reported were made on "dry" lipid extracts, but also on isolated membranes, and were of microviscosity. We now use the convenient term fluidity to describe bilayer lipid motion (as did Sinenksy) but are stuck with viscous in the term homeoviscous.

There are several ways in which the composition of a lipid bilayer can be changed to increase its fluidity. An increase in the proportion of cis-unsaturated fatty acyl chains is particularly widespread in offsetting the ordering effect of low temperature, because of the 40-degree kink associated with the double bond. Another fluidising effect is to reduce the cholesterol level and to introduce phospholipids with a bulky headgroup, such as phosphatidylglycerol, and reduce phosphatidylethanolamine (Cossins 1994; Ernst et al. 2016).

In some microorganisms, and rarely in animals, lipids with branched chains increase fluidity by disrupting the packing of the bilayer, offsetting the ordering effect of low temperature. Their effectiveness in fluidising bilayers is demonstrated by the behaviour of liposomes made of a single branched-chain lipid, analogous to the effect of cis double bonds. Earl (1986), a colleague in Aberdeen, synthesised branched-chain lipids with the branch at selected positions along the acyl chain. Liposomes made of pure DPPC, 2 MeHPC, 3 MeHPC, 4 MeHPC and 5 MEHPC were labelled with the fluorescent probe DPH and subjected to a temperature ramp, from below 0 °C to above 50 °C. The fluorescence anisotropy of DPH showed the characteristic abrupt change at 41 °C in the case of DPPC liposomes, defining its transition temperature. In the case of 2 MeHPC liposomes, the transition occurred at 28 °C and, in the case of 3 MeHPC, at 20 °C. However, liposomes made of 4 MeHPC had a "transition" spread over a temperature range of over 10 °C but centred on 12 °C, whilst that of 5 MeHPC liposomes was even more spread out and centred on 0 °C. Thus, a single methyl branch in an acyl chain has a fluidising effect on the packing of a bilayer which increases as the position of the branch approaches the middle of the chain. In nature, this fluidising effect would work in a mixture of lipids.

The significance of this measured fluidity has to be qualified. All the methods used provide an average for the bilayer as a whole. But as was described earlier, the fluidity of boundary lipids influences the performance of enzymes and other proteins embedded in the bilayer and they only constitute a small fraction of the bilayer lipids. So, does the fluidity of the considerable area of bilayer which is free of proteins have any physiological significance? In both eukaryote and prokaryote cells, the permeability of the bilayer to the diffusion of ions and small uncharged molecules has significance, but in the prokaryotes, the surface area of the membrane relative to the volume of the cell is huge, hence the diffusion of ions through the membrane is particularly important. The homeoviscous adaptation of the membrane bilayer of *Bacillus subtilis* to temperature provides a particularly interesting development of Sinensky's original experiment. Liposomes prepared from cells grown at low (13 °C) and high (40 °C) temperature had the same permeability to hydrogen ions, i.e. protons when tested at the temperature at which the cells were grown. "Homeoproton permeability adaptation" is undoubtedly physiologically important to the cell (Van de Vossenberg et al. 1999).

Homeoviscous adaptation of membrane lipids implies cells have a means of detecting bilayer fluidity and regulating it to a "set point" value by adjusting the composition of the lipid mix in the membranes (Ernst et al. 2016). There is evidence that these regulatory processes exist in *Bacillus subtilis*, other prokaryotes and in Eukaryotes; yeast (*Saccharomyces cerevisiae*) and in fish, the Carp (Tiku et al). A key element in regulating the degree of unsaturation in phospholipids is the class of enzyme called desaturases which can insert a double bond in a carbon chain, at a specific position along its length.

We have to leave the problem of regulating fluidity with that tantalising information and return to the question of adapting membranes to high pressure, which can now be rephrased as "does homeoviscous adaptation to high pressure occur?". Remember that pressure reduces the fluidity of lipid bilayers to the extent that 50 MPa has an effect equivalent to a reduction in temperature of 10 °C. Lipid bilayers in hadal organisms functioning at 100 MPa and a temperature of say, 2 °C, are at the equivalent of minus 18 °C, the temperature in a deep freeze. Clearly, adaptation to high pressure is needed. Does homeoviscous adaptation occur, by adjusting the composition of membrane bilayers, increasing their fluidity to counteract the ordering effect of pressure?

In the following section, we shall consider the evidence for such adaptation to high pressure in organisms living in the cold deep sea. Those occupying the hydrothermal vents are separately dealt with in Chapter 9 because the physical conditions are different.

## 7.2    Membranes from Deep-sea Fish

This section is written in a personal style because it is the outcome of collaboration between the author and Andrew Cossins of the Liverpool University. Apart from the science, it provides an example of how scientists go about their business, which often entails collaboration.

I contacted Andrew at a meeting of the Society of Experimental Biology because he was a leading light in the study of homeoviscous adaptation to temperature. It is rare in Biology to be able to make firm and sweeping predictions but we were soon able to frame two, in the form of testable hypotheses about the membranes of deep-sea organisms, and of fish in particular. First, they would have a greater bilayer fluidity, measured in standard conditions, than shallow water equivalents, the fluidity arising from the absence of the ordering effect of their ambient high pressure. The second hypothesis was the obvious sequel—the extra fluidity would be achieved by an increase in the proportion of unsaturated fatty acids in the membrane lipids, and there could well be additional "fluidising" factors present. An important caveat was that the homeoviscous adaptation we were predicting would not perfectly compensate for the ordering effect of pressure. Even in temperature studies, imperfect compensation is found (Macdonald and Cossins 1985). We were focussing on deep-sea fish living in cold waters; the problems presented by the then newly discovered hydrothermal vent animals which experience high pressure combined with high temperature were for another day.

### 7.2.1    The Fluidity of the Lipid Bilayer in Deep-sea Fish Membranes

We were delighted and not a little relieved to receive the necessary research grant and ship time from the UK Natural Environment Research Council (UK), to test our hypotheses. We were allocated short cruises on RV Challenger to work in the familiar Porcupine Sea Bight which, as already mentioned, provides trawling grounds at depths of around 4000 m, conveniently close to the UK. I was as confident as one can be, when working at sea, that we would be able to secure suitable fish, and thus their membranes, for our purpose.

Cell membranes were prepared from the brain and the liver of large fish such as Coryphaenoides, using established methods that were simplified for work at sea. We needed a suspension of well-defined membranes in order to measure their bilayer fluidity using fluorescence anisotropy (Chapter 2; Box 2.4). The ship's cold laboratory and a gimbal table to accommodate a miniature centrifuge proved invaluable. The quality of the three membrane fractions was later assessed in a land laboratory using an electron microscope, and it was found to be satisfactory. A brain myelin fraction comprised myelin whorls and membranous vesicles, with little contamination from mitochondria. Myelin is a mixture of lipids making up the membrane of special cells which are wrapped around nerve cells and provide electrical insulation. The brain synaptic fraction was free of myelin whorls, consisting of a variety of membranous vesicles. From the liver, a mitochondrial fraction was prepared which was as free of droplets of liver oil as possible. These membrane fractions proved to be close to the quality used in studies of more accessible fish, processed in a laboratory on land. Furthermore, the membrane fractions prepared from the shallow and deep-sea fishes we collected appeared to be of similar purity. Deep-sea fish often have large livers full of oil and we were concerned to ensure the samples were not biased by such factors. The fish we used were subsequently identified by experts at the Natural History Museum, London and at the Marine Laboratory, Aberdeen.

The steady-state fluorescence spectrometer, normally used in Andrew's laboratory, was mounted on an optical bench which was then bolted to a wooden laboratory bench. The motion of the ship nevertheless exerted a small (1–2%) effect on the signal from each photomultiplier, which was smoothed by computer processing. Checks were made on the spectrometer's performance, using standard procedures. Because we had not yet acquired the apparatus required to carry out the anisotropy measurements at high pressure we used conventional equipment, making the measurements at normal pressure and 4 °C, a temperature sufficiently low to suit the fish but one which caused condensation on the flat-sided quartz container which held the membrane suspension! This was eventually eliminated by a stream of dry nitrogen gas, after some initial confusion. Humidity seemed to be remarkably high at sea. The fluidity of membrane fractions, especially the mitochondria of the shallow species, varied with temperature in the same way as similar membranes from freshwater fish, for example, the Green Sunfish (*Lepomys cyanellus*) (Cossins et al. 1984).

We studied membranes from 15 species of fish, trawled from depths of 200–4000 m. The DPH anisotropy measurements, made at atmospheric pressure, indicated a bilayer fluidity greater than would be the case at the animals' normal high pressure. The fluidity of brain myelin membranes from three species increased with the depth from which the animal was trawled (In the case of a mitochondrial-rich membrane fraction from the liver, the relationship was less clear, for technical reasons.) The more fluid state of the membrane bilayers from the deepwater fish is offset by the high pressure at which they live. The membranes from *Coryphenoides rupestris* (collected from 900 m; species depth range 400–2000 m) are compared to those from *C. armatus* (collected from 4000 m; see Fig. 5.3b). When the fluidity of the latter, at atmospheric pressure, is adjusted for the effect of pressure (40 MPa), it becomes similar to that of *C. rupestris* at its ambient pressure. This is good evidence for the homeoviscous adaptation of deep-sea membranes to high pressure. The adaptation is partial, offsetting about half the ordering effect of pressure (Cossins and Macdonald 1984).

This conclusion was confirmed with brain myelin membranes prepared during the second cruise to the same area, the membranes being stored at minus 20 °C, for transport home and then subsequently at minus 80 °C for up to a year. Time-resolved anisotropy measurements were obtained at high pressure using an apparatus fitted

**Table 7.1** Comparison of the Order Parameter (P2) for DPH in brain myelin membranes from deep-sea and a shallow water fish. (Modified from Behan et al. 1992)

| Species and depth | Order Parameter (P2) | |
|---|---|---|
| | From measurements at 4 °C and atmospheric pressure | Calculated value at the membrane's ambient temperature and pressure |
| *Pleuronectes platessa* <2 m | 0.829 | 0.807 (9 °C, 0.1 MPa) |
| *C. rupestris* 900 m | 0.799 | 0.811 (4 °C, 9 MPa) |
| *C. armatus* 4000 m | 0.794 | 0.813 (3 °C, 40 MPa) |

with special high-pressure windows in conjunction with the synchrotron radiation source at the Daresbury Laboratory, UK. Our collaborator at the laboratory was Gareth Jones and the author was simply the provider of the membranes. Whereas steady-state spectroscopy provides a measure of the "wobbliness" of DPH, i.e. the range of movement which the molecule has, the time-resolved technique provides a measure of the rate at which the probe moves, oscillating in a partially rotary fashion. From these measurements, the order parameter called (P2) was calculated. This is a parameter that can also be obtained by other forms of spectroscopy. In *C. armatus* membranes (P2) increases linearly with pressure whereas the same type of membranes, obtained from fish that live at lesser depths such as *C. rupestris* and the shallow water Plaice, show (P2) increasing very steeply over the first 50 MPa. This means that the compressibility of the *C. armatus* membranes is less than that of the other two, presumably because of differences in the composition of the membranes (Behan et al. 1992).

Table 7.1 sets out the important concluding data, showing that the fluidity of the membranes from these three species are very similar at their respective ambient temperature and pressure. Note that in the first column the difference between the measured values for the Plaice (*Pleuronectes platessa*) and *C. armatus* membranes is 0.035 and in the second column, which provides the value for the membranes at their normal pressure and temperature, it is only 0.006. They have a similar fluidity in their normal conditions, in other words, homeoviscous adaptation to pressure. How that adaptation is achieved is the subject of the next section, Section 7.2.2.

## 7.2.2  The Lipid Composition of Deep-sea Fish Membranes

Our second prediction was that membranes from deep-sea animals would contain a greater proportion of unsaturated lipids and other molecular components known to fluidise bilayers, than the equivalent membranes from species living at shallow depths. This would take any difference in their normal ambient temperatures into account.

Plasma membranes and intracellular membranes have a distinctive composition in any given organism so it was essential that we compared like with like, i.e. the same membranes from similar species that live at different depths. It turned out that of the membrane fractions we were able to prepare during the Challenger cruises in 1981 and 1982, only the mitochondrial-rich fraction from the liver was obtained in sufficient quantity to enable us to carry out the exhaustive analysis required to identify and quantify the fatty acids in the membrane lipids. We obtained samples from 12 species of fish, trawled from depths of 200–4000 m. Using the mitochondrial-rich fraction is paradoxical as you may recall that this membrane fraction yielded inconclusive results in the fluorescence anisotropy measurements described earlier. The technical reasons mentioned then included the strong possibility that pigment compounds from the livers may have contaminated the samples sufficiently to obscure the results, a problem also encountered with shallow water fish material processed in more favourable conditions than we had at sea. This contamination would not be a problem when analysing the composition of the membranes.

The principal fatty acids in the following phospholipids were the focus of our attention: ethanolamine (PE), serine (PS), choline (PC), inositol (PI) and cardiolipin. The question of how to "score" the degree of unsaturation in a long list of constituents, to provide an index of composition, is a potentially tricky one. For example, a fatty acid with 16 carbon atoms linked by simple carbon–carbon bonds is a saturated fatty acid whilst one with 22 carbons linked by bonds which include 6 double bonds, 22:6, is polyunsaturated. Fortunately, the simple index of the saturation ratio has served well in previous studies, and we use it here. It is the ratio of the weight percent saturated to unsaturated fatty acids. The saturation ratio for PE decreases with the increase in the depth of the species. Statistically, the trend is highly significant. The saturation index for PC shows the same trend which is also statistically significant. These results are consistent with the predictions made for homeoviscous adaptation to high pressure (Cossins and Macdonald 1986). It is probably the long-term, evolved "genomic" adapted state. Interestingly the data are similar to that in a short-term, cold-adapted shallow water fish, the Green Sunfish, in which the PE and PC from the same membranes show the same homeoviscous trend in the saturation ratio. In the latter case the adaptation is seasonal, the reversible regulation of gene expression (Cossins et al. 1984).

The saturation ratio varies in membranes from individuals of the same species trawled from the same depth. This reminds us that the detailed composition of the membranes reflects not just the physical conditions in which the animals live (temperature, pressure) but also other factors in the animals' life, such as food sources and activity levels.

Homeoviscous theory prompts the question, mentioned earlier, what exactly is the fluidity of the bilayer for? We can now provide a partial answer. Temperature acclimation studies have provided some evidence showing that enzymes situated in membranes are influenced by their surrounding lipids in a way that is consistent with their homeoviscous adaptation. Evidence that the same occurs in homeoviscous adaptation to high pressure comes from the biochemical expertise of Gibbs and Somero.

The enzyme they studied was the $Na^+/K^+$ -ATPase, which is embedded in the membrane bilayer of cells in the gill tissue of teleost fish. Its function is to transfer sodium ions, $Na^+$, and potassium ions, $K^+$, across the membrane bilayer, thus regulating the ionic composition of the animal's blood. Marine fish live in a $Na^+$ rich solution (seawater) yet their blood and body fluids generally contain a lesser concentration of $Na^+$ ions, due to the activity of the enzyme in their gills which transfers (pumps) $Na^+$ ions from the blood to the higher concentration in seawater. The ATPase part of the enzyme's name refers to the fact that it hydrolyses ATP as it transfers the ions. ATP is commonly used in this way, feeding energy into enzymatic pumping processes. A reasonable way to imagine what goes on during ion pumping is that the enzyme, which spans the bilayer, combines with $Na^+$ ions on one side of the bilayer, undergoes a rotation and probably a shape change, and then releases the $Na^+$ on the other side of the bilayer. The frictional resistance of the boundary lipid molecules which affects the enzyme's movement and the release of the $Na^+$ ion into a high concentration of the same ion means that energy is required. If, when subject to cooling, the lipids surrounding the ATPase are too stiff and inflexible then the ion pumping process slows down and the composition of the blood deteriorates. An important question is whether the enzyme's boundary lipids undergo homeoviscous adaptation to compensate for the low temperature, sufficient to retain adequate fluidity for the enzyme's function, or does the enzyme, a protein, also change in some way to compensate.

Following Moon (1975) and DeSmedt et al. (1979), Gibbs and Somero (1989, 1990) used membrane fragments prepared from gill tissue from *Coryphaenoides* and other species. The fluidity of the membranes was measured using DPH anisotropy, as previously described, and the activity of the ATPase enzyme was measured in the same membrane suspension. It was the rate at which ATP was hydrolysed that was measured, not the rate at which $Na^+$ ions were pumped. Species of fish from different depths provided tissue, for example *Coryphaenoides armatus*, from depths of 1885–4815 m and 2–4 °C; *C. leptolepis* from 2300 to 4600 m, 2–4 °C; *C. acrolepis*, from 700 to 1820 m, 4–8 °C; the Sablefish *Anoplopoma fimbria*, from 1 to 1500 m, 4–8 °C; and the Barracuda, *Sphyraena barracuda*, from 1 to 20 m, 24 °C. The gill tissue was stored at minus 80 °C immediately after dissection from the specimens at sea. In the home laboratory, the membrane fragments underwent further processing before the activity of the enzyme was measured over a range of pressures.

The results showed, first, that the enzyme from deepwater species was less inhibited by high pressure than the same enzyme from shallow water. Second, the membranes responded to changes in pressure and temperature in the same way, irrespective of their depth of origin, with high pressure reducing both bilayer fluidity and enzyme activity. Conversely, increasing the temperature increased fluidity and enzyme activity. An increase of 15–25 °C offsets the effect of 100 MPa on both enzyme activity and bilayer fluidity. The correlation between the two is strong evidence, short of proof, that adapting the lipid bilayer also adapts the enzyme's activity. But does the fluidity actually determine the activity of the enzyme? This was answered by ingenious experiments using a "lipid substitution" technique using membranes from *C. armatus* and the Barracuda. This introduces selected lipids to the bilayer surrounding the enzyme, mimicking changes that take place in homeoviscous adaptation. Substituting phosphatidylcholine increases the pressure sensitivity of the enzyme in both membranes, but significantly, an interspecific difference persists. This suggests that the enzyme from each species, when surrounded by the same lipids, retains its individual activity. Thus, the enzyme's boundary lipid does not totally dictate the enzyme's activity, implying that the structure of the individual enzyme (protein) plays a part. When both the enzymes have the same boundary lipids, the enzyme from the deepwater *C. armatus* is less inhibited by pressure than that from the Barracuda, implying that the enzyme per se from the former is better adapted to high pressure.

This type of ATP-ase has been the subject of many high-pressure studies, discussed by Winter and Jeworrek (2009) and by Winter (2015).

Although these experiments are a significant step in understanding homeoviscous adaptation to high pressure, more physiological measurements are desirable. Measuring the ATPase activity of the enzyme is not the same as measuring ion transport across the gill, and the latter is the critical function whose "fitness" is presumably selected for in evolution. In the case of other membrane-bound ATPases, the effect of pressure has been interpreted in other ways. Nevertheless, the above example is a good one and has yet to be reinterpreted. We also need to know more about ionic regulation in the different species. For example, a slow rate of Na transfer (per enzyme molecule) might be sufficient for an animal's needs if there are many enzymes distributed over a large area of gill tissue. Furthermore, comparisons with more fully understood membrane-bound ATPases such as that in the mammalian kidney would be productive. In the latter, for example, the ATPase activity is reduced by pressures comparable to those used in the Gibbs and Somero experiments, by ordering the boundary lipids which constrain the enzymes' flexibility. One-third of the enzyme is embedded in the bilayer and two-third projects into the aqueous phase.

There is a great deal more to be learned in this area. More of the same kind of data would be worthwhile but how is the regulation of membrane lipids in deep-sea fish related to the individual's depth–pressure range in life? Can the long-term homeoviscous adaptation be distinguished from short-term, reversible adaptation mentioned earlier? Some deep-sea species have a significant vertical range, suggesting that homeoviscous adaptation may be an ongoing homeostatic process, as it is in freshwater fish in response to temperature. Have those species which occupy a deep but limited depth zone lost their short-term adaptive capacity? Investigating other animal groups is clearly desirable, to test the assumption that homeoviscous adaptation is a general phenomenon. The fatty acids from amphipods collected from hadal depths in the north and the south Pacific have been determined. Unfortunately, membranes were not isolated to provide convincing data and we are left with inventories of whole-body fatty acids which are likely to include fatty acids from membranes, lipid stores and other sources.

Does pressure act through the same fluidity sensor as temperature? Have the important desaturase enzymes, central to regulating the balance between saturated and unsaturated fatty acids, adapted to high pressure? Examination of the genome of certain species might answer some of these questions, but generally, there is plenty of scope for future experimenters.

It might have been expected that microorganisms and microbiologists would lead the way in enquiring if homeoviscous adaptation had a role in deep-sea biology. However, the next section shows they came a close second. As we have seen, the phenomenon was first established in sea-level bacteria, and ever since research into the homeostasis of bacterial lipids has been very active. We should remember that Gram-negative bacteria, like *E. coli*, have an inner and an outer membrane which may differ in composition and the Gram-positive bacteria (e.g. *Bacillus subtilis*) have just the inner membrane. In what follows the complexity of the synthesis and modification of membrane lipids will be ignored. We will simply focus on what membrane lipids are present and the extent to which they are consistent with homeoviscous theory.

## 7.3 Homeoviscous Adaptation in the Membranes of Deep-sea Bacteria

DeLong and Yayanos (1985), at the Scripps Institution of Oceanography, USA, showed that the deep-sea Vibrio CNPT3, which grows over the range atmospheric pressure to 68 MPa at 2 °C, contained lipids with a conveniently limited number of fatty acids, six. The ratio of unsaturated:saturated fatty acids increased with the growing pressure. At normal atmospheric pressure, it was 1.9 and at 68 MPa it was 3. In the latter case, the growth rate was the same as at atmospheric pressure but the proportion of the dominant fatty acid, 16:1 was 56%. At atmospheric pressure, it was 40%. This bacterium is Gram-negative, with an inner and an outer membrane, from which the phospholipids certainly came. These data are consistent with homeoviscous adaptation to pressure, but, as the authors pointed out, the physical state of the membrane bilayer was not investigated. Numerous other studies have reported the same result. A year later (1986), the authors published further data on the fatty acid composition of the membrane lipids of various deep-sea bacteria grown at pressures up to 57 MPa. For example, Vibrio marinus MP-1 and a strain known as PE 36 both increased the proportion of unsaturated fatty acids at high pressure.

Yano and colleagues, at the Japanese National Research Institute of Fisheries Science, approached the same problem by isolating two new species of deep-sea bacteria. One (strain 16C1) was from the gut of *Corypheanoides armatus*, caught at a depth of 3100 m and the other (Strain 2D2) from the gut of *C. yaquinae*, at 6100 m. Both grew well at 20–40 MPa at low temperature. Their dominant lipids were PE and PG, and their fatty acids varied with the growth pressure, at 5 °C. Strain 16C1 was grown at atmospheric pressure 21 and 41 MPa and showed a clear reduction in the proportion of saturated fatty acids at the higher pressure and conversely an increase in the proportion of unsaturated fatty acids at high pressure. This pattern was also seen in Strain 2D2 over the pressure range 21–62 MPa. These results are consistent with homeoviscous theory.

More examples from different laboratories can be cited. (1) Alteroma, isolate F1A, was collected from a depth of 4900 m and grown at 2 °C at normal atmospheric pressure and 40 MPa by Wirsen et al. (1987). The ratio of unsaturated:saturated fatty acids as used above was 1.6 and 2.3, respectively. (2) Two species of Halomonas from hydrothermal vent sediments, both of which produced more unsaturated fatty acids when grown at 45 MPa than at atmospheric pressure (Kaye and Baross 2004). (3) A third example is Pseudomonas sp BT1, from sediments at 4420 m depths in the Japan trench. Its fatty acids are limited to only four. Growth at 10 °C and pressures up to 20 MPa increased the unsaturated:saturated ratio which levelled off at 30 MPa (Kaneko et al. 2000). These, like the previous examples, are also Gram-negative bacteria, but the Gram-positive *Sporosarcina*, from sediments at a depth of 6500 m in the Japan Trench also appears to show a homeoviscous response to pressure (Wang et al. 2014). It grows best at normal atmospheric pressure and only very slowly at 60 MPa, both at the high temperature of 35 °C. Presumably it is dormant in trench conditions. However, when the fatty acids are analysed after growth in these conditions, the results are intriguing. The membrane is rich in branched-chain fatty acids and at high pressure, the proportion of anteiso branched chains increases whilst that of the iso branched chains decreases. This would cause a fluidising change in the membrane as the methyl branch is further from the end of the chain than is the case in iso branched chains. In other experiments, measuring the ability of benthic bacterial communities occupying the sediment surface at 1000 m depths to grow at 20 MPa or more, an increase in the proportion of branched-chain fatty acids was found. In none of these experiments was the fluidity of the membrane bilayers measured. From mud samples collected by Li and colleagues at a depth of 10,898 m in the Mariana Trench, two barophilic bacteria were isolated, DB21MT-2 and DB21MT-5, which grew well at 70 and 80 MPa, respectively. In both, 70% of the membrane lipids were unsaturated fatty acids which is consistent with homeoviscous adaptation.

In the context of numerous examples of deep-sea bacteria containing membrane lipids consistent with homeoviscous adaptation, the detailed performance of *Photobacterium profundum* SS9 is particularly interesting. It was isolated from Amphipod material collected from 2550 m in the Sulu Trough, (see Chapter 12, Section 12.1.3 for details). Strains of SS9 long ago proved to be good experimental cells in, for example, Bartlett's laboratory (Bartlett 1999; Allen et al. 1999). The temperature and pressure optima for growth are, respectively, 15 °C and 25 MPa but it can grow well outside these conditions. When grown at low temperature or high pressure, it produces an increased proportion of monounsaturated fatty acids (MUFAs) and polyunsaturated fatty acids (PUFAs). These would be located in the inner and outer membranes, consistent with homeoviscous adaptation. Their presence and implied membrane fluidity are important for growth, as demonstrated by the following experiments. With the synthesis of MUFAs inhibited (but not PUFAs or saturated fatty acids) by a specific chemical, the growth rate and yield were reduced at low temperature and high pressure, a condition reversed by adding MUFAs to the cultures. Furthermore, mutants (EA3) were discovered which are deficient in MUFAs and they performed in the same way as the chemically inhibited cells. The effect could be reversed by the addition of the fatty acid 18:1 to the culture.

The gene primarily responsible for regulating 18:1 production is fabF, which generates an enzyme system that is sensitive to pressure. High pressure increases the rate of catalysis of the enzyme, perhaps by reducing the Km. Transcriptional regulation, i.e. special protein or mRNA synthesis, is not involved.

Abe and Usui (2013) and Usui et al. (2012) have made a major contribution in this field by using time-resolved fluorescence spectroscopy at high pressure (Chapter 2; Box 2.4). Recall that this technique measures fluidity and also provides an order parameter, in this case of the fluorescent probe TMA-DPH, which obligingly dissolves in the inner membrane of Gram-negative bacteria. Here the deep-sea piezophile *Shewanella violacea*, strain DSS12, was studied, originally isolated from a depth of 5110 m in the Ryukyu Trench. It grows best at 30 MPa, 8 °C (Many deep-sea bacteria are named after scientists and this one honours James Shewan of the Torry Research Station, Aberdeen). The probe positions itself at the polar, outer, region of the membrane bilayer. The order parameter revealed the membrane of *S. violacae* to be surprisingly little affected by pressure, up to 100 MPa at 10 °C. This means the membrane bilayer is not very compressible, rather like a simple lipid bilayer in its gel state. In contrast, the order parameter for the equivalent membrane in *E. coli* was greatly increased by the same pressure.

The question arises: how is it that *S. violacea*'s membrane is so ordered, despite its fatty acid composition indicating otherwise? A mutant strain of the bacterium which is deficient in the fatty acid EPA provides a plausible explanation. EPA is short for eicosapentaenoic acid which has 20 carbons in its chain with five double bonds. It is a PUFA, various forms of which occur in psychrophilic and deep-sea bacteria. EPA is present in the membrane of the normal form of *S. violacea* which, as just mentioned, is little affected by pressure. The order parameter of a mutant's inner membrane (lacking EPA) increases considerably with pressure, similar to *E. coli*'s membrane. This suggests that the highly unsaturated EPA fails to fluidise the membrane bilayer of normal *S. violacea*, but instead renders it well ordered and relatively insensitive to pressure. EPA, with its long carbon chain made highly flexible by its five double bonds, coexists with the tight packing of the bilayer lipids. This is contrary to homeoviscous theory but it may be an adaptation. It seems to create a tightly packed bilayer that is insensitive to pressure, as measured by the order parameter, and EPA may play a role in providing membrane proteins with suitable boundary conditions. Other deep-sea bacteria have been found to require specific fatty acids for certain physiological functions, such as motility or, in the case of *S. violacea*, cell division, described below. The presence of such fatty acids in the membrane bilayer complicates the idea of homeoviscous adaptation. But then we are dealing with a complex metabolism using a simple idea of regulating membrane fluidity, an orthodox scientific approach but clearly a limited one. Wang et al. (2009) have shown that many strains of Shewanella have the genome to synthesise EPA but many do not, at least not in readily detectable amounts. How fatty acid synthesis is regulated in these bacteria, and particularly in those growing at high pressure, is not known.

Bacteria that metabolise hydrocarbons are important in the recovery of the oceans following oil spills, such as the catastrophic case of the Deep Water Horizon,

Chapter 9. Understandably most investigations have dealt with shallow water bacteria but a deep-sea bacterium, *Marinobacter hydrocarbonoclasticus* strain 5, has produced some interesting results. It was collected from 3745 m in the western Mediterranean and when cultured on hexadecane in the laboratory its growth was found to be the same at 35 MPa as at atmospheric pressure. Its metabolism of hexadecane was also much the same at the two pressures. This is clearly a pressure-tolerant bacterium of some importance. Does it show homeoviscous membrane adaptation to high pressure? The answer is, marginally, but the report, by a team led by a Grossi, focussed on total cell lipids. Electronmicrographs show the cell contains lipid inclusions which appear to be smaller in cells grown at 35 MPa. However, their composition is affected by pressure. Cells grown at atmospheric pressure have a high proportion of wax esters and small proportions of triacylglycerols and phospholipids. Cells grown at 35 MPa have a much smaller proportion of wax esters and higher proportions of triacylglycerols and phospholipids. The cell membrane contributes to the last category and the degree of unsaturation was slightly more in cells grown at 35 MPa. However, the proportion of unsaturated chains in the lipids generally was much greater in the cells grown at 35 MPa. Temperature is known to drive the homeoviscous regulation of lipid stores.

### 7.3.1 Some Physiological Functions Related to the Membranes of Marine Bacteria

An example was demonstrated by DeLong and Yayanos in 1987. They investigated the mechanism by which glucose is transported into deep-sea bacteria. The ordinary marine bacterium *Vibrio marinus* was compared with a vibrio, PE-36, isolated from 3584 m depths. It grows well at 20 MPa, 2 °C in contrast to *V. marinus* which grows best at normal atmospheric pressure. Both cells take up glucose from the external environment by a similar mechanism, a process that involves a complicated series of reactions best understood in laboratory bacteria. Measuring the rate of glucose uptake showed that PE-36 was far more pressure adapted than *V. marinus*. The transport system of the former functioned maximally at 20 MPa and adequately at up to 80 MPa whereas that of *V. marinus* was inhibited by increased pressure and became negligible at 80 MPa. Membrane fluidity may be unimportant here as these two bacteria have remarkably similar membrane fatty acids, so the adaptation in PE-36 probably lies in the coupled reactions driving the transport, and therefore more directly in the enzymes involved.

Cell division in deep-sea bacteria has more recently been studied, following progress made with the laboratory bacterium *E. coli*. In this cell, filamentous forms appear when it is grown at high pressure. A late stage in the formation of a septum to partition the elongated cell into two is inhibited. High pressure depolymerises a tubulin-like protein which forms a ring (Z ring) at the site where the septum is to form. In the deep-sea *Shewanella violacea* DSS12, mentioned above, the synthesis of EPA is, in some way, a prerequisite for the proper function of the Z ring and septum forming process. This was demonstrated by the use of the mutant, also mentioned earlier, which is incapable of synthesising the normal amount of EPA (Kawamoto et al. 2011). At normal atmospheric pressure, its growth and form were normal, but at 30 MPa it grew poorly and formed filaments, a condition which was reversed when EPA was provided in the growth medium. Thus, in the absence of EPA the filamentous *S. violacea* resembled *E. coli* under high pressure, in which the Z ring protein is depolymerised. EPA might function to accommodate certain proteins in the membrane bilayer which are involved in septum organisation.

In summary, homeoviscous adaptation of deep-sea bacterial membranes is apparent in experimental evidence, ranging from substantial to fragmentary, from numerous deep-sea bacteria. However, the adaptation of their membrane lipids involves more than changes in bilayer fluidity. More direct measurements of membrane fluidity are needed. Other membrane functions obscure and complicate the simplicity of homeoviscous theory. An account of the genetic basis of membrane adaptation to pressure and the synthetic pathways involved has not been attempted here but knowledge is growing from favourable cells such as *Photobacterium profundum* SS9.

## 7.4     Bacteria in the Deep Biosphere

Few experiments have been carried out on bacteria collected from this environment, which is not surprising given the practical difficulties in obtaining samples.

### 7.4.1   Sample Collection

Some of those practical difficulties of collecting samples are described in Chapter 8.

### 7.4.2   Membrane Lipids in Bacteria From the Deep Biosphere

However, the lipids in bacteria isolated from the Deep Biosphere in the Nankai Trough, off the coast of Japan, have been analysed. The bacterium, designated LT25 and closely related to *Vibrio diazotrophicus*, was a motile Gram-negative rod. It originated from a depth of around 297 m in sediments lying below a water depth of 4790 m and hence from a pressure of around 50 MPa. The temperature was not recorded but it would have been warm. In any event, the bacterium grew well at 37 °C at atmospheric pressure, less well at higher pressures and hardy at all above 40 MPa. To investigate the effects of pressure on its lipids the cells were grown at 37 °C at 25 MPa and compared with those grown at normal pressure. The analysis was very detailed and the results, with the predictions of homeoviscous theory in mind, not easy to assess. On the one hand, cells grown at 25 MPa had an increased proportion of monounsaturated fatty acid side chains attached to phosphoglycerol, which can be a potent fluidising ingredient of membranes. On the other hand, there were many other changes that may have cancelled out any overall fluidising effect. Mangelsdorf and his colleagues (2005) concluded that, overall, the pressure-dependent changes did maintain membrane fluidity at high pressure. There is clearly a need to measure bilayer fluidity directly to confirm this.

A more recent study was of the Gram-negative *Desulfovibrio indonesiensis* strain P23, collected from a depth of 260 m in sediments on the Juan de Fuca Ridge, in a seawater depth of 2656 m (Fichtel et al. 2015). The bacteria were thus living at around 30 MPa and an estimated 61 °C. At 45 °C growth was demonstrated at both 40 MPa and normal atmospheric pressure. In the latter conditions, the cells were typical vibrio rods, 1–1.7 u in length. At 40 MPa they were filamentous and about ten times longer, which implies that cell division was inhibited. In turn, this suggests that the cells were ill-adapted to the conditions. The lipids were investigated in considerable detail and although the authors concluded that growth at pressure was accompanied by a decrease in the proportion of saturated fatty acids (consistent with homeoviscous adaptation), the picture was clouded by chain length and the presence of branched-chain fatty acids.

These investigations demonstrate that the lipid metabolism of these cells is significantly disturbed by high pressure. Are the changes pathological? Perhaps, but probably not, and only perseverance will clarify the situation (Chapter 8).

## 7.5    Homeoviscous Adaptation: Other Aspects

It has been pointed out earlier that membrane lipids do not just provide a molecular barrier and support, as they are metabolically active in, among other roles, cellular signalling. This is an aspect that is not dealt with here, as limits have to set even in a wide-ranging book such as this. However, see Siebenaller and Garrett (2002) and Koyama and Aizawa (2002). And, as was mentioned at the beginning of this chapter, the homeoviscous adaptation of Archaeal membranes is dealt with in Chapter 9, Hydrothermal Vents.

## 7.6    Do Shallow Water Cells Show Homeoviscous Adaptation When Exposed to High Pressure?

*Tetrahymena pyriformis*, strain NT-1, not a bacterium but a Eukaryotic unicellular "laboratory" organism previously mentioned, fails to show any homeoviscous change in its membrane lipids or in the DPH anisotropy of its membranes when subjected to pressures of up to 26 MPa for 13 hours (Macdonald 1984). Previously, Nozawa et al. (1974) and Martin and Thompson (1978) had shown that a similar strain manifested a degree of homeoviscous adaptation to temperature changes. Similarly, *E. coli* (the cell which gave rise to the concept of homeoviscous adaptation of membrane bilayers to temperature) in the form of strain S116 does not show a homeoviscous response when grown at high pressure, 30 MPa at 37 °C (Allen and Bartlett 2000). On the other hand, different strains of *E. coli* manifest homeoviscous adaptation to pressure in longer-term laboratory evolution experiments described in Chapter 12. In the same chapter, mention is made of the fluidity change which the gill cells of the eel undergo after a month at 10 MPa (Section 12.2.4).

# References[1]

Abe F (2013) Dynamic structural changes in microbial membranes in response to high hydrostatic pressure analysed using time-resolved fluorescence anisotropy measurement. Biophys Chem 183:3–8. *

Abe F, Usui K (2013) Effects of high hydrostatic pressure on the dynamic structure of living *Escherichia coli* membrane: a study using high pressure time-resolved fluorescence anisotropy measurement. High Pressure Res 33:278–284

Alberts B et al (2014) Essential cell biology. Garland Science, USA & UK. *

Allen EE, Bartlett DH (2000) FabF is required for piezoregulation of cis-Vaccenic acid levels and piezophilic growth of the deep sea bacterium Photobacterium profundum strain SS9. J Bacteriol 182:1264–1271

Allen EE, Facciotti D, Bartlett DH (1999) Monounsaturated but not polyunsaturated fatty acids are required for growth of the deep sea bacterium Photobacterium profundum SS9 at high pressure and low temperature. Appl Environ Microbiol 65:1710–1729

Ballweg S, Ernst R (2017) Control of membrane fluidity: the OLE pathway in focus. Biol Chem 398:215–228. *

Bartlett DH (1999) Microbial adaptations to the psychrosphere/piezosphere. J Mol Microbiol Biotechnol 1:93–100

Bartlett DH, Lauro FM, Eloe EA (2007) Microbial adaptation to high pressure. In: Gerday C, Glansdorff N (eds) Physiology and biochemistry of extremophiles. ASM Press, Washington. USA, pp 333–348. *

Behan MK, Macdonald AG, Jones GR, Cossins AR (1992) Homeoviscous adaptation under pressure: the pressure dependence of membrane order in brain myelin membranes of deep sea fish. Biochim Biophys Acta 1103:317–323

Behan-Martin MK, Jones GR, Bowler K, Cossins AR (1993) A near perfect temperature adaptation of bilayer order in vertebrate brain membranes. Biochim Biochem Acta 1151:216–222. *

Boone DR et al (1995) Bacillus infernus sp. nov., an Fe (III) and Mn (IV) reducing anaerobe from the deep terrestrial subsurface. Int. J. Systematic Bact 45:441–448. *

Boonyaratanakornkit BB, Clark DS (2008) Physiology and Biochemistry of Methanocaldococcus jannaschii at elevated pressure. In: Michiels C, Bartlett DH, Aertsen A (eds) High pressure microbiology. ASM Press, Washington, USA, pp 293–304. *

Cario A, Grossi V, Schaeffer P, Oger PM (2015) Membrane homeoviscous adaptation in the piezo-hyperthermophilic archaeon Thermococcus barophilus. Front Microbiol 6:1–12. *

Casadei MA, Manas P, Niven G, Needs E, Mackey BM (2002) Role of membrane fluidity in pressure resistance of *Escherichia coli* NCTC 8164. Appl Environ Microbiol 68:5965–5872. *

Chong PL-G (2010) Archaebacterial bipolar tetraether lipids: physico-chemical and membrane properties. Chem Phys Lipids 163:253–265. *

Chong PL-G, Cossins AR (1983) A differential polarised phase fluorometric study of the effects of high hydrostatic pressure upon the fluidity of cellular membranes. Biochemistry 22:409–415. *

Chong PL-G, Fortes PAG, Jameson DM (1985) Mechanisms of inhibition of (Na, K)—ATPase by hydrostatic pressure studied with fluorescent probes. J Biol Chem 260:14484–14490. *

Chong PL-G, Sulc M, Winter R (2010) Compressibilities and volume fluctuations of Archaeal tetraether liposomes. Biophys J 99:3319–3326. *

---

[1]Those marked * are further reading.

Chong PL-G, Ayesa U, Daswani VP and Hur EC (2012) On physical properties of Tetraether lipd membranes: effects of cyclopentane rings. Archaea 2012:1–11. Hindawi Publishing Corporation, article 138439. *

Cossins AR (1994) Homeoviscous adaptation of biological membranes and its functional significance. In: Cossins AR (ed) Temperature adaptation of biological membranes. Portland Press, London, pp 63–76

Cossins AR, Macdonald AG (1984) Homeoviscous theory under pressure II. The molecular order of membranes from deep sea fish. Biochim Biophys Acta 776:144–150

Cossins AR, Macdonald AG (1986) Homeoviscous adaptation under pressure III. The fatty acid composition of liver mitochondrial phospholipids of deep sea fish. Biochim Biophys Acta 860:325–335

Cossins AR, Macdonald AG (1989) The adaptation of biological membranes to temperature and pressure: fish from the deep and cold. J Bioenerg Biomembr 21:115–135. *

Cossins AR, Kent J, Prosser CL (1984) A steady state and differential polarised phase fluorimetric study of the liver microsomal and mitochondrial membranes of thermally acclimated green sun fish (Lepomis cyanellus). Biochim Biophys Acta 599:341–358

De Long EF (2006) Archaeal mysteries of the deep revealed. Proc Natl Acad Sci U S A 103:6417–6418. *

De Long E, Yayanos AA (1985) Adaptation of the membrane lipids of a deep sea bacterium to change in hydrostatic pressure. Science 228:1101–1103

De Long E, Yayanos AA (1986) Biochemical function and ecological significance of novel bacterial lipids in deep sea procaryotes. Appl Environ Microbiol 51:730–737

De Long E, Yayanos AA (1987) Properties of the glucose transport system in some deep sea bacteria. Appl Environ Microbiol 53:527–532. *

De Smedt H, Borghgraef R, Ceuterick F, Henans K (1979) Pressure effects on lipid protein interactions in (Na+K+)—ATPase. Biochim Biophys Acta 556:479–489

Ding W, Palaiokostas M, Shahane G, Wang W, Orsi M (2017) Effects of high pressure on phospholipid bilayers. J Phys Chem B 121:9597–9606

Earl CRA (1986) Physical properties of membranes formed from phospholipids having monomethyl branched fatty acyl chains. In: Klein RA, Schmitz B (eds) Topics in lipid research. Royal Soc. Chemistry, UK

Ernst R, Ejsing CS, Antonny B (2016) Homeoviscous adaptation and the regulation of membrane lipids. J Mol Biol 428:4776–4790

Fichtel K, Logemann J, Fichtel J, Rullkotter J, Cypionka H, Engelen B (2015) Temperature and pressure adaptation of a sulphate reducer from the deep sub surface. Front Microbiol 6:1078

Gibbs A, Somero GN (1989) Pressure adaptation of the Na+ / K+ -ATPase in gills of marine teleosts. J Exp Biol 143:475–492

Gibbs A, Somero GN (1990) Pressure adaptation of gill Na+/K+—adenosine triphosphate: role of the lipid and protein moieties. J Comp Physiol B 160:4341–4439

Grossi V et al (2010) Hydrostatic pressure affects membrane and storage lipid compositions of the piezotolerant hydrocarbon-degrading Marinobacter hydrocarbonoclasticus strain 5. Environ Microbiol 12:2020–2033. *

Kamimura K, Fuse H, Takimura O, Yamaoka Y (1993) Effects of growth pressure and temperature on fatty acid composition of a barotolerant deep sea bacterium. Appl Environ Microbiol 59:924–926. *

Kaneko H, Takami H, Inoue A, Horikoshi K (2000) Effects of hydrostatic pressure and temperature on growth and lipd composition of the inner membrane of barotolerant Pseudomonas sp. BT11 isolated from the deep sea. Biosci Biotechnol Biochem 64:72–79

Kaneshiro SM, Clark DS (1995) Pressure effects on the composition and thermal behaviour of lipids from the deep sea thermophile Methanococcus jannaschii. J Bacteriol 177:3668–3672. *

Karmer MB, DeLong EF, Karl D, M. (2001) Archaeal dominance in the mesopelagic zone of the Pacific Ocean. Nature 409:507–510. *

Kato M, Hayashi R, Tsuda T, Taniguchi K (2002) High pressure-induced changes of biological membranes. Eur J Biochem 269:110–118. *

Kawamoto J et al (2011) Favourable effects of eicosapentaenoic acid on the late step of the cell division in a piezophilic bacterium, Shewanella violacea DSS12, at high hydrostatic pressure. Environ Microbiol 13:2293–2298

Kaye JZ, Baross JA (2004) Synchronous effects of temperature, hydrostatic pressure and salinity on growth. phospholipid profiles, and protein patterns of four Halomonas species isolated from deep sea hydrothermal vent and sea surface environments. Appl Environ Microbiol 70:6220–6229

Koyama S, Aizawa M (2002) PKC-dependent IL-6 production and inhibition of IL-8 production by PKC activation in normal human skin fibroblasts under extremely high hydrostatic pressure. Extremophiles 6:413–418

Lacey CN (2015) The ecology and physiology of trench amphipods from bathyal to hadal depths. PhD Thesis. University of Aberdeen. UK *

Lane N (2016) The vital question. Profile Books Ltd.. *

Lauro FM et al (2008) Large-scale transponson mutagenesis of Photobacterium profundum SS9 reveals new genetic loci important for growth at low temperature and high pressure. J Bacteriol 190:1699–1709. *

Li L, Kato C, Horikoshi K (1999) Microbial communities in the world's deepest ocean bottom–the Mariana Trench. In: Ludwig H (ed) Adv high pressure bioscience and biotechnology. Springer, pp 18–20. *

Lombard J, Lopez-Garcia P, Moreira D (2012) The ealy evolution of lipd membranes and the three domains of life. Nat Rev Microbiol 10:507–515. *

Macdonald AG (1984) Homeoviscous theory under pressure, 1. The fatty acid composition of Tetrahymena pyriformis NT-1 grown under high pressure. Biochim Biophys Acta 775:141–149

Macdonald AG, Cossins AR (1985) The theory of homeoviscous adaptation of membranes applied to deep sea animals. In: Laverack MS (ed) Symposia of the society for experimental biology, vol 39. The Company of Biologists, Cambridge, UK, pp 301–302

Mangelsdorf K, Zink K-G, Birrien J-L, Toffin L (2005) A quantitative assessment of pressure dependent adaptive changes in the membrane lipids of a piezosensitive deep sub- seafloor bacterium. Org Geochem 36:1459–1479

Marietou A, Nguyen ATT, Atten EE, Bartlett DH (2015) Adaptive laboratory evolution of Escherichia coli K-12 MG1655 for growth at high hydrostatic pressure. Front Microbiol 5:1–8. *

Martiensson VT et al (1999) Thermococcus barophilus sp. nov., a new barophilic and hyperthermophilic archaeon isolated under high hydrostatic pressure from a deep sea hydrothermal vent. Int J Syst Bacteriol 49:351–358. *

Martin CE, Thompson GA (1978) Use of fluorescence polarization to monitor intracellular membrane changes during temperature acclimation. Correlation with lipid compositional and ultrastructural changes. Biochemistry 17:3581–3586

de Mendoza D (2014) Temperature sensing by memebranes. Annu Rev Microbiol 68:101–116. *

Moon TW (1975) Effects of hydrostatic pressure on gill Na-K-ATPase in an abyssal and a surface-dwelling teleost. Comp Biochem Physiol B 52:59–65

Nozawa Y, Iida H, Fukushima H, Ohki K, Ohnishi S (1974) Studies on Tetrahymena membranes: temperature- induced alterations in fatty acid composition of various membrane fractions in Tetrahymena pyriformis and its effect on membrane fluidity as inferred by spin-label study. Biochim Biophys Acta 367:134–147

Oger PM, Cario A (2013) Adaptation of the membrane in Archaea. Biophys Chem 183:42–56. *

Perrone FM, Croce ND, Dell'anno A (2003) Biochemical composition and trophic strategies of the amphipod Eurythenes gryllus at hadal depths (Atacama trench, South Pacific). Chem Ecol 19:441–449. *

Roer RD, Bekman MY, Shelton MG, Brauer RW, Shvetzov SG (1985) Effects of changes in hydrostatic pressure on Na transport in gammarids from Lake Baikal. J Exp Zool 233:65–72. *

Scoma A et al (2019) Reduced TCA cycle rates at high hydrostatic pressure hinder hydrocarbon degradation and obligate oil degraders in natural deep sea microbial communities. The ISME J 13:1004–1018. *

Siebenaller JF, Garrett DJ (2002) The effects of the deep sea environment on transmembrane signaling. Comp Biochem Physiol B 131:675–694

Silliakus MF, van der Oost J, Kengem SWM (2017) Adaptations of archaeal and bacterial membranes to variations in temperature, pH and pressure. Extremophiles 20:721–740. https://doi.org/10.1007/s00792-017-0939-x. *

Sinensky M (1974) Homeoviscous adaptation–A homeostatic process that regulates the viscosity of membrane lipids in *Escherichia coli*. Proc Natl Acad Sci 71:522–525. *

Tiku PE, Gracey AY, Macartney AI, Beynon RJ, Cossins AR (1996) Cold-induced expression of Delta 9 -desaturase in carp by transcriptional and post translational mechanisms. Science 271:815–818. *

Usui K et al (2012) Eicosapentaenoic acid plays a role in stabilizing dynamic membrane structure in the deep sea piezophile Shewanella violacea: a study employing high pressure time-resolved fluorescence anisotropy measurement. Biochim Biophys Acta 1818:574–583

Valentine RC, Valentine DL (2004) Omega-3 fatty acids in cellular membranes: a unified concept. Prog Lipid Res 34:383–402. *

Van de Vossenberg JLCM, Driessen JM, da Costa MS, Konings WN (1999) Homeostasis of the membrane proton permeability in Bacillus subtilis grown at different temperatures. Biochim Biophys Acta 1419:97–104

Vezzi A et al (2005) Life at depth: Photobacterium profundum genome sequence and expression analysis. Science 307:1459–1461. *

Wang F, Xiang X, Ou H-Y, Gai Y, Wang F (2009) Role and regulation of fatty acid biosynthesis in the response of Shewanella piezotolerans WP3 to different temperatures and pressures. J Bacteriol 191:2574–2584

Wang J et al (2014) Alterations in membrane phospholipid fatty acids of Gram-Positive piezotolerant bacterium Spororsarcina sp. DSK25 in response to growth pressure. Lipids 49:347–356

Winter R (2015) Pressure effects on artificial and cellular membranes. In: Akasaka K, Matsuki H (eds) High pressure bioscience, sub cellular biochemistry, vol 72. Springer Science+ Business Media, Dordrecht

Winter R, Jeworrek C (2009) Effect of pressure on membranes. Soft Matter 5:3157–3173

Wirsen CO, Jannasch HW, Wakeham SG, Canuel EA (1987) Membrane lipids of a psychrophilic and barophilic deep sea bacterium. Curr Microbiol 14:319–322

Yano Y, Nakayama A, Ishihara K, Saito H (1998) Adaptive changes in membrane lipids of barophilic bacteria in response to changes in growth pressure. Appl Environ Microbiol 64:479–485. *

Zhai Y et al (2012) Physical properties of Archaeal tetraether lipid membranes as revealed by differential scanning and pressure perturbation calorimetry, molecular acoustics and neutron reflectometry: effects of pressure and cell growth temperature. Langmuir 28:5211–5217. *

Zhang Y-M, Rock CO (2008) Membrane lipid homeostasis in bacteria. Nat Rev Microbiol 6:222–233. *

# Prokaryotes at High Pressure in the Oceans and Deep Biosphere

**8**

In Chapter 1 the point was made that the study of microorganisms was a major part of Deep-Sea Biology, and the Deep Biosphere is a huge high-pressure environment for microorganisms. Here we focus on Prokaryotes as such: Bacteria and Archaea. But deep-sea bacteria dominate the proceedings. They were, after all, studied by Certes (1884) and subsequently in the early twentieth century by Portier, Benecke, and then by ZoBell, who participated in the Galathea Expedition of 1950–1952, and by Morita. However, in 1968, scientists at the Woods Hole Oceanographic Institution, USA made a notable contribution:

The preservation of sandwiches.

In 1968, the research submersible Alvin, operated by the Woods Holes Oceanographic Institution, was lost from its mother ship, sinking to the seafloor at a depth of 1540 m. Fortunately, the crew escaped but a sandwich lunch was not so lucky. It remained on board for 10 months, and when Alvin was salvaged, the food became an object of intense microbiological interest. It was remarkably well preserved, by it would appear, 15 MPa and 4 °C. Bacterial and fungal activity seems to have been arrested by 15 MPa, as there is little doubt that at 4 °C and atmospheric pressure the food would deteriorate much more. The episode provided a good example of the unexpected effects of moderate deep-sea pressures, and further justification, if that was needed, for the existence of Alvin.

The above comes from Jannasch et al. (1971) and the microbiological observation was followed up in 1973 and in other papers.

This chapter deals with the collection and growth of microorganisms from the deep sea and then moves on to the Deep Biosphere. Eventually, we consider the adaptation to the high pressure of some of the microbial proteins. This latter part complements the account of the pressure adaptation of proteins in deep-sea fish (Chapter 6).

© Springer Nature Switzerland AG 2021
A. Macdonald, *Life at High Pressure*, https://doi.org/10.1007/978-3-030-67587-5_8

## 8.1    Microorganisms in the Deep Sea and Their Collection

We begin with experiments which have been carried out on microorganisms collected from deep sea water with, and without, decompression, and then proceed to the equivalent from the Deep Biosphere, from deep sediments and rock. Chapter 13 describes the equipment designed to collect microorganisms from high-pressure environments without decompression (isobaric collection).

In this subject area, the growth of microorganisms is basic to just about all experimental work, and much non-experimental work. Growing bacteria at high pressure is not particularly difficult to do simply and crudely, but it is difficult to do in a fully controlled way which, fortunately, is not always necessary. Similarly, measuring the metabolic or biochemical activity of microorganisms can be done simply if that is all that is required, but to closely mimic natural conditions, which is important for certain studies, is more demanding.

In bulk seawater, deep or shallow, pelagic bacteria grow on dilute nutrients in solution, made available by cell lysis, decomposition, and excretion. They are oligotrophic. Bacteria also adhere to, and live off, particulate material, "marine snow", which it solubilises. The number of bacteria in seawater is typically a hundred million per litre in shallow water and a million per litre in the deep sea. Measuring growth (multiplication) in natural conditions with such low cell and nutrient concentrations is a general problem for marine microbiologists (Giovannoni and Stingl 2007). In 1978, Carlucci and Williams of the Institute of Marine Resources, California, published a salutary report of such measurements. Specific strains were isolated and returned to the natural seawater from whence they came. Shallow water strains, from 1 and 75 m depths and 18–22 °C had a doubling time of 14 h or less. Strains decompressed from 1500 and 5550 m depths and recompressed to their original pressures in natural deep seawater had a doubling time of 145 and 210 h, respectively (2.5–1.5 °C). Similarly, the strains from shallow water consumed oxygen many hundreds of times faster than those from deep water. As decompression and recompression seemed to have no effect on the deepwater strains (the authors were not very detailed on this point), we have to conclude that the metabolic state of shallow water bacteria can be very different from that of the deepwater strains.

One of the first attempts to measure the metabolic activity of deep-sea bacteria in seawater, in situ, was reported by Jannasch and Wirsen (1973). They used the submersible Alvin to fill pre-sterilised bottles containing enriched media with top sediment water, thus inoculating the enriched media. These were left to incubate for a year on the seafloor at a depth of 1830 m and 4 °C. The metabolic activity was very slow in these experiments, but subsequent work has found otherwise. If mixed populations of deep-sea bacteria are provided with an increased concentration of dissolved nutrients (enrichment), they often respond vigorously. This probably applies to bacteria freely suspended in water as well as the many which are stuck to organic particles, and slowly sinking in the water column. These pelagic bacteria are probably in a metabolic state rather different from that of sedimentary bacteria which exist in high cell concentrations.

Many bacteria can form spores, also called endospores, which are extremely resistant to just about everything, and indeed they arise from growing cells when conditions become unfavourable. They are very important in food technology and feature in the Deep Biosphere research literature more than they do in ocean research.

## 8.2 Effects of Decompression on Deep-Sea Bacteria

Recall that when an abyssal amphipod is brought to the surface with decompression it looks dead and soon will be if it is not restored to its normal high pressure. Deciding whether decompressing deep-sea bacteria has any adverse effect on them is not always simple. Neither is deciding if a particular species is adapted to a given high pressure and therefore likely to be an indigenous species (autochthonous). (The meaning of species in this context is given in Chapter 1; Section 1.5.) Bearing in mind that growth performance is a basic criterion in this field, the point made by the Russian microbiologist Kriss many years ago should not be overlooked. Pressure-tolerant species of bacteria can be isolated from garden soil!

### 8.2.1 Bacteria Decompressed During Collection and Recompressed

We start with some celebrated examples of bacteria which, when collected from the deep sea, were subjected to decompression. The most recent and spectacular case is *Colwellia marinimaniae*, whose optimum pressure for growth is 120 MPa, the world record (Kusube et al. 2017). Many years ago, from Yayanos' laboratory, came the important and productive MT-41, a strain isolated from a dead Amphipod collected from a depth of 10,476 m in the Mariana Trench. MT-41 proved to be an obligate piezophile, unable to grow at less than 40 MPa and routinely batch cultured at 101 MPa and 2 °C (Yayanos et al. 1981). Serial transfers to fresh culture media, exposing the cells to atmospheric pressure for less than an hour, is tolerated. The lethal effect of more prolonged exposure to atmospheric pressure was quantified by testing the ability of dilute suspensions of cells so held for periods of many hours to grow when transferred to 101 MPa. No cells survived for more than 90–170 h at atmospheric pressure at 0 °C. A high percentage died within 6 h. The fine structure of the cells was examined under an electron microscope. Cells fixed immediately after decompression from 103 MPa appeared no different to cells fixed just before decompression, but after several hours at atmospheric pressure unhealthy looking changes appeared, beginning with a significant increase in the number of intracellular vesicles (Chastain and Yayanos 1991).

Deep-sea bacteria have been isolated from other deep-sea animals, for example from deep-sea Holothurians (sea cucumbers) and a mycoplasma from a deep-sea isopod (Bathynomus) with which it had a symbiotic relationship.

In Yayanos' lab the effects of decompression on another hadal bacterium, collected with decompression, was also investigated, namely strain CNPT-3, which is a psychrophile and a piezophile (optimum growth between 10 and 50 MPa) but not an obligate piezophile like MT-41, which could not grow at less than 40 MPa (Yayanos and Dietz 1982). It was shown to grow well at 10 °C and at the pressure corresponding to its depth of origin, 58 MPa, but grew better at 40 MPa. It died out over a period of a few hours when incubated at atmospheric pressure, at 10 and 20 °C. In fact, this cell was regarded as a typical deep-sea bacterium whose tolerance to decompression is such that brief exposures to atmospheric pressure for practical purposes are tolerated. (To complete the piezo terminology, piezotolerant cells grow well above atmospheric pressure but not above 10 MPa, Fang et al. 2010.)

The general idea about the isobaric collection expressed by Yayanos and Dietz in 1982 was that the "majority of marine microbiology is approachable with samplers thermally insulated and with cultivation at high pressure". That is, isobaric collection is not essential but of course, "approachable" is a rather vague term.

## 8.2.2  Isobaric Samples

Isobaric collection is necessary to find out what microorganisms exist at depth because decompression can kill sensitive species which will therefore fail to grow in subsequent experiments. The isobaric procedure has to be carefully designed to avoid damaging the organisms. Experience has also shown that isobaric collection is important in preserving the natural mix of organisms. For example, La Cono and colleagues demonstrated the undesirable effects of decompression by monitoring a bathypelagic sample for 3 days after collection. The archaeal population decreased markedly but the bacterial fraction doubled.

The first systematic use of isobaric bacterial sampling equipment was by Jannasch et al. (1973, and other colleagues at Woods Hole, Chapter 13; Section 13.8). They were followed by Colwell's group, Tabor et al. (1981) at the University of Maryland and then by Bianchi and Garcin (1993) and Bianchi et al. (1999) in Marseille, France. The initial observation was that radiolabelled substrate utilisation was generally slower at in situ pressures than in decompressed samples. Acetate incorporation was less affected by decompression than the incorporation of amino acids (Jannasch and Wirsen 1977, 1982). Intuitively this seems to imply that no "damage" was inflicted by decompression, but the slower rate of utilisation may be the real natural condition. Interpreting these data is complicated by not knowing the composition of the bacterial populations involved. Undoubtedly there were true deep-sea species present, along with surface survivors that happened to be present in the depths.

(Historians of science may one day establish who, in fact, was the first to work on isobarically collected deep-sea bacteria. The 1972 paper of Gundersen and Mountain, of the University of Hawaii, which describes their equipment (Chapter 13) implies they have a serious claim.)

When Bianchi and Garcin (1993) deployed their isobaric equipment in the relatively warm waters of the eastern Mediterranean, they obtained rather different results. Water samples, containing in the region of 100,000 cells per ml, were collected from 1100 m depths with and without decompression. Subsequent incubation showed that the decompressed bacteria incorporated radiolabelled glucose more slowly (at normal atmospheric pressure) than those which were not decompressed from 11 MPa. Subsequent experiments with samples from 2000 m depths using the radiolabelled amino acid leucine to quantify protein production by the microorganisms present (i.e. Bacteria and Archaea) showed a big difference. The sample held at the in situ pressure incorporated twice as much leucine as the decompressed sample. If the mixed population of cells behaved uniformly then the decompression certainly had a significant physiological effect, with ecological significance. If the population was not uniform, perhaps some cells were unaffected by decompression whilst others were inhibited, then there is another question, what are the physiological differences between the two groups? A proportion of bacteria in deep water are there by chance and merely surviving rather than growing as an autochthonous species. The selection of nutrients and the concentrations used in these experiments present difficult choices, and for how long should the incubations be?

For those who wish to know the metabolic activity of the mixed population of microorganisms in situ, i.e. at depth, it appears that data from isobaric samples are to be preferred. Tamburini and colleagues (2013) have made this point strongly, further confirmed by Garel et al. (2019). Measurements made on decompressed samples subsequently recompressed to the pressure of origin are probably better than those simply made on decompressed samples. For those who wish to know about the performance of individual species then isolates have to be prepared from the mixed population. That entails enrichment and an isobaric technique for this was developed by Jannasch and colleagues and is described in Chapter 13; Section 13.8.

Seawater also contains enzymes in solution. These enzymes, ectoenzymes, are synthesised and released by microorganisms to break down organic particles. Viruses also rupture cells in the water column, releasing enzymes and other organic molecules (Corinaldesi et al. 2019). Typically, the ectoenzymes cleave proteins and other polymers. Tamburini and his colleagues (2002) reported the use of the isobaric Serial Sampler (Bianchi et al. 1999; Chapter 13) to measure the activity of ectoenzymes. The activity of aminopeptidase and phosphatase in undecompressed samples from as deep as 2000 m was about twice that measured after decompression. This extended the measurements of Kato and colleagues of the activity of the ectoenzyme serine ectoprotease, released by *Sporosarcina sp.* at a depth of 6500 m in the Pacific Ocean. Its activity at atmospheric pressure was half that at 65 MPa. Thus, ectoenzymes are adapted to their ambient high pressure.

## 8.2.3 Bacteria from the Hadal Zone

Kato and his colleagues (Yanagibayashi et al. 1999; Kyo et al. 1991; Kato et al. 1999), using the submersible Shinkai 6500 and a remarkable isobaric collecting and transferring system called Deep Bath (Chapter 13), have carried out some important experiments with hadal bacteria. Among other things they too have demonstrated that isobaric collection is essential if the natural mix of species is to be accessible. The manipulating arm of the submersible scooped up samples of sediment from the surface and transferred it to the isobaric holder. This was at a depth of 6292 m in the Japan Trench. Back in a land laboratory, the isobaric diluting system was used to transfer, dilute and culture the microorganisms present in the sediment. Unlike the Jannasch method (Chapter 13), this is a purely hydraulic operation, carried out at 65 MPa and 10 °C. DNA was also extracted from the samples of the sediment. After 4 days of growth at 10 °C and 65 MPa, monitored by the optical devices integrated in the equipment, the stationary phase was reached and the culture was transferred a further four times in like fashion. The bacteria which became dominant in these mixed cultures were primarily the piezophiles *Shewanella* and *Moritella*, but the control incubations, carried out on decompressed samples, produced mainly *Pseudomonas*, which is widely distributed in the ocean waters. The cultures yielded DNA fragments which encoded for an operon ORF1.2, known to be expressed at high pressure in *Shewanella*. In control cultures incubated at atmospheric pressure, ORF1.2 became attenuated in a few generations, confirming its pressure dependency. From the sediments, fragments of DNA encoded for the same operon were also detected, at low concentrations (Kato et al. 1999; Yanagibayashi et al. 1999).

However, it is possible to obtain certain living bacteria from the greatest depths without isobaric techniques. In 1976, Morita published an account of the collection and the isolation of *Pseudomonas bathycetes* from the Mariana Trench, with decompression. It grew very slowly at 100 MPa and 2 °C, and also at normal atmospheric pressure, and accordingly is considered a survivor and not a permanent member of the Trench flora (Yayanos et al. 1981).

Kato and his colleagues (1998) and Kato (2011) have sampled, with decompression, shallow muds at 10,897 m depths in the Mariana Trench using the unmanned submersible Kaiko and isolated some very interesting bacteria. Two strains, DB21MT-2 (similar to *Shewanella*) and DB21MT-5 (related to *Moritella*) proved to be obligate piezophiles, as they failed to grow at less than 50 MPa. Another, *Thermaerobacter marianensis*, was found to be aerobic and strangely sensitive to pressure, with a generation time of 6 h at 5 MPa and failing to grow at pressures above 20 MPa. At a suitably low pressure, its optimum growth temperature was 75 °C, at which its generation time was 1.5 h (Takai et al. 1999). Like the *Pseudomonas*, it is a survivor and not well adapted to its environment.

## 8.3    Growing Bacteria at High Pressure

Samples of bacteria collected by an isobaric method need to be transferred to a new medium for growth and isolation of individual species, which can then be identified. Growth requires nutrients, and if the sample is isobarically transferred to a new solution containing a high concentration of nutrients (an enrichment culture), then the result will probably be vigorous growth, with some species doing better than others. This provides plenty of cells to isolate and to work with but an unnatural mixture of species growing unnaturally fast.

The traditional method of growing bacteria and other unicellular microorganisms in the laboratory is to use a test tube or larger flask containing a solution of nutrients and initiate growth by adding a small volume of an existing culture, the inoculum. Prior sterilisation precludes unwanted contaminants. After an initial delay, the lag phase, the cells begin to grow in size and divide in two at a rapid rate, which then levels off as nutrients or some other factor begins to limit metabolism. Cells in the lag phase are adapting to their new environment and differ from those in the rapidly growing phase, which in turn are different to those in the late stages of the culture, the stationary phase. In the extended stationary phase, mutants can arise which use the breakdown products of spent cells. This phase can last for up to a year and might have some similarity to the ultra-slow activity of microorganisms in the Deep Biosphere, of which more soon. These characteristics and limitations of batch cultures are well known. To study the effect of a single factor, such as a particular nutrient, acidity, temperature or pressure, it is important that varying it exerts no significant effect on anything else. This is the concept of the single independent variable and one way of cleanly exposing such a variable to growing cells is to use a method called continuous culture, which is explained in Section 8.3.1.

High hydrostatic pressure is a difficult variable. For example, it affects the ionisation of certain chemicals used to stabilise the acidity of a culture (i.e. buffer the pH, Chapter 2), so pressure-stable pH buffers have to be used in high-pressure experiments. A classic study of the growth of the thermophilic Archaeon *Methanococcus thermolithotrophicus*, published by Bernhardt and a group of expert chemists and microbiologists in 1988, provides an example of how to proceed. The growth conditions included high temperatures, high hydrostatic pressures, high pressures of hydrogen and carbon dioxide and numerous other factors, all controlled, see Section 8.7.4. To that can be added the observation of Nelson et al. (1991). The growth of the Archeon ES4 at high temperature and high pressure in a stainless steel vessel was markedly different (inhibited) from that in a vessel lined with glass and ES4 is not unique. Trace amounts of metals such as chromium, nickel and iron are thought to be leached from the steel.

Growth may be measured by one or more criteria; cell number, doubling time, dry weight of cells or of DNA or protein. Indirect measurements are also used, such as optical density, which is convenient at high pressure but not precisely related to the previous criteria.

## 8.3.1  Continuous Culture Methods

The technique was developed independently by Szilard and Novick and by Monod in the 1950s. A growing population of cells is held in a container that is supplied with a continuous flow of fresh culture solution. The inflow of fresh solution matches the outflow of "spent" solution, producing a steady-state population of cells growing in constant conditions. The technique was designed for several reasons. One was to study the rate of selection of mutants. Another was to enable a single independent variable such as the effect of the concentration of a specific nutrient to be studied. It is a technique that can mimic the oligotrophic natural seawater environment in which growth is often limited by the concentration of nutrients. Needless to say, the continuous culture apparatus, also called a chemostat, has become very complicated (Nelson et al. 1992; Hoskisson and Hobb 2005; Houghton et al. 2007; Foustoukos and Ileana 2015).

One of the first high-pressure chemostats was developed by Jannasch et al. (1996), particularly to study the growth of piezophiles at low concentrations of nutrients, typical of deep ocean water. It was designed for use at up 70 MPa and at low temperatures. The growth chamber (nowadays called a reactor) was made of titanium, coated with nylon and held 500 ml of stirred culture solution through which fresh solution was passed at rates of from 0.01 to 10 ml per minute. Pressure fluctuated ±1%. Samples of the growing culture could be decompressed into a special chamber (volume 10 ml) fitted with a piston which allowed a controlled volume to be withdrawn without shear forces rupturing the cells. Typically, the concentration of cells was low and determined by direct microscopic counts. The number of viable cells could also be assessed by plating out (by depositing the culture on the surface of agar gel which supports the growth of discrete colonies). One measurement, which seems very simple but which the traditional batch culture technique could only achieve in a limited way, was of the rate growth of a naturally occurring population of bacteria in the seawater in which they were sampled and at their original ambient temperature and pressure. This rate of growth was thought to reflect the extent to which naturally occurring nutrients limit the growth of the population (at high pressure and low temperature).

So, bacteria were filtered from seawater collected with decompression, from a depth of 4400 m and subsequently reintroduced into the chemostat filled and primed with seawater from the same depth. The experiment was run at 3 °C and 45 MPa with a very slow dilution rate. Growth was hardly detectable but when the nutrient level in the culture was raised to 0.33 mg C/L, growth became measurable. A steady-state concentration of cells at around a million cells per ml was achieved. Setting aside the possible effects of decompressing the sample, the conclusion is that in situ deep-sea bacteria grow very slowly but can respond to a small increase in the availability of nutrients (Wirsen and Molyneaux 1999). Another caveat is that this applies to bacteria in suspension in seawater, but not necessarily to bacteria adhering to particulate material in suspension.

Much information about deep-sea bacteria comes from experiments in which the cells are grown on high concentrations of nutrients which might well affect their tolerance of, and apparent adaptation to, their normal ambient pressure and temperature. In fact, this possibility seems to be rare. Individual species isolated from the seawater samples which are piezophilic in batch cultures with high concentrations of nutrients are often also piezophilic when grown in the chemostat in nutrient-limiting conditions.

## 8.4    Pressure Experienced by Sinking Particles

The surface of the sediments in the deep ocean is covered by freshly accumulated particulate material. The steady descent of the material down through the water column appears to deliver material that is relatively unaltered from its state in shallow water. The slowly sinking, negatively buoyant particles, sometimes called marine snow, appear to support a lot of eukaryotic microorganisms which slowly degrade the organic material at significant depths (Bochdansky et al. 2017).

The relatively fast-sinking material comprises aggregated faecal pellets, fragments of organisms and prokaryotes. Chemically it is a complex mix of inorganic "ballast" (calcium carbonate and silicates) and organic material such as chitin, cellulose, pigments and the familiar components of organisms; protein, lipids and carbohydrate, with for good measure, strange chemicals of which most biologists have never heard. Both bacteria and archaea are present. Sinking rates vary a lot and in part depend on the density difference between the particles and the water in which they are suspended. Taking a middling rate of 200 m per day it is clear that the particles experience a gradual increase in pressure and decrease in temperature spread over several weeks. This is a strange but important incubation in which numerous processes are affected, but the general idea is that degradation, by various means, is slowed by the increase in pressure. Incubating particles at high pressure and at slowly increasing pressures which mimic the gradual descent has provided experimental evidence for this conclusion and has also shown that the initial mixture of Bacteria and Archaea found on the particles in shallow water becomes dominated by Bacteria.

## 8.5    Eukaryotic Microorganisms

The limited amount of experimental work that has been carried out on deep-sea Eukaryote microorganisms has been discussed in Chapter 5. For a long time, they have been a promising source of illuminating experiments.

## 8.6    Viruses

The American marine microbiologist Claud Zobell was one of the first to detect viruses in the sea and in 1955 the first marine virus, a bacteriophage (explained below) was isolated. We now know that the numbers of viruses and other "mobile genetic elements" (MGE) in a given volume of seawater can be very significant (Forterre 2013; Biller et al. 2014).

One of the first thorough attempts to find viruses in deep water was in the North Pacific. Setting aside the technicalities, which are considerable, viruses were found to be present at concentrations of around a million per millilitre. The concentration decreased over the initial few hundred metres descent in the water column but thereafter was relatively constant, down to the maximum depth sampled, 5000 m.

The ratio of viruses to bacteria varied from 2:1 to 5:1, but in other oceans, it can be much higher. A high proportion of viruses in the sea are bacteriophages and they cause the release of proteins and nucleic acids from the bacteria they lyse. Other viruses attack Archaea (Rohwer and Thurber 2009; Prangishvili 2013). More recently the difficult problem of estimating the number of viruses in deep-sea sediments has been studied. In the case of sediment at a depth of 4800 m in the North Atlantic, virus concentrations of 100,000 per gram dry weight were typical. Separately, the number of species (technically viral genotypes) has been estimated at an amazing 5000 per 100 litres of bulk seawater or a million in a kilogram of sediment. In such sediments a ratio of virus: prokaryote of ten is common and in anoxic sediments, it can reach a hundred. In the recycling of organic material which goes on in sediments, virus material is significant.

Exactly what is a virus and how small is it? Viruses are essentially particles made of a protein capsule, called a capsid, which encloses either DNA or RNA. A big virus, such as the herpes virus, is about 200 nm in diameter (and contains DNA) whilst a small one, for example the virus that causes poliomyelitis in humans, is about 30 nm in diameter (and contains RNA, but that has nothing to do with its smallness). The detailed structure of the capsid of many viruses is known. For example, the Tobacco mosaic virus, which is particularly well studied, has a rod-shaped capsid with protein wrapped around a helical shaft of RNA. The protein molecule consists of 158 amino acids and is wrapped around the RNA, which comprises 6395 nucleotides. The remarkable feature of the virus capsid is that it self-assembles into a highly regular, quasi-crystalline structure.

Some capsids are enclosed by a lipid membrane and others are naked. As viruses fuse with cells and inject or transfer their nucleic acid, the presence of a membrane influences the details of this process. The fusion is preceded by binding to a receptor on the surface of the host cell, the receptor molecule might be any one of a variety of molecules, a glycolipid or a proteoglycan for example, but specific for a given fusion. Universally the outcome of fusion is the replication of the virus and replicated viruses often acquire a gene that can be passed on to a subsequent host. Replication is frequently concluded by the destruction of the host cell, lysis, and viruses thus release all sorts of organic molecules into the environment. Sediments in all deepwater localities, including hypersaline basins in the Mediterranean Sea (3000 m) and the Mariana Trench, acquire many refractory chemical compounds as a result, including DNA. It has been suggested that 20% of the ocean biomass is killed by lysis each day (Suttle 2007; Danovaro et al. 2008; Corinaldesi et al. 2019). A spectacular example of this was seen in northern Europe in 1988 and 2002, when Harbour seal populations were decimated by a distemper virus.

Viruses are too small to sink and their distribution in the water column is the result of current flow and the balance between replication and decay. The decay of viruses collected from deep Atlantic water was assessed by Parada and colleagues by measuring their number, over several days, in samples incubated at the temperature of collection. The data allowed the virus turn over time to be calculated. For samples collected from 4000 m depths, this turned out to be about 39 days and for viruses from shallower water, it was around 12 days. However, the measurements ignored pressure, which at the level of 40 MPa might have a significant effect.

Vesicles, mentioned earlier as another form of MGE, are also suspended in ocean water at about the same concentration as the bacteria which generate them. For example, Gram-negative marine bacteria produce vesicles from a swelling of their outer membrane, which pinches off to create a vesicle some 70–100 nm in diameter containing a variety of proteins, lipids, DNA and RNA. They have various functions and consequences, one of which is to perform horizontal gene transfer although how the vesicle fuses with a host cell is not clear.

From experiments on "laboratory" viruses, we can predict that deep-sea pressures must affect deep-sea viruses, and indeed MGEs. Their self-assembly and the numerous molecular interactions they indulge in, such as binding to a receptor molecule and replication, are typically pressure labile (Bispo et al. 2012; Lou et al. 2015).

## 8.7    Microorganisms in the Deep Biosphere

In Chapter 1, some of the salient features of the Deep biosphere were mentioned. Gold's seminal, 1992, paper entitled "The deep hot biosphere" reminds us that it is a recently recognised environment. Here a comment from a 2012 paper by Roy and colleagues alerts us to its strangeness: For example, sedimentary rock containing trapped organic material, such as coal beds, can support a microbiological flora for millions of years. Other, "abiotic" sources of energy are also available. The marine sediments, which reach a maximum depth of 10 km, the subsurface under kilometres of ice (Parnell and McMahon 2016) and the basement "hard rock" are all inhabited by microorganisms and viruses (Anderson et al. 2013). The Hydrothermal vent and cold seep environments also merge with these and are separately considered in Chapter 9.

A very significant proportion of the Earth's biomass is in the Deep Biosphere. The scale of the Deep Biosphere has been confirmed by Bar-On et al. (2018) and by D'Hondt and colleagues in a 2019 review. In terms of volume, the marine sediments are 260 million and the igneous basement rock 1000 million cubic km. The latter is restricted to rock below a temperature of 122 °C, commonly regarded as the upper-temperature limit for life, but see Barras (2021). But organisms occupy the moist interstitial spaces in this material. In the case of the sediments and basement they are each about 80 million cubic km or 6% of the volume of the ocean, i.e. 12% combined. The Deep Biosphere is frequently compared with extraterrestrial environments.

### 8.7.1    The Presence and Activity of Microorganisms in the Deep Biosphere

Collecting organisms from the deep sea is not a simple process but from the Deep Biosphere, it is much more difficult and restricted. Some of the practical difficulties are illustrated in a paper that reported the isolation and subsequent growth experiments carried out with a typical Deep Biosphere bacterium, *Bacillus infernus*, mentioned in Chapter 7. It was obtained from the continental crust which is more accessible than the oceanic crust.

Access to the Oceanic crust usually involves a specialised ship capable of drilling through the sedimentary layers and beyond. The flux of oxygen from seawater into marine sediments can, in nutrient-poor conditions, extend to many metres depth and in rare instances, down to the basement rock. This is due to the low consumption of oxygen by the small population of cells in the sediment. These oxic sediments (oxygen present) typically have cell densities of no more than 10 million cells per gm and often much less. In contrast, anoxic sediments have cell densities typically ten to a hundred times greater. They arise from a greater sedimentation rate in the water above and the great number of cells consuming the oxygen. However, both types of sediment have a similar ratio of Bacteria : Archaea, very approximately one, an abundance of viruses, fungi and diverse extracellular enzymes (Hoshino and Inagaki 2019).

A rather different environment, ice, several kilometres depth of it, can support, or preserve, populations of cells, as in the Antarctic (see Fig. 1.2, Chapter 1) (Parnell and McMahon 2016).

Recycling of organic material is a marked feature of the Deep Biosphere and so is the mutually beneficial activities of different organisms, syntrophy. This "obligatory mutualistic metabolism" is well developed in regions in which organic compounds are degraded, which means marine sediments and much of the Deep Biosphere. It is not unique to these geographical scale environments; the human intestine harbours a syntrophic population of bacteria. For would-be experimenters with Deep Biosphere cells it means that creating reaction conditions which match the natural state can be tricky, as "the product of one conversion step is the substrate for the next organism in the chain" (Dolfing 2014).

Sediments become sedimentary rock which together with igneous and metamorphic rock comprise the potential environment for Deep Biosphere organisms. Evidence for organisms occupying these deep places, such as cell counts and traces of DNA, is not lacking, but proof of ongoing living processes rather than of past activity is important to establish, and the cores made available by the Deep Drilling Program have produced some decisive evidence. Using these cores, Schippers and a team of six produced evidence based on the CARD-FISH technique which targets ribosomal RNA, that "a large fraction of the sub sea floor prokaryotes is alive, even in very old (16 million yr) and deep (> 400 m) sediments". All the living cells were Bacteria.

An example of bacteria from deep sediments (before they become sedimentary rock) is the strictly anaerobic, sulphate-reducing *Desulfovovibrio indonesiensis* strain P23, studied by Fichtel and colleagues, also mentioned in Chapter 7; Section 7.4.2. It came from a sediment on the San Juan Ridge (International Oceanographic Drilling Project site U1301). The ambient pressure was estimated to be 30 MPa and the estimated temperature a maximum of 62 °C. The sediment was composed of very dense clay which blocked water movement from the basaltic aquifer up into the sediments. Here we will consider its growth, best measured from the protein synthesised or sulphide produced. It was fastest at 45 °C and 30 MPa, even after the cells used as the inoculum had been cultured at 20 °C and atmospheric pressure for 3 years. The maximum temperature for growth at atmospheric pressure, 48 °C, was increased to 50 °C at 20°MPa. The growth experiments suggest the estimate of the environmental temperature was high, a point discussed by the authors of the paper. This bacterium appears to have arisen from cells or spores present in the initial sediments.

Much of the Deep Biosphere is much deeper than the above. For example, in one investigation of sedimentary rock, coal, organisms have been detected at a depth of 2.5 km (details coming soon), and even deeper in waters in the fracture zones of other types of strata. In oil reservoirs, there is evidence of microbial activity down to 4000 m below the surface (Head et al. 2003). However, staying with very deep sedimentary environments, it is possible to learn something of the inhabitant's physiology. From the numbers of cells present deep in oxic sediments and the influx of oxygen, it is possible to calculate that the respiration rate, per cell, is exceedingly low, a thousand times less than the sort of rate measured in cultures in the lab. The state of the cells is more like that of cells in long-term stationary phase batch culture than in an active growth phase. In fact, it is more extreme than that. Populations of cells exists in a state in which their rate of metabolism sustains a negligible growth rate for thousands of years. Other studies of the steady-state carbon flux through microbial populations in deep sediments has led to the conclusion that the "turnover" time, to be imagined as the time it takes for a cell to replace its organic material without dividing in two, or in growth terms, the time it takes for one cell to grow into two of similar size, is also exceptionally slow, in the region of 100–3000 years. In this study the bulk of the cells were thought to be living, not dead nor in some state of suspended animation, whatever that is.

A separate argument has been constructed by Parkes and colleagues to determine the growth rate of bacteria in organic-rich layers of deep sediment called sapropels, in the Mediterranean. The layers can be dated and older layers have higher cell densities because of cell multiplication. From the age–cell density relationship the doubling (generation) time can be calculated. It is 920,000 years. To reassure the reader this is not a printing error the figure used in the original paper is 0.92 mya. Microscopic examination of the cells showed the proportion of cells in the process of dividing was such that the actual division process must take 105,000 years. These figures are, of course, to be thought of as averages for a diverse population and even if the data from which they have been derived is only approximate, they imply that an utterly different physiological time scale applies in the metabolism of the cells. Remember that *E. coli* famously divides once every 20 min in lab culture. Intuitively, it seems necessary to postulate inhibition must be at present as it is difficult to imagine reaction kinetics being so intrinsically slow. The phenomenon of life proceeding slowly in circumstances of an extremely limited energy supply is a new subject in biology and the significance of high pressure is unknown.

The concept of "basal power requirement" (Hoehler and Jorgensen 2013) is the minimum energy required to sustain metabolically active cells. Ultimately the basal requirement is to repair macromolecules as they suffer an intrinsic physical tendency to undergo molecular change. In the case of proteins, the constituent amino acids are of the L form, in contrast to the D form. The L and D forms relate to the optical properties of the underlying structure of the amino acids, revealed in the way they rotate the plane of polarisation of light. The naturally present L form spontaneously and slowly change to the D form, a process called racemization, which is speeded up by an increase in temperature. Such a change would impair the function of the protein, so metabolic energy is required to repair the background damage. The point here is that racemization probably incurs no volume change and is therefore unaffected by pressure. In an analogous fashion nucleotides undergo depurination and although the effect of temperature on the process is rather less than on racemization, there is probably no volume change and so the process is probably also not affected by high pressure. That said, the very slow growth rates thought to occur in the Deep Biosphere are well above the ultra-slow rate required to repair macromolecules damaged by physical processes (Jorgensen and Marshall 2016).

The average biologist, learning of this remarkable world of minimal metabolic activity, will probably ask "how on earth (!) do you experiment with such slow-motion life?" Answers have been provided by several investigations and here we shall consider one, a major international project studying the deep coal beds off Japan. The Integrated Ocean Drilling Program Expedition 337, employing the drillship Chikyu at Site C00200, succeeded in drilling to a depth of 2500 m below the seafloor, in a water depth of 1180 m (Fang et al. 2017). Cores were obtained from lignite coal seams and shale of mid-Miocene age (about 20 million years old). The temperature range was 40–60 °C and the maximum pressure about 30 MPa. Intact microbial cells were found over the full depth range. The cores were carefully sampled and stored for subsequent analysis. Contamination checks were rigorous so we can ignore those details and focus on the results. The bacterial communities (Archaea were virtually absent) in the deeper regions were rather different to those in the shallow zones. Many types were identified but we can skip the taxonomic names. The origin of the deeper population was freshwater/terrestrial in contrast to the shallow, more marine community. The depth profile of the cell densities was striking. Greater numbers than expected were present in the upper cores and fewer (100–1000 cells per ml) than expected in the deep cores. Evidence of past microbial activity was provided by the nature of the carbon isotope in methane ($CH_4$) and the presence of the biomarker F430, indicating the cellular presence of a specific enzyme in the sequence of reactions leading to the synthesis of methane. Separate growth experiments using a continuous flow system at 10 °C and normal atmospheric pressure demonstrated the synthesis of methane after 289 days. Concerning the sparsity of cells in the deeper cores, it was suggested that this might be the result of the energy cost of macromolecular repair in warm conditions (40 °C +). This implies the microbial community would be limited by the supply of energy and not by temperature, pressure or other comparable factors.

Experiments have been carried out on bulk samples and on individual cells in an attempt to measure the rate of metabolic processes which, from other studies were predicted to be very slow. The samples experienced decompression during collection and processing. Bulk suspensions of microorganisms were prepared from small pieces of coal incubated in special solutions, which included isotopically labelled methanol and methylamine, for 30 months at 45 °C, and at normal atmospheric pressure. The results showed that metabolism was present and, in particular, that methane was produced by the microorganisms present and not by inorganic means. Furthermore, the biomass generation (replacement) time was, according to conditions, in the range of "a few months to more than 100 years". This is an order of magnitude, or more, faster than the generation times calculated from geochemical processes, mentioned earlier, but it is still exceedingly slow compared to normal microbial activity. The experiments were not carried out at high pressure, 30 MPa would have been appropriate, which might have altered the results significantly. This brief summary hardly does justice to a major collaborative project. (One of the first papers to report the work, in 2012, had 46 authors, headed by Inagaki.)

Growing mixed enrichment cultures at high pressure enabled piezophiles to be isolated (Fang et al. 2017). Gram-positive, endospore-forming bacteria were conspicuous, able to grow in the presence or absence of oxygen. A specific example, 19R1-5, grew best at 42 °C, atmospheric pressure, and at 46 °C grew best at 20 MPa. The corresponding environment (source) conditions were 46 °C and 53 MPa. The mismatch between the two is not understood. The deep community of bacteria in these coal beds most probably have arisen from the original population of bacteria present in the sediments. But is it a community of active cells or of endospores, which respond in enriched cultures to produce physiologically active cells? Spores are visible in sediment samples. As the depth of the sediments grew over time (20 million years), those strains which could grow best and form spores predominated.

Endospores in the deep (400 m) sediments beneath the ocean floor (5000 m maximum depth) off Peru have been investigated by Lomstein et al. (2012). The number of spores was similar to the number of vegetative ie growing, bacterial cells, about ten million per ml sediment. This was estimated by using the chemical markers; dipicolinic acid for spores, muramic acid for cells and spores and acridine orange stain for recently dead but intact cells. The age of the sediment occupied by the bacteria was estimated from racemization, the change in the ratio of the D form of the amino acid aspartic acid to the L form. In growing cells that ratio is about 0.02, a low value because the L form is the normal currency for living cells. On death, racemization proceeds at a known rate. Ratios as high as 0.8 were found in sediments 150–250 m below the seafloor, indicating their age was 10 million years. In these sediments, the original deposition of organic material was sustaining the population of slowly metabolising bacteria (and Archaea), for that length of time. At the time of deposition, the pressure would have been 4.2 MPa in the shallow region rising to 50 MPa in the deepest region, and the temperature typical of the cold deep sea. Over 10 million years the pressure increased only slightly as the depth of sediment grew but the temperature would have increased considerably.

The role of spores in these circumstances is not clear. In well studied "laboratory" bacteria spores are formed when conditions are unfavourable, typically when nutrients are limited. Spores are a survival mechanism, as they are highly resistant, metabolise at a negligible rate yet remain "in touch" with the environment sufficiently to respond to chemical cues, such as amino acids, to start germination which results in normal vegetative cells. It is a curious fact that cues other than nutrients can stimulate germination and one is a hydrostatic pressure of one to several hundred MPa. In favourable conditions, spores survive a very long time, the record being

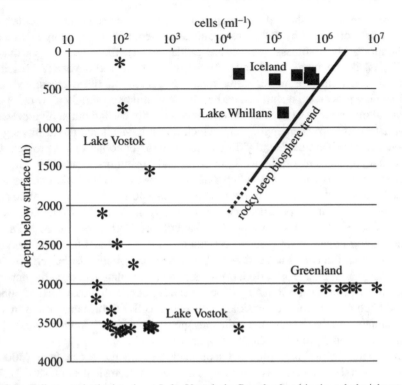

**Fig. 8.1** Cell counts in the ice above Lake Vostok, in Greenland and in the subglacial waters of Iceland and Lake Whillans (Antarctica). The asterisks denote counts from ice samples and the squares from subglacial liquid water. The line relates to cell counts in continental aquifers. (Parnell and McMahon 2016)

250 million years (Vreeland et al. 2000). In the present case, it would be interesting to know if the viability of the spores declined with the age of the surrounding sediment.

Older geological formations supporting microbial activity are known. For example the Canadian Tar Sands are the result of 35 million years of bacterial biodegradation of oil and the Cretaceous black shales (Equatorial Atlantic, off the coast of French Guyana) which supports microorganisms, are 100 million years old (Arndt et al. 2006). Sediments 1700 m below the Newfoundland Margin support bacteria and Archaea at high temperature and pressure and are 111 million years old (Parkes et al. 2014a).

There is a special case of a glacial and sub-ice Deep Biosphere which Parnell and McMahon have described in a 2016 paper. Antarctica, Greenland, the Canadian Arctic, Iceland and even the Alps in both Europe and New Zealand have examples. The very large and deep subglacial Lake Vostok in the Antarctic and the ice above it have been investigated, incompletely, due to the severe conditions, but nevertheless to good effect (Chapter 1; Section 1.2.2). Both the ice and the water in the lake contain bacteria, a significant proportion of which are metabolising (Christner et al. 2006). Figure 8.1 provides some extraordinary data. This is a high pressure but cold environment, with a host of other interesting features, especially sources of energy and nutrients.

## 8.7.2   Growth in High-Pressure Gases

Microorganisms are exposed to significant gas pressures in the Deep Biosphere and in certain laboratory techniques, as we have seen. High-pressure gases present special problems for the engineer and for the biologist they simultaneously present three distinct factors, requiring this digression. They are the partial pressure of the gas, the hydrostatic pressure, which might be the same as the partial pressure or it might be greater, and the chemical properties of the gas. The gas partial pressure determines the concentration of gas that dissolves in the various compartments of cells and tissues. The chemical reactivity of the specific gas is obviously important, and its effects increase with its concentration.

The effects of high pressure of an inert gas such as helium on lipid bilayers provides a simple example to start with. In the following experiment, the partial pressure and the hydrostatic pressure are equal and chemical reactivity is absent. Recall that the phase transition of a pure phospholipid bilayer is a quasi-melting process that occurs at a specific temperature, dependent on the length of the carbon chain. In the case of DPPC, it is 42 °C and the increase in the molar volume of the bilayer at the transition means that high hydrostatic pressure opposes it, raising the temperature at which it occurs (Chapter 2). In the presence of a high pressure of helium, the concomitant hydrostatic pressure increases the transition temperature whereas the helium dissolved in the bilayer fluidises it and lowers the transition temperature. The net result is helium pressure increases the transition temperature of DPPC bilayers by 0.21 C per MPa whereas hydrostatic pressure increases it by 0.24 °C per MPa. The pressure of nitrogen, much more soluble than helium, increases the transition temperature by only 0.06 °C per MPa, meaning that pressure and the fluidising effect of nitrogen nearly cancel out (MacNaughtan and Macdonald 1980). Unfortunately, methane has not been used in this context but we can reasonably predict that its bilayer fluidising effect would offset the accompanying hydrostatic pressure. Methane is an important gas in the Deep Biosphere.

Just as helium can dissolve in a lipid bilayer so other inert gases (methane is included here) can dissolve (bind) in the hydrophobic interior of proteins without entering into chemical reactions, of which more soon. The classical inert gases, in increasing order of their feeble ability to react chemically, are: helium, followed by hydrogen (surprisingly), neon, argon, krypton, xenon and, for good measure, nitrogen. They all have weak anaesthetic properties, which do not involve chemical reactions, manifest in human divers and experimental mammals (Chapters 11 and 4, Section 4.3). Once helium was suspected of being so weak that its "anaesthetic" effect would be masked by the hydrostatic pressure component of its partial pressure. However, it does have some weak biological effects and so too do hydrogen and methane which may have unexpected effects in the Deep Biosphere. These effects are generally referred to as narcotic effects at the cellular level and in air-breathing animals they, and many others, can cause, or contribute to, inert gas narcosis which is a form of anaesthesia (Chapter 11).

### 8.7.3   High Partial Pressures of Helium, Hydrogen and Methane

Experiments have shown that high pressures of helium and of hydrogen affect ordinary living cells in a way distinct from the concomitant hydrostatic pressure (Chapter 6; Section 6.2.1). These inert gases thus counteract the effects of hydrostatic pressure in dividing cells (as the more potent gases do in whole animal experiments, see also Chapter 4, Section 4.3; Chapter 11, Sections 11.2, 11.4.3). At the time of the experiments (Macdonald 1975), the preferred interpretation was that the gases weakly interacted (bound) at molecular sites in such a way as to counteract the pressure effect, but now it is understood that their presence, as apolar inert molecules, might conceivably affect the structure of water close to the sites primarily affected by hydrostatic pressure—a solvent effect. Similar high pressures of helium counteract the inhibition of growth by hydrostatic pressure in *Acholeplasma laidlawii* (30 MPa) (MacNaughtan and Macdonald 1982) and the piezotolerant marine bacterium EP-4 (50 MPa) (Taylor 1987). Somewhat lower pressures of helium (less than 14 MPa) offset the inhibition of the growth of *Saccharomyces cerevisiae* (Thom and Marquis 1984) and even lesser pressures (2–4 MPa) actually stimulates the growth of *Streptococcus faecalis*. To add to a confusing picture, the thermophilic Archaeon ES4, originally isolated from a hydrothermal vent, grew rapidly in a hydrostatic pressure of 50 MPa at 95 °C but much less rapidly when the same pressure was applied with helium (Nelson et al. 1992). However, the difference was absent at 25 MPa. How might these results be transferred to conditions in the Deep Biosphere? Perhaps the best answer is they alert us to the unpredictability of pressure–gas–temperature interactions at the cellular level.

Methane's cellular narcotic or toxic effects have not been fully explored (Narcotic implies reversible and benign, toxic irreversible and lethal). Methane is grouped with other saturated short-chain hydrocarbon gases, ethane, n-propane and cyclopropane, the last being a clinical general anaesthetic years ago. Methane at a partial pressure of 0.2 MPa protects mice from electroshock convulsions and around 10 MPa blocks nerve conduction in rat sciatic nerve. About 20 MPa severely inhibits the growth of yeast cells. 4 MPa methane blocks mitosis in onion root cells, compared to 0.15 MPa n-propane and it specifically binds to sickle cell haemoglobin, reversing the sickled state (Chapter 2). It probably binds to hydrophobic regions of proteins in general and to the interior of lipid bilayers (Macdonald and Wann 1978). It is a hazard in mining, tunnelling, deep drilling and construction work on landfill sites as it is potentially inflammable and explosive. The unsaturated hydrocarbon gases such as ethylene, propylene and n-butene are much more toxic (for example, to yeast cells), but it is not clear how significant they are in the Deep Biosphere. They have the interesting property, probably shared with saturated short-chain hydrocarbons, of dissolving in the yeast cell membrane and expanding its area. From these laboratory experiments, we should conclude that for cells to thrive in hydrocarbon environments is rather surprising, irrespective of the temperature or pressure.

## 8.7.4   Gases Important in the Deep Biosphere

Subsurface microorganisms produce and/or consume hydrogen, hydrogen sulphide, carbon dioxide, nitrogen, ammonia and methane, typically in a warm or hot high-pressure environment.

Hydrogen gas contributes to the "dark energy" available to microorganisms remote from sunlight or the products of photosynthesis. Methanogens use hydrogen to synthesise methane according to: $4H_2 + CO_2 = CH_4 + 2H_2O$. For example, the fracture waters in the deep mines of South Africa support this process. Ethane ($C_2H_6$) and propane ($C_3H_8$) are also produced by microorganisms. Chemolithotrophs are organisms that secure their energy from oxidising inorganic compounds, here hydrogen, a primaeval biochemical reaction.

The Archaeon *Methanococcus thermolithotrophicus* manifests an optimum growth yield at 65 °C and 50 MPa hydrostatic pressure, fuelled by hydrogen gas, as in the above equation (Bernhardt et al. 1988). The partial pressure of the dissolved hydrogen was 0.08 MPa and the experimenters ensured the culture tube, made of nickel in place of plastic, was impermeable to hydrogen. At pressures higher than 50 MPa lysis occurred and at 75 MPa the shape of the cells turned from the normal cocci to large elongated cells, presumably because cell division was impaired.

The strange acronym SLiME comes from Subsurface Lithoautotrophic Microbial Ecosystems, which are interesting as they can generate organic material on which chemoorganotrophs thrive, under moderate hydrostatic pressure. This is a plausible scenario for the early stages in the evolution of life. Hydrogen is available in the deep biosphere, but other energy sources, such as organic compounds photosynthetically produced, can be carried in by water from lesser depths in specific regions. Chemoorganotrophs metabolise this organic material as well as the material produced by "dark energy".

Methane is an important gas both in the Deep Biosphere and in the deep ocean. It may exist as a free gas, and as gas in solution or as a hydrate, which looks like frozen snow and indeed the crystals are lighter than water. The term clathrate applies to the special form of hydrate in which methane (four molecules) and water (23 surrounding molecules, a cage) exist in a stable crystalline structure. Other gases form clathrates, such as carbon dioxide and nitrogen. At 5 °C and 5 MPa hydrostatic pressure, seawater saturated with methane forms hydrate crystals. Crystals form at 0 °C and about 3 MPa and at 20 °C a pressure of 29 MPa stabilises the crystal. In sediments, the hydrate is likely to incorporate methane which has been produced by microorganisms, but the hydrate is also formed from thermogenic methane.

In order to measure the three different states in which methane exists in sediments, free gas, dissolved gas and clathrate, it is necessary to retrieve isobaric sediment cores and "degas" them in controlled conditions. The first report of this achievement was by Heeschen and his team in 2007. Sediment cores from 500 and 1000 m depths in the Gulf of Mexico were retrieved by two rather different pieces of equipment, the Multi-Autoclave Corer and the Dynamic Autoclave Piston Corer, which have a depth capability of 1400 and 3000 m, respectively (Chapter 13; Section

13.9.1). They are rather different in their mode of operation but both pull the sediment core into a cylindrical pressure vessel which retains the ambient pressure as it is brought to the surface. On board the ship the pressurised vessel is connected to high-pressure plumbing and decompressed in a way that allows the free gas to be collected. The methane is accompanied by small amounts of ethane and propane. X-Ray Computer Tomography (medical equipment) was used to detect the volume of free gas in the vessel before decompression. The results can be left to the geologists to unravel, a good grasp of physical chemistry is needed, and biologists can enjoy the direct demonstration of this strange high pressure, tripartite sedimentary environment which organisms inhabit.

Globally, methane hydrate is important as a potential source of energy (good) but it is a greenhouse gas (bad), so its deliberate release into the atmosphere is contentious (Adam 2002).

Many microorganisms can anaerobically oxidise methane in the presence of sulphate according to $CH_4 + SO^-_4 = HCO^-_3 + HS^- + H_2O$. This is a major source of metabolically useful energy in the Deep Biosphere. Methane is a reactant like any other and the idea that its concentration might drive the rate of sulphide production in the above reaction stimulated a number of studies. They also illustrate the development of high-pressure techniques for investigating the biochemistry and geochemistry of the Deep Biosphere.

Nahaus and colleagues (2002), based in Bremen, Germany, obtained mixed samples of microorganisms, Archaea and Bacteria, from sediment cores taken at a temperature of 4–5 °C and a depth of 780 m on the Hydrate Ridge, in the north Pacific Ocean. In situ, the methane partial pressure was 0.1 MPa but incubating the samples in a pressure vessel with a methane partial pressure increased to 1.1 MPa greatly increased the rate of sulphide production.

A relatively simple apparatus for incubating rock or sediment samples at high pressure in either batch mode or continuous flow was developed by Sauer et al. (2012), at the University of Potsdam, Germany. It was used to monitor the oxidation of methane by microorganisms in mud collected from 1020 m depths on the Isis mud volcano, off the River Nile delta (Mediterranean Sea). A sample incubated for 18 days at 23 °C and 10 MPa in the presence of a very low methane concentration produced sulphide at a constant and low rate. Samples similarly incubated, but in the presence of a partial pressure of 4 MPa methane, showed sulphide production increasing after 8 days and continuing at a high rate for the next 10 days. Part of this increase might be due to microbial growth but most is due to the kinetic effect of methane concentration.

A final example is provided by Rajala et al. (2015) and Rajala and Bomberg (2017), using samples from a depth of 500 m in a drill hole in Fennoscandian bedrock, Finland. The mixed population of microorganisms, a community of Archaea and Bacteria, were not metabolically active in situ, where the dissolved methane concentration was quite low. When it was increased by about fourfold, both respiration and sulphide production (as above) and transcription were stimulated in 60% of cells, at normal atmospheric pressure.

The last three experiments used material collected with decompression, so now we need to consider organisms isobarically collected from the Deep Biosphere.

## 8.8   Isobaric Samples from the Deep Biosphere

Several isobaric sampling devices have been described which are capable of collecting water samples from the Deep Biosphere but they have been used for chemical rather than biological purposes (Chapter 13; Section 13.9). Isobaric sampling in the Deep Biosphere for microbiological experiments is a difficult proposition but it has been achieved by several groups.

A comprehensive system for culturing microorganisms from the Deep Biosphere has been developed by Parkes at the University of Cardiff, UK and his collaborators (2009). They have combined the isobaric retrieval of drill cores, using a system called Hyacinth, with an isobaric store for drill cores and a core cutting and processing apparatus called Press. The pressurised and refrigerated samples can be transported from the drill site to the laboratory in the UK where microbiological processing can take place. The isolation of pure strains uses an isobaric method similar to that developed by Jannasch and colleagues. Isobaric transfers also enable a wide range of enrichment cultures to be grown over many weeks, at selected pressures. The system can perform its isobaric functions at pressures up to 25 MPa and the batch cultures can be conducted at 100 MPa. All the subsurface sediments studied contained methane hydrates and an abundance of microorganisms. For example, cores were obtained from the Cascadia Margin, in the Pacific west of Vancouver Island. They were from 170 m below the seafloor in a water depth of 1315 m and isobarically handled at 15 MPa in anaerobic conditions. Batch cultures (14 weeks) in one enriched culture medium produced maximum growth at 14 MPa but in other media, growth was best at atmospheric pressure and declined with an increase in pressure. Growth was measured by microscopic counts of stained cells. In another major drilling project off the Indian east coast, sediment cores were secured from a water depth of 1049 m and a sediment depth of 77 m and subsequently handled at a constant 12.5 MPa. The variously enriched cultures (6 weeks) yielded bacterial species similar to those seen in the Cascadia sediments. Over a longer time three groups were conspicuous: *Carnobacterium*, *Acetobacterium* and *Clostridium*. Because enrichment cultures grown at 80 MPa for 560 days also grew at atmospheric pressure the isolation of pure cultures was carried out at atmospheric pressure.

Considering the pressures from which the cores were retrieved it is not surprising that many cultures proved the be piezotolerant, a few were peizophilic and none were obligately piezophilic. Broad characterisation of the metabolism of some of the cells (anaerobic, aerobic, etc.) was reported and genomic analysis was also carried out, but the results are not closely related to the question of high-pressure tolerance and adaptation. However, Archaea were absent from all sediments examined, which is interesting. The sediment depths which were sampled imply that the flora living there is permanent rather than recently arrived from the surface fall out. This work provides a much-improved picture of what is going on at the top of the deep Biosphere.

Case and a team (2017) used an isobaric sediment retrieval system deployed by an ROV, in an attempt to investigate Deep Biosphere organisms more economically than by using deep drilling rigs. The system, a development of an earlier JAMSTEC apparatus, is called HP-Core. It was used to collect material from a depth of 985 m on the Joetsu Knoll, off the north coast of Japan. The mother ship R V Natsushima operated the ROV Hyper-Dolphin, whose robotic arm scraped bacterial mats growing on the surface of rock—like methane hydrate, along with sediment, into the receiving pressure vessel. During the transit of the ROV from depth to the mother ship, there was an accidental drop in the pressure in the sample vessel to 2 MPa but once on board, the pressure was restored to the original 10 MPa, which was held thereafter. Over a 40-day incubation the anaerobic oxidation of methane, mentioned earlier, proceeded and when oxygen was added to the culture methane was consumed aerobically. Microbial diversity was assessed. It was considerable but of no obvious relevance to the question of high-pressure adaptation. However, this system is clearly capable of carrying out isobaric experiments on organisms from deep sediments.

### 8.8.1  CORKs and FLOCS

The use of boreholes in the ocean floor for continuous monitoring and study has proved possible by the development of retrofit kits (Circulation Obviation Retrofit Kits, CORKs) which contain passive samplers. These provide samples for analysis in shore laboratories. To tackle questions about microbial colonisation and growth in deep borehole waters, Flow-through Osmo Colonization Systems, FLOCS, have been developed. These provide substrates, such as small pieces of rock, which are colonised by microorganisms over many months. They are deployed and retrieved by a manned submersible. Thus far the results are primarily demonstrating what grows and roughly how quickly. As it was not the intention, it is not surprising that they have not yet provided biochemical or physiological information relevant to high-pressure adaptation, but they might. The use of these microbial observatories and other in situ equipment such as landers is described in Toffin and Alain (2014) and in Chapter 13.

### 8.8.2  Pollution in Deep Environments: Hydrocarbons and Other Toxic Compounds

The Deep Water Horizon catastrophe in the Gulf of Mexico arose from the rupture of a wellhead situated on the seafloor at a depth of 1544 m and 4 °C. Oil and gas flowed into the sea, creating a buoyant plume and then a descending mass of dispersed oil which eventually covered 3000–8000 square km of seafloor. About 30% of the released gas and oil was bio-degraded in the water column. Sedimentary microorganisms also responded. Certain bacteria are recognised as specialists in this activity, for example *Marinobacter hydrocarbonoclasticus* (Grossi et al. 2010). Not all contaminant hydrocarbon compounds are toxic. Methane, propane and ethane are respired by bacteria, creating changes in the natural communities. Pressure experiments have been carried out with specific bacteria, demonstrating a variety of metabolic and biochemical effects in the response to hydrocarbons but no overriding, dramatic outcome emerged. Generally, the deeper the pollution the slower it is processed (Fasca et al. 2017). Although oil spills are serious for humans and disturb the environment, they are not all that different from natural oil seepages. Some have even described the outpourings of hydrothermal vents (next chapter) as natural pollution.

Some attempts have been made to construct a thermodynamic view of the effectiveness of hydrocarbon pollutants in the deep sea, stretching narcosis theory, and in particular the pressure reversal of anaesthesia (Chapter 11) beyond proper limits. The general conclusion that high pressure mitigates the potency of hydrocarbon pollutants is inconsistent with two facts: high pressure acts additively with anaesthetics in certain Crustacea and in dividing cells (Youngson and Macdonald 1967; Kirkness and Macdonald 1972), and hydrocarbon pollutants contain toxic compounds which are not unreactive narcotics (Paquin et al. 2018).

The development of deep-sea mining techniques has created a major problem of toxic pollution which is set out in an excellent paper by Hauton et al (2017). The paper explains the scale of the problem and provides information about the organisations concerned with regulating the industry (See also more recent papers by Santos et al. 2018 and by Lelchat et al. 2020).

## 8.9 Adaptations of the Enzymes of Deep Sea and Deep Biosphere Microorganisms to High Pressure

We start with an example of a pressure-adapted bacterium discovered in particularly unusual circumstances. An underwater telescope, in the form of an array of light detectors, is positioned at a depth of 2745 m in the Mediterranean Sea. It is designed to detect neutrinos but inevitably detects luminescence emitted by animals (Priede et al. 2008) and by bacteria. One species of the latter has been isolated by a French team of scientists (Ali et al. 2010), *Photobacterium phosphoreum*. They reported it to be piezophilic and that its bioluminescence at 22 MPa was much brighter than at normal atmospheric pressure. Bioluminescence is a complex enzymatic process that featured in the early studies of the kinetics of high pressure (Johnson et al. 1954), which will make the details of this phenomenon, when they are unravelled, particularly interesting. A start has been made by Martini et al. (2013).

In Chapter 6, the pressure adaptation of the proteins (and enzymes) of deep-sea fish was classified as intrinsic or extrinsic, so the same distinction is adopted here.

## 8.9.1　Intrinsic Adaptation of Microbial Proteins to High Pressure

Research in bacterial adaptation to high pressure benefits from molecular genetics to a greater extent than is the case with other organisms. *Photobacterium profundum* SS9 was the first deep-sea piezophilic bacterium to have its genome sequenced, followed by *Shewanella piezotolerans* WP3. Many more species have now that status. This is fundamental progress. The molecular geneticist sees adaptation to the environment as a change in the genome, a change in coded information. That can tell us what types of enzyme and other proteins a given organism might produce, but it cannot yet provide the fine detail required to understand the adaptation of a protein to specific conditions. Recent progress in understanding protein folding suggests this may change, see Deep Mind and Alpha folding on the Web. The biochemist/ physiologist sees adaptation at the "cutting edge", in the molecules in the phenotype whose fine details and coordinating factors perform, as a whole, the function whose effectiveness is selected for in Nature. That affects the survival of the individual and of the critical genes involved. This book is primarily concerned with the molecular, and other, adaptations of the phenotype and not with coded information. It is concerned with energy, the P. $\Delta$ V term in thermodynamics and its kinetic consequences, and not only at the molecular level (Chapters 10 and 11).

Motility is one of the most pressure-sensitive aspects of *E. coli* and also of the pressure-sensitive *P. profundum*, strain 3TCK, isolated from shallow seas (See Chapter 12; Section 12.1.3 for details). The piezophilic *P. profundum* SS9, in contrast, is motile at high pressures and has two motile mechanisms. One is a rotary motor that drives a polar flagellum, powered by a gradient of Na ions. This drives swimming through water and is effective at high pressure. In contrast, the equivalent mechanism in strain 3TCK functions best at atmospheric pressure. The other is a rotary motor powered by a hydrogen ion gradient which drives lateral flagella, enabling the cells to swarm over a moist substrate at high pressure. The genetic basis for these systems is well documented but the adaptation of the mechanisms in *P. profundum* to high pressure is not understood (Eloe et al. 2008). The deep-sea *Shewenalla piezotolerans* WP3 is similarly equipped (Wang et al. 2008).

High pressure affects gene expression. Examples of this are thought to play a role in adapting a deep-sea bacterium to high pressure. In *P. profundum* strain SS9, pressure affects the transcriptional regulation of genes that code for proteins that become positioned in the cells' outer membrane and function as channels, known as porins. Porins are trimers and each of the three subunits is a short, wide cylinder, traversing the membrane as a tightly bound triplet. They are positioned in the outer membrane of Gram-negative bacteria where they function as pores, connecting the periplasm with the external environment. Solutes and water diffuse through the pores which are gated (open-closed) by various little-understood mechanisms. Recall the outer membrane of these bacteria is unusual in that its outer monolayer is largely lipopolysaccharide, to which the porin is coupled by an elaborate assembly process. Bartlett and colleagues (Bartlett and Welch 1995; Bartlett et al. 1995) found certain porins to be pressure regulated; outer membrane protein OmpH, whose gene is ompH, and OmpL (gene ompL). The ratio of OmpL to OmpH depends on the degree of bilayer order produced by pressure (Welch and Bartlett 1998). The transcription regulator, ToxR, is a protein that is attached to the inner membrane and whose cytoplasmic domain binds to the genes in question. The fluidity of the inner membrane bilayer seems to affect the way the regulator allows transcription to proceed. When the bilayer is relatively fluid, the synthesis of OmpL is favoured over OmpH. This occurs at high temperature and low pressure. On the other hand, when the bilayer is well ordered, the transcriptional regulation determined by ToxR favours the synthesis of OmpH. In the presence of an inert anaesthetic, which fluidises the bilayer, ToxR favours the synthesis of OmpL, even at high pressure. OmpH is presumably adapted to function at high pressure which leads to its expression. It has a larger pore than OmpL but it is not clear why a large number of the OmpH porins are required in the outer membrane at high pressure. The picture is still incomplete at the physiological level but this pressure regulation of gene expression is assumed to be "ecologically meaningful".

An interesting footnote to the work on OMPs is the dissociation of the ToxR dimer by 20–50 MPa, which is a rare example of a moderate pressure dissociating a membrane–protein complex (Linke et al. 2008).

Porins can be extracted from bacteria and reconstituted in the phospholipid membrane of liposomes (Chapter 4; Section 4.5.3.1). This may seem surprising in view of the peculiar outer membrane from which they come, but, even more surprisingly, they can then be investigated with a patch clamp technique which not only confirms the porins are stable and functional in the liposome, but they are all orientated in the same way. At least that is the case with OmpC extracted from *E. coli*. This porin is incorporated such that its normally outer end is positioned at the outside face of the liposome bilayer. This fortuitously convenient property does not end here, as the patch of liposome bilayer formed at the tip of the patch pipette is, like a patch of a natural membrane, stable at high pressure (Chapter 13; Fig. 13.4).

OmpC from *E. coli* reconstituted in this way appears to be gated by voltage and preferentially permeable to cations in normal conditions. Using high-pressure apparatus (Chapter 4), it was shown by the author in collaboration with Martinac and Bartlett, that whilst 30 and 90 MPa, at room temperature, did not have a dramatic effect on the gating kinetics, pressure favoured the opening of the porin without affecting its conductance. The experiments also showed that the activity of the porins OmpH and OmpL expressed in *P. profundum* might be measured at high pressure, perhaps revealing some pressure adapted feature in OmpH (Macdonald and Martinac 1999; Macdonald et al. 2002). However, the porins in *P. profundum* function in contact with seawater and periplasm, the latter probably being rather different to that of *E. coli*, which normally lives in the mammalian gut. The process of porin extraction and reconstitution in liposomes was based on *E. coli*, so caution about using the same procedure for a marine psychrophilic bacterium was sensible. However, the fortuitously convenient nature of this area seemed to extend to *P. profundum*, as the established procedure seemed to work well. Patch clamp recordings of OmpH and OmpL demonstrated typical porin-like behaviour, despite being in contact with alien solutions. With one such solution on either side of the patch, and holding potentials of between 10 and 100 mV, high pressure (30 and

**Table 8.1**  Pressure stability of the protein SSB in strains of *Shewanella*

| Strain | Depth of origin<br>meters | optimum growth pressure<br>MPa (at optimum temperature) | P 1/2 MPa |
|---|---|---|---|
| *Shewanella hanedia* | 1 | 0.1 | 52 |
| ..        SC2A | 2000 | 13.6 | 65 |
| ..        PT99 | 8600 | 68 | 75 |

P1/2 is the pressure which dissociates half of the SSB oligomers. A stable oligomer is required for normal function and the differences in the stability lies, ultimately. in the primary sequence of amino acids.

83 MPa) appeared to favour the opening of OmpH rather more than OmpL. The conditions are utterly unphysiological but, on the assumption that appropriate changes to the solutions can be made to mimic those prevailing in vivo, the prospects for investigating OmpH and OmpL at high pressure are good.

The adaptation of the respiratory pathway in *P. profundum* SS9 to high pressure has been studied and compared with that of *Shewanella violacea*. This is a compli-cated, multi-component system to compare, but interesting differences have been identified. Whereas *S. violacea* growing at a higher than optimum pressure of 50 MPa has cytochromes somewhat different to those at lower pressure (Oger and Jebbar 2010), *P. profundum* does not change the gene expression of its cytochromes in corresponding conditions. One idea put forward is that *P. profundum*'s ability to modify its membrane lipids when grown at high pressure may, in some way, assist its respiratory enzymes to perform. The speculative idea seems to be that *S. violacea*, which is adapted to higher pressure than *P. profundum*, may have the enzyme components of its respiratory pathway adapted intrinsically, and *P. profundum* gets by with extrinsically adapted enzymes (Tamegal et al. 2012).

In contrast, a much simpler example of pressure adaptation is seen in the repair and replication of single-stranded DNA. This is accomplished by a protein with the acronym SSB and its oligomeric structure is understood in some detail. The pressure stability of that structure was investigated by Chilikuri and colleagues (2002) and Chilikuri and Bartlett (1997), at the Scripps Institution of Oceanography, California using strains of *Shewanella*. Three strains were selected and the pressure stability of their SSB was determined. Table 8.1 summarises the correlation between the pressure/depth of origin of the strains, their optimum growth pressure and the pressure which dissociates half the SSB oligomers, a measure of its stability. The correlation is not particularly strong, but surely ecologically meaningful.

Other proteins extracted from deep-sea bacteria that have intrinsic high-pressure features might be more interesting to chemists than biologists.

One intriguing enzyme, dihydrofolate reductase, otherwise DHFR, is involved in the synthesis of nucleotides (Murakami et al. 2011; Ohmae et al. 2012, 2013a, b). DHFR in the shallow water *Shewanella oneidensis* has the anomalous property of being activated by high pressure. The same enzyme from several deep-sea bacteria, including *Shewanells violacea*, *S. benthica* and a different genus *Moritella profunda*, showed a biphasic effect of high pressure; activity increasing up to 50–100 MPa and subsequently decreasing at higher pressures. The DHFR in other deep-sea bacteria, *M. japonica* and *P. profundum*, are only inhibited by pressure over the same range. These differences are quite unexpected but they keep us alert! In one paper the intriguing statement is made: "DHFRs from ambient atmospheric species are not necessarily incompatible with deep sea environments, and interestingly, DHFRs from deep-sea *Shewanella* species are not necessarily tolerant to high pressure".

Deep-sea bacterial DHFR has been intensively studied by a host of techniques, following the detailed work carried out on the catalytic reaction of DHFR from the standard laboratory bacterium, *E. coli*.

Consider, for example, some experiments on DHFR from the bacterium *Moritella profunda*, collected from a depth of 2815 m and, for comparison, the laboratory *E. coli*. Their names are abbreviated to mpDFHR and ecDHFR (Ohmae et al. 2012). The reduction of dihydrofolate to tetrahydrofolate requires the ligand NADH, and the reaction can be followed optically at high pressure. mpDHFR shows the biphasic effect of pressure, previously mentioned, which is interpreted as a change in the rate-determining step (and hence thermodynamic parameters) at around 50–100 MPa. The experiments were carried out at 25 °C so both the temperature and the critical pressure are unlike those experienced by the enzyme in nature. In contrast, the activity of ecDHFR is steadily reduced by pressure, although that of an active site mutant in which aspartic acid is substituted by glutamic acid (designated D27E) showed a biphasic effect of pressure, similar to that of mpDHFR, above. As there are many, at least five, steps in the catalytic reaction of ecDHFR it is perhaps not surprising that the experiments with mpDHFR and the D27E mutant shows the complication of a change in the rate-limiting step. Whether the difference between these enzymes has any adaptive significance is not clear. Clearly, the D27E mutant, mimicking the deep-sea enzyme, shows the potency of a single amino acid substitution in the enzyme's active site to offset its sensitivity to high pressure (Ohmae et al. 2013a).

Changes in the structure of the enzymes under high pressure were measured optically, using a well-established fluorescence technique, which detects the unfolding of the protein. The unfolding of ecDHFR under pressure involved a much greater volume decrease than that of mpDHFR. Such volume decreases can only arise from a decrease in the void volume of the protein, or from an increase in the electrorestriction of solvent water at the surface, or a combination of the two. Other measurements showed that mpDHFR has a smaller partial molar volume (displacement volume) than ecDHFR, yet it has a greater aggregate void volume than ecDHFR. As their surface areas are similar we have to conclude that: (a) mpDHFR is has a more loosely packed structure than ecDHFR, and (b) it has an intensely hydrated surface which keeps its partial molar volume small. Exactly how the characteristic feature of mpDHFR relates to its effective function at high pressure is not explicitly stated by the authors of this work. It is difficult to imagine that there is no such connection and significantly an American group (Huang et al. 2017; also Schnell et al. 2004), approaching the problem using molecular simulations, have suggested that the loose structure of mpDHFR means that at high-pressure water enters the void volumes without distorting the structure of the enzyme, and hence affecting its activity. This mpDHFR is pressure stable. However, in assessing the "fitness" of an adapted enzyme it is essential to focus on the kinetic conditions (concentration of ligands, substrate and other factors) which prevail in physiological conditions.

The structure of another enzyme, 3-isopropylmalate, one which plays a part in the synthesis of the amino acid leucine, has been investigated by kinetic, mutational and crystallographic methods. It is conveniently abbreviated to IPMDH and the enzyme from the deep-sea *Shewanella benthica* (sbIPMDH) was compared with the same enzyme from the shallow water *Shewanella oneidensis* (soIPMDH). In vivo the enzyme is a dimer, with the two subunits joined by a hinge region which allows movement between an "open" and a more "closed" state. Pictures of its structure derived from crystallographic analysis show the two subunits to be a tangle of coiled polypeptides within which numerous cavities, void volumes, exist. These are significant because under high-pressure molecules of water from the surrounding solution enter the cavities, causing its partial molar volume, to decrease, in accordance with Le Chatelier's Principle. The soIPMDH has 32 such cavities in its subunits whereas sbIPMDH has 39 and the total volume of the cavities is 1426 A and 1537 A, respectively (Nagae et al. 2012). The difference is considerable and clearly sbIPMDH has a looser structure which, it is argued, would allow it to compress under pressure with less distortion to its native structure. The additional void volume is described as a sort of damper which renders the catalytic activity of the enzyme much less susceptible to high pressure than that of soIPMDH.

The kinetics of the enzymes were investigated in a separate study, and the activity of sbIPMDH remained high up to 100 MPa, revealing it to be a remarkably pressure-resistant enzyme. In contrast, the activity of soIPMDH decreased steadily with increases in pressure, typical of a shallow water enzyme. Studies using mutant enzymes established that the central region of the dimer is the critical part of the molecule where adaptation to pressure, in the form of a change in the amino acid sequence, has led to the creation of void volumes. In particular, the substitution of alanine for serine at position 266 is particularly important in conferring the pressure resistance of sbIPMDH. This conclusion goes a long way to justify the title of the major concluding paper: "Pressure adaptation of 3-isopropylmalate dehydrogenase from an extremely peizophilic bacterium is attributed to a single amino acid substitution" (Hamajima et al. 2014, 2016). A final comment on this work is that these high-pressure studies have extended to pressure well beyond those of immediate biological significance. Deep-sea organisms provide us with proteins that are not only of practical use (Chapter 6; Section 6.1.1.2) but of considerable theoretical interest to chemists.

Finally, it is interesting to note that a fundamentally different approach to understanding the adaptation of proteins to high pressure is being pursued by Martinez et al. (2016). Using a version of a neutron scattering technique they assess the dynamic state of water molecules on and around molecular surfaces in the cell. Differences are seen between a piezophilic archaeon, *Thermococcus barophilus* and the pressure-sensitive *T. kodakarensis* at normal and high (40 MPa) pressure. It is too early to conclude that these differences are significant, but the approach is potentially powerful.

## 8.9.2    Extrinsic Adaptations of Proteins

The role of compatible solutes in protecting pressure labile proteins in deep-sea fish was described in Chapter 6. The question arises, do deep-sea microorganisms similarly protect their proteins against high pressure?

Evidence that "ordinary" bacteria can be protected from the lethal effects of high pressure by the presence of compatible osmolytes comes from a surprising source, the use of high pressure to sterilise food. A potentially dangerous and quite common bacterium found in such foods as cooked meats and soft cheese is *Listeria monocytogenes*. In laboratory tests, *Listeria* can be killed by a 5-minute pulse of 400 MPa, but if the cells contain betain and, or, L-carnitine, the lethal effect is much reduced. A protein protective mechanism is very likely to be at play here, but effects on membrane fluidity are also possible (Smiddy et al. 2004).

A number of Archaea and Methanogenic bacteria are known to accumulate osmolytes such as beta glutamate, which has no role in metabolism (Robertson et al. 1999), and which in other organisms appears to stabilise proteins against high pressure, so it was thought that such osmolytes might be accumulated in deep-sea bacteria to stabilise their proteins. Martin et al. (2002) accordingly used *P. profundum* to see if osmolytes accumulated when it was grown at high pressure. They found that an unusual compound, beta-polyhydroxybutyrate (PHB) accumulated when the cells were grown at 28 MPa. It also accumulates when the cells are grown in a high salt concentration. Thus, beta-PHB apparently functions both as a piezolyte and as an osmolyte. (Recall that beta-amino acids are not used in normal proteins and they are consequently metabolically inactive.)

Other studies with *P. profundum* SS9 and with *P. phosphoreum* (mentioned earlier) but here strain ANT-2200 show that TMAO is involved in piezophilic metabolism (Campanaro et al. 2005; Zhang et al. 2016). Another example is *Vibrio fluvialis* strain QY27, which was isolated from a depth of 2500 m, by Yin and a team of scientists from several Chinese laboratories (2018). At 10 °C its growth rate is reduced by high pressure but in the presence of TMAO, provided as a nutrient, the growth rate is much less inhibited. Using the criterion of biomass at stationary phase, the optimum pressure for growth increased from atmospheric pressure to 30 MPa in the presence of 1% TMAO. The cell becomes a piezophile. It appears that the cells' ability to respire anaerobically when so required comes into action. TMAO becomes an electron acceptor in anaerobic respiration, a change that involves the induction of TMAO reductase by high pressure. It appears that TMAO contributes to the piezophilic metabolism and may also protect proteins from the effects of high pressure. Its concentration in the cell is similar to the concentration which pressure-protects enzymes in vitro (Yancey and Siebenaller 1999).

If the reader, like the author, was looking for clear cut examples of TMAO stabilising proteins against the adverse effects of high pressure, then microorganisms have failed to deliver. It is hard to imagine that the presence of TMAO is without such benefit, and clearly, this is an attractive hypothesis to test, using extracted proteins.

The adaptation of the membranes of deep-sea microorganisms is dealt with in Chapter 7, Section 7.3.

## 8.10   Conclusion

Deep-Sea and Deep Biosphere microbiology is a huge body of knowledge and its high-pressure aspects have only been lightly touched on here. Isobaric collection and subsequent processing is important for several reasons but is not required for many purposes. The adaptation of Prokaryote proteins to high pressure by intrinsic means mirrors that in deep-sea animals, and extrinsic adaptation appears very likely. The microorganisms and enzymes of The Deep Biosphere seem to be unique in the extent of their slow turnover. Their adaptation to high pressure is self-evident, if still largely obscure, and it is not clear if pressure plays a role in their life in slow motion.

## References[1]

Abegg F, Hohnberg H-J, Pape T, Bohrmann G, Freitag J (2008) Development and application of pressure-core-sampling systems for the investigation of gas and gas-hydrate—bearing sediments. Deep Sea Res. I 55:1590–1159. *

Adam D (2002) Fire from ice. Nature 415:913–914

Ali BA et al (2010) Luminous bacteria in the deep sea waters near the ANTARES underwater neutrino telescope (Mediterranean Sea). Chem Ecol 26:57–72

Anderson RE, Brazelton WJ, Baross JA (2013) The Deep Viriosphere: assessing the viral impact on microbial community dynamics in the deep subsurface. Rev Mineral Geochem 75:649–675

Arndt S, Brumsack H-J, Wirtz KW (2006) Cretaceous black shales as active bioreactors: a biogeochemical model for the deep biosphere encountered during ODP Leg 207 (Demerara Rise). Geochim Cosmochim Acta 70:408–425

Arunmansee W et al (2016) Gram-negative trimeric porins have specific LPS binding sites that are essential for porin biogenesis. Proc Natl Acad Sci U S A 113:E5034–E5043. *

Bakelar J, Buchanan SK, Noinaj N (2016) The structure of the b-barrel assembly machinery complex. Sciencve 351:180. *

Barras C (2021) Deepest land-dwelling microbes found at bottom of 5km hole in China. New Scientist, 19 Feb, p 14

Bar-On YM, Phillips R, Milo R (2018) The biomass distribution on earth. Proc Natl Acad Sci U S A 115:6506–6511

Bartlett DH, Welch TJ (1995) ompH Gene expression is regulated by multiple environmental cues in addition to high pressure in the deep sea bacterium Photobacterium species strain SS9. J Bact 177:1008–1016

Bartlett DH, Kato C, Horikoshi K (1995) High pressure influences on gene and protein expression. Res Microbiol 146:697–706

Bernhardt G et al (1988) Effect of carbon dioxide and hydrostatic pressure on the pH of culture media and the growth of methanogens at elevated temperature. Appl Microbiol Biotechnol 28:176–181

---

[1]Those marked * are further reading.

Bernhartdt G, Jaenicke R, Ludemann H-D (1987) High pressure equipment for growing microorganisms on gaseous substrates at high temperature. Appl Environ Microbiol 53:1876–1879. *

Bianchi A, Garcin J (1993) In stratified waters the metabolic rate of deep sea bacteria decreases with decompression. Deep Sea Res I 40:1703–1710

Bianchi A, Garcin J, Tholosan O (1999) A high pressure serial sampler to measure microbial activity in the deep sea. Deep Sea Res I 46:2129–2142

Biddle JF et al (2006) Heterotrophic Archaea dominate sedimentary subsurface ecosystems off Peru. Proc Nat Acad Sci 103:3846–3851. *

Biller SJ et al (2014) Bacterial vesicles in marine ecosystems. Science 343:183–186

Bispo JAC et al (2012) Entropy and volume change of dissociation in Tobacco Mosaic virus probed by high pressure. J Phys Chem B 116:14817–14828

Bochdansky AB, Clouse MA, Herndl GJ (2017) Eukaryotic microbes, principally fungi and labyrinthulomycetes, dominate biomass on bathypelagic marine snow. The ISME J 11:362–373

Boone DR et al (1995) Bacillus infernus sp. nov., an Fe(III)—and Mn(IV)—reducing anaerobe from the deep terrestrial subsurface. Int J Systemat Bacteriol 45:5441–5448. *

Campanaro S et al (2005) Laterally transferred elements and high pressure adaptation in Photobacterium profundum strains. BMC Genomics 6:122

Carlucci AF, Williams PM (1978) Simulated in situ growth rates of pelagic marine bacteria. Naturwissenschaften 65:541–542. *

Carpenter FG (1954) Anesthetic action of inert and unreactive gases on intact animals and isolated tissues. Am J Phys 178:505–509. *

Case DH et al (2017) Aerobic and anaeroibic methanotrophic communities accocioated with methane hydrates exposed on the sea floor: a high pressure samp-ling and stable isotope incubation experiment. Front Microbiol 8:art.2569

Certes A (1884) Sur la culture, a l'abri des germes atmospheriques des eaux et des sediments rapports par les expeditions du Travailleur et du Talisman. CR Acad Sci 98:690–693

Chastain RA, Yayanos AA (1991) Ultrastructural changes in an obligately barophilic marine bacterium after decompression. Appl Environ Microbiol 57:1489–1497

Chilukuri LM, Bartlett DH (1997) Isolation and characterization of the gene encoding single stranded DNA binding protein (SSB) from four marine Shewanella strains that differ in their temperature and pressure optima for growth. Microbiology 143:1163–1174

Chilukuri LN, Bartlett DH, Fortes PAG (2002) Comparison of high pressure induced dissociation of single stranded DNA binding protein (SSB) from high pressure sensitive and high pressure adapted marine Shewanella species. Extremophiles 6:377–383

Christner BC et al (2006) Limnological conditions in subglacial Lake Vostok, Antarctica. Limnol Oceanogr 51:2485–2501

Colwell FS, D'Hondt S (2013) Nature and extent of the deep Biosphere. Rev Mineral Geochem 75:547–574. *

Corinaldesi C et al (2019) High rates of viral lysis stimulate prokaryote turnover and C recycling in bathypelagic waters of a Ligurian canyon (Mediterranean Sea). Prog Oceanog 171:70–75

D'Hondt S, Pockalny R, Fulfer VM, Spivack AJ (2019) Subseafloor life and its biogeochemical impacts. Nat Commun 10:3519

Danovaro R et al (2008) Major viral impact on the functioning of benthic deep sea ecosystems. Nature 454:1084–1097

Dolfing J (2014) Syntrophy in microbial fuel cells. The ISME J 8:4–5

Edwards JK, Wheat CG, Sylvan JB (2011) Under the sea: microbial life in volcanic ocean crust. Nat Rev Microbiol 9:703–712. *

Eloe EA, Lauro FM, Vogel RF, Bartlett DH (2008) The deep sea bacterium Photobacterium profundum SS9 utilizes separate flagellar systems for swimming and swarming under high pressure conditions. Appl Environ Microbiol 74:6298–6305

Escudero C, Oggerin M, Amils R (2018) The deep continental subsurface: the dark biosphere. Int Microbiol 21:3–14. *

Fang J, Zhang L, Bazylinski DA (2010) Deep sea piezosphere and piezophiles: geomicrobiology and biogeochemistry. Trend Microbiol 18:413–422

Fang J et al (2017) Predominance of viable spore-forming piezophilic bacteria in high pressure enrichment cultures from 1.5 to 2.4 km deep coal bearing sediments below the ocean floor. Front Microbiol 8:article 137

Fasca H et al (2017) Response of marine bacteria to oil contaminants and to high pressure and low temperature deep sea conditions. Microbiol Open e550:7

Fichtel K et al (2015) Temperature and pressure adaptation of a sulfate reducer from the deep subsurface. Front Microbiol:6, art 1078. *

Forterre P (2013) The virocell concept and environmental microbiology. ISME J 7:233–236

Foustoukos DI, Ileana P-R (2015) A continuous culture system for assessing microbial activities in the piezosphere. Appl Environ Microbiol 81:6850–6856

Garel M et al (2019) Pressure-retaining sampler and high pressure systems to study deep sea microbes under in situ conditions. Front. Microbiol:10, art 453

Giovannoni S, Stingl U (2007) The importance of culturing bacterioplankton in the "omics" age. Nat Rev Microbiol 5:820–826

Gold T (1992) The deep, hot biosphere. Proc Natl Acad Sci U S A 89:6045–6049

Grossi V et al (2010) Hydrostatic pressure affects membrane and storage lipid compositions of the piezotolerant hydrocarbon-degrading Marinobacter hydrocarbonoclasticus strain 5. Environ Microbiol 12:2020–2033

Gundersen KR, Mountain CW (1972) A temperature and pressure preserving sampling, transfer and incubation system for deep sea microbiological and chemical research. In: Brauer RW (ed) Barobiology and the experimental biology of the deep sea. University of North Carolina, USA. *

Hamajima Y et al (2014) Pressure effects on the chimeric 3-isopropylmalate dehydrogenase of the deep sea piezophile Shewanella benthica and the atmospheric pressure-adapted Shewanella oneidensis. Biosci Biotechnol Biochem 78:469–471

Hamajima Y et al (2016) Pressure adaptation of 3-isopropylmalate dehydrogenase from an extremely piezophilic bacterium is attributed to a single amino acid substitution. Extremophiles 20:177–186

Hauton C et al (2017) Identifying toxic impacts of metals potentially released during deep sea mining—a synthesis of the challenges to quantifying risk. Front Marine Sci:4, art. 368

Hazael R, Meersman F, Ono F, McMiillan PE (2016) Pressure as a limiting factor for life. Life 5:34. *

Head IM, Jones DM, Larter SR (2003) Biological activity in the deep subsurface and the origin of heavy oil. Nature 426:344–352

Heeschen KU et al (2007) In situ hydrocarbon concentrations from pressurized cores in surface sediments, Northern Gulf of Mexico. Mar Chem 107:498–515

Hinrichs K-U, Inagaki F (2012) Downsizing the deep biosphere. Science 338:204–205. *

Hoehler TM, Jorgensen BB (2013) Microbial life under extreme energy limitation. Nature Rev Microbiol 11:83–94

Holters C, van Almsick G, Ludwig H (1999) High pressure inactivation of anaerobic spores from Clostridium pasteurianum. In: Ludwig H (ed) Advances in high pressure science and biotechnology. Springer, Berlin, pp 65–68. *

Hoshino T, Inagaki F (2019) Abundance and distribution of Archaea in the subsurface sedimentary biosphere. ISME J 13:227–231

Hoskisson PA, Hobb G (2005) Continuous culture—making a comeback? Microbiology 151:3153–3159

Houghton JL, Seyfried WE, Banta AB, Reysenbach A-L (2007) Continuous enrichment culturing of thermophiles under sulfate and nitrate-reducing conditions and at deep sea hydrostatic pressures. Extremophiles 11:371–382

Huang Q, Rodgers JM, Hemley RJ, Ichiye T (2017) Extreme biophysics: enzymes under pressure. J Comput Chem 38:1174–1182

Imachi H et al (2011) Cultivation of methanogenic community from subsea floor sediments using a continuous flow bioreactor. The ISME J 5:1913–1925. *

Inagaki F et al (2015) Exploring deep microbial life in coal-bearing sediment down to—2.5 km below the ocean floor. Science 349:420–424. *

Jannasch HW, Taylor CD (1984) Deep sea microbiology. Ann Rev Microbiol 38:487–314. *

Jannasch HW, Wirsen CO (1973) Deep sea microorganisms: in situ response to nutrient enrichment. Science 180:641–643

Jannasch HW, Wirsen CO (1977) Retrieval of concentrated and undecompressed microbial populations from the deep sea. Appl Environ Microbiol 33:642–646

Jannasch HW, Wirsen CO (1982) Microbial activities in undecompressed and decompressed deep seawater samples. Appl Environ Microbiol 43:1116–1124

Jannasch HW, Wirsen CO, Farmanfarmaian A (1971) Microbial degradation of organic matter in the deep sea. Science 171:672–675

Jannasch HW, Wirsen CO, Winget CL (1973) A bacteriological pressure-retaining deep sea sampler and culture vessel. Deep Sea Res 20:661–664

Jannasch HW, Wirsen co, Taylor CD (1982a) Deep sea bacteria: isolation in the absence of decompression. Science 216:1315–1317. *

Jannasch HW, Wirsen CO, Taylor CD (1982b) Deep sea bacteria: isolation in the absence of decompressin. Science 216:1315–1317. *

Jannasch HW, Wirsen CO, Doherty KW (1996) A pressurised chemostat for the study of marine barophilic and oligotrophic bacteria. Appl Environ Microbiol 62:1593–1596

Johnson FH, Eyring H, Pollisar MJ (1954) The kinetic basis of molecular biology. John Wiley & Sons, NY

Jorgensen BB, Boetius A (2007) Feast and famine—microbial life in the deep sea bed. Nat Rev Microbiol. 5:770–761. *

Jorgensen BB, Marshall IPG (2016) Slow microbial life on the seabed. Annu Rev Mar Sci 8:311–332

Kato C (2011) Distribution of Piezophiles. Chapter 5.1 in Horikishi K (ed), Extremophiles handbook. Springer

Kato C et al (1998) Extremely barophilic bacteria isolated from the Mariana Trench, Challenger Deep, at a depth of 11,000 meters. Appl Environ Microbiol 64:1510–1513

Kato C, Yanagibayashi Y, Nogi LL, Horikoshi K (1999) Changes inn the microbial community in deep sea sediment during cultivation without decompression. In: Ludwig H (ed) Advances in high pressure bioscience and biotechnology. Springer, pp 11–15

Kawachi S, Arao T, Suzuki Y, Tamura K (2010) Effects of compressed unsaturated hydrocarbon gases on yeast growth. Ann N Y Acad Sci 1189:121–126. *

Kirkness CM, Macdonald AG (1972) Interaction between anaesthetics and hydrostatic pressure in the division of Tetrahymena pyriformis W. Exp Cell Res 75:329–336

Koebnik R, Locher KP, Van gelder P (2000) Structure and function of bacterial outer membrane proteins: barrels in a nutshell. Molec. Microbiol. 37:239–253. *

Kumar A, Hajjar E, Ruggerone P, Ceccarelli M (2010) Structural and dynamical properties of the porins OmpF and OmpC: insights from molecular simulations. J Phys Conden Matter 22:454125. (11p) *

Kusube M et al (2017) Colwellia marinimaniae sp. nov., a hyperpiezophilic species isolated from an amphipod within the Challenger Deep, Mariana Trench. Int J Syst Evolut Microbiol 67:824–831

Kyo M, Tuji Y, Usui H, Itoh T (1991) Collection, isolation and cultivation system for deep sea microbes study: concept and design: 419–423. https://doi.org/10.1109/Oceans.1991.613968

La Cono V et al (2015) Shifts in the meso- and bathypelagic archaea communities composition during recovery and short term handling of decompressed deep sea samples. Environ Microbiol Rep 7:450–459. *

Lelchat F et al (2020) Measuring the biological impact of drilling waste on the deep sea floor: an experimental challenge. J Hazard Mater 389:1221

Lever MA et al (2015) Life under extreme energy limitation: a synthesis of laboratory—and field—based investigations. FEMS Microbiol Rev 39:688–728. *

Linke K et al (2008) Influence of high pressure on the dimerizatiion of ToxR, a protein involved in bacterial signal transduction. Appl Environ Microbiol 74:76231–77623

Lomstein BA, Langerhuus AT, D'Hondt S, Jorgensen BB, Spivack AJ (2012) Endospore abundance, microbial growth and necromass turnover in deep sub-seafloor sediments. Nature 484:101–104

Lou F, Neetoo H, Chen H, Jianring L (2015) High hydrostatic pressure processing: a promising non-thermal technology to inactivate viruses in high risk foods. Annu Rev Food Sci Technol 6:589–409

Luo M et al (2018) Benthic carbon mineralisation in Hadal Trenches: insights from in situ determination of benthic oxygen consumption. Geophys Res Letters 45:2752–2760. *

Macdonald AG (1975) The effect of helium and of hydrogen at high pressure on the cell division of *Tetrahymena pyriformis* W. J Cell Physiol 85:511–528

Macdonald AG, Martinac B (1999) Effect of high hydrostatic pressure on the porin OmpC from *Escherichia coli*. FEMS Microbiol Letters 173:327–334

Macdonald AG, Wann KT (1978) Physiological aspects of anaesthetics and inert gases. Academic Press, London, p 308

Macdonald AG, Martinac B, Bartlett DH (2002) High pressure experiments with the porins from the barophile Photobacterium profundum SS9. In: Hayashi R (ed) Trends in high pressure bioscience and biotechnology. Elsevier Science. B V

Macdonald AG, Martinac B, Bartlett DH (2003) Patch-clamp experiments with porins extracted from a marine bacterium (Photobacterium profundum, strain SS9) and reconstituted in liposomes. Cell Biochem Biophys 37:157–167. *

MacNaughtan W, Macdonald AG (1980) Effects of gaseous anaesthetics and inert gases on the phase transition in smectic mesophases of dipalmitoyl phosphatidylcholine. Biochim Biophys Acta 597:193–198

MacNaughtan W, Macdonald AG (1982) Effects of pressure and pressure antagonists on the growth and membrane bound ATP-ase of Acholeplasma laidlawii B. Comp Biochem Physiol 72 A:405–414

Mangelsdorf K, Zink K-G, Birrien J-L, Toffin L (2005) A quantitative assessment of pressure dependent adaptive changes in the membrane lipids of a piezosensitive deep sub-sea floor bacterium. Org Geochem 36:1459–1479. *

Marietou A et al (2018) The effect of hydrostatic pressure on enrichments of hydrocarbon degrading microbes from the Gulf of Mexico following the Deepwater Horizon oil spill. Front Microbiol 8:808. *

Marquis RE, Thom SR, Crookshank CA (1978) Interactions of helium, oxygen and nitrous oxide affecting bacterial growth. Undersea Biomedical Res 5:189–198. *

Martin DD, Bartlett DH, Roberts ME (2002) Solute accumulation in the deep sea bacterium Photobacterium profundum. Extremophiles 6:507–514

Martinez N et al (2016) High protein flexibility and reduced hydration water dynamics are the key pressure adaptive strategies in prokaryotes. Sci Reports 6:32816–32827

Martini S et al (2013) Effects of hydrostatic pressure on growth and luminescence of a moderately piezophilic luminous bacteria Photobacterium phosphoreum ANT-2200. PLoS One:8 e66580

Mason OU et al (2010) First investigation of the microbiology of the deepest layer of ocean crust. PLoS One 5:e15399. *

Masui N, Kato C (1999) New method of screening for pressure-sensitive mutants at high hydrostatic pressure. Biosci Biotechnol Biochem 63:235–237. *

Miyazaki J et al (2017) WHAT-3: an improved flow through multi-bottle fluid sampler for deep sea geofluid research. Front Earth Sci:5 art 45. *

Morono Y et al (2011) Carbon and nitrogen assimilation in deep subseafloor microbial cells. Proc Natl Acad Sci U S A 108:18295–18300. *

Murakami C et al (2011) Comparative study on dihydrofolate reductase from Shewanella species living in deep sea and ambient atmospheric pressure environments. Extrempophiles 15:165–175

Murayama M (1973) Sickle cell haemoglobin: Molecular basis of sickling phenomenon: theory and practice. Crit Rev Biochem 1:461–499. *

Nagae T, Kato C, Watanabe N (2012) Structural analysis of 3-isopropylmalate dehydrogenase form the obligate piezophile Shewanella benthica DB21MT-2 and the nonpiezophile Shewanella oneidensis MR-1. Acta Cryst F 68:265–268

Nagata T et al (2010) Emerging concepts on microbial processes in bathypelagic ocean-ecology, biogeochemistry, and genomics. Deep Sea Res II 57:1519–1536. *

Nauhaus K, Boetius A, Kruger M, Widdel F (2002) In vitro demonstration of anaerobic oxidation of methane coupled to substrate reduction in sediment from a marine gas hydrate area. Environ Microbiol 4:296–305

Nelson CM, Schuppenhauer MS, Clark DS (1991) Effects of hyperbaric pressure on a deep sea Archaebacterium in stainless steel and glass-lined vessels. Appl Environ Microbiol 57:3576–3580

Nelson CM, Schuppenhauer MR, Clark DS (1992) High pressure, high temperature bioreactor for comparing effects of hyperbaric and hydrostatic pressure on bacterial growth. Appl Environ Microbiol 58:1789–1793

Oger PM, Jebbar M (2010) The many ways of coping with pressure. Res Microbiol 161:799–809

Ohmae E et al (2012) Pressure dependence of activity and stability of dihydrofolate reductase of the deep sea bacterium Moritella profunda and *Escherichia coli*. Biochim Biophys Acta 11824:511–519

Ohmae E, Miyashita Y, Kato C (2013a) Thermodynamic and functional characteristics of deep sea enzymes revealed by pressure effects. Extremophiles 17:701–709

Ohmae E et al (2013b) Solvent environments significantly affect enzymatic function of *Escherichia coli* dihydrofolate reductase: comparison of wild type protein and active site mutant D27E. Biochimica et Bipohysica Acta 1834:2782–2794. *

Ohmae E, Gekko K, Kato C (2015) Environmental adaptation of dihydrofolate reductase from deep sea bacteria. Chapter 21, in Akasaka K, Matsuki H (ed) High pressure bioscience. Springer Science + Business Media, Dordrecht, p 423–421 *

Orcutt B, Wheat CG, Edwards KJ (2010) Subseafloor ocean crust microbial observatories: development of FLOCS (Flow-through osmo colonization system) and evaluation of borehole construction materials. Geomicrobiol J 27:143–157. *

Paquin PR, McGrath J, Fanelli CJ, Di Toro DM (2018) The aquatic hazard of hydrocarbon liquids and gases and the modulating role of hydrostatic pressure on dissolved gas and oil toxicity. Mar Pollut Bulletin 133:930–942

Parada V, Sintes E, van Aken HM, Weinbauer MG, Herndl GJ (2007) Viral abundance, decay, and diversity in the meso- and bathypelagic waters of the North Atlantic. Appl Environ Microbiol 73:4429–4438. *

Parkes RJ et al (1994) Deep bacterial biosphere in Pacific Ocean sediments. Nature 371:410–413. *

Parkes RJ, Cragg BA, Wellsbury P (2000) Recent studies on bacterial populations and processes in subseafloor sediments: a review. Hydrogeol J 8:11–28. *

Parkes RJ et al (2009) Culturable prokaryotic diversity of deep, gas hydrate sediments: first use of a continuous high-pressure anaerobic, enrichment and isolation system for subseafloor sediments (DeepIsoBUG). Environ Microbiol 11:3140–3153

Parkes RJ et al (2014a) A review of prokaryotic populations and processes in sub-seafloor sediments, including biosphere: geosphere interactions. Mar Geol 352:409–425

Parkes RJ et al. (2014b) Studies on prokaryotic populations and processes in subseafloor sediments. In Microbial life of the deep biosphere, De Gruyter Inc. *

Parnell J, McMahon S (2016) Physical and chemical controls on habitats for life in the deep sub surface beneath continents and ice. Phil Trans R Soc A 374:20140293

Patching JW, Eardly D (1997) Bacterial biomass and activity in the deep waters of the eastern Atlantic—evidence of a barophilic community. Deep Sea Res. I 44:1655–1670. *

Prangishvili D (2013) The wonderful world of Archaeal; viruses. Annu Rev Microbiol 67:565–585

Priede IG, Jamieson A, Heger A, Jessica C, Zuur AF (2008) The potential influence of biolumines-cence from marine animals on a deep sea underwater neutrino telescope array in the Mediterra-nean Sea. Deep Sea Res. I 55:1471–1483

Rajala P, Bomberg M (2017) Reactivation of deep subsurface microbial community in response to methane or methanol amendment. Front Microbiol. 8 article 431

Rajala P et al (2015) Rapid reactivations of deep subsurface microbes in the presence of C-1 compounds. Microorganisms 3:17–33

Reineke K, Mathys A, Knorr D (2011) The impact of high pressure and temperature on bacterial spores: inactivation mechanisms of Bacillus subtilis above 500 MPa J. Food Sci 76:M189–M197. *

Robertson DE, Roberts MF, Belay N, Stetter KO, Boone DR (1999) Occurrence of B-Glutamate, a novel osmolyte, in marine methanogenic bacteria. Appl Environ Microbiol 56:1504–1508

Rohwer F, Thurber RV (2009) Viruses manipulate the marine environment. Nature 459:207–212

Roussel EG et al (2008) Extending the sub-sea floor biosphere. Science 320:1046. *

Roy H et al (2012) Aerobic microbial respiration in 86 million year old deep sea red clay. Science 336:922–925. *

Saad-Nehme J, Silva JL, Meyer-Fernandes JR (2001) Osmolytes protect mitochondrial Fo F1-ATPase complex against pressure inactivation. Biochim Biophys Acta 1546:164–170. *

Santos MM et al (2018) The last frontier: coupling technological developments with scientific challenges to improve hazard assessment of deep sea mining. Sci Tot Environ 627:1505–1514

Sauer P, Glombitza C, Kellmweyer J (2012) A system for incubations at high gas partial pressure. Front Microbiol. 3 article 25, p 1–9

Schippers A et al (2005) Prokaryotic cells of the deep sub-sea floor biosphere identified as living bacteria. Nature 433:861–864. *

Schnell JR, Dyson HJ, Wright PE (2004) Structure, dynamics and catalytic function of dihydrofolate reductase. Annu Rev Bipohys Struct 33:119–140

Scoma A et al (2016) Microbial oil-degradation under mild hydrostatic pressure (10 MPa): which pathways are impacted in piezosensitive hydrocarbonoclastic bacteria? Sci Reports 6:23526. *

Seewald JS, Doherty KW, Hammar TR, Liberatore SP (2001) A new gas tight isobaric sampler for hydrothermal fluids. Deep Sea Res I 49:189–196. *

Smiddy M, Sleator RD, Patterson MF, Hill C, Kelly AL (2004) Role for compatible solutes glycine, betaine and L-carnitine in Listerial barotolerance. Appl Environ Microbiol 70:7555–7557

Suttle CA (2007) Marine viruses—major players in the global ecosystem. Nature Rev Microbiol 5:801–812

Tabor PS et al (1981) A pressure retaining deep ocean sampler and transfer system for measurement of microbial activity in the deep sea. Microb Ecol 7:51–65

Takai K, Inoue A, Horikoshi K (1999) Thermaerobacter mariansis gen. nov. sp. nov., an aerobic extremely thermophilic marine bacterium from the 11, 000 m deep Mariana Trench. Int J System Bact 49:615–628

Takami H, Inoue A, Fuji F, Horikoshi K (1997) Microbial flora in the deepest sea mud of the Mariana Trench. FEMS Microb Lett 152:279–285. *

Tamburini C, Garcin J, Ragot M, Bianchi A (2002) Biopolymer hydrolysis and bacterial production under ambient hydrostatic pressure through a 2000 m water column in the NW Mediterranean. Deep Sea Res II 49:2109–2123

Tamburini C et al (2009a) Distribution and activity of Bacteria and Archaea in different water masses of the Tyrrhenian Sea. Deep Sea Res II 56:700–712. *

Tamburini C et al (2009b) Effects of hydrostatic pressure on microbial alteration of sinking fecal pellets. Deep Sea Res I 56:1533–1546. *

Tamburini C, Boutrif M, Garel M, Colwell RR, Deming JW (2013) Prokaryotic responses to hydrostatic pressure in the ocean—a review. Environ Microbiol 15:1262–1274. *

Tamegal H, Nishikawa S, Haga M, Bartlett DH (2012) The respiratory system of the piezophile Photobacterium profundum SS9 grown under various pressure. Biosci Biotechnol Biochem 76:1506–1510

Taylor CD (1987) Solubility properties of oxygen and helium in hyperbaric systems and the influence of high pressure oxy-helium upon bacterial growth, metabolism and viability. In Jannasch HW, Marquis RE, Zimmerman AM (eds) Incurrent perspectives in high pressure biology. Academic Press

Thom SR, Marquis R (1984) Microbial growth modification by compressed gases and hydrostatic pressure. Appl Environ Microbiol 47:780–787

Toffin L, Alain K (2014) Cultivation of marine subseafloor microorganisms: state of the art solutions and major issues. In Kallmeyer J, Wagner D (ed), Microbial life in the deep biosphere, Vol. 1, De Gruyter, p 83–99

Trembath-Reichert E et al (2017) Methyl-compound use and slow growth characterize microbial life in 2 km deep subsea floor coal and shale beds. Proc Natl Acad Sci U S A 114:E9206–E9215. *

Valentine DL et al (2010) Propane respiration jump-starts microbial response to a deep oil spill. Science 330:208–211. *

Vreeland RH, Rosenzweig WD, Powers DW (2000) Isolation of a 250 million year old halotolerant bacterium form a primary salt crystal. Nature 407:897–900

Wang F et al (2008) Environmental adaptation: genomic analysis of the piezo tolerant and psychrotolerant deep sea iron reducing bacterium Shewenella piezotolerans WP3. PLoS One 3:e1937

Wang Y et al (2016) Genomic characteriuzation of symbiotic mycoplasmas from the stomach of deep sea isopod Bathynomus sp. Environ Microbiol 18:2646–2659. *

Welch TJ, Bartlett DH (1998) Identification of a regulatory protein required for pressure responsive gene expression in the deep sea bacterium Photobacterium species strain SS9. Molec Microbiol 27:977–985

Wirsen CO, Molyneaux SJ (1999) A study of deep sea natural microbial populations and barophilic pure cultures using a high pressure chemostsat. Appl Environ Microbiol 65:5314–5321

Wuytack EY, Soons J, Poschet F, Michiels CW (2000) Comparative study of pressure- and nutrient-induced germination of Bacillus subtilis spores. Appl Environ Microbiol 66:257–261. *

Xu Y, Huangmin G, Fang J Biogeochemistry of hadal trenches: recent developments and future perspectives. Deep Sea Res. II 155:19–26. *

Yanagibayashi M, Nogi Y, Lina L, Kato C (1999) Changes in the microbial community in Japan Trench sediment from a depth of 6292 m during cultivation without decompression. FEMS Microbiol Lett 170:271–279

Yancey PH, Siebenaller JF (1999) Trimethylamine oxide stabilises teleost and mammalian lactate dehydrogenase against inactivation by hydrostatic pressure and trypsinolysis. J Exp Biol 202:1597–1603

Yayanos AA, Dietz AS (1982) Thermal inactivation of a deep sea barophilic bacterium, isolate CNPT-3. Appl Environ Microbiol 43:1481–1489

Yayanos AA, Dietz AS (1983) Death of a hadal deep sea bacterium after decompression. Science 220:497–498. *

Yayanos AA, Dietz AS, Von Boxtell R (1981) Obligately barophilic bacterium from the Mariana trench. Proc Natl Acad Sci U S A 78:5212–5215

Yin Q-J et al. (2018). High hydrostatic pressure inducible Trimethylamine N oxide Reductase improves the pressure tolerance of piezosensitive bacteria Vibrio fluvialis. Front Microbiol. 8 art 2646 *

Youngson AF, Macdonald AG (1967) Interaction between halothane and hydrostatic pressure. Brit J Anaesthesia 42:801–802

Zhang Y, Arends JBA, de Weile TV, Boon N (2001) Bioreactor technology in marine microbiology: from design to future application. Biotechnol Adv 29:312–321. *

Zhang S-D et al (2016) Genomic and physiological analysis reveals versatile metabolic capacity of deep sea Photobacterium phosphoreum ANT-2200. Extremophiles 20:301–310

Zhang Z, Wu Y, Zhang X-H (2018) Cultivation of microbes from the deep sea environment. Deep Sea Res II 155:34–43. *

Zivaljic S et al. (2017) Survival of marine heterotrophic flagellates isolated from the surface and the deep sea at high hydrostatic pressure: literature review and own experiments. Deep Sea Res II. April *

Zobell CE, Kim J (1972) Effects of deep sea pressures on microbial enzyme systems. The effects of pressure on organisms. Symp Soc Exp Biol 26:125–146. Cambridge University Press, UK *

# Hydrothermal Vents: The Inhabitants, Their Way of Life and Their Adaptation to High Pressure

## 9.1    Hydrothermal Vents: Physical Conditions

In the 1970s, oceanographers and geologists were exploring the Pacific Ocean floor and the waters immediately above it. They were pursuing, among other things, anomalous warm patches of deep water and volcanic fissures in the seafloor, using both towed instruments and the manned submersible, Alvin. One of the clues that volcanic activity might be disturbing the seafloor was the occasional presence of dead benthic fish floating on the surface. In 1976, the crew of Alvin discovered hot fluids pouring out of vents in the seafloor, on the Galapagos Rift, creating a volume of warm water some 8 to 40 metres above the bottom. The hot water was seawater, which had penetrated the basaltic crust, where it was heated and subsequently discharged back into the ocean, carrying additional chemicals dissolved as it passed through the crust. Photographs of the seafloor included dramatic pictures of clusters of live animals close to the vents at depths of 2488–2925 m. A report was published in 1977 by Lonsdale who carefully noted that "This report of phenomena at hydrothermal vents in the Galapagos Rift is preliminary in that teams of scientists are presently diving there aboard DSRV Alvin and are sampling the hydrothermal precipitates and benthic organisms". Among the living animals, there were numerous large bivalve Molluscs and the photographs clearly showed many of their shells lying empty. This was evidence of recent death. Also recorded were sea anemones, sea pens, bivalves and a crab. In 1979, a fuller report was published by 11 authors, headed by Corliss, in the journal *Science*. Many of the animals and bacteria, fungi, Archaea and associated viruses found around the vents, together with their biochemistry and physiology were new to science. They proved to be a stimulus for a range of geological, chemical and biological investigations which are continuing to the present day. They have extended our knowledge of life deep in the Earth's crust and our ideas of how life may have arisen on Earth and, perhaps, on other planets. The discovery of the hydrothermal vents and their inhabitants was surely one of the most exciting and important episodes in the history of Biology, and indeed of

© Springer Nature Switzerland AG 2021
A. Macdonald, *Life at High Pressure*, https://doi.org/10.1007/978-3-030-67587-5_9

Science. (Readers can benefit from the Website offerings on many aspects of hydrothermal vents.)

Investigating the vents requires manned and unmanned deep submersible vehicles, and is, therefore, an expensive business, as such vehicles require a mother ship as well as a support team of engineers and technicians. Careful collecting methods have to be used and that includes isobaric methods. Submersibles also enable in situ experiments to be carried out. Hydrothermal vents are typically on mid-ocean ridges, in deep water, so the inhabitants must be adapted to high pressure. The vents are named exotically, for example, Snake Pit, depth 3600 m; Lucky Strike, 1700 m; Endeavour, 2250 m; and Mariana, 3600 m. The deepest vent is Beebe, 4964 m, in the Cayman Trough (Dalmasso et al. 2016a). More than 600 hot vents have been discovered. They are a paradise for a select group of organisms, and that certainly includes biologists (Van Dover 2000; Desbruyeres 2007).

There are also cold seeps, waters which emerge from geological faults and which contain hydrogen sulphide and, or, methane. They also create an environment in which chemosynthesis supports an animal community usually dominated by a mussel, Bathymodiolus. The seeps occur over a range of depths, the greatest being in the hadal zone.

A good starting point for this necessarily brief discussion of the hot vent organisms is the temperature of the venting fluids. Often it exceeds the normal boiling point of water. That is defined as the temperature at which the vapour pressure of water is equal to that of the environment. In everyday life, it is 100 C, or perhaps a little lower if you are at high altitude, say in Denver or if you are lucky enough to be trekking in the high Himalaya, where the atmospheric pressure is reduced. As liquid water vaporises there is a large enthalpy change and a very large increase in volume. Thus, increased pressure raises the boiling point of water, a classic example of Le Chatelier's Principle. When liquid water is superheated under high pressure the molecules acquire so much energy they separate, losing cohesion, becoming a supercritical fluid. In the case of pure water, this occurs at a pressure close to 22 MPa and a temperature of 374 °C, respectively, the Critical Pressure and Critical Temperature. Supercritical water does not boil, i.e. does not undergo a phase change to a vapour at any temperature. The salts in seawater make a small difference to these values and seawater at 22 MPa reaches its boiling point close to 360 °C. Some vent waters nearly reach 400 °C. Scientists aboard Alvin have extended a temperature probe, using its mechanical manipulator, into supercritical seawater. Presumably, it would have been a bad idea for Alvin to get in closer contact! In this respect, Alvin was behaving like some of the active vent animals, in keeping clear, just, from lethal temperatures. These very hot waters are, of course, devoid of life, but when cooled by the cold ocean water the result is very warm oceanic water, say 20–80 °C, but which, nevertheless, supports specially adapted organisms. Deepwater in the western Mediterranean Sea is unusually warm, for example 13 °C at 2500 m depths in the Tyrrhenian sea, but this does not arise from hydrothermal vents. The variety of chemicals in the vent waters are a source of biochemical primary production and the heat merely assists the biochemical reactions. The chemicals also provide sensory information for animals and, for simple organisms, a directional stimulus. Vents are usually too deep for sunlight to be detected but bioluminescence, as in the cold abyss, is commonplace, so organisms use it in various ways. Additionally, the temperatures encountered, 350 °C or more, are high enough to emit "black body" radiation, and it seems to be sufficient in the 400–1050 nm band to stimulate special photoreceptors in certain organisms.

So, hot vents support organisms living at high temperatures, say up to 80 °C and in the most extreme case, 120 °C, of which more later. Other vent waters are at a lower temperature, say 20 °C, still significantly higher than typical abyssal temperatures of 2–4 °C. Adaptation to high pressure is thus combined with adaptation to either extremely hot or warm temperatures.

The typical hydrothermal vent, on the basalt of a mid-ocean ridge, emits water containing many dissolved chemicals: lithium, magnesium, calcium, barium, silica, aluminium, nickel and manganese along with the dissolved gases, methane, hydrogen, carbon dioxide and hydrogen sulphide. The spectacular Black Smoker vents produce chimneys several metres high made of precipitated black metal sulphides. White smokers emit light coloured particles of minerals containing calcium and barium. A less common type of vent occurs in older ocean crust, away from the mid-ocean ridge spreading centres. These are the alkali vents, which are not volcanic and are heated by an exothermic reaction between water and bedrock containing various minerals, including olivine in particular. The process, a set of chemical reactions, is called serpentinisation, as one of its products is serpentinite. The heat released causes the expansion of rock and water, the latter usually attains a temperature of less than 100 °C. Magnesium hydroxide and carbonate minerals form an alkaline solution, which on cooling precipitate to form pale coloured chimneys, which do not act as chimneys but as porous pinnacles through which water permeates. The water also contains a lot of dissolved hydrogen and methane which is produced by the serpentinisation reactions. These gases are an important source of energy for the rich biofilms of mainly archaea but also bacteria, which clad the pinnacles of the alkali vents.

Both types of vent exist in the darkness of the depths and appear dramatically out of the murk when the light from an approaching submersible reaches them or when filmed by a remotely operated camera. The first alkaline vent was discovered in 2000 by a team led by Kelley. It was named Lost City because of the numerous pale coloured towering pinnacles (chimneys) seen by the crew of the submersible Alvin, at a depth of 700 m in the Atlantic. Both types of vent support a fascinating population of microorganisms and animals at a significant pressure but the features of alkaline vents, such as Lost City, have a special significance as they illustrate a scenario for the origin of life, of which more below.

Before 1977 we all accepted that most life on earth ultimately depended on photosynthesis. This amazing chemical process has evolved to convert (reduce) carbon dioxide to form organic molecules such as sugars, with oxygen as a by-product.

$$CO_2 + H_2O + \text{Energy from sunlight} \rightarrow O_2 + CH$$

The discovery of the hot vent communities changed our thinking as they are independent of photosynthesis, at least in the short term. Vent microorganisms harness the energy from the oxidation of, for example, sulphides and use the energy to incorporate carbon into organic molecules.

$$S + CO_2 + O_2 + H_2O + \text{energy} \rightarrow SO_4 + CH_2O$$

Methane, ammonia and, in particular, hydrogen sulphide, are also used to provide energy:

$$H_2S + CO_2 + H_2O + O_2 \rightarrow H_2SO_4 + CH_2O$$

In the absence of oxygen in the water, a not unusual occurrence in sediments, the oxidation of dissolved hydrogen gas and the reduction of nitrates, sulphates and carbon dioxide leads to the production of organic molecules. Indeed it was the discovery of obligate anaerobic archaea which established the vents as sites where life was strictly independent of sunlight.

Also, in the environment of the alkaline vents and the deep granite aquifers (Chapter 8), gaseous hydrogen and carbon dioxide are a source for the synthesis of the organic compounds, methane and acetic acid, which are thought to have been important in the origin of life.

$$4H_2 + CO_2 \rightarrow CH_4 + 2H_2O$$

$$4H_2 + 2CO_2 \rightarrow CH_3COOH + 2H_2O$$

Before 1977 it was also accepted that animals depended on other animals and ultimately, organic material produced by photosynthesis, for food. But many of the microorganisms of the hydrothermal vents derive energy and synthesise organic matter independent of sunlight and provide food for animals. Through symbiosis, they also directly support the nutrition of animals. Unravelling the reactions involved in this chemosynthesis (also called chemolithoautotrophy) has taken time and the specific effects of pressure on them have not yet been studied to any great extent.

The presence of well-populated hydrothermal vents on the otherwise desert-like deep ocean floor has been compared to oases in hot dry deserts. Although many hot vent sites have been discovered, their aggregate biochemical activity is very small compared to that of the deep oceans generally.

Hot hydrothermal vents are ephemeral volcanic habitats. Alkali vents, for example Lost City, exist for many thousands of years, as the process by which they form tends to be self-perpetuating. The death of the volcanic vents is marked by the cessation of flow and recent deaths are apparent from the debris of empty mollusc shells left behind. Remarkably, the birth of several vents has been observed. The interesting question arises, how are they colonised? In many cases, the answer is by planktonic larvae, which then raises further questions about the pressure tolerance of the eggs and larvae, and indeed their tolerance to a whole lot of factors which do not exist in the vents. Some interesting experiments have been carried out in this context, and they will be described soon.

## 9.2   Geobiochemistry and the Origin of Life

The discovery of the hydrothermal vents provided a huge stimulus to the subjects of geobiochemistry and the origin of life. Serious thought and some effort had already gone into the study of oil well brine microorganisms, which exist at high pressures and temperatures. The existence of hydrothermal vent organisms made the idea of life penetrating the earth's crust and its fluids more plausible. Remember that pressure increases by 10 MPa for every kilometre depth of water. In contrast to hydrostatic pressure, lithostatic pressure, created by the head of water and the weight of porous sediments, increases by rather more (Chapter 8), but not enough to preclude life. The temperature of the earth's crust increases with depth by about 30 °C per kilometre and 125 °C, generally regarded as the upper limit for life, is attained at a depth of around 4 km.

Current ideas about the origin of life have moved on a great deal from Darwin's "warm little pond" and have focussed on a source of energy which can be harnessed by reacting systems. There is an extensive literature in this subject and the following brief summary comes from a 2010 paper by Lane, Allen and Martin, published in BioEssays. I offer my thanks and apologies.

The conditions present in alkaline vents, where acidic seawater is separated from alkaline vent water by thin, porous mineral barriers, creates a flux of hydrogen ions. This "proton motive force" is remarkably similar to that present in certain living cells and thought to be an ancient feature of the hypothetical organism LUCA. This stands for last universal common ancestor, in this case, ancestor to Archaea and Bacteria, but not Eukaryotes. This hypothetical cell, like contemporary primitive cells, contained a fluid with a low concentration of hydrogen ions separated by a membrane from the external fluid which had a high concentration of hydrogen ions. The ensuing flux of hydrogen ions, the proton motive force, is postulated to have driven synthetic reactions. In the precellular stage of evolution the porous, vesicular, mineral structures are postulated to have behaved similarly, synthesising organic molecules. There was a continuous supply of alkaline vent water and an infinitely large supply of acidic seawater so the hydrogen ion gradient did not equilibrate. The progression from static mineral vesicles to free cells might have been achieved by hydrocarbon like compounds being deposited on the interior of the vesicle and behaving like a supported cell membrane. Presumably one day they may have broken free.

This meeting of alkaline and acidic solutions in the porous fabric of precipitated mineral pinnacles is very different from what goes on in a warm pond. Specific temperatures and pressures are not central to these ideas but vents can occur at significant pressure, such as Lost City, 7 MPa. There is no reason to suppose that pressure would have been an obstacle to these early processes and, on the contrary, some scientists have argued that high pressure could have had a significant role. Woese's view that LUCA was very probably hyperthermophilic is consistent with a deep subsurface origin. Others, Hazen et al. (2002), Trevors (2002), Nickerson (1984), and Wächtershäuser (2006), have variously incorporated a high-pressure environment in their ideas about the origin of life. Most specifically, Giulio argues that the evolution of the genetic code took place under high pressure.

There is no particular type of reaction postulated in an early form of life which would be favoured by high pressure. Amino acids result from ammonia reacting with simple organic acids and various organic molecules can be synthesised in normal laboratory conditions, as for example in the electrochemical reactor constructed to simulate conditions in alkali hydrothermal vents by Herschy and colleagues (2014). A high-pressure version might prove to be interesting as it could provide a much higher partial pressure of hydrogen. Chemists have demonstrated the polymerisation of amino acids in vitro. Two groups have pursued this route, one led by Imai et al. (1999) and the other by Alargov. The latter used a solution of the amino acid glycine which was injected in a supercritical state at 39.5 MPa and 400 C, into a cold receiving chamber at the same pressure. The glycine molecules bonded to form diglycine and triglycine. This process is thought to simulate conditions in a hot deep-sea hydrothermal vent and the result was described as a possible "prebiotic" reaction, subsequently taken over by ribosomal synthesis. It certainly demonstrates the abiotic formation of peptide bonds. This type of creative chemistry has implications for ideas about extraterrestrial life, see Box 9.1.

Some readers may be wondering why, at the start of this section, was LUCA defined to exclude Eukaryotes. The reason is Eukaryotes are thought to have arisen later, from an Archaeal-like cell which evolved mitochondria from symbiotic bacteria (Chapter 1, Williams et al. 2013).

---

**Box 9.1 Extraterrestrial Life**

Ideas about the geochemical origins of biochemical reactions naturally lead to the question, has life arisen on other planets? The answer is either maybe, or probably, depending on your disposition. Either way, serious studies are examining the issues carefully. Confining our thoughts to life based on liquid water still allows plenty of scope because several planets in the solar system have liquid water to support familiar chemistry. Europa, one of Jupiter's moons, is a fascinating prospect. It has an ice–water "ocean" estimated to be 125–300 km, deep overlying a rock floor (Carsey et al. 1999/2000). Whatever ideas there may be about the geochemistry there, pressure is likely to be significant, although Europa's low gravity generates less pressure in the immense depths than one might have anticipated. Another of Jupiter's moons, Callisto, may also have an ocean. Enceladus, one of Saturn's moons, has a 50-km-deep subsurface salty ocean in which a certain amount of organic

---

(continued)

**Box 9.1** (continued)

chemistry has taken place. Additionally, hydrogen gas has been detected in quantities suggesting its steady production, probably by a hydrothermal process. Mars has been more fully studied and estimates for the existence of liquid water led to the possibility that a "deep hot biosphere" might extend some 300 km or more in depth (Jones et al. 2011). There, liquid water, perhaps as a film in porous rock rather than in bulk, exists at a pressure of approximately 1000 MPa at around to 100 °C or less.

## 9.3    The Inhabitants of the Vent Sites

All grades of organism have been found in and around hydrothermal vents: Archaea, Bacteria, Eukaryotes and, inevitably, viruses. We are concerned with their adaptation to high pressure at high or warm temperatures. There are two contexts. One is the permanent adult, in the case of animals, and the steady-state population, in the case of prokaryotes. And because hydrothermal vents are transient structures which die and reappear elsewhere, their inhabitants are capable of colonising both existing and new sites, usually through a pelagic larval stage in the case of animals. In this respect, they have much in common with other deep-sea benthic animals. So, the second context is the migrant larval stage of the hydrothermal animal, whose pressure tolerance must be considerable and similar to that of the larvae of the benthos in the cold abyss (Van Dover 2000; Van Dover et al. 2002; Van Dover and Lutz 2004).

### 9.3.1    Animals

When the first hydrothermal vent was seen by scientists aboard Alvin it was the profusion of sedentary animals which attracted the most attention, but vents in general support plenty of mobile animals. Access to them is not easy, requiring submersible craft, manned or unmanned. The main features of interest to experimenters are symbiosis, by which bacteria support the nutrition of active and sedentary animals, the ability to cope with the toxic chemicals in the vent waters, life history and colonisation and, of course, the high temperatures and pressures present. Actually, there are no uninteresting aspects to the lives of vent animals, so competition for experimental access is severe.

#### 9.3.1.1 The Tubeworm Riftia

A spectacular vent animal is the tubeworm *Riftia pachyptila* which grows in clusters. The reader is recommended to access websites for illustrations of vent animals and indeed the vents themselves. Each worm-like creature, which can be 2 metres in length, is housed in a chitinous tube anchored to the seafloor. At the top end, a red plume of fine tentacles protrudes (Van Dover and Lutz 2004). This is richly supplied with blood vessels, the blood containing haemoglobin in solution. Its ability to bind sulphide is one of the mechanisms by which *Riftia* is protected from the toxic concentration of $H_2S$ in which it lives. The plume is the organ which absorbs carbon dioxide, facilitated by the enzyme carbonic anhydrase which converts $CO_2$ to bicarbonate, $HCO_3^-$ (Goffredi et al. 1999), and also sulphides, for onward transport to the trophosome lower down the length of the body. The trophosome accommodates the intracellular symbiotic bacteria which sustain the animal, which, accordingly, has no mouth or digestive system. *Riftia* is classified as follows: Phylum Pogonophora, Class Obturata, Order Vestimentifera, and it is confined to hydrothermal vents in the Pacific. The animals' remarkable nutrition and metabolism has attracted a lot of interest and in situ observations have shown that, although it does not occupy particularly warm water, it can grow fast, gaining about 80 cm in length each year. This contributes to its ability to colonise its habitat rapidly, an advantageous property of vent organisms. We will deal with its life history in due course. *Riftia*'s occurrence in moderately warm waters is reflected in the chemistry of its collagen, an extracellular protein whose molecular properties have been investigated across the animal kingdom, throwing light on adaptation to habitats and broader evolutionary relationships. Collagen extracted from bovine calf skin denatures at 38 °C, i.e. just above the animal's body temperature whilst that from *Riftia pachyptila* denatures at 29 °C. These data apply to measurements at atmospheric pressure. The denaturation of calf skin collagen involves a small positive volume increase such that at 20 MPa the denaturation temperature is increased by only 1.4 °C. The denaturation temperature of *Riftia*'s collagen is negligibly affected by pressure, implying an even smaller volume increase (Auerbach et al. 1994). The actual growth of the collagen has been measured in animals restored to their ambient pressure of 21 MPa and 26 MPa at 8 °C after collection with decompression by submersibles, from the East Pacific Rise (Gaill et al. 1997).

### 9.3.1.2 Other Worms

Perhaps the star animal performer on the hydrothermal stage is *Alvinella pompejana*, otherwise known as the Pompeii worm (Desbruyeres et al. 1998; Le Bris et al. 2005). Unlike *Riftia* it is an active worm (Classification: Phylum Annelida, Class Polychaeta, Order Terebellida), about 6 cm long and has a furry appearance due to a coating of filamentous bacteria on its dorsal surface. It spends a lot of its time confined to a thin-walled tube built on the chimney walls of hot vents but does venture out to graze on bacteria. It can survive very hot conditions, which are difficult to quantify (Cary et al. 1998; Chevaldonne et al. 2000; Henscheid et al. 2005). Its habitat range is typically 2–40 °C and the vents it inhabits on the East Pacific Rise are 2600 m deep, giving a pressure of 26 MPa. The unstable swirling currents of hot water discharging from the vents are liable to erratically affect many of the organisms in the vicinity. It has been claimed that *Alvinella* can survive brief exposures to 105 C and even a gradient of 60 °C along its length, 20 °C at the head end and 80 °C at the posterior, for a period of hours, at 26 MPa. However, the isobaric collection (or nearly so in some cases) and transfer of *Alvinella* at its pressure of origin, 25 MPa, on the East Pacific Rise enabled Shillito's team to observe the animals at controlled temperatures. At 55 °C they behave abnormally and died within 2 hours, but 42 °C was survived for an hour or so (Ravaux et al. 2013; Chapter 13, Figs. 13.25–13.27). The conclusion is that *Alvinella* is very heat tolerant at its normal pressure but perhaps it is not world champion. That distinction probably belongs to a desert mite, *Paratarsotomus macropapis*, which routinely forages in direct sunlight at more than 55 °C (Wu and Wright 2015). The waters of hot springs in Oregon, also 55 °C, are home to the Ostracod *Potamocypris*, another contender. These two live at normal atmospheric pressure.

Generally, the experience of experimenters is that *Alvinella pompejana* is a delicate creature, less tolerant to extreme heat than was first thought, but able to withstand brief decompression. For example, Shillito et al. (2004) were able to collect 26 *Alvinella*, with decompression, from vent chimneys on the East Pacific Rise, 2600 m depth, using the remotely operated vehicle (ROV) Victor 6000. They were restored to 26 MPa at 15 °C on the mother ship. From 6 to 20 hours post recompression about 40% were alive and well, suggesting that a healthy equilibrium might have been reached, during which time meaningful experiments could probably be carried out, although full isobaric treatment is obviously preferable.

More to the point is understanding *Alvinella*'s adaptation to pressure. Dahhoff and Somero (Dahlhoff and Somero 1991) selected a cytoplasmic version of the enzyme malate dehydrogenase (MDH) to study and showed that the Km of its cofactor NADH is a function of pressure. The Km is a measure of the binding of NADH to the enzyme. It was little affected by pressure at 5 °C, as was that of *Riftia pachyptila*. This shows both animals are adapted to function at high pressure. However, the effect of temperature on the Km, at environmental pressures differed, with the Km for *Riftia*'s enzyme cofactor showing a much greater increase at 25 °C than was the case for *Alvinella*'s enzyme. This means that *Alvinella*'s MDH–NADH system is the more temperature tolerant of the two, at the normal high pressure.

Studies of a wide variety of species, using the relationship between the maximal sustained body temperature and the Km for the MDH cofactor NADH, strongly indicates that *Alvinella pompejana* is well adapted to 30 °C and *Riftia pachyptila* to 23 °C, at their normal pressure of 26 MPa. This is consistent with the denaturation temperatures of their collagens; *Riftia* 29 °C (mentioned above) and *Alvinella* 46 °C. Both are little affected by pressure. The thermal stability of *Alvinella*'s proteins have been studied in a number of cases but as pressure was not taken into account their value is questionable.

A rather more robust Polychaete, *Hesiolyra bergi*, which lives close to *Alvinella*'s tubes, has been collected with decompression but, within 2 hours, restored to its normal pressure of 26 MPa and 15 °C. In such conditions, the worms survived for 48 hours, during which their oxygen consumption was typical of "normal" worms of their kind. They were healthy animals and their thermal tolerance at 26 MPa was assessed (Shillito et al. 2001). Gradually increasing the temperature caused death at 46 °C. Severe stress was obvious at 38 °C and other evidence points to the temperature of above 35 °C being avoided. An interesting pathological detail is that heat caused the cuticle to separate from the underlying epithelial cells. Generally, *Hesiolyra* is less tolerant of heat than *Alvinella* and probably *Paralvinella sulfincola*, and similar to that of the vent crab *Bythograea thermydron*.

### 9.3.1.3 Crabs and Mussels

Mickel and Childress (1982a, b) experimented with *Bythograea thermydron*, probably the first vent animal to be collected for shipboard experiments. Observations carried out on the Galapagos vents using DSRV Alvin established that this active crab usually occupied water no warmer than 22 °C and often as cool as 2 °C, at depths of around 2500 m. Its coordination was impaired after collection with decompression and some individuals survived for 5 days at 5 °C. atmospheric pressure. When recompressed to their normal pressure survival for 18 months has been achieved at 8 °C. Electrodes were implanted, during a brief decompression, to monitor its heart rate, which proved remarkably unaffected by pressure up to 34 MPa, at 5 °C (Airriess and Childress 1994). Higher pressures reduced the frequency markedly. At 20 °C increasing the pressure to 14 MPa increased the rate, beyond which higher pressures, up to 50 MPa, were without effect. All this seems compatible with its normal pressure and temperature.

Separate experiments on the kinetics of its muscle MDH–NADH system, similar to those described above for *Alvinella*, showed that the Km for NADH was unaffected by high pressure, clear evidence of its adaptation to pressure. Other experiments showed that the composition of its blood (haemolymph) was little affected over a period of 13 hours when the animal was subjected to a range of salinities at 26 MPa, or at normal atmospheric pressure. In its normal life among the vents, it occupies water with a low concentration of oxygen and a high concentration of hydrogen sulphide, the toxicity of which it effectively neutralises. *Bythograea's* oxygen consumption at its normal environmental pressure was similar to that of subtidal crabs and other benthic crustacea in their normal environment.

The first hydrothermal vents to be discovered were occupied by many bivalve Molluscs, such as clams and mussels. They are interesting because of their symbiotic bacteria and also because their shells, incorporating calcium carbonate, are produced in an unusual environment. Both high temperature and high pressure enhance the solubility of calcium carbonate, with the aragonite form being more rapidly dissolved than calcite.

The structure and growth of the molluscan shell presents a number of interesting processes, in which high pressure is likely to be a significant factor, for example the organisation of extracellular structural proteins and the crystallisation of minerals extracted from seawater. With this in mind *Bathymodiolus azoricus* has been collected, with decompression, from vents along the Mid Atlantic Ridge using the ROV Victor 6000 (Kadar et al. 2008a, b). At its most shallow site, Menez Gwen, the depth is 850 m and at the deepest site, Broken Spur, it is 3000 m. The adult form is about 5 cm long and dark in colour, not unlike familiar shallow-water mussels. It was found that decompression from depths of 850–2300 m led to a deterioration of the layer in the shell known as nacre, which is composed of organised deposits of aragonite. This is crystalline calcium carbonate, and within a day at normal atmospheric pressure it looked "as if it had been subjected to acidic dissolution treatment". The outer prismatic layer, made up of calcite embedded in protein fibres, was unaffected. The nacre may have been affected by an increase in the acidity of the fluids bathing the underlying mantle epithelium, as anaerobic respiration generated more lactic and acetic acids. Maintaining the mussels at the relatively low pressure of 2 MPa reduced the effect, which was therefore regarded as a decompression stress. Accordingly, subsequent investigations of the growth and repair of the shell ensured the animals were re-compressed to their normal pressures. The experiments took the form of drilling a small hole in the shell and monitoring the response, primarily of the haemocytes in the blood. Many of these contained particulate calcium carbonate, which appeared to be added to the damaged shell, a process which was regarded as a sort of first aid response rather than a resumption of the normal growth. The repair sites clearly showed signs of freshly deposited crystals, but only in the animals kept at their normal ambient pressure. Controls kept at atmospheric pressure failed to show any such signs of repair.

*Bathymodiolus azoricus* is sustained by symbiotic bacteria in its gill epithelial cells where the bacteria oxidise either methane or sulphur with the resultant energy driving the fixation of carbon in the first step of chemosynthesis. It also has an alimentary canal and can digest filtered particulate food. It is thus mixotrophic, and in the short term, it is not dependent on its symbionts. In an initial study of the symbionts, the mussels were collected by two different methods, from 830 and 2300 m depths. One method entailed decompression during collection, as in the work described earlier (Kadar et al. 2005). The second method was isobaric (Szafranski and colleagues 2015), in which the animals' pressure was kept close to their original ambient pressure, by means of PERISCOP (see Chapter 13, Figs. 13.25, 13.26). In both methods, the animals were transferred to their normal pressure once safely on board ship, but the isobarically collected animals experienced only a brief, 10 minute, period of decompression and re-compression. As a means of securing the symbiotic bacteria for subsequent experimental work with the mussel host, the isobaric method was superior, especially in the case of mussels from the deeper vents. The main idea emerging from the work is that as the substrates available in the vent fluids vary considerably in time and place, a versatile symbiotic population is advantageous in ensuring a steady supply of energy for the host. Other species, *B. thermophilus* and *B. childressii*, have predominately sulphur-oxidising and methane-oxidising symbionts, respectively (Duperron et al. 2005).

*Bathymodiolus azoricus* from 850 m survives in aquaria at normal atmospheric pressure and in the absence of sulphides. Kadar and colleagues have shown that the symbionts living in the gill tissue disappear over 30 days but reappear in animals kept in the company of infected animals. Presumably, this form of cross-infection occurs as the young colonise new sites (Laming et al. 2015).

### 9.3.1.4 Shrimps

The vents on the Mid Atlantic Ridge often have swarms of active shrimps, *Rimicaris exoculata*, apparently exposed to temperatures as high as 40 °C and in close proximity to water at 70 °C, at pressures of 20 MPa or more. Their nutrition is provided by a variety of bacteria, epibonts, which live on the mouthparts and in the gill cavity. Moulting eliminates bacteria living on the cuticle and quite separately, the dominant types of bacterial epibionts change as the animal progresses from the early stages of its life cycle to the juvenile–adult stages (Guri et al. 2012).

Experiments have been carried out to try to clarify what temperatures these shrimps can tolerate, at their normal pressure of 20 or more MPa. Ravaux and colleagues (2003), using the ROV Victor 6000 on the Rainbow Vent of the Mid Atlantic Ridge, collected shrimps with decompression. They survived for 2 hours before being restored to their normal pressure of 23 MPa, and 15 °C. In these conditions, an exposure to 25 °C for about 15 minutes, with the necessary increase and decrease in temperature above 15 °C occupying an hour or so, appeared to elicit the synthesis of heat shock proteins, hsp 70. This might indicate that the shrimps' real-life thermal tolerance, at 23 MPa, is rather less than observations on the vents might suggest. Further studies have been made (Ravaux et al. 2007, 2009, 2019), broadly confirming that view. However, decompression preceded the analysis of the heat shock proteins in all cases so it is not clear what the state of the process is prior to decompression. Stable, slowly turning over proteins probably undergo little change on decompression, but stress proteins are different. Some are normally present as part of the cell's homeostasis but others are, as their name suggests, induced by specific stimuli. In the shallow water shrimp *Palaemonetes* a marked increase is seen in the one type of heat shock protein within 30 minutes of treatment, which is interesting but hardly resolves our problem (Cottin et al. 2010). The decompression of some deep-sea protists appears to elicit a stress response and so too does the application of high pressure to ordinary protists, so the experimental design, in this case, is critical. Freezing the specimens promptly after decompression may be quick enough, but the reader needs reassurance. When the effects of pressure on cellular structure is investigated fixation prior to decompression is carried out (Chapter 8; Section 8.2.1).

The isobaric collection of *Rimicaris*, using the PERISCOP apparatus (Shillito et al. 2008; Chapter 13, Figs. 13.25, 13.26), from 2300 m depths on the Mid Atlantic Ridge enabled its healthy, active swimming, similar to that seen at the vent sites, to be observed on board ship. Animals collected with decompression were disorientated, often upside down, sluggish and exhibiting the occasional convulsion. Some years later, Auguste and a large team (2016) isobarically collected *Rimicaris*, again by the ROV Victor 6000, from the Trans-Atlantic Geotraverse site (TAG) in the same region. This time the collecting depth was 3630 m and the purpose was to examine the significance of the locally high concentration of copper and other potentially toxic salts. After a brief decompression and recompression on board the mother ship the animals were exposed to copper, but no other potentially toxic salts, in the high-pressure aquarium IPOCAMP, at 30 MPa and 10 C for 72 hours (Chapter 13). The animals survived this, with some signs of a loss of vigour, and subsequent analysis of the animals' tissues, post decompression, showed that metallothioneins were present, a known defence against the copper. The presence of the copper increased the concentration of reactive oxygen species so the response to that was of interest. Lipid peroxidation was not increased so presumably the antioxidant defence enzymes were in good order. Although the activity of superoxide dismutase (SOD) was not increased that of catalase was. On these vent sites, *Rimicaris* is a healthy animal, its natural defence mechanism coping with the water's significant concentration of copper and other salts. If mining operations change the environment seriously that happy state may not persist. The presence of potentially toxic salts in vent waters is normal and the organism's defence mechanisms are, equally, a normal state of affairs, of considerable interest. Globally the development of deep-sea mining is getting underway, and our knowledge of the tolerance of deep-sea organisms, including those in the vents, to high concentrations of toxic salts, is insufficient for sound planning purposes.

And in case readers are curious about the animal's name, *Rimicaris exoculata* is indeed eye-less. The normal shrimp eye is much modified and contains a visual pigment sensitive to the radiation emitted at 350 C. Pelli and Chamberlain (1989) have argued that the radiation would so affect the pigment that the signal to background noise ratio would provide the animal's nervous system with a useful stimulus. A similar argument concluded that the dark-adapted human eye would just fail to detect a signal from the same source, which is consistent with what observers in submersibles report.

### 9.3.1.5 Fish

There are a few species of fish endemic to vents and they too have an active lifestyle, moving rapidly about, making it difficult to know exactly what their temperature preference and tolerances are. *Thermarces andersoni* and *Bythites hollisi* occupying depths centred on 2500 m are two whose thermal tolerance has been assessed by Dahlhoff et al. (1990), examining the kinetics of their muscle malate–NADH systems (as above) and comparing them with that of the abyssal *Coryphaenoides armatus* which typically lives much deeper, in cold water. The latter's enzyme differed markedly from that of *T. andersoni* in thermal tolerance, indicating that *T. andersoni* was capable of withstanding much warmer conditions and that *C. armatus* was probably unlikely to access the vent waters for food. The case of *Bythites hollisi* was particularly interesting because its enzyme appeared similar in its thermal sensitivity to that of *C. armatus*. This was consistent with it living in a vent site significantly cooler than the one occupied by *T. andersoni* and being a more continuous swimmer. To know exactly what temperature these mobile animals actually experience probably requires the use of recording tags, but other techniques exist (Rinke and Lee 2009; Bates et al. 2010).

Footnote: A variety of vent animals are kept alive in a public aquarium in Japan, at normal atmospheric pressure but in a controlled chemical and temperature environment. The deepest came from depths of 1500–1200 m: the shrimp *Opaepele sp.* and the barnacle *Neoverruca sp.* (Miyake et al. 2007; Watanabe et al. 2004). Similarly, the gastropod *Alvinoconcha sp*, collected from a depth of 2400–3200 m from a hydrothermal vent in the Indian ocean remains in good condition at atmospheric pressure (Sigwart and Chen 2018).

## 9.3.2    Microorganisms

### 9.3.2.1 Eukaryote Microorganisms

As fungi are found in the cold ocean depths and the sediments, it seems likely that they would colonise hot vent habitats. The waters from hydrothermal vents contain few fungi but the vent animals, particularly shrimps and mussels, have proved a source of numerous species. Yeasts, like all fungi, are heterotrophs, that is they use organic material for their source of energy. Many are saprotrophs, using the remains of dead organisms. A largely French group, Burgaud and colleagues (2010, 2015), has made a major contribution in this field, collecting material from numerous hydrothermal vents and their fauna. Numerous species of yeasts proved capable of growing at pressures matching their original source, around 25 MPa, in warm but not hot conditions, suggesting that they truly belong in the vents and are not just surviving. An example is *Rothia mucinlagiosa*, a Basidiomycete yeast which has been isolated from cold abyssal waters, but in this case was collected from the waters of the Rainbow vent, depth 2300 m and also from shrimps and mussels living there. It grew at normal atmospheric pressure and at pressures as high as 60 MPa, 25 °C. Other yeasts manifested abnormal filamentous growth at high pressure, which was regarded as different to the filamentous growth seen in bacteria. Generally, the Basidiomycete yeasts tolerated high pressure (60 MPa) better than Ascomycete yeasts (Fig. 9.1).

### 9.3.2.2 Protists

Flagellates collected with decompression by the submersible Alvin from vents at a depth of 2500 m on the East Pacific Rise were isolated and grown at a range of high pressures. These experiments required bacterial cultures to provide food for the flagellates, which seemed to cope well at normal atmospheric pressure. Two species were of special interest: *Rhynchomonas nasuta* and *Caecitallus parvulus*. Pressures above atmospheric reduced their growth rate but at 25 MPa and 20 °C, they proved capable of growing moderately well and thus qualify as piezotolerant. Both species formed cysts at pressures higher than 25 MPa, and these produced abundant cells when restored to normal atmospheric pressure for a few days. These species are pelagic and widely distributed and are thought to be carried on sinking particles to the seafloor. The growth of the same species isolated from shallow sources was also measured at high pressure and found to be less vigorous than the cells from the deep vents. Perhaps some adaptation to pressure takes place when cells survive at depth, and perhaps the cysts are redistributed to the pelagic zone by plumes rising from the vents. These experiments were carried out over 20 years ago (Atkins et al. 1998), since when little interest has been shown in vent protists. A 2016 paper reports the discovery of a variety of ciliated protists in the Okinawa Trough, 1000 m depth, by Zhao and Xu, but these were detected from DNA and RNA extracted from sediment samples and nothing is known of them as cells.

The ROV Victor 6000, mentioned earlier, deploying its slurp gun bottle on the Ashadze site on the Mid Atlantic Ridge, collected, with decompression, mites (*Copidognathus*) colonised by *Suctorians* (Protist), at a depth of 4090 m. These

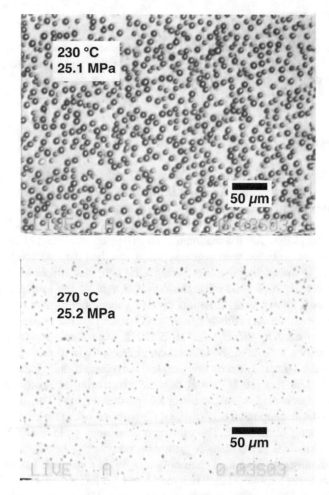

**Fig. 9.1** Yeast cells at high pressure and 230 and 250 °C. The cells of the Basidimycete yeast *Cryptococcus sp* No. 6 are structurally stable at 230 °C and 25.1 MPa, but when the temperature was increased to 250 C, the cellular structure is destroyed. The yeast was isolated from sediments collected by the manned submersible Shinkai, working at 4500–6500 m depths, and 1–2 °C, in the Japan Trench. It grows well at atmospheric pressure and has been studied because of its exceptional tolerance to the heavy metal copper. Subjecting cells to supercritical conditions was a technical exercise in the development of a high-pressure microscope flow cell for use in supercritical conditions. Reproduced with permission from Deguchi and Tsujii (2002)

single cells contain a giant nucleus enclosed in a tulip-like cell body about 30 μm across and held above the substrate by a stalk. Fine protuberances, called tentacles, decorate the top of the tulip. The new species, *Corynophyra abyssalis*, described by Bartsch and Dovgal (2010), has the same appearance as the suctorians colonising freshwater crustacea which I learned about as an undergraduate. It is a good example of the appearance of a deep-sea organism masking its molecular adaptation to extreme conditions.

## 9.4    Prokaryotes

Prokaryotes are present as biofilms, in sediments and in suspension in the waters around the vents. A submersible is required for collecting and high concentrations of cells can be obtained, but the original community is not preserved.

### 9.4.1   Archaea

Over 60 species of Archaea have been isolated from hot vent habitats. They include *Thermococcales, Methanococcales* and *Desulfurococcales*. The hydrothermal vents are not in the deepest parts of the ocean, being absent in the hadal regions, so these microorganisms are not the deepest living organisms, but they include some spectacular extremophiles. An international team, (Birrien et al. 2011) enriched with French scientists, collected with decompression and then isolated and characterised an Archaeon, *Pyrococcus yayanosi* (strain CH1). It was isolated from sulphide samples from a depth of 4100 m on the Ashadze vent on the Mid Atlantic Ridge, then the deepest known vent. Isolation involved anaerobic growth at 42 MPa and 95 C, using the extinction to dilution method (Chapter 13). The purity of the ensuing culture was checked by cloning the 16 S rRNA. The Archeon was named after Aristides Yayanos, a distinguished American microbiologist. It proved to be both an obligate anaerobe (i.e. is killed by oxygen) and a hyperpiezophile (unable to grow at a pressure of less than 10 MPa and capable of growing at up to 120 MPa, an unusually high pressure) as well as being a hyperthermophile. This means it grows well at very high temperatures, 60–108 °C, with an optimum of 98 °C and an optimum pressure of 52 MPa. The optimum conditions yield a doubling time of 50 minutes, which is pretty fast. It grows on complex carbon substances and is therefore formally classified as a chemoorganotrophic organism. Under the microscope, it appears as a spherical cell, between 1 and 1.5 μm in diameter, with a tuft of flagella which make it motile. Its unique identity was defined by its 16S tRNA gene sequence. The molecular adaptations which underlie CH1's performance are largely unknown. The growth of other species at high pressure is summarised in Table 9.1 (compiled from Zeng et al. 2009). They were grown in standard conditions and their

**Table 9.1** Growth of Vent Archaea ai 85–95 C

| Species | Depth of origin | Good growth (MPa) | No growth (MPa) |
|---|---|---|---|
| *Pyrococcus furiosus* | shallow | 1–20 MPa | 35 |
| *P. horikoshii* | 1400 m | 1–35 | 50 |
| *P. abyssi* | 2200 m | 1–40 | 55 |
| *P. glycovorans* | 2650 m | 1–40 | 50 |
| *Thermococcus barophilus* | 3550 m | 1–80 | 85 |
| *P. yayanosi* | 4110 m | 30–120 | No growth <15 MPa |
| | | | Or at 130 MPa |

significant growth yields were measured over a range of pressures. The results tally nicely with their depth of origin.

The growth yield of *P. yayanosi* CH1 peaked at 50–60 MPa and uniquely it was zero at a pressure of less than 15 MPa and it was also zero at 130 MPa.

Some interesting experiments have been carried out on *P. horikoshii* collected from a vent at a depth of 1395 m in the Okinawa Trough (Zeng et al. 2009; Gonzalez et al. 1998). The growth of the organism was optimal at 95 °C at pressures up to 15 MPa. At higher pressure, 30 MPa, the temperature optimum was shifted to 100 °C. This is one of several examples, another is *P. abbyssi*, of high pressure increasing thermal tolerance. At the molecular level, high pressure can enhance the thermal tolerance of enzymes extracted from thermophiles. *P. horikoshii* also provided experimenters (Roenbaum et al. 2012), with an enzyme, a peptidase, abbreviated to TET3, which proved to be remarkably pressure tolerant. X-ray techniques showed its structure and oligomeric state was unaffected by pressures up to 300 MPa, at 25–90 °C. Its kinetics, monitored optically at high pressure, were also unaffected by high pressure. To be teleological, this is a case of an apparently "over designed" enzyme, for which there is doubtless a good explanation, as yet unknown. It alerts us to the fact that a neat relationship between the pressure at which an organism lives and the pressure tolerance of an isolated component such as an enzyme, may not always exist. Microorganisms are complicated enough to accommodate a host of features susceptible to selective pressures working in diverse ways. How is it, for example, that the hyperthermophile Thermococcus piezophilus, which has optimal growth at 75 °C and 50 MPa, can grow, to some extent, over an enormous pressure range, atmospheric pressure up to 130 MPa (Dalmasso et al. 2016a, b).

The growth of these vent Archaea at high temperatures and pressures has dramatically changed our ideas about life in extreme environments. *P. yayanosi* probably holds the world record for growing at high pressure, 120 MPa, and we should note that another Archaeon, *Methanopyrus kandleri* strain 116, holds the record for growth at high temperature, in the presence of significant pressure. This organism was known from collections at various vents but a team of Japanese scientists carried out some remarkable high-temperature experiments using a strain collected from the Kairei hydrothermal vents on the Central Indian Ridge, at a depth of 2420–2450 m (Takai et al. 2008). Under the microscope the cells appeared rod-shaped, 0.5–0.7 μm wide and about 7 μm long. They are hyperthermophilic hydrogenotrophic methanogens, and anaerobic for good measure. They oxidise hydrogen according to: $CO_2 + 6H_2 \rightarrow (CH_2O)$ organic compound $+ CH_4 + 3H_2O$. Their culture conditions are demanding, with a partial pressure of 0.4 MPa hydrogen gas to provide significant dissolved hydrogen, carbon dioxide, ammonium or nitrate to provide a source of nitrogen, acidic conditions with additives such as iron. At a pressure of 40 MPa, both growth and methane production were apparent at 122 °C. The cells could survive 130 °C for three hours in these conditions. We should not regard 122 °C as the highest possible temperature for growth, as culture conditions might be varied fortuitously to enable growth at even higher temperatures. And other organisms may be discovered with an even higher heat tolerance. We cannot, however, rationally predict a precise upper limit to the temperature at which life processes can continue, but a common hunch is that it might be as high as 150 °C.

Shallow cores of the hydrothermal sediments obtained by the submersible Alvin in the Guaymas Basin at a depth of 2013 m had a maximum temperature of 200 °C. The rate at which sulphate was reduced and methane anaerobically oxidised were measured by Kallmeyer and Boetius (2004). The sediments contained a diverse population of Protists, but the point is some activity was detectable at up to 130 °C and 22 MPa. In the case of sulphate reduction, it peaked at 100 °C at 22 MPa, its ambient pressure.

## 9.4.2  Bacteria

More than a hundred species of Bacteria have been isolated from the waters, chimneys, sediments and macrofauna of the hot vents.

### 9.4.2.1  Bacteria Growing on Substrates

According to a 2015 review, the Domain of the Bacteria includes major occupants of the vents such as *Desulfurbacteriales* and *Acidobacteria* but is dominated by *Epsilonprotobacteria*. Bacteria such as these form biofilms several centimetres thick on chimney surfaces. Some bacteria synthesise a polysaccharide cement and others grow long filaments. In one example, Loki's Castle vent (2400 m), detailed examination revealed small nematode worms living in the fabric of the biofilm, whilst polychaete worms, shrimps, crabs and gastropods lived on and around it. The animals were also coated with episymbiotic bacteria, and some hosted endosymbiotic bacteria. In the sediments, bacterial populations live surprisingly mobile lives alternating with an attachment to sediment particles.

### 9.4.2.2  Bacteria in Water Samples

An isobaric water and gas sampler has been used to collect vent water samples containing a natural population of bacteria, dissolved gases and salts. The equipment is described in Chapter 13, Fig. 13.32. It was deployed on the East Pacific Rise, to sample the waters venting from Crab Spa at a depth of 2506 m and a temperature of 24 °C. The operation was carried out by a team from Woods Hole led by McNichol and involved the ROV Jason II and its mother ship R V Atlantis. One of the aims of the work was to measure the rates of biochemical reactions in a natural population of bacteria, mostly *Epsilonproteobacteria*, incubated at their in situ pressure (presumably 25 MPa), and temperature (24 °C), with only minor enrichments. For example, the consumption of hydrogen sulphide, oxygen and hydrogen were measured and found to be equal to, and often much higher than, similar measurements made by other workers using decompressed samples. A 2018 paper confirmed the subseafloor microorganisms are active primary producers.

*Desulfovibrio hyrothermalis* AM13 was collected from a hot vent on the East Pacific Rise at a depth of 2600 m by Amrani et al. (2014). The bacterium is a sulphate reducer, the reduction of sulphate being a form of respiration which generates ATP. Experiments revealing gene expression were carried out using cells grown anaerobically at 30 C and at atmospheric pressure, 10 and 26 MPa, the latter being the pressure from which the bacteria were retrieved. It was found that 65 genes were differentially expressed, according to the growth pressure. They fell into several groups, such as genes involved in energy production, and glutamate metabolism. In the case of the latter, the upregulated genes implied that the net concentration of glutamate in the cells should be increased at high pressure, and indeed, direct analysis showed this to be the case. Cells grown at 26 MPa had more than twice the intracellular concentration of glutamate than those grown at normal atmospheric pressure. This is significant because glutamate can act as a compatible solute to protect proteins from adverse effects of high pressure (Chapter 6). Other examples of this are the methanogenic bacteria *Methanococcus thermolithotropicus* and *M. jannaschii*, whose optimum growth temperature is in the region of 60–85 °C. and contain significant concentrations of beta-glutamate. *Photobacterium profundum*, mentioned in Chapter 8, which lives in cold abyssal waters, produces a high intracellular concentration of another neutral and protective solute, beta-hydroxybutarate, when grown at high pressure.

As with abyssal bacteria, details of the growth techniques are important. In the case of hot vent bacteria, the usual batch culture methods work well enough but continuous culture techniques have been used by, for example, Houghton et al. (2007). Their continuous culture bioreactor was designed to simulate vent conditions but also to allow enrichment growth experiments to be carried out. The high-pressure chemostat developed by Foustoukos and Rodriguez (2015) at the Carnegie Institute in Washington DC operates at pressures up to 69 MPa over the notable temperature range of 20–120 °C. The hot vent bacterium *Marinitoga piezophilia*, originally obtained from the East Pacific Rise at a depth of 2630 m, was shown to perform well in the apparatus during a growth experiment lasting 352 hours, when its growth at 30 and 40 MPa was followed.

The adaptation of the membranes of vent bacteria is of particular interest and is dealt with in Section 9.6.

## 9.5   Viruses and Other Mobile Genetic Elements

An abundance of viruses and other MGEs are found in hydrothermal vent waters (Anderson et al. 2013, 2014). In the hottest, pure vent waters it is generally thought that viruses are absent, but in the mixed waters, the concentration of virus-like particles is up to about 10,000,000 per ml, with rather fewer in the neutrally buoyant plume waters (Ortmann and Suttle 2005). Virus concentrations are typically ten times those of bacteria. Six "species" of bacteriophage have been identified in vent waters, of which four lyse their host cells. Archaea and bacteria are both subject to virus infections. A good example is the piezophilic, anaerobic, hyperthermophile called *Marinitoga piezophilia* mentioned above, which grows at 65 °C on the walls of a Black Smoker at a depth of 2630 m in the East Pacific Rise. It is infected with a DNA virus MPV1, which is non-lytic and also caries plasmids. Vesicles are also found in the vent waters and those derived from Archaea have a membrane composed of characteristic Archaean lipids (Roine and Bamford 2012). The collective term for this population of genetically active elements is mobilome (Lossouarn et al. 2015). Its existence implies a high degree of adaptation to the ambient temperatures and pressures, but little experimental effort has gone into exploring this, so far. The resistance of viruses to deep-sea conditions seems to be taken for granted by many workers (Silva et al. 1996; Williamson et al. 2008). No reports of high-pressure experiments with marine viruses, including those to be found in the deep sea, seem to be available (Chapter 8; Section 8.6).

## 9.6    Homeoviscous Adaptation in Hydrothermal Vent Organisms

It will be recalled that increasing pressure and temperature cancel each other's effects on the fluidity of lipid bilayers (Chapter 7). As many organisms tend to regulate the fluidity of their bilayers it is clear that a high pressure–high temperature environment is radically different to the high pressure–low temperature of the abyss. The bilayer ordering effect of 10 MPa is counteracted by a temperature increase of approximately 2 °C. The deepest hydrothermal vent, Beebe, lies at a depth of 4964 m thus providing a hydrostatic pressure of about 50 MPa. Even if deeper vents are found in the future it is unlikely, for geological reasons, that they will be much deeper. Thus, homeoviscous adaptation to pressure in hydrothermal vent organisms is limited to 50 MPa, whose effect can theoretically be offset by a temperature increase of around 10 °C, which is small in the vent environment. For hyperthermophiles living at 80 °C and 40 MPa, we can predict that their membrane bilayers are at the equivalent of 72 °C and normal atmospheric pressure. However, predictions cannot be made for organisms living in wildly fluctuating temperatures.

Finally, as previously pointed out, the theoretical implications of homeoviscous theory should not be overstated. It provides us with a strategy for studying membranes in the cold depths. Here we shall see how it works out in vent organisms.

### 9.6.1    Animals

Quite early in the investigation of the hydrothermal vent fauna, Taghon (1988) analysed the lipids of the polychaete *Paralvinella palmiformis* collected from the Juan de Fuca Ridge vent at a depth of 2200 m. Some specimens came from water at 17 °C and others from 30–40 °C. A comparison of the fatty acids extracted from the whole body showed that the animals from the warmer waters had a greater proportion of saturated fatty acids and a lesser proportion of unsaturated fatty acids than animals from the cooler water. As most fatty acids come from membranes this is consistent with homeoviscous theory. Some years later a major study of hydrothermal vent lipids was led by Phleger. *Alvinella pompejana* from the East Pacific Rise was studied. As previously mentioned, these animals often have a significant temperature gradient along their body, the anterior being cooler than the posterior, sometimes by as much as 60 °C. The membranes from each part of the body were not isolated but a comparison of the fatty acids from the anterior and posterior parts of the body was made. Thus, at 25 MPa the anterior part of the worm has a slightly higher proportion of polyunsaturated fatty acids than the posterior part. Whilst this may encourage a project to isolate membrane fractions it also suggests that sedentary animals, preferably from deeper vents, might be more suitable for the precise measurement of membrane fatty acids which testing the hypothesis demands. Mussels are common in hydrothermal vent waters and a great deal is known about their lipids and fatty acids in particular. They have also been studied to understand the contribution which endosymbiotic bacteria make to their nutrition. Methane- and sulphur-oxidising bacterial symbionts produce characteristic fatty acids which become incorporated in the mussels' cells. Although this is a complication there is no reason why a host mussel should not incorporate "appropriate" fatty acids into its membrane bilayers irrespective of their source. Natural selection could drive the use of symbiont fatty acids in the direction of homeoviscous adaptation.

## 9.6.2  Bacteria

The bacteria *Halomonas axialensis* and *H. hydrothermalis* were isolated by Kaye and Baross (2004) from warm vent waters in the Pacific at depths of 1533 m and 27 °C and 2580 m and 9 °C, respectively. They showed a difference in their fatty acids when grown at normal atmospheric pressure and 45 MPa. In both cells, the fatty acids were predominantly 18:1 and 16:1 (unsaturated) and 16:0 (saturated), and in both cells growth at high pressure caused similar changes. The proportion of monounsaturated fatty acids increased and that of the saturated 16:0 approximately halved. This is consistent with homeoviscous theory, but in the absence of fluidity measurements, this is as far as we can go. However, that conclusion is somewhat clouded by the data for terminally branched saturated fatty acids. Although present in very small amounts they can exert a strong influence on bilayer fluidity. In the case of *H. axialensis* their proportion doubled at pressure but in *H. hydrothermalis* it was halved!

Another case is the bacterium identified as Strain RS103, isolated from surface sediments in the Okinawa Trough, 1350 m depth and 17 °C. It contained significant amounts of branched-chain fatty acids. The effect of temperature at atmospheric pressure on the cells' fatty acids was first investigated, growth taking place over the range 10–40 °C. At elevated temperature, the proportion of unsaturated saturated fatty acids decreased markedly, but the branched chains iso C 15:0 and iso C:17 actually increased. Growth at pressures up to 40 MPa at 30 °C produced cells with increased proportions of unsaturated fatty acids and the branched-chain iso C17:1. A slight decrease was seen in the iso C15:0. This is consistent with homeoviscous adaptation, but as is often the case, bilayer fluidity measurements are lacking.

## 9.6.3   Archaea

There are abundant Archaea in the cold deep oceans and in the hydrothermal vent sites. It is estimated that they are almost as numerous as Bacteria, a further example of the strange similarity between these two Domains. Under the light microscope, they are indistinguishable, their distribution overlaps a great deal and for billions of years they have co-existed. But crossing the Lipid Divide to the Archaea (Chapter 1), means we need to understand their different lipids and membrane structure before we can think about their adaptation to high pressure. Recent papers by Silliakus et al. (2017) and by Caforio and Driessen (2017) are instructive in this regard.

### 9.6.3.1 Archaeal Lipids

Bacterial lipids are based on phosphatidic acid, which has two hydrocarbon chains joined (by ester bonds) to the 1 and 2 position on glycerol 3-phosphate. In contrast, archaeal membranes have, as a core lipid, molecules based on archaetidic acid. This has hydrocarbon chains of methyl branched isoprenoids connected (by ether bonds) to the 2 and 3 positions on glycerol 1-phosphate. As mentioned in Chapter 1, these glycerol backbones are mirror images, enantiomers, but here we need not be too concerned with their chemistry. Figure 9.2 shows the important point, which is the similarity in the shape of the molecules, and how their polarity enables them to form stable bilayers in water.

A second important characteristic of Archaeal membranes is that in addition to containing molecules which form bilayers, such as archaeol (Fig. 9.3a), they also have very long molecules which have polar groups at each end. Such molecules were once thought to be unique to Archaea but it is now known they occur in certain Bacteria. An example of a bipolar, monolayer forming Archaeal lipid has the acronym GDGT. (This stands for glycerol dialkyl glycerol tetraether lipid). From its structure (Fig. 9.3b) it is easy to see why it is thought to arise from the end to end fusion of two shorter molecules. Other lipids incorporate five-sided and six-sided rings (Fig. 9.3c). The presence of monolayers, serial methyl groups and ring structures suggests that Archaeal membranes are rather dense and well-packed structures, which is confirmed by experiments. Additionally, they have a high protein content compared to membranes on the other side of the Lipid Divide. Nevertheless, they perform the same role as Bacterial and Eukaryote membranes: providing a permeability barrier which supports proteins with many functions including the transport of molecules across the membrane. The right amount of molecular motion is required for these functions so the details of the membrane lipids are important. That motion can be measured in situ and in liposomes made from extracted lipids.

The membrane lipids of the Archaeon *Solfolubus acidocaldarius*, grown at 78 C, were used by Chang in a study of unilamellar liposomes made largely of the bipolar

# Bacteria

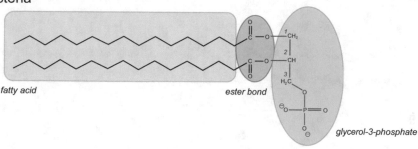

*fatty acid*                                        *ester bond*

glycerol-3-phosphate

glycerol-1-phosphate

# Archaea

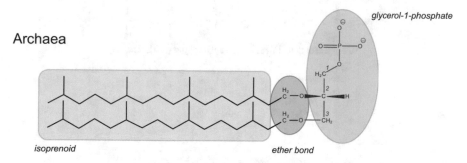

*isoprenoid*                                        *ether bond*

**Fig. 9.2**  Characteristic Bacterial and Archaeal membrane lipids. Reproduced with permission from Caforio and Dreissen (2017)

lipids GDGT and GDNT, the latter being glycerol dialkyl calditol tetraether. These are known as polar lipid fraction (PLF) membranes and they contained a variable number of cyclopentane rings. Their properties were compared to those of liposomes made of the familiar membrane phospholipid DPPC (Chapter 2, Section 2.4.1). The Archaeal liposomes, sometimes called archeosomes, were found to be very impermeable to dyes and to hydrogen ions, and the lipid packing was tight with very little lateral diffusion. The membrane lipids of the Archaeon *Solfolubus acidocaldarius*, grown at 78 C, were studied by Chang (1994), Elferink et al. (1994), and Jarrell et al. (1998). Valentine has argued (2007) that the archaeal membrane minimises ion leakage in sustaining the cells' chemiosmotic potential, making the cells very energy efficient. The liposomes have been shown to be very stable at high temperatures. Phase transitions were detected at temperatures below the growth temperature, and they had small the $\Delta H$ and $\Delta V$ values, compared to the analogous transitions in DPPC liposomes. The compressibility of the membrane is much smaller than that of DPPC bilayers (Chong et al. 2010), and in a subsequent paper, neutron reflectivity measurements on a supported PLFE membrane showed little effect of 180 MPa on its structure (Zhai et al. 2012). Significantly, spectroscopic measurements of the fluidity of lipid dispersions (similar to liposomes) showed a significant decrease on

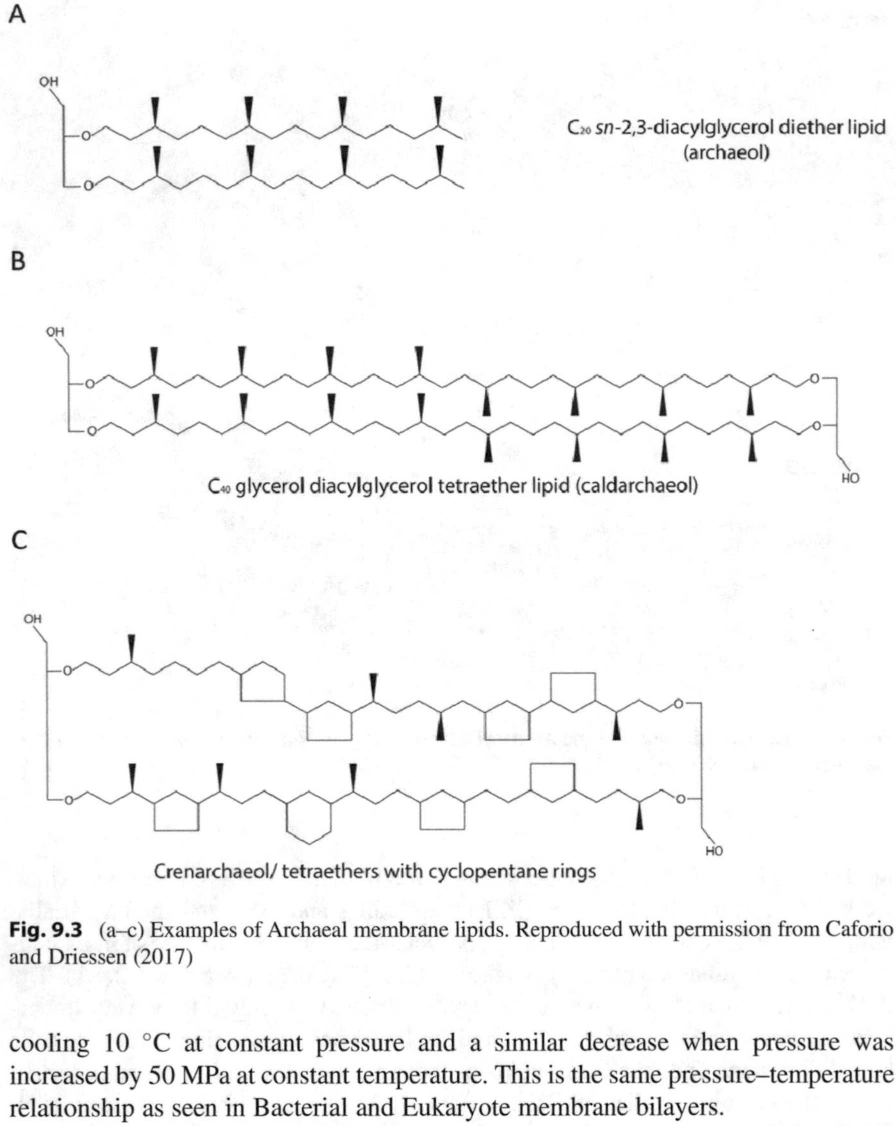

A

C₂₀ sn-2,3-diacylglycerol diether lipid
(archaeol)

B

C₄₀ glycerol diacylglycerol tetraether lipid (caldarchaeol)

C

Crenarchaeol/ tetraethers with cyclopentane rings

**Fig. 9.3** (a–c) Examples of Archaeal membrane lipids. Reproduced with permission from Caforio and Driessen (2017)

cooling 10 °C at constant pressure and a similar decrease when pressure was increased by 50 MPa at constant temperature. This is the same pressure–temperature relationship as seen in Bacterial and Eukaryote membrane bilayers.

Growth experiments demonstrated that the number of cyclopentane rings in the tetraether lipids decreased with a decrease in temperature, consistent with the rings ordering the membrane. This is a form of homeoviscous adaptation to temperature. Generally, the Archaeal membrane is a well-ordered structure and impermeable because of the tight packing. In this regard, it is rather similar to the DPPC liposome bilayer below its main phase transition temperature and unlike the more fluid membranes of Bacteria.

We might speculate that the Archaeal membrane is adapted to function in a highly ordered state and might therefore be pre-adapted to high pressure. Oger (2015) has considered their homeoviscous adaptation and cited 12 examples of adaptation to temperature and only 1 to high pressure.

### 9.6.3.2 Methanococcus janaschii

Compared to deep-sea Bacteria only a few deep-sea Archaea have been studied. For example, *Methanococcus jannaschii* and *M. thermolithotrophicus* (Phylum Euryarchaea) have been shown to be piezophilic. The former, named after the distinguished microbiologist Holger Jannasch, is a flagellated cocci which was isolated from samples collected by the submersible Alvin in 1982 from the White Smoker vent on the East Pacific Rise, at 2600 m. It is a remarkable cell, an anaerobic methanogen, which "fixes" nitrogen, i.e. it uses it to make amino acids. That requires energy in the form of ATP, which it generates from reactions involving hydrogen gas and carbon dioxide, producing a waste product, methane, natural gas.

The overall reaction is: $4 H_2 + CO_2 = CH_4 + 2 H_2O$ . Chemiosmotic coupling drives ATP synthase to produce ATP. The details of this process closely resemble ATP synthesis in various "modern" aerobic cells. *M. jannschii* is probably similar to the earliest form of cells which existed more than 3 billion years ago (Lane 2016). A complicated high-pressure bioreactor was developed by Boonyaratanakornkit and Clark (2008) for growth experiments with *M. jannaschii* in well-controlled conditions. For example, in the presence of the gases hydrogen, nitrogen and carbon dioxide, at 86 C *M. jannaschii* has a doubling time of 30 minutes at 36 MPa and of 15 minutes at 75 MPa. From what we know of Archaeal membrane lipids in liposome form such pressures are expected to reduce the fluidity of an already viscous membrane.

The composition of the membrane lipids has been measured in cells grown at high pressure. A simple prediction of homeoviscous theory is that membrane lipids should change in the same way when growth takes place at reduced temperature (constant pressure) and at increased pressure (constant temperature). The cells grown at 86 °C at atmospheric pressure, 25 and 50 MPa showed shifts in lipid composition inconsistent with this prediction. For example, at increased pressure, the proportion of macrocyclic archaeol was increased and not decreased. However, Oger (2015) concluded that *M. jannascherii* grown at 50 MPa did show a homeoviscous response in its membrane lipids, and he may be right. The lack of spectroscopic data on the fluidity of the membranes of *M. jannascherii* is serious. It renders the assessment of fluidity from the proportions of different membrane lipids rather speculative.

### 9.6.3.3 Thermococcus barophilus

The archaeon *Thermococcus barophilus* (Phylum Crenarchaea) appears to show a degree of homeoviscous adaptation in its membrane lipids. This cell was isolated by a multinational team working with the submersible Alvin and a French ship, the Jean Charcot, on the Mid Atlantic Ridge in 1993. From an unusually deep hydrothermal vent, Snake Pit, 3550 m, chimney rock and water samples were collected and subsequently subjected to a complicated and rigorous procedure which resulted in the isolation of *T. barophilus*. It is an irregular, mobile cocci, 0.8–2 u across and it metabolises sulphur anaerobically at temperatures and pressures which qualify it as both a hyperthermophile and piezophile. The optimum temperature for growth is 85 °C, at both 40 and 0.3 MPa. Growth above 95 °C requires a pressure of 15–17.5 MPa. Subsequent investigation of its membrane properties was carried out by Cario et al. (2015a). The membrane lipids were classed as either diether or tetraether and the ratio of the two was determined following growth at either constant temperature and variable pressure or vice versa. Table 9.2 summarises the results.

On the basis that the tetraether lipids decrease membrane fluidity and the diether lipids increase its fluidity, the following conclusions are clear. Growth at constant pressure produces a smaller proportion of tetraether lipids at low temperature, consistent with homeoviscous theory (Table 9.2). The same trend should be seen at increased pressure, at constant temperature, which is the case.

*T. barophilus* also contains non-polar lipids which probably reside in the membrane, performing some unknown function. It is interesting that they also show a homeoviscous response to temperature and pressure (Right half of Table 9.2) The number of double bonds increases with a reduction in growth temperature and an increase in growth pressure. So, the conclusion is that *T. barophilus* demonstrates homeoviscous adaptation to high pressure, but direct measurements of membrane fluidity are needed to confirm this provisional conclusion.

When focussing on membranes we should not lose sight of the healthy growth of the whole cell. In the course of developing a defined growth medium, one which contains the minimum specific ingredients for the good growth of *T. barophilus* (from Snake Pit), Cario et al. (2015b) singled out the amino acids required for growth. At the optimum pressure of 40 MPa and 85 °C, they differed from those required at normal atmospheric pressure. An amino acid which is essential for growth under all conditions implies the cell lacks functional genes for its synthesis. At normal atmospheric pressure only three amino acids are essential; glutamic acid, cysteine and tyrosine. These are energy demanding to synthesise. At 40 MPa all amino acids are essential except three; alanine, proline and glutamine, which are relatively simple to synthesise. The cells do not lack the ability to absorb the amino acids from the growth medium but are deficient in their synthesis. In its growth requirements, *T. barophilus* at 40 MP and 85 °C is clearly very different to *T. barophilus* at normal atmospheric pressure, something which would not be

**Table 9.2**  Membrane lipids of *T. barophilus* grown at different temperatures and pressures

|            | Diether | Tetraether | Average number of double bonds in non-polar lipids |
|------------|---------|------------|----------------------------------------------------|
| At 85 °C   |         |            |                                                    |
| 70 MPa     | 37%     | 63%        | 3.7                                                |
| 40         | 16      | 84         | 2.9                                                |
| 0.1        | 12      | 88         | 2.5                                                |
| At 40 MPa  |         |            |                                                    |
| 90 °C      | 8       | 92%        | 2.2                                                |
| 85 °C      | 16      | 84         | 2.9                                                |
| 75 °C      | 435     | 55         | 3.8                                                |

apparent when the cells are grown on a medium containing diverse ingredients. Gene expression probably plays a role here.

## 9.7 The Dispersal from, and the Colonisation of, Hydrothermal Vents

In the second part of this chapter, we will consider the pressure tolerance required of vent organisms colonising established and newly formed vents. Unlike the vast floor of the deep oceans, hydrothermal vents are ephemeral, volcanic structures, some with a life as short as a hundred years or so. Movement of organisms from one vent to another involves passage through cold abyssal waters. A few mobile animals such as crabs may be able to walk or swim to a neighbouring vent, but for most animals, migration entails a larval planktonic stage, with exposure to considerable changes in depth and temperature. Those animals which are nutritionally dependent on symbiosis have also to re-establish their symbionts which, it is generally thought, do not infect eggs. This view has been confirmed in the case of *Rimicaris exoculata*, the interior of whose eggs appear to be devoid of bacteria although plenty are present on the exterior surface. Prospective symbionts and other planktonic microorganisms presumably disperse and colonise vents sites by chance. The few hundred vents known to exist on the ocean floor provide a sparsely distributed collection of tiny landing sites.

Sedentary abyssal animals and microorganisms occupying the vastness of the abyssal plains also disperse and colonise new territory. Animals achieve this by means of a planktonic larval stage in their life cycle. Although they are liable to experience some temperature change they are also exposed to large changes in pressure. Like the planktonic larvae from the hot vents, their pressure tolerance is important to their survival, but unlike the vent fauna, they have a vast landing site awaiting them although detailed requirements for a given species may restrict survival.

### 9.7.1 Dispersal from Hydrothermal Vents

Here are some examples of dispersal which can be treated as natural experiments in demonstrating the pressure tolerance of the participants.

The iconic animal of the hydrothermal vents, the giant vestimentiferan Tubeworm *Riftia pachyptila*, has attracted a lot of study. van Dover (1994) observed the spawning of this extraordinary animal whilst she was aboard the submersible Alvin on the East Pacific Rise. She reported seeing clouds of gametes (eggs or sperm) being released and propelled upwards by the rapid withdrawal of the worm into its tube. Dense groups of the worms, males and females, colonise the vents, their red plumes making a striking sight. Fertilisation may occur internally in some way, before the release of the eggs, or externally, and then the eggs sink before being swept by currents away from the warm waters of the vent. Experiments carried out by Brooke and Young (2009) have shown that the eggs contain nutrients sufficient to sustain the developing embryo for 30 or so days. This was established by measuring the respiration of the embryos at 2 °C and 25 MPa. In a subsequent series of very impressive experiments, Alvin collected (with decompression) mature worms from a depth of 2500 m on the East Pacific Rise to enable in vitro fertilisation to be carried out at surface pressure. The embryos were subsequently incubated in pressure vessels and in containers placed by the vents, and it was shown that they developed normally at 24 MPa and 2 °C, but were drastically arrested by 10 MPa or less. The eggs' buoyancy at 25 MPa and 2 °C is positive, suggesting a rate of passive ascent of 2 m per day, and probably slightly more at lesser pressures. The warm water plume rising from the vent becomes neutrally buoyant at 2300 m so, if the embryos were being carried in the plume they would only ascend further at the rate set by their buoyancy. After about 20 days the microscopic ciliated larval stage, called a trochophore, is attained which can sustain its level in the water column. Trochophores are commonly seen in polychaete development and usually develop a mouth and gut, but this does not happen in the present case. Nothing seems to be known about the settlement and early growth of the trochophore. It probably does not contain symbionts and presumably, these are acquired from the surrounding water at an early stage. Just how the infection takes place is also unknown, but unless it does the animal is doomed.

*Alvinella pompejana*, as famous as *Riftia*, is active on the walls of vent chimneys and tolerates the higher temperature range of 20–80 °C. Normally fertilisation of the eggs takes place internally, but the larvae have never been seen. Pradillon and her team (2005) have carried out remarkable experiments using in vitro fertilisation which has enabled the pressure tolerance of the early development of the eggs to be assessed, using a special microscope pressure chamber (Pradillon et al. 2004). Eggs from 2500 m depths on the East Pacific Rise vents developed well at 25 MPa and 10 C but not at 2 C. Abyssal pressures are required for most eggs to develop normally. In situ experiments in which eggs were incubated close to a White Smoker chimney, where the fluctuating temperature, the pH and dissolved sulphides had an adverse effect, show that the eggs have to get clear of the adult environment. *Alvinella* does not need to acquire symbionts but at some stage it acquires the "hairy" episymbionts, presumably from the environment, which gives the animal its characteristic appearance and possibly some advantages. There is direct evidence that the planktonic larvae of mussels (e.g. *Idas simpsoni*), once settled on a substrate, rapidly acquire symbiotic bacteria in their gill filaments. These are essential for the animals' nutrition.

A prolonged planktonic existence, exposing the larvae to significant pressure changes, is made possible by food reserves and the ability to feed by the development of a mouth and alimentary tract. However, the larvae of shrimps such as *Rimicaris exoculata* and *Mirocaris fortunata*, from vents on the Mid Atlantic Ridge, are equipped with special lipid food stores which enable them to survive for a long time without feeding. Tyler and Dixon (2000) worked on *M. fortunata* larvae and found them particularly interesting. When reared in vitro on board ship, they were rendered moribund by 30 MPa at the adults' normal temperature of

10 C. At the same temperature and lesser pressures, they were active, but at higher temperatures they were moribund. The idea is their survival at depth is limited by pressure and the extent of their effective ascent into shallow water by temperature. The depth range of the adult population is 820–3000 m on the Mid Atlantic Ridge vents, which is rather a good match.

The Blood Red Limpet *Shinkailepas myojinensis* lives on the sulphide chimney walls of vents at depths of 442–1227 m in the N W Pacific. Its shell is about 1 cm across and being a limpet, it grazes on the bacteria encrusting the sulphite. Its planktonic larvae survive and grow at surface pressures where there is abundant food and eventually, they descend and colonise the specialised sulphide environment of the vents (Yahagi et al. 2017). Some adaptation to a moderate pressure seems likely in this life cycle, which also encompasses a number of drastic changes, notably sensory and nutritional.

### 9.7.2   Abyssal and Benthic Animals Also Produce Planktonic Larva

First, for comparison, consider the pressure tolerance of the larvae of some shallow water Molluscs commonly seen on temperate shores, *Mytilus edulis*, the mussel and *Crepidula fornicata*, the slipper limpet. The trochophore stage of the mussel is a tiny planktonic organism equipped with cilia, and it develops rapidly at 20 °C after the eggs are fertilised in seawater. Mestre et al. (2009, 2013) used a method of fertilising the eggs at high pressure and assessed the rate of development at selected pressures. At 10 °C, normal pressure, many eggs reached the early blastula stage in 24 hours, but at 10 MPa very few did. Clearly, this moderate pressure exerts some influence on the animals' eventual distribution. *Mytilus* is a typical shallow-water mussel but has been found at depths of more than 100 m, living on a substrate. Experiments with *Crepidula* used active veliger larvae removed from the capsule in which they normally develop. The pressure tolerance of the early stage veliger was compared to that of the later stage. At 10 °C a high proportion the early stage veligers succumbed at pressures above 20 MPa, but more of the late-stage veligers survived. This is an unspectacular example of a marked increase in pressure tolerance in the development of the animal, but it does not seem to have any ecological meaning.

The early stages of another Mollusc, the shallow water whelk *Buccinum undatum*, have also been studied, and they too develop within an egg capsule and not freely in the plankton. A juvenile (miniature adult) emerges from the capsule. The process by which the eggs develop into veliger larval stages, utilising food stores in the capsule, takes many days, and normally would take place in shallow water. When incubated at high pressure the process is significantly slowed, effectively ceasing at 30 MPa, which is not surprising. The oxygen consumption of normal veligers is, however, little affected by high pressure unlike that of the juveniles which is reduced by pressures of 20 MPa or more. This curious ontogenetic reduction in pressure tolerance also lacks obvious ecological meaning.

In Chapter 2, Marsland's experiments on cell cleavage, using the eggs from the shallow water sea urchin *Arbacia* were described. The eggs were artificially fertilised to initiated cell division. In normal conditions, a succession of cell cleavages creates a ball of cells which grows into a free-swimming larva. The same procedure was used by Young and his colleagues (Young and Tyler 1993; Young et al. 1996) to assess the pressure tolerance of cleavage in the eggs of the deep sea star fish, *Plutonaster bifrons*, and the sea urchin *Echinus affinis* both in the phylum Echinodermata. The starfish were collected from a seafloor depth of 2200 m in the N. Atlantic and were fertilised at normal pressure and 4 °C., thus demonstrating a tolerance to decompression. After incubation at selected pressures and 4 °C for 12 hours the embryos were decompressed and assessed. A much higher proportion of those held at 20 MPa had divided normally compared to those held at higher and lower pressures. Only those held at 20 MPa achieved the 32-cell stage in 24 hours. 25 MPa and 10 MPa impaired cleavage. The eggs of *Plutonaster*, which do not float, presumably undergo normal cleavage at the pressure in which they are normally released, i.e. at depths of 1000–2500 m. Pressures outside that depth band markedly inhibited cleavage.

Similar experiments were carried out with the eggs from the sea urchin *Echinus affinis*, whose vertical distribution is 1750–2450 m. The specimens used were trawled from a seafloor depth of 2000 m. The cleavage of the eggs, artificially fertilised as before and held at selected pressures and 6 °C, was closely followed. Significant numbers at 20 MPa reached the 8-cell stage in 12 hours. Recompression after the observations had been made enabled further development to proceed, especially so at 20 MPa. None held at 10 MPa or less achieved cleavage. These sea urchin eggs are tolerant to the high pressure they would normally experience but require a significant pressure, more than 10 MPa, to cleave at all. The larvae are thought to be planktonic, implying they would float up from the seafloor and presumably encounter less than 10 MPa.

A different species, *E. esculentus*, as an adult lives no deeper than 100 m. It is common on European beaches and it breeds in shallow waters. However, its planktonic larvae are found as deep as 2000 m, apparently healthy. Experiments reported by Tyler and Young (1998) indicate that larval *E. esculentus* has a degree of tolerance to a significant pressure. The larvae of *Psammechinus miliaris*, the starfish *Asterias rubens* and the Antarctic sea urchin *Sterechinus neumayeri* similarly tolerate a significant pressure, greater than experienced by the adult form, presumably an advantage in their planktonic distribution.

## 9.7.3    Pelagic Adults and Larvae

In 1984, George published some particularly interesting experimental observations made in the Antarctic, on the effects of temperature and hydrostatic pressure on the developmental stages of krill, a major presence in the plankton of polar waters. Krill is in fact *Euphausia superba*, belonging to a Class of Crustacea we have met before. It thrives at between 2 and −0.5 C. The adult form lives in fairly shallow water, at around 200 m and not deeper than 400 m, whereas the eggs released by the females sink sufficiently fast that their cleavage and subsequent development might begin at a depth of 2000 m (20 MPa). Cleavage was, in fact, found to be inhibited by a pressure of 20 MPa, which is not unexpected. Presumably, cleavage normally proceeds at depths of less than 2000 m. In fact, George and Stromberg (1985) found that 5–20 MPa accelerated cleavage. The larva, called a nauplius, is microscopic in size and ascends as it develops through a series of stages. George tested each stage by holding them at selected pressures for 24 hours. The first nauplius stage survived 20 MPa but not 25 MPa. The second stage survived 15 MPa but not 20 MPa. The next stage, the post-larval krill, which is about 20 mm long, survive 10 MPa but not 15 MPa and the mature krill survived 2.5 MPa but not 5 MPa. This progressive increase in sensitivity to pressure broadly matches the vertical distribution of the developmental stages. As the larvae develop and rise in the water column their tolerance to pressure changes. Is this driven by a genetic clock or a gradual acclimation type of process, responding to environmental conditions? Presumably, those larvae which find themselves in the "wrong" depth zone at the wrong time fail to survive.

The eggs of the copepod *Calanus sinicus* also appear to be sensitive to the pressure they are liable to meet in their development. Female copepods typically receive sperm packaged in a spermatophore and their fertilised eggs develop within an egg sac which remains attached to the adult. The nauplius larva emerges from the egg and passes through several stages of moulting its exoskeleton, producing a succession of nauplius larvae. They then undergo a major moult to become adult-like in form, at normal surface pressure. Yoshiki et al. (2011) collected *C. sinicus* from less than 300 m depths using plankton nets and selected healthy females to obtain freshly spawned eggs. When these were incubated at 5 °C and 1 MPa they failed to hatch, although the nauplius developed at a normal rate within the transparent egg case. Cleavage was presumably unaffected. Many of those which did hatch, at higher temperatures, for example, were deformed. The biological implication is that eggs only hatch in the sea at depths of less than 100 m and 5 °C. It appears that some process involved in hatching is inhibited by 1 MPa, which is very little hydrostatic pressure to exert such an effect. In contrast to *C. sinicus*, species of *Neocalanus* which live and spawn in much deeper waters, produce eggs which develop and hatch normally at 4 C and 10 MPa.

The observations made by Koyama and colleagues (2005) on the larval development of the shrimp *Alvinocaris*, collected from a cold seep at a depth of 1157 m in the north Pacific, are something of a curiosity. Traps were placed on the seafloor and subsequently the trapped shrimps were transferred by the submersible Shinkai to a pressure aquarium vessel and taken to the surface (i.e. isobaric collection). Three shrimps were then subject to a very slow decompression to atmospheric pressure, over 63 days. Two individuals appeared healthy and contained fertilised eggs which subsequently hatched, producing larvae which developed over 74 days. Their pressure of origin, 11.5 MPa, is not particularly high, but nevertheless it is interesting that atmospheric pressure was tolerated so well. Perhaps this observation belongs with other acclimatisation experiments (Chapter 12).

## 9.7.4    Colonisation

Colonisation of new vent sites can proceed rapidly. In 1991, observers on board Alvin examined a region of the East Pacific Rise at a depth of 2500 m soon after new lava flows had appeared. Superheated fluids were seen flowing from fresh fissures in the exposed basalt and whilst no animals were to be seen, thick mats of bacteria had become established on the ground. A year later an extensive colony of the vestimentiferan tube worm *Tevnia jerichonana* was present, but *Riftia pachyptila* was absent. Over the following year, a dramatic change took place. A team led by Lutz et al. (1994) reported that the giant *Riftia* had established itself and was crowding out the Tevnia. Some of the *Riftia* were over 1.5 m in length and subsequent measurements recorded growth rates in excess of 85 cm per year. Another example is the 2005 eruption on the East Pacific Rise which wiped out certain vent communities. They did not renew themselves but were replaced, initially at any rate, by a conspicuous immigrant Gastropod, *Ctenopelta porifera*, which most likely came from a population 300 km away (Mullineaux et al. 2010). Colonisation is not just rapid, the planktonic distribution of larvae seems to make it random, but the range of pressure tolerated plays an important role in dispersal (McVeigh et al. 2017).

## References[1]

Airriess C, Childress JJ (1994) Homeoviscous properties implicated by the interactive effects of pressure and temperature on the hydrothermal vent crab Bythograea thermydron. Biol Bull 187:208–214

Alargo, D.K., Deguchi, S., Tsujii, K. and Horikoshi, K. (2002) Oligomerization of glycine in supercritical water with special attention to the origin of life in Deep sea hydrothermal system. In Trends in High pressure Biosc. Biotech. ed. Hayashi, R., (pp. 631–636). *

Allen CE, Tyler PA, van Dover CL (2001) Lipid composition of the hydrothermal vent clam *Calyptogena pacifica* (Mollusca: Bivalvia) as a trophic indicator. J Mar Biol Assoc UK 81:817–821. *

Amend JP, Teske A (2005) Expanding frontiers in deep subsurface microbiology. Palaeogeogr Palaeoclimatol Palaeoecol 219:134–155. *

Amrani A et al (2014) Transcriptomics reveal several gene expression patterns in the piezophile Desulfovibrio hydrothermalis in response to hydrostatic pressure. PLoS One 9:e106831

Anderson RE, Brazelton WJ, Baross JA (2013) The deep viriosphere: assessing the viral impact on microbial community dynamics in the deep subsurface. Rev Mineral Geochem 75:649–675

Anderson RE, Sogin M, Baross JA (2014) Evolutionary strategies of viruses, bacteria and archaea in hydrothermal vent ecosystems revealed through metagenomics. PLoS One 9:e109696

Aquina-Souza R, Hawkins SJ, Tyler PA (2008) Early development and larval survival of *Psammechinus miliaris* under deep sea temperature and pressure conditions. J Mar Biol Assoc UK 88:453–461. *

Atkins MS, Anderson OR, Wirsen CO (1998) Effect of hydrostatic pressure on the growth rates and encystment of flagellated protozoa isolated from a deep sea hydrothermal vent and a deep shelf region. Mar Ecol Prog Ser 171:85–95

---

[1]Those marked * are further reading.

Auerbach G, Gaill F, Jaenicke R, Schulthess T, Timpl R, Engel J (1994) Pressure dependence of collagen melting. Matrix Biol 14:589–592

Auguste M et al (2016) Development of an ecotoxicological protocol for the deep sea fauna using the hydrothermal vent shrimp Rimicaris exoculata. Aquat Toxicol 175:277–285

Bartsch I, Dovgal IV (2010) A hydrothermal vent mite (Halacaridae, Acari) with a new Corynophyra species (Suctoria, Ciliophora), description of the suctorian and its distribution on the halacarid mite. Eur J Protistol 46:196–3203

Bates AE, Lee RW, Tunniclffe V, Lamare MD (2010) Deep sea hydrothermal vent animals seek cool fluids in a highly variable thermal environment. Nat Commun 1:14–19

Beaulieu SE, Baker ET, German CR, Maffei A (2013) An authoritative global database for active submarine hydrothermal vent fields. Geochem Geophys Geosyst 14:4892–4905. *

Birrien JC et al (2011) Pyrococcus yayanosii sp. Nov. obligate piezophilic hyperthermophilic archaeon isolated from a deep sea hydrothermal vent. Int J Syst Evol Microbiol 61:2827–2831

Boetius A (2005) Lost City life. Science 307:1420–1422. *

Boonyaratanakornkit BB, Clark DS (2008) Physiology and biochemistry of Methanocaldococcus jannaschii at elevated pressures. In: Michiels C, Bartlett DH, Aertsen A (eds) High pressure microbiology. ASM Press, Washington, pp 293–304

Bradley AS (2009) Expanding the limits of life. Sci. American December 62–67 *

Brazelton W (2017) Hydrothermal vents. Curr Biol 27:R431–R510. *

Brazelton WJ et al (2010) Archaea and bacteria with surprising micro diversity show shifts in dominance over 1,000 year time scales in hydrothermal chimneys. Proc Natl Acad Sci U S A 107:1612–1617. *

Brooke SD, Young CM (2009) Where do the embryos of Riftia pachyptila develop? Pressure tolerances, temperature tolerances and buoyancy during prolonged embryonic dispersal. Deep-Sea Res II 56:1599–1606

Burgaud G et al (2010) Marine culturable yeasts in deep sea hydrothermal vents: species richness and association with fauna. FEMS Microbiol Lett 73:121–133

Burgaud G et al (2015) Effects of hydrostatic pressure on yeasts isolated from deep sea hydrothermal vents. Res In Microbiol 166:700–709

Caforio A, Driessen AJM (2017) Archaeal phospholipids: structural properties and biosynthesis. Biochim Biophys Acta 1862:1325–1339

Cario A, Gross V, Schaeffer P, Oger PM (2015a) Membrane homeoviscous adaptation in the piezo-hyperthermophilic archaeon Thermococcus barophilus. Front Microbiol 6:article 1152

Cario A, Lormieres F, Xiang X, Oger P (2015b) High hydrostatic pressure increases amino acid requirements in the piezo-hyperthermophilic archaeon Thermococcus barophilus. Res In Microbiol 166:710–716

Carsey FD, Chen D-S, Cutts J, French L (1999/2000) Exploring Europa's ocean: a challenge for marine technology of this century. Mar Technol Soc J 33:5–12

Cary CS, Shank T, Stein J (1998) Worms bask in extreme temperatures. Nature 391:545–546

Chang EL (1994) Unusual thermal stability of liposomes made from bipolar tetraether lipids. Biochem Biophys Res Commun 202:673–679

Chevaldonne P, Fisher CR, Childress JJ, Desbruyeres D, Jollivet D, Zal F, Toulmond A (2000) Thermotolerance and the 'Pompeii worms'. Mar Ecol Prog Ser 208:293–295

Chong PL-G (2010) Archaebacterial bipolar tetraether lipids: physico-chemical and membrane properties. Chem Phys Lipids 163:253–265. *

Chong PL-G, Sulc M, Winter R (2010) Compressibilities and volume fluctuations of Archaeal tetraether liposomes. Biophys J 99:3319–3326

Chong, PL-G, Ayesa U, Daswani VP, Hur EC (2012) On the physical properties of tetraether lipid membranes: effects of cyclopentane rings. Archaea, article 138439. Hindawi Publishing Corp. *

Chyba CF, Phillips, C.B. (2001) Possible ecosystems and the search for life on Europa. Proc Natl Acad Sci U S A 98:801–804. *

Corliss JB et al (1979) Submarine thermal springs on Galapagos Rift. Science 203:1073–1083

Cottin D et al (2010) Comparison of heat shock responses between the hydrothermal vent shrimp Rimicaris exoculata and the related coastal shrimp *Palaemonetes varians*. J Exp Mar Biol Ecol 393:9–16

Craddock C, Lutz RA, Vrijenhoek RC (1997) Patterns of dispersal and larval development of archaeogastropod limpets at hydrothermal vents in the eastern pacific. J Eptl Mar Biol Ecol 210:37–51. *

Dahlhoff E, Somero GN (1991) Pressure and temperature adaptation of cytosolic malate dehydrogenases of shallow and deep living marine invertebrates: evidence for high body temperatures in hydrothermal vent animals. J Exp Biol 159:473–487

Dahlhoff E, Schneidemann S, Somero GN (1990) Pressure – temperature interactions on M4-Lactate dehydrogenase form hydrothermal vent fishes: Evidence for the adaptation to elevated temperatures by the Zoarcid *Thermarces andersoni*, but not by the Bythitid, Bythites hollisi. Biol Bull 179:134–139

Dalmasso C et al (2016a) Thermococcus thermophilus sp. nov., a novel hyperthermophilic and piezophilic archaeon with a broad pressure range for growth, isolated from a deepest hydrothermal vent at the Mid-Cayman Rise. Syst Appl Microbiol 39:440–444

Dalmasso C et al (2016b) Complete genome sequence of the hyperthermophilic and piezophilic Thermococcus piezophilus CDGS, able to grow under extreme hydrostatic pressure. Genome Announc, American Society for Microbiology, 4, e00610–16

Daniel I, Oger P, Winter R (2006) Origins of life and biochemistry under high pressure conditions. Chem Soc Rev 35:858–875. *

Deguchi S, Tsujii K (2002) Flow cell for in situ optical microscopy in water at high temperatures and pressures up to supercritical states. Rev Sci Instrum 73:3938–3941

DeLong EF (2006) Archaeal mysteries of the deep revealed. Proc Natl Acad Sci U S A 103:6417–6418. *

Desbruyeres D (2007) Hydrothermal vents. In: Nouvain C (ed) The deep. Chicago University Press

Desbruyeres D et al (1998) Biology and ecology of the 'Pompeii worm' (Alvinella pompejana Desbruyeres and Laubier) a normal dweller of an extreme deep sea environment: a synthesis of current knowledge and recent developments. Deep-Sea Res II 45:383–422

Di Giulio M (2005) A comparison of proteins from Pyrococcus furiosus and Pyrococcus abyssi: barophily in the physicochemical properties of amino acids and in the genetic code. Gene 346:1–6. *

Di Giuo M (2005) The ocean abysses witnessed the origin of the genetic code. Gene 346:7–12. *

Duperron S et al (2005) Dual symbiosis in a Bathymodiolus sp. mussel from a Methane seep on the Gabon Continental Margin (Southeast Atlantic): 16S rRNA phylogeny and distribution of the symbionts in gills. Appl Environ Microbiol 71:1694–1700

Elferink MGL et al (1994) Stability and proton—permeability of liposomes composed of archaeal tetraether lipids. Biochim Biophys Acta 1193:247–254

Foustoukos DI, Rodriguez HP (2015) A continuous culture system for assessing microbial activities in the Piezosphere. Appl Environ Microbiol 81:6850–6856

Fujikura K et al (1999) The deepest chemosynthesis-based community yet discovered from the hadal zone, 7326 m deep, in the Japan Trench. Mar Ecol Prog Ser 190:17–26. *

Gabriel JL, Chong PL-G (2000) Molecular modelling of archaebacterial bipolar tetraether lipid membranes. Chem Phys Lipids 105:193–200. *

Gaill F, Shillito B, Menard F, Goffinet G, Childress JJ (1997) Rate and process of tube production by the deep sea hydrothermal vent tubeworm *Riftia pachyptila*. Mar Ecol Prog Ser 148:135–143

George RY (1984) Ontogenetic adaptations in growth and respiration of *Euphausia superba* in relation to temperature and pressure. J Crustac Biol 4:252–262

George RY, Stromberg J-O (1985) Development of eggs of Antarctic krill *Euphausia superba* in relation to pressure. Polar Biol 4:125–133

Gliozzi A et al (1983) Effect of isoprenoid cyclization on the transition temperature of lipids in thermophilic archaebacteria. Biochim Biophys Acta 735:234–242. *

Goffredi SK, Girguis PR, Childress JJ, Desaulniers NT (1999) Phjysiological functioning of carbonic anhydrase in the hydrothermal vent tubeworm *Riftia pachyptila*. Biol Bull 196:257–264

Gold T (1992) The deep hot biosphere. Proc Natl Acad Sci U S A 89:6045–6049. *

Gonzalez JM et al (1998) Pyrococcus horikoshii sp. nov., a hyperthermophilic archeon isolated from a hydrothermal vent at the Okinawa Trough. Extremophiles 2:123–130

Guri M et al (2012) Acquisition of epibiotic bacteria along the life cycle of the hydrothermal shrimp Rimicaris exoculata. ISME J 6:597–609

Hazen RM et al (2002) High pressure and the origins of life. J Phys Condens Matter 14:11489–11494

Henscheid KL, Shin DS, Cary SC, Berglund JA (2005) The splicing factor U2AF65 is functionally conserved in the thermotolerant deep sea worm Alvinella pompejana. Biochim Biophys Acta 1727:197–207

Herschy B et al (2014) An origin-of-life reactor to simulate alkaline hydrothermal vents. J Mol Evol 79:213–227

Higashi Y et al (2004) Microbial diversity in hydrothermal surface to surface environments of Suiyo seamount, Izu-Bonin Arc, using a catheter-type in situ growth chamber. FEMS Microbiol Ecol 47:327–336. *

Holm NG, Oze C, Mousis O, Waite JH, Guilbert-Lepoutre A (2015) Serpentinization and the formation of $H_2$ and $CH_4$ on celestial bodies (planets, moons, comets). Astrobiology 15:587–600. *

Houghton JL, Siefried WE, Banta AB, Reysenbach AL (2007) Continuous culturing of thermophiles under sulphate and nitrate-reducing conditions and at deep sea hydrostatic pressures. Extremophiles 11:371–382

Huber C, Wachterauser G (2003) Primordial reductive amination revisited. Tetrahedron Lett 44:1695–1697. *

Imai E-I, Honda H, Hatori K, Brack A, Matsuno K (1999) Elongation of oligopeptides in a simulated submarine hydrothermal system. Science 283:831–833

Jarrell HC, Zukotynski KA, Sprott DG (1998) Lateral diffusion of the polar lipids from Thermoplasma acidophilum in multilamellar liposomes. Biochim Biophys Acta 1369:259–266

Jebbar M, Franzetti B, Girard E, Oger P (2015) Microbial diversity and adaptation to high pressure in deep sea hydrothermal vent prokaryotes. Extremophiles 19:721–740. *

Jennings DH, Lysek G (1996) Fungal Biology. BIOS Scientific Publishers Ltd, Oxford, UK. *

Jones EG, Linewcaver CH, Clarke JD (2011) An extensive phase space for the potential Martian biosphere. Astrobiology 11:1017–1033

Kadar E et al (2005) Experimentally induced endosymbiotic loss and re-acquirement in the hydrothermal vent bivalve Bathymodiolus azoricus. J Exp Mar Biol Ecol 318:99–110

Kadar E, Guerra IT, Checa A (2008a) Post–capture hyperbaric simulations to study the mechanism of shell regeneration of the deep sea hydrothermal vent mussel Bathymodiolus azoricus (Bivalvia: Mytilidae). J Exp Mar Biol Ecol 364:80–90

Kadar E, Checa AG, Oliviera ANDP, Machado JP (2008b) Shell nacre ultrastructure and depressurisation dissolution in the deep sea hydrothermal vent mussel Bathymodiolus azoricus. J Comp Physiol B 178:123–130

Kallmeyer J, Boetius A (2004) Effects of temperature and pressure on sulfate reduction and anaerobic oxidation of methane in hydrothermal sediments of Guaymas basin. Appl Environ Microbiol 70:1231–1233

Kaneshiro SM, Clark DS (1995) Effects on the composition and thermal behaviour of lipids from the deep sea Methanococcus jannaschii. J Bacteriol 177:3668–3672. *

Karmer MB, DeLong EF, Kari DM (2001) Archaeal dominance in the mesopelagic zone of the Pacific Ocean. Nature 409:507–510. *

Kaye JZ, Baross JA (2004) Synchronous effects of temperature, hydrostatic pressure and salinity on growth, phospholipid profiles and protein patterns of four Halomonas species isolated from deep sea hydrothermal vent and surface environments. Appl Environ Microbiol 70:6220–6229

Kelley DS et al (2001) An off-axis hydrothermal vent field near the mid-Atlantic ridge at 30 degrees N. Nature 412:145–149. *

Kennish MJ, Lutz RA (1999) Calcium carbonate dissolution rates in deep sea bivalve shells on the East Pacific Rise at 21 N: results of an 8 year in-situ experiment. Palaeaogeog Palaeaoclimat & Palaeao Ecol 154:293–299. *

Koga Y (2012) Thermal adaptation of the Archaeal and Bacterial lipid membranes. Archaea Article 789652. *

Koyama S et al (2005) Survival of deep sea shrimp (Alvinocaris sp.) during decompression and larval hatching at atmospheric pressure. Mar Biotechnol 7:272–278

Laming SR et al (2015) Adapted to change: the rapid development of symbiosis in newly settled, fast maturing chemosymbiotic mussels in the deep sea. Mar Environ Res 112:100–111

Lane N (2016) The vital question. Profile Books

Lane N, Martin WF (2012) The origin of membrane bioenergetics. Cell 151:1406–1416. *

Lane N, Allen JF, Martin W (2010) How did LUCA make a living? Chemiosmosis in the origin of life. BioEssays 32:271–280

Le Bris N, Zbinden M, Gaill F (2005) Processes controlling the physico-chemical microenvironments associated with Pompeii worms. Deep Sea Res I 52:1071–1083

Lombard J, Lopez-Garcia P, Moreira D (2012) The early evolution of lipid membranes and the three domains of life. Nat Rev Microbiol 10:507–515. *

Lonsdale P (1977) Clustering of suspension feeding macrobenthos near abyssal hydrothermal vents at oceanic spreading centres. Deep Sea Res 24:857–863

Lossouarn J et al (2015) An abyssal mobilome: viruses, plasmids, and vesicles from deep sea hydrothermal vents. Res Microbiol 166:742–752

Lutz RA et al (1994) Rapid growth at deep sea vents. Nature 371:663–664

Marsh AG, Mulllineaux LS, Young CM, Manahan DT (2001) Larval dispersal potential of the tubeworm *Riftia pachyptila* at deep sea hydrothermal vents. Nature 411:77–80. *

Marteinsson VT et al (1997) Physiological responses to stress conditions and barophilic behaviour of hyperthermophilic vent Archaeon Pyrococcus abyssi. Appl Environ Microbiol 63:1230–1236. *

Martiensson VT et al (1999) Thermococcus barophilus sp. nov., a new barophilic and hyperthermophilic archaeon isolated under high hydrostatic pressure from a deep sea hydrothermal vent. Int J Syst Bacteriol 49:351–358. *

Mathai JC, Sprott GD, Zeidel ML (2001) Molecular mechanisms of water and solute transport across archaebacterial lipid membranes. J Biol Chem 276:27266–27271. *

McNichol J, Sylva SP, Thomas F, Taylor CD, Sievert SM, Seewald JS (2016) Assessing microbial processes in deep sea hydrothermal systems by incubation at in situ temperature and pressure. Deep Sea Res I 115:221–232. *

McNichol J et al (2018) Primary productivity below the seafloor at deep sea hot springs. Proc Natl Acad Sci U S A 115(26) *

McVeigh DM, Eggleston DB, Todd AC, Young CM, He R (2017) The influence of larval migration and dispersal depth on potential larval trajectories of a deep sea bivalve. Deep Sea Res I 127:57–64

Mestre NC, Thatje S, Tyler PA (2009) The ocean is not deep enough: pressure tolerances during early ontogeny of the blue mussel *Mytilus edulis*. Proc R Soc B 276:717–726

Mestre NC, Brown A, Thatje S (2013) Temperature and pressure tolerance of larvae of *Crepidula fornicata* suggest thermal limitation of bathymetric range. Mar Biol 160:743–750

Michous G, Jebbar M (2016) High hydrostatic pressure adaptive strategies in an obligate piezophile Pyrococcus yayanosi. Sci Rep 6:27289. *

Mickel TJ, Childress JJ (1982a) Effects of pressure and temperature on the EKG and heart rate of the hydrothermal vent crab Bythograea thermydron (Brachyura). Biol Bull 162:70–82

Mickel TJ, Childress JJ (1982b) Effects of temperature, pressure and oxygen concentration on the oxygen consumption rate of the hydrothermal vent crab Bythograea thermydron (Brachyura). Physiol Zool 55:199–207

Miyake H, Kitada M, Tsuchida S, Okuyama Y, Nakamura K-I (2007) Ecological aspects of hydrothermal vent animals in captivity at atmospheric pressure. Mar Ecol 28:86–92

Moore D (2013) Fungal biology in the origin and emergence of life. Cambridge University Press, Cambridge, UK. *

Mullineaux LS, Adams DK, Mills SW, Beaulieu SE (2010) Larvae from afar colonise deep sea hydrothermal vents after a catastrophic eruption. Proc Natl Acad Sci U S A 107:7829–7834

Munro C, Morris JP, Brown A, Hauton C, Thatje S (2015) The role of ontogeny in physiological tolerance: decreasing hydrostatic tolerance with development in the Northern Stone Crab *Lithodes maja*. Proc R Soc B 282:20150577. *

Nickerson K (1984) A hypothesis on the role of pressure in the origin of life. J Theoret Biol 110:487–499

Oger PM (2015) Homeoviscous adaptation of membranes in Archaea. Chapter 19 in Akasaka K, Matsuki H (eds) High pressure bioscience. Springer Science+Business

Oger PM, Cario A (2013) Adaptation of the membrane in Archaea. Biophys Chem 183:42–56. *

Ortmann A, Suttle C (2005) High abundance of viruses in a deep sea hydrothermal vent system indicates viral mediated microbial mortality. Deep-Sea Res 52:1515–1527

Pelli DG, Chamberlain SC (1989) The visibility of 350 C black body radiation by the shrimp Rimicaris exoculate and man. Nature 337:460–461

Phleger CF et al (2005a) Lipid biomarkers of deep sea hydrothermal vent polychaetes–*Alvinella pompejana*, *A. caudata*, *Paralvinella grasslei* and *Hesiolyra bergii*. Deep Sea Res I 52:2333–2352. *

Phleger CF et al (2005b) Lipid composition of deep sea hydrothermal vent tubeworm *Riftia pachyptila*, crabs Munidopsis subsquamosa and Bythograea thermydron, mussels Bathymodiolus sp. and limpets Lepetodrilus spp. Comp Biochem Physiol 141B:196–210. *

Picard A, Daniel I (2013) Pressure as an environmental parameter for microbial life—a review. Biophys Chem 183:30–41. *

Pond DW et al (1998) Stable carbon isotope composition of fatty acids in hydrothermal vent mussels containing methanotrophic and thiotrophic bacterial endosymbionts. Appl Environ Microbiol 54:370–375. *

Pond DW et al (2000) Unusual fatty acid composition of storage lipids in the bresilioid shrimp Rimicaris exoculata couples the photic zone with MAR hydrothermal vent sites. Mar Ecol Prog Ser 198:171–179. *

Pradillon F, Shillito B, Young CM, Gaill F (2001) Developmental arrest in vent worm embryos. Nature 413:698–699. *

Pradillon F, Shillito B, Chervin J-C, Hamel G, Gaill F (2004) Pressure vessels for in vivo studies of deep sea fauna. High Pressure Res 24:237–246

Pradillon F et al (2005) Influence of environmental conditions on early development of the hydrothermal vent polychaete Alvinella pompejana. J Exp Biol 208:1551–1561

Raghukumar C, Damare S (2008) Deep sea fungi. In: Michiels C, Bartlett DH, Aertsen A (eds) High pressure microbiology. ASM Press, Washington, DC, pp 265–291. *

Ravaux J et al (2003) Heat-shock response and temperature resistance in the deep sea vent shrimp Rimicaris exoculata. J Exp Biol 206:2345–2354

Ravaux J et al (2007) First hsp 70 from two hydrothermal vent shrimps, Myrocaris fortunate and Rimicaris exocultata: characterization and sequence analysis. Gene 386:162–172

Ravaux J et al (2009) Hydrothermal vent shrimps display low expression of the heat-inducible hsp70 gene in nature. Mar Ecol Prog Ser 396:153–156

Ravaux J et al (2013) Thermal limit for metazoan life in question: in vivo heat tolerance of the Pompeii worm. PLoS One 8:e64074

Ravaux J, Leger N, Hamel G, Shillito B (2019) Assessing a species thermal tolerance through a multiparameter approach: the case study of the deep sea hydrothermal vent shrimp Rimicaris exoculata. Cell Stress Chaperones 24:647–659

Reysenbach A, Longnekker K, Kirshtein J (2000) Novel bacterial and archaeal lineages from an in situ growth chamber deployed at a Mid Atlantic Ridge hydrothermal vent. Appl Environ Microbiol 66:3798–3806. *

Rinke C, Lee RW (2009) Macro camera temperature logger array for deep sea hydrothermal vent and benthic studies. Limnol Oceanogr Methods 7:527–534

Robb F, Clark DS (1999) Adaptation of proteins from hyperthermophiles to high pressure and high temperature. J Mol Microbiol Biotechnol 1:101–106. *

Robertson DE, Roberts MF, Belay N, Stetter KO, Boone DR (1990) Occurrence of Beta-glutamate, a novel osmolyte, in marine methanogenic bacteria. Appl Environ Microbiol 56:1504–1508. *

Roenbaum E et al (2012) Effects of hydrostatic pressure on the quaternary structure and enzymatic activity of a large peptidase complex from Pyrococcus horikoshii. Arch Biochem Biophys 517:104–110

Roine E, Bamford DH (2012) Lipids of Archaeal viruses. Archaea 212:art 384919

Russell MJ, Hall AJ, Martin W (2010) Serpntinization as a source of energy at the origin of life. Ceobiology 8:355–n371. *

Seewald JS, Doherty KW, Hammar TR, Liberatore SP (2001) A new gas tight isobaric sampler for hydrothermal fluids. Deep Sea Res I 49:189–190. *

Shillito B et al (2001) Temperature resistance of Hesiolyra bergi, a polychaetous annelid living on deep sea vent smoker walls. Mar Ecol Prog Ser 216:141–149

Shillito B, Le Bris N, Gaill F, Rees JF, Zal F (2004) First access to live Alvinellas. High Pressure Res 24:1–9

Shillito B et al (2008) Live capture of megafauna from 2300 m depth using a newly designed pressurised recovery device. Deep Sea Res I 55:881–889

Shrestha UR et al (2015) Effects of pressure on the dynamics of an oligomeric protein from deep sea hyperthermophile. Proc Natl Acad Sci U S A 112:13886–13891. *

Sicot F-X et al (2000) Molecular adaptation to an extreme environment: origin of the thermal stability of the Pompeii worm collagen. J Mol Biol 302:811–820. *

Sigwart J, Chen C (2018) Comparative oxygen consumption of gastropod holobionts from the deep sea hydrothermal vents in the Indian Ocean. Biol Bull 235:102–112

Silliakus MF, van der Oost J, Kengen SWM (2017) Adaptations of archaeal and bacterial membranes to variations in temperature, pH and pressure. Extremophiles 20. https://doi.org/10.1007/s00792-017-0939-x

Silva JL, Foguel D, Da Posian AT, Prevelige PE (1996) The use of hydrostatic pressure as a tool to study viruses and other macromolecular assemlages. Curr Opin Struct Biol 6:166–175

Smith KE, Thatje S (2012) The secret to successful deep sea invasion: does low temperature hold the key? PLoS One 7:e51219. *

Sojo V, Herschy B, Whicher A, Camprubi E, Lane N (2016) The origin of life in alkaline hydrothermal vents. Astrobiology 16:181–197. *

Soler N, Gaudin M, Marguet E, Forterre P (2011) Plasmids, viruses and virus-like membrane vesicles from Thermococcales. Biochem Soc Trans 39:36–44. *

Steel EL, Davila A, McKay CP (2017) Abiotic and biotic formation of amino acids in the Enceladus Ocean. Astrobiology 17:862–875. *

Stokke R et al (2015) Functional interactions among filamentous Epsilonprotobacteria and Bacteroidetes in a deep sea hydrothermal vent biofilm. Environ Microbiol 17:4063–4077. *

Szafranski KM, Piquet B, Shillito B, Lallier FH, Duperron S (2015) Relative abundance of methane- and sulfur-oxidising symbionts in gills of the deep sea hydrothermal vent mussel Bathymodiolus azoricus under pressure. Deep Sea Res I 101:7–13

Taghon GL (1988) Phospholipid fatty acid composition of the deep sea hydrothermal vent polychaete Paralvinella palmiformis (Polychaete—Ampharetidae): effects of thermal regime and comparison with two shallow water confamilial species. Comp Biochem Physiol 91B:593–596

Takai K et al (2008) Cell proliferation at 122 C and isotopically heavy CH4 production by a hyperthermophilic methanogen under high pressure cultivation. Proc Natl Acad Sci U S A 105:10949–10954

Tamura Y, Gekko K, Yoshioka K, Vonderviszt F, Namba K (1997) Adiabatic compressibility of flagellin and flagellar filaments of Salmonella typhimurium. Biochim Biophys Acta 1335:120–126. *

Trevors JT (2002) The subsurface origin of microbial life on the earth. Res Microbiol 153:487–491

Tyler PA, Dixon DR (2000) Temperature/pressure tolerance of the first larval stage of Mirocaris fortunata from Lucky Strike hydrothermal vent field. J Mar Biol Ass UK 80:739–740

Tyler PA, Young CM (1998) Temperature and pressure tolerances in dispersal stages of the genus echinus (Echinodermata:Echinoidea): perquisites for deep sea invasion and speciation. Deep-Sea Res II 45:253–277

Tyler PA, Young CM, Clarke A (2000) Temperature and pressure tolerances of embryos and larvae of the Antarctic sea urchin Sterechinus neumayeri (Echinodermata: Echinoidea): potential for deep sea invasion from high latitudes. Mar Ecol Prog Ser 192:173–180. *

Valentine DL (2007) Adaptations to energy stress dictate the ecology and evolution of the Archaea. Nat Rev Microbiol 5:316–323

Van Dover CL (1994) In situ spawning of hydrothermal vent tubeworms (Riftia pachyptila). Biol Bull 186:134–135

Van Dover CL (2000) The ecology of hydrothermal vents. Princeton University Press

Van Dover CL, Lutz RA (2004) Experimental ecology at deep sea hydrothermal vents: a perspective. J Exp Mar Biol Ecol 300:273–307

Van Dover CL, Szuts EZ, Chamberlain SC, Cann JR (1989) A novel eye in 'eyeless' shrimp from hydrothermal vents of the Mid Atlantic ridge. Nature 337:458–460. *

Van Dover CL, Reynolds GT, Chave AD, Tyson JA (1996) Light at deep sea hydrothermal vents. Geophys Res Lett 23:2049–2052. *

Van Dover CL, German CR, Speer KG, Parson LM, Vrijenhoek RC (2002) Evolution and biogeography of deep sea vent and seep invertebrates. Science 295:1253–1257

Von Damm KL et al (1995) Evolution of East Pacific rise hydrothermal vent fluids following a volcanic eruption. Nature 375:47–50. *

Wächtershäuser G (2006) From volcanic origins of chemoautotrophic life to bacteria. Archaea and Eukarya Phil Trans R Soc B 361:1787–1808

Watanabe H et al (2004) Larval development and intermoult period of the hydrothermal vent barnacle Neoverruca sp. J Mar Biol Assoc UK 84:743–745

Williams TA, Foster PG, Cox CJ, Embley TM (2013) An Archaeal origin of eukaryotes supports only two primary domains of life. Nature 504:231–236

Williamson SJ et al (2008) Lysogenic virus host interactions at deep sea diffuse flow hydrothermal vents. ISME J 2:1112–1121

Wu GC, Wright JC (2015) Exceptional thermal tolerance and water resistance in mite Paratars otomus macropalpis (Erythracaridae) challenge prevailing explanations of physiological limits. J Insect Physiol 82:1–7

Yahagi T, Watanabe HK, Kojima S, Kano Y (2017) Do larvae from deep sea hydrothermal vents disperse in surface waters? Ecology 98:1424–1534

Yoshiki T, Toda T, Yoshida T, Shimizu A (2006) A new hydrostatic pressure apparatus for studies of marine zooplankton. J Plankton Res 28:563–570. *

Yoshiki T, Ono T, Shimizu A, Toda T (2011) Effect of hydrostatic pressure on eggs of Neocalanus copepods during spawning in the deep-layer. Mar Ecol Prog Ser 430:63–70

Young CM, Tyler PA (1993) Embryos of the deep sea echinoid Echinus affinis require high pressure for development. Limnol Oceanogr 38:178–181

Young CM, Tyler PA, Gage JD (1996) Vertical distribution correlates with pressure tolerance of early embryos in the deep sea asteroid Plutonaster bifrons. J Mar Boil Assoc UK 76:749–757

Zeng X et al (2009) Pyrococcus CH1, an obligate piezophilic hyperthermophile: extending the upper pressure-temperature limits for life. ISME J 3:873–876

Zhai Y et al (2012) Physical properties of Archaeal tetraether lipid membranes as revealed by differential scanning and pressure perturbation calorimetry, molecular acoustics and neutron diffraction reflectometry: effects of pressure and cell growth temperature. Langmuir 28:5211–5217

Zhao F, Xu K (2016) Molecular diversity and distribution pattern of ciliates in sediments from deep sea hydrothermal vents in Okinawa trough and adjacent sea areas. Deep Sea Res I 116:22–32

# Buoyancy at Depth

**10**

There is a curious, intuitive fascination about buoyancy at great depths in the ocean. It was reported of ancient mariners that they feared burial at sea because they thought that the weighted body would only sink so far and no further. In some confused way, they thought the depths prevented the body sinking further, so it would drift with the currents. Not a proper burial. We are now clearer about Archimedes Principle, probably the most widely known law of physics.

An object, a fish or a rock, in water experiences two forces. Its weight creates a downward force and the volume of water it displaces also has a weight which acts as an upward, buoyant force. When the two are equal (the condition feared by the mariners), a state of neutral buoyancy exists. When the upward force exceeds the downward force, positive buoyancy exists, and the object will rise. Conversely, it will sink when the downward force prevails, as in the case of a weighted corpse buried at sea. Cells and animals are composed of proteins, carbohydrates and skeletal materials all of which are more dense than seawater and they tend to sink. Their lipids, which are less dense, are normally present in quantities too small to render the organism buoyant.

Buoyancy has evolved independently on numerous occasions in both big and small organisms. Generally, on purely energetic grounds, buoyancy is economic in slow swimming animals, but for fast movers, muscle power generating lift is more favourable (Pelster 1997; Webber et al. 2000). The role of hydrostatic pressure in adaptations to buoyancy ranges from the negligible to one of remarkably interesting physical chemistry.

## 10.1 Densities of Sea Water and Cell Constituents

Seawater, at 0 °C, of standard salinity and at atmospheric pressure has a density of 1028.3 kg m$^{-3}$, or in more familiar units of gm per cubic centimetre, 1.028. Its density is increased by 2.1% at 50 MPa and by 4.1% at 100 MPa (Libes 1992). Decreased salinity and increased temperature reduce density (Table 10.1).

© Springer Nature Switzerland AG 2021
A. Macdonald, *Life at High Pressure*, https://doi.org/10.1007/978-3-030-67587-5_10

**Table 10.1**   The densities of typical cell constituents at 0 °C, in units of kg m$^{-3}$

| Carbohydrate | 1550 | Cartilage and muscle | 1100 approximately |
|---|---|---|---|
| Protein | 1300 | Lipid | 860 |

Although the densities of lipids and oils are increased by high pressure, approximately by 3% at 50 MPa, they remain less dense than seawater at comparable pressures and temperatures.

## 10.2   Small Planktonic Organisms

Small aquatic organisms are greatly affected by the viscosity of water, because of their relatively large surface area in relation to the power they can generate, so the factors underlying their mobility and buoyancy differ from those of macroscopic animals. But both the viscosity and, as just mentioned, the density of seawater vary little over the pressure range in the oceans. Cyanobacteria float because of their unique gas vacuole and phytoplankton in general, which needs to be sufficiently buoyant to remain in the shallow photosynthetic layer of the sea, regulates the composition of the cytoplasm using ion transport and metabolic adjustments to reduce their intrinsic density. A giant diatom, *Ethmodiscus rex*, for example, incorporates trimethylammonium ions (Me3NH+), which have a large partial molar volume and thus form solutions of low density, to become buoyant (see Section 10.3.1). The carbohydrate reserves of *Ethmodiscus* make them sink, but when they are metabolised, buoyancy returns, enabling the cells to ascend at a velocity of 10 m per day. All this goes on in depths where the pressure is unimportant (Boyd and Gradmann 2002). *Noctiluca*, a Dinoflagellate which lives (and luminesces spectacularly) in shallow waters, is also buoyant because it accumulates ammonium ions, and there is evidence that deeper living Crustacea do the same. For example, the shrimp *Notostomus gibbosus* (400–1000 m depths) accumulates high concentrations of ammonium and trimethylammonium ions, making it positively buoyant, but other active Crustacea in the plankton do not (Sanders and Childress 1988).

Many eggs in the plankton contain sufficient fatty yolk to float, and as we have seen, some will have ascended from the deep benthos (Chapter 9). A number of gelatinous planktonic animals, Siphonophores such as jellyfish, Ctenophores (*Beroe*) and Tunicates (*Cyclosalpa*), have acquired neutral buoyancy by reducing both their protein content and their heavy sulphate (S04) ion concentration (Bidighare and Biggs 1980). All three can swim weakly; jellyfish by contracting its umbrella structure, *Beroe* uses sets of cilia and *Salpa* propels itself by jet propulsion, i.e. it contracts to create a jet of water. They occupy deeper water than phytoplankton but experience no significant pressure.

Planktonic Crustacea, Copepods, various shrimp-like creatures and diverse larvae, can swim to avoid sinking and their long spines probably reduce their tendency

**Table 10.2**  Low-density solutions of ammonium compounds and of urea

| | Partial molal volume $10\text{-}6\ m^3\ mol^{-1}$ | Density, in solution Of 1000 mol m$^{-3}$kg m$^{-3}$ |
|---|---|---|
| NH4+ | 12.86 | 1004.3 |
| Methyl ammonium (MeNH3+) | 31.11 | 1000.1 |
| Trimethyl ammonium (Me3NH+ | 66.51 | 992.7 |
| Tetramethyl ammonium (Me4N+) | 84.57 | 988.8 |
| Urea | 54.1 | 1015.5 |

to sink. Many, however, have a store of lipid which contributes buoyancy as well as providing an energy store.

The Copepod *Calanus acutus*, an important member of the plankton in the Southern Ocean, is made buoyant by a store of wax esters, the result of feeding on diatoms in the summer. When the wax esters attain 50% unsaturation, the animals actively swim down to considerably deeper than 500 m, 5 MPa and to temperatures of 0–2 °C. It appears that the wax esters freeze, reducing their buoyancy, so the animals become neutrally buoyant. They enter a dormant state called diapause which can last for many months, and which is thought to be an effective over-wintering strategy. The polyunsaturated wax esters are selectively metabolised, slowly, so their buoyancy changes and the animals ascend from diapause after winter. Biochemical evidence suggests that the carbon chain in the alcohol part of the wax ester is the site of unsaturation and thus of the pressure sensitivity of the mixture. A further aspect of this buoyancy scenario is that high concentrations of ammonium ions have been found in *C. acutus*. They will contribute to buoyancy but may also play a role in the mechanism of diapause (Pond 2012, Pond and Tarling 2011, Pond et al. 2012; Sartoris et al. 2010). Interestingly, many other Calanus-like copepods do not descend the water column or enter diapause but their buoyancy is provided by triacylglycerol stores (Table 10.2).

## 10.3  Buoyancy in Larger Animals Living at Significant Pressures

Buoyancy is created at depth just as it is in shallow water, in three ways. Surprisingly perhaps, a gas phase can be created and sustained, but also oils and/or lipids can be accumulated and, as mentioned earlier, the ionic composition of body fluids is regulated, selecting ions with a large partial molar volume, hence low density. Combinations of these options occur and they can be supplemented by the reduction of dense tissues.

Muscle-powered swimming prevents many animals from sinking and it is often aided by buoyancy. A reduction of dense tissues, muscle protein and bone is not uncommon in deep-sea fish which presumably reduces the energy cost of generating uplift. Complex organs such as the swim bladder in fishes and the osmotic systems of cephalopods specifically create buoyancy, which can be adjusted on a fairly short time scale. Pressure plays a conspicuous and crucial role in these organs. Any buoyancy device decreases the animal's density by increasing its volume and thus increases the volume of the water displaced and also the resistance the animal encounters as it moves through the water. There is an interesting trade-off between the energy used in muscle-powered lift and the energy required of buoyancy mechanisms, but there are other, complicating, factors at work. To be neutrally buoyant permits a static and virtually silent presence.

Gas provides buoyancy efficiently. The gas-filled swim bladder of fish generally increases the volume of the animal by only 6% (Alexander 1966). The gas-filled rigid shell of certain Cephalopods, cuttlefish, increases their volume by about 9%, but the coiled shell of *Nautilus* adds 30% to its volume. These are celebrated, complex buoyancy organs whose mechanisms have intrigued physiologists for years. If bulk lipids or oils are used to provide buoyancy, their presence also increases the volume of the animal by around 30%, which clearly incurs a significant increase in the resistance to movement through the water. Finally, if an aqueous solution of low-density salts provides buoyancy, without creating osmotic complications, then the animal's volume increases by 300%. Such a buoyant design corresponds to the Bathyscaphe whereas we shall see that the rigid Cephalopod shell corresponds to the buoyancy system of a submarine. The fish swim bladder has no obvious engineering equivalent, but the occasional use of gas-filled lifting bags in salvage work comes close. These are similar to the device adopted by a rare oceanic Cephalopod, the argonaut *Argonauta nodosus*. The females use their shell to entrap air which enables them to be neutrally buoyant at a depth of a few metres. This behaviour is confined to the females. The males are diminutive creatures, two orders of magnitude smaller (Finn and Norman 2010). A pelagic octopod, *Ocythoe sp.* also traps air as a buoyancy aid (Packard and Wurtz 1994).

## 10.3.1  Ionic Regulation in Bathypelagic Squid

Species of bathypelagic squid can be divided into those which are heavier than water, and thus use muscle power to avoid sinking and those which generate buoyancy by accumulating ammonium ions. Others combine both features.

*Helicocranchia pfefferi*, a species belonging to the Cranchid Family of Cephalopods, has been shown to float in the water column. It is buoyed by the low density of the fluid filling the body cavity which comprises two-thirds of the animal (Denton et al. 1969; Clarke et al. 1979). The fluid has a very high concentration of ammonium ions, replacing sodium ions, with chloride being the main counter anion. Ammonia, $NH_3$, is a common product of metabolism. It combines with H ions or with $H_2O$ to form $NH_4^+$ and $OH^-$.

From Boyd and Gradmann (2002) we learn of a series of ammonium compounds which produce very low-density solutions. Some are listed, with urea, in Table 10.2.

The point of the table is to illustrate the relative densities; we need not be too concerned about the precise definitions of the terms, which are not simple. All the examples are, in practical situations, less dense than seawater.

Ions which have a small or negative partial molar volume and therefore produce dense solutions are H+, $Na^+$, Cl− and SO4−.

The osmotic pressure of the *Heliocranchia*'s body fluid matches that of seawater. The concentration of ammonium ions in the body cavity fluid is also vastly greater than that of the blood and other tissues. The selective transport of hydrogen ions into the body cavity fluid is thought to be "captured" by $NH_3$, an end product of nitrogen metabolism and essentially an excretory product, forming the $NH_4^+$ ion. The fluid is accordingly very acidic. The relevant membranes are not permeable to the ammonium ion, so the diffusive "leak" is small, but nevertheless has to be offset by the continuous transport of $H^+$. If it were otherwise the animal would be poisoned, its nerves, for example, would be unable to function. Although *Helicocranchia* is a small animal, its relatively large buoyancy "tank" is clearly an obstacle to active swimming. Other pelagic squids, some much larger, have low-density ammonium-rich fluids distributed in special tissues throughout their body. Generally, squid buoyed by ammonium rich fluid are present in large numbers in moderately deep water where they provide a significant proportion of the diet of sperm whales. The well-known giant squid, *Architeuthis*, is one such and its pockets of ammonium-rich fluids have been analysed. It is also an active muscular predator, recorded in one case attacking a baited camera suspended at a depth of 900 m. That particular specimen was only 8 m in length, well short of the biggest recorded specimen which was 18 m in length. The list of unidentified deep-sea squid photographed from submersibles or ROVs at depths from 1940 to 4735 m, published by Vecchione et al. (2003), suggests that significant numbers exist in the depths, and escape capture. The important point is that the ion transport process and permeabilities which support the ammonium buoyancy mechanism can function harmoniously over a significant pressure range, well over 10 MPa.

According to Clarke et al. (1979), there is just one intriguing species of buoyant squid, *Gonatus fabricii*, which does not use ammonium ions. Instead, it uses low-density oil, not unlike certain fish.

## 10.4  Gas Phase Buoyancy

Gas phase buoyancy devices have evolved independently in various deep-sea animals, taxing the ability of physiologists to understand how they work. They come in two forms. In one, the gas is at a pressure close to atmospheric and confined in a rigid container which withstands the ambient pressure, like a submarine. In the other, the gas is confined in a flexible bladder, in equilibrium with the ambient pressure, which can be very high.

### 10.4.1  Cephalopods

Cephalopods have evolved an improbable buoyancy system based on a rigid, pressure-resistant shell. It contains modified air at, or close to, atmospheric pressure. The shell comes in two forms, flat or coiled.

In Cuttlefish (*Sepia officinalis*), described by a distinguished marine physiologist as "amongst the most remarkable of all animals", the buoyancy shell is an internal organ, a flat, elongated oval shape. In contrast, the shell of *Spirula* is a coil of chambers, but also internally situated. The third example is *Nautilus*, whose shell is also a coiled series of chambers but it lies externally (see papers by Ward et al. and Greenwald and Ward 2010). The largest, and most recently formed chamber is occupied by the animal whose squid-like tentacles extend conspicuously (Fig. 10.1). *Nautilus* is related to the now-extinct *Ammonites*. The shells of *Nautilus*, often some 15 cm in diameter, are frequently sold as ornaments, sectioned to reveal the elegant chambers arranged in what mathematicians call a logarithmic series. The shells of Sepia, on the other hand, are often found on the seashore as dried out chalky white oval plates, which are given to canaries in their cages to groom their beaks.

These three animals are normally close to being neutrally buoyant. Nautilus, for example, has been weighed in seawater and found to be very close to neutral buoyancy whilst their shells alone were positively buoyant in proportion to the extent to which their watery fluid had been removed by the animal. So how does their buoyancy mechanism work?

Despite the structural differences between the shells their buoyancy mechanism is essentially the same, as first shown by Denton and Gilpin-Brown, working at the Marine Biological Laboratory, Plymouth, England. The structure of Nautilus is shown in Fig. 10.1.

The siphuncle is a tube carrying blood vessels and is supported by a strong but porous tube. It prevents the soft living tissue of the siphuncle being forced into the gas-filled chamber but the seawater does seep in. That process is opposed by an osmotic pressure across the siphuncle, i.e. between the fluid in the chamber and the animal's blood. $Na^+$ ions are transported by the epithelial cells lining the siphuncle, from the chamber fluid into the blood, creating a local osmotic pressure, so an osmotic flow of water follows, evacuating the chamber. The water is transferred from the animal to the external environment by the kidneys. The space created by its

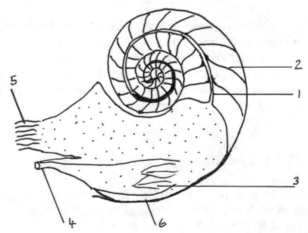

**Fig. 10.1** A sketch of *Nautilus sp.* showing the main body parts. 1, siphuncle, 2, septa which divides the shell. The animal occupies the most recently formed shell. 3, gills, 4, funnel, through which a jet of water can propel the animal, 5, tentacles, 6, mantle wall

removal from the chamber is occupied by humid air. The maximum osmotic pressure which can exist across the siphuncle is set by the maximum osmolarity of the blood versus pure water, which is 2.4 MPa, equivalent to the hydrostatic pressure at a depth of 240 m. This is, therefore, the maximum depth that the simple osmotic mechanism can provide buoyancy (D.

So, thus far we can imagine that as Nautilus grows and adds a new chamber to its shell, its additional weight is offset by the osmotic emptying of the water in the new chamber. Emptying is a slow process. It is interesting to consider the depth distribution of *Nautilus* (Dunstan et al. 2011). *N. belauensis* has been tracked at a depth of 500 m and probably spends most of its time between 200 and 300 m. *N. pompilius* has been recorded taking bait at 703 m depth (7 MPa). Spirula lives deeper, for example, it has been collected from a depth of 1200 m off the Canary Islands. Sepia probably spends most of its time no deeper than 200 m (2 MPa). The following shell implosion pressure have been experimentally determined: 2 MPa (*Sepia officinallis* and other species; Sherrad 2000), 6.2 MPa (*Nautilus macromphalus*) and 3–7 MPa (*N. pompilus*; Saunders and Wehman 1977), 12 MPa (*Spirula spirula*). These are probably biased on the low side because dried shells, which were often used, tend to be weaker than shells which are still part of the animal. For living *N. pompilius* there is better data. The shells of live animals lowered in traps collapsed at 7.5–9 MPa and the shell of one animal, observed in a pressure vessel, imploded and shattered at a pressure of 8.05 MPa (Kanie et al. 1980). The pressure resistance of the shells matches, with a safety margin, the observed depth distributions of the animals. The behavioural response to compression has been observed by Jordan and colleagues in the case of young *N. pompilius* adapted to shallow aquarium depths. Stepwise (100 kPa) compression to the equivalent of a depth of 200 m elicited short bursts of upward swimming, implying that the danger of excessive depth was detected.

Nautilus and Spirula certainly spend time at depths greater than 240 m, 24 MPa, which is the critical depth limit for the simple osmotic mechanism just described. For these animals we have to invoke a supercharged version of the simple osmotic mechanism (see Box 10.1).

**Box 10.1 Solute Coupled Water Transport in Nautilus**

Evidence that a supercharged osmotic system exists in *Nautilus* lies in the fine structure of the epithelia across which the osmotic process takes place. It is similar to the epithelia in other animals, including mammals, which is involved in fluid transport. In Nautilus, the epithelia is about 100 u thick and made of closely packed cells full of mitochondria. It lies along the siphuncle, the tissue which is supported by the rigid but permeable tube which runs through the length of the coiled chambers. One side of the epithelia adheres to the rigid tube and the other side is perfused by blood vessels. The water osmotically removed from the shell passes into the blood stream at this juncture. The surface of the cells in contact with the blood vessels is deeply invaginated, and

(continued)

**Box 10.1**  (continued)

finer tubes penetrating the cell branch from the main invagination, called the drainage channel. A high concentration of NaCl arises in the drainage channel from the activity of the many ion pumps situated along the extensive area of cell membrane, powered by ATP produced by mitochondria. The salt is conveyed in solution along the drainage channel by both osmotic flow and diffusion to the blood vessels. The same arrangement occurs in numerous transport epithelia, a good example being the mammalian kidney, which uses osmosis to transport water. The solution thus produced is either of the same osmotic concentration as the source (the chamber fluid in the case of Nautilus) or it is more concentrated. The process is called solute-coupled water transport, and its properties have been mathematically analysed by Diamond and Bossert whose paper enthusiasts may wish to read.

A significant point about this buoyancy mechanism is that the chamber fluid, from which salt is transported to drive the water flow, can be of a very limited volume, just enough to cover the siphuncle. It not only enables *Nautilus* and *Spirula* to prevent their shells from being flooded at depths greater than 240 m, 2.4 MPa, the osmotic pressure of sea water. It also provides more rapid buoyancy adjustment than a simple osmotic arrangement.

The system is primarily suited to provide neutral or near-neutral buoyancy leaving the animal to swim economically. Tests with *N. pompilius* in a pressure aquarium showed it responded to a pressure increase of 100 kPa with an alarm response and upward swimming, but did not respond to a similar pressure decrease. In the diurnal vertical migration of *Nautilus pompilius* in the tropical Pacific, active swimming moves the animal and changes in buoyancy play no part. Nautilus scavenges on the sea floor and can, in other circumstances, reduce its metabolic rate to less than 8% of normal and survive hypoxic conditions for a day, during which time it may draw on oxygen stored in the shell gas.

Let us return to Sepia, which is not as exotic as the other two species but nonetheless is a remarkable animal. Unlike Nautilus the genus Sepia (i.e. cuttlefish) has evolved many species, which either live in shallow water (e.g. *S. officinalis*) or in deeper water (*S. orbignyana*). This is reflected in the pressure at which their cuttlebones implode. In the case of adult *S. officinalis*, it implodes at 1.5–2.0 MPa, i.e. at depths of 150–200 m. The cuttlebone in young animals is weaker, imploding at 0.5–1.0 MPa. The implosion pressure for the cuttlebone in *S. orbignyana* is 5.5–6.0 MPa. The flat laminated structure of the cuttlebone is an unlikely compression-resistant shape, but the septa between the laminae and their separation clearly contribute to its strength under compression. It is therefore unsurprising that the deeper living species have cuttlebones in which the septa are thicker and more closely spaced than their counterparts in shallow living species. It would be interesting to know if, during the development of *S. officinalis*, cuttlebone strength is enhanced by exposure to mechanical stress (depth), as in vertebrate bone. Rearing the animals in aquaria might shed some light on this point.

Some species of *Sepia* exist in sufficient numbers in coastal waters to support commercial fishing. Typically, they have a high metabolic rate, and use their buoyancy for poise, stealth and efficient swimming. The repertoire of the genus also includes an amazing talent for camouflage and, a sophisticated array of senses supported by an "advanced" brain, features which the physiologist quoted earlier had in mind.

## 10.4.2  The Gas-Filled Swim Bladder of Bony Fishes

Dissect a bony fish and the swim bladder is apparent as a silver white oblong in the mid-dorsal region of the body, comprising about 6% of its volume. In many shallow water fish, the swim bladder is connected to the oesophagus by a duct, and the animal gulps air to fill the bladder. Here we are concerned with swim bladders which are not connected to the oesophagus and which contain gas which comes from the blood. Typically, the gas is air enriched with oxygen, and thus a mixture of nitrogen and oxygen in which the pressure of oxygen (called its partial pressure) is greater than it is in air.

The swim bladder is situated within the coelomic cavity, below the spinal cord and provides the fish with buoyancy and a stable orientation (Fig. 10.2). Some swim bladders are connected to the oesophagus (physostomatous) but the one shown is not so connected (physoclistous).

Against all reasonable expectations, gas-filled swim bladders are found in many deep-sea fish. They are common in bentho-pelagic species, for example *Coryphaenoides*, whose swimming behaviour is consistent with being close to neutral buoyancy. Pelagic species living deeper than 2000 m rarely have a functional swim bladder. *Bassogigas profundissimus* is often quoted as the deepest known example, which was collected by Nielsen and Munk (1964) from a depth of more than 7000 m, which means the pressure of gas in the bladder was in excess of 70 MPa. Their paper considers the evidence that the fish actually came from 7000 m depths, and it ignores the puzzling feature of the swim bladder being intact after decompression from such a pressure, and suggests it would have contributed little

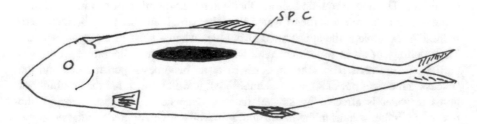

**Fig. 10.2**  Bony fish swim bladder, shown in black, below the spinal cord, sp.c

buoyancy. Generally, bony fish are not found deeper than 8400 m (Yancey et al. 2014). However, there is plenty of evidence that functional swim bladders exist at considerable depths, i.e. around 5000 m. This is a remarkable state of affairs. It means that the gas in the bladder is "pumped" up a huge pressure gradient and somehow confined in the bladder. Pumped is a convenient term but a misleading one. The gas actually diffuses down a partial pressure gradient into the bladder. The clever stuff lies in creating the gradient.

Consider a hypothetical fish with a swim bladder filled with gas, making the animal neutrally buoyant. At the surface the partial pressure of air in the water is 100 kPa, of which 20% is oxygen (20 kPa) and 80% nitrogen (80 kPa), in round numbers. The surface sea water circulates and our fish, which we will assume is at a depth of 100 m (1000 kPa, 1 MPa), passes it over its gills, where the dissolved gas diffuses into the circulating blood. Nitrogen, at a partial pressure of 80 kPa, dissolves in the blood just as it does in water and so does 20 kPa of oxygen. But oxygen also combines with haemoglobin, in great quantities, forming oxyhaemoglobin. The amount of oxygen thus combined is many times the amount physically dissolved in the blood, but its partial pressure (pO2) is still 20 kPa. The blood circulates to the swim bladder and through the gas gland in the bladder wall which secretes gas into the bladder. At a depth of 100 m (1000 kPa, 1 MPa), the combined partial pressure of nitrogen and oxygen equals 1000 kPa, ten times the partial pressure of the gas in the surrounding sea water. Thus, the gas has been pumped up to a partial pressure ten times that of the ambient sea water, i.e. a tenfold pumping ratio). Explaining how this comes about has attracted physiologist for years. We have a good understanding of some of the factors at play, but a detailed account of the extreme cases such as Bassogigas has not yet been achieved.

Here is an account of our present understanding, organised in seven stages.

First, the tenfold pumping ratio in our hypothetical swim bladder at a depth of 100 m is not strictly accurate. The reason is that hydrostatic pressure affects the partial pressure of dissolved gas. This is an often misunderstood point, but not really a difficult one.

The quantity of gas in a solution is the product of its partial pressure and the solubility coefficient for the liquid in question, here water. This is Henry's Law.

$$\text{Quantity of gas dissolved} = \text{Partial pressure} \times \text{Solubility coefficient.}$$

The solubility coefficient is a technical way of saying solubility in standard conditions. The above equation shows that when the solubility of gas in solution is decreased in a discrete volume of water, then the partial pressure must be increased. A familiar example is the appearance of bubbles when a glass of tap water warms. The solubility of air in water decreases with an increase in temperature so the gas partial pressure rises. The solution of air in water becomes supersaturated and the "excess" air comes out of solution as bubbles. It is well known that the solubility of gases in water is affected by several factors: temperature, the gas pressure, the presence of other solutes in the liquid and, less well known, by hydrostatic pressure.

The last point means that if water, containing gas in solution, is hydrostatically pressurised in the absence of a gas phase, the partial pressure of the dissolved gas will increase because the hydrostatic pressure decreases its solubility. The nitrogen partial pressure of 80 kPa acquired in the surface waters of the sea is increased by the hydrostatic pressure at great depth. This means that, at significant depth, the gas in a swim bladder has been pumped from a somewhat higher partial pressure than might have been expected; hence the pumping ratio is less than might have been expected. Exactly how much the pumping ratio changes with increase in depth has been calculated from thermodynamics and confirmed experimentally by a simple and elegant experiment. It happens to be one of the author's all-time favourite high pressure experiments and is described in Box 10.2.

---

**Box 10.2 Hydrostatic pressure affects gas solubility**

Imagine sea water, containing dissolved gas at a given partial pressure, being hydrostatically pressurised, with no gas phase present. Thermodynamic theory predicts that the gas solubility coefficient is decreased by the pressure, so the gas partial pressure must be increased. An experiment which demonstrates this is simplicity itself.

The experiment was carried out by Enns, Scholander and Bradstreet of the Scripps Institution of Oceanography, California, and published in 1965. They used a Teflon tube of 0.3 mm internal diameter and a wall thickness of 0.23 mm. A 30-cm length of such a tube was mounted in a water-filled pressure vessel and connected to the exterior such that the inside of the tube was at normal atmospheric pressure. Inside the vessel the tube withstood the water pressure. The plastic was permeable to gas but not to water. The water in the pressure vessel contained dissolved gas which, at the start of the experiment, was in equilibrium with the air in the tube. When the hydrostatic pressure was increased, the dissolved gas diffused through the wall of the plastic tube and was measured outside the pressure vessel. The higher the hydrostatic pressure the more the gas diffused through the plastic tube. The effect was small, but convincing. The maximum pressure the experiment could handle was 10 MPa, as the plastic tube tended to collapse at higher pressures, but 10 MPa was sufficient to provide crucial data. It increased the tendency for nitrogen and for oxygen to diffuse from the pressurised water, through the wall of the plastic tube, by 14%. That means the partial pressures of the gases in the pressurised water was similarly increased. Other gases behaved in the same way. The partial pressure of dissolved helium increased by 13% and carbon dioxide by 16%. The partial pressure ("escaping pressure") of the dissolved gas exponentially increases with increase in hydrostatic pressure, which in the sea increases linearly with depth. The data extrapolate to a 1.5-fold increase in the partial pressure of the dissolved gases at 30 MPa, 3000 m and a fourfold increase at 100 MPa, 10,000 m depth. The data are consistent with

---

(continued)

**Box 10.2** (continued)
thermodynamic theory and show that the effective partial pressure of oxygen
and nitrogen in deep water is higher than anticipated from the simple view we
took at the start of this account. Thus, at significant depths the pumping ratio
for deep-sea swim bladders is less than expected.

The partial pressures of dissolved oxygen and nitrogen are increased fourfold by
hydrostatic pressure as seawater descends from the surface to 10,000 m depths. This
assists the swim bladder buoyancy mechanism only marginally but we should note
the existence of this phenomenon. It also implies that a higher than expected partial
pressure of oxygen exists in the depths, the implications of which are discussed by
Ludwig and Macdonald (2005).

The second point to be considered in understanding the high-pressure swim
bladder is the effectiveness of compressed gas as a buoyancy fluid. The density of
gases increases with pressure but Preide has recently carefully examined the data and
concluded that oxygen, the densest of the relevant gases, in the deepest possible case,
could still provide more buoyancy than lipids. For example, at 60 MPa (as in
Bassogigas) the density of oxygen is 672 kg m$^3$, a very light oil, see Table 10.2.

For the third stage in this account, we return to Henry's Law, and the way in
which gas solubility can be decreased, simultaneously increasing the partial pressure. Here the context is blood circulating through capillaries influenced by gas
gland (Fig. 10.3).

For reasons which will soon be apparent the gas in deep-sea swim bladders has a
high concentration of oxygen, typically 90%. A fish at 4000 m depths would have
swim bladder gas comprising 4 MPa nitrogen and 36 MPa oxygen. The partial

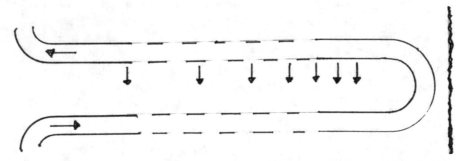

**Fig. 10.3** The countercurrent multiplier mechanism in the rete. A capillary in the form of a hairpin
loop is depicted, one of many clustered in the rete, lying in the wall and surrounding tissue of the
swim bladder. The swim bladder lumen is on the right and the gas gland, not shown as such, is at the
apex of the loop. The rete has an arterial entry and a venous exit, respectively, feeding afferent
(entry) and efferent (exit) capillaries. The flow of blood entering and leaving the hairpin loop is
shown. Also shown is the diffusion of lactate and of gas, crossing from efferent to afferent
capillaries. The dashed lines indicate that the length of the loops vary: long (c. 25 mm) in deep-
sea fish and short (1–2 mm) in shallow water fish

pressure of nitrogen dissolved in the blood entering the capillaries is close to that of seawater, i.e. about 80 kPa. It is increased by the secretion of ions, mainly lactate ions, in the gland. These decrease the solubility of gas, i.e. nitrogen and also oxygen, physically dissolved in the blood, a process known as "salting out", and the consequence is their partial pressure is increased, but only a little.

The curious arrangement of the capillaries in the swim bladder wall is called a rete. This is the fourth stage in our stepwise account. The rete is a bundle of intermingled capillaries in a hairpin loop, with arterial blood entering and venous blood leaving the bladder. A loop is shown in Fig. 10.3. Lactate ions are secreted by the cells in the gas gland at the apex of the loop and adjacent to the bladder lumen. Lactate thus diffuses into the efferent (exit) capillaries of the rete where, in accordance with Henry's Law, they increase the partial pressure of the dissolved gases by decreasing their solubility. The increased lactate concentration in the efferent capillaries in the rete also leads to its diffusion into the afferent capillaries. This boosts the concentration of lactate in the rete. Similarly, the increased gas partial pressure in the efferent capillaries causes the diffusion of gas across to the afferent (entry) capillaries of the rete. By a continuous process, a high partial pressure of gas is created and conserved in the gas gland and the gas diffuses into the lumen of the bladder.

The rete acts as a countercurrent multiplier (Fig. 10.3). It creates the high partial pressure of nitrogen found in deep-sea swim bladder gases by a "salting out" process. For example, Scholander found the nitrogen partial pressure to be 0.93 MPa in the bladder gas of the ratfish, *Macrourus bairdii*, trawled from a depth of 1240 m, 12 MPa. The balance was made up of oxygen, of which more soon. Long retes are a feature of deep-sea swim bladders, and their length increases with the depth at which they function. This is particularly marked over the depths down to 2000 m, beyond which the effect is moderate.

Countercurrent multipliers are found in other physiological contexts, a well-known example being the loop of Henle in the mammalian kidney. Its function is to create a region of high osmotic pressure through which the collecting ducts pass. Fluid moving along the ducts is osmotically concentrated, producing the concentrated solution called urine once it is stored in the bladder. The process conserves water for terrestrial mammals, including humans. The longer the loop of Henle the greater the concentration of urine produced. The loop of Henle in desert rodents is much longer than in ordinary mammals, enabling the rodents to produce highly concentrated urine and conserve water. The length of the countercurrent multiplier of the rete in deep-sea swim bladders and in desert rodents is an adaptation, clever chemical engineering. In deep-sea fish, it is a structural adaptation to high pressure whilst molecular adaptations to pressure presumably lie in the transport and permeability processes involved (Pelster 2001; Rummer and Brauner 2015).

The fifth point concerns the properties of oxyhaemoglobin, which is the source of oxygen in the swim bladder. Whilst oxygen physically dissolves in blood just like nitrogen, most is associated (bound) with haemoglobin. This accounts for oxygen's very high partial pressure in deep-sea swim bladders. Acidity, a high concentration of hydrogen ions, favours the dissociation of oxygen from fish haemoglobin and it severely limits the amount of oxygen which can combine with it, the Root effect. The blood in the rete is acidified (Fig. 10.3) causing a lot of oxygen to dissociate from the haemoglobin, thus increasing its partial pressure. This is the main source of gas in swim bladders.

A number of mathematical models of the countercurrent multiplier have been created, allowing various parameters to be varied, such as the length of the rete, the through-flow of blood and the rate of diffusion of ions and other solutes (Kuhn et al. 1963; Enns et al. 1967; Steen and Sund 1977; Kobayashi et al. 1989). As we have seen, the multiplier works, not just on the physically dissolved nitrogen and oxygen, but more importantly, on the oxygen dissociating from oxyhaemoglobin. The model of Gerth and Hemmingsen (1982) can generate a partial pressure of oxygen in excess of what any deep-sea fish might require, suggesting that the problem of understanding how deep-sea swim bladders acquire their high partial pressure of gas is solved, in principle at any rate. But that is not enough. To solve the problem properly needs evidence that selected parameters actually prevail in real fish, and that requires experimental measurements on healthy fish. Isobaric collection and maintenance of suitable specimens would appear necessary. Attempts to do this are described in Chapter 13. Furthermore, there appears to be variation in important features, such as the way in which oxygen is dissociated from oxyhaemoglobin, the amount of nitrogen in the bladder gas and the detailed structure of the retes, which mean there are interesting variations in the primary mechanism of generating the high partial pressures required.

The sixth stage is to consider the retention of gas within the bladder. Gas is lost in two ways. There is a controlled diffusion to the bloodstream, in order to regulate buoyancy. This generally occurs at a specific site in the bladder wall, the oval window, which can be restricted by a sphincter muscle. In the case of farmed Atlantic cod, which are kept in cages, raising them to reduce pressure by 40% is tolerated without injury but 10 h is required for the new buoyancy equilibrium to be achieved. Secreting gas (pumping) is an even slower process. To achieve neutral buoyancy after a doubling of pressure requires 24 h or more. In deep-sea fish, a given vertical movement causes relatively small changes in bladder volume and we can assume any buoyancy adjustment is similarly slow. All the mathematical models mentioned earlier predict very slow fluxes of gas.

Gas is also lost by continuous diffusion through the bladder wall, a process which increases significantly with depth. Shallow water swim bladders are rendered impermeable to gas by the deposition of crystal plates of guanine and hypoxanthine in the multi-layered wall. A characteristic feature of many deep-sea fish bladders is the additional presence of a great deal of lipid on the inside surface of the bladder wall. This feature has attracted much interest and it is one area of traditional deep-sea physiology where the importance of isobaric collection has been appreciated, but not solved (Chapter 13). On retrieval in a trawl, deep-sea fish are usually found with their swim bladder hugely swollen, often rupturing the whole animal. Collecting satisfactory gas samples from such animals is problematical but studying the bladder lipids is even more difficult. They are a mass of bubbles, a foam, and possible chemically altered as well. The foam consists of a mixture of phospholipids and a high proportion of cholesterol. The barrier to the diffusion of dissolved gas which the lipids present is very significant but only in situ, when they are solidified by the combination of low temperature and high pressure. The gas has a low solubility in this form of lipid through which it diffuses very slowly. This is another example of a phase transition, but this time the frozen state is physiologically important (Lapennas and Schmidt-Nielsen 1977; Ross 1979; Wittenberg et al. 1980). There is a lot of microscopical detail in the cells where the gas is released into the bladder which needs to be understood (Prem et al. 2000).

We tend to think of a swim bladder as an inert container, but its walls are metabolically active. Phleger, who has studied buoyancy lipids in detail, and colleagues (1971) have demonstrated the synthesis of cholesterol and lipids in the gas secreting patch of the swim bladder of *Coryphaenoides* collected from a depth of 2000 m. The excised bladder tissue was incubated in a solution with the very high partial pressures of oxygen experienced by deep-sea fish. The incorporation of a radiolabel showed that cholesterol synthesis was unaffected by pressure, at 4 C. The interesting point is that synthesis actually proceeds at such high partial pressures of oxygen.

The seventh and final stage in understanding deep-sea swim bladders is the problem of how the cells are protected from the toxic effects of oxygen. Highly reactive free radicals and other reactive species occur at high partial pressures of oxygen, as we have seen. This is a fundamental problem which has influenced evolution and continues to be of clinical importance. Over 50 years ago D'aoust confirmed the prediction that the high partial pressures of oxygen tolerated in swim bladders are lethal to the intact fish. An oxygen partial pressure of 0.5 MPa is lethal for *Sebastodes miniatus*, and *Anguilla* only survived it for 25 h (Barthelemy et al. 1981). The phenomenon is well known but little studied; a 2015 review paper reported no significant progress in this non-existent field.

The deep-sea swim bladder is a remarkable complex physiological system which works well for numerous species at high pressure. A recent phylogenetic analysis of its evolution concluded that it emerged a hundred million years after a simpler oxygen secreting system in the otherwise poorly vascularised eye of fish was established (Berenbrink et al. 2005).

### 10.4.3 Other Functions of the Swim Bladder

In shallow water fish, the swim bladder can function as a change-of-depth gauge, signalling the change in tension in the bladder wall to the central nervous system, but it is not known if this occurs in deep water fish. It has been suggested that the swim bladder could provide fish with the information to navigate in the vertical dimension. Although it would be a complicated process, it may be of more use in the dark abyss than in depths where sunlight penetrates.

In many fish, the swim bladder is intimately associated with producing and detecting sound. In the deep sea, sound waves travel effectively, their velocity increases slightly with increased pressure as a consequence of the reduction in the compressibility of water. However, the importance of sound in the deep sea is more to do with the limitations of other sensory modes than sound velocity. The same can be said for the acoustic aspects of marine mammals and especially the spermaceti organ of the sperm whale.

Historically, dried swim bladders have been a source of isinglass, dried collagen, which is used to clarify wine and beer by aggregating yeast and other particulate material. A more modern use of the collagen from swim bladders, which are available in quantity from fish processing plants, is in the fabrication of bio-piezo-electric nanogenerators (Maitra et al. 2017).

## 10.5   Buoyancy Provided by Oil and Lipid

The buoyancy of tiny planktonic organisms is generated by low-density oils and lipids, (Section 10.2), which are also effective in big-bodied fish and whales. They are metabolically available and less dense than seawater, and, as we shall see, they remain so at high pressure.

The term wax refers to substances formed from the combination of long-chain fatty acids (C14–C36) with long-chain alcohols (esterification). Waxes have melting points in the range 60–100 C, depending on the chain length and number of double bonds present. Their buoyancy role is often linked to their role as a food store. Triacylglycerols, also called fats, are esters, with three fatty acids combined with glycerol. Squalene is a 30 carbon chain with six double bonds and known in mammalian and fish biochemistry as a precursor of cholesterol. Spermaceti is a mixture of triacylglycerols and wax, highly unsaturated, so it is liquid at mammalian body temperatures, but as we shall see, it has interesting temperature–pressure properties. Pristane is a long chain alkane, $C19H40$, which is found in small quantities in the liver of certain sharks. It is listed in Table 10.3, not because of its role in buoyancy, which is uncertain, but because of its unusually low density. The synthesis of these buoyancy compounds is complicated and mercifully need not concern us here, but we should note that energy is required in their synthesis and it can be released when they are metabolised.

**Table 10.3** Buoyancy generated by lipids and other substances which have a low density. Density, kg m$^{-3}$, at approximately room temperature and atmospheric pressure, and where appropriate, with similar chain lengths. Compiled from various sources

| Pristane | 780 | Seawater | 1029 |
|---|---|---|---|
| Wax | 860 | Bone | 1300–1500 |
| Squalene | 860 | Muscle | 1100 |
| Spermaceti | 890 (853 at 37 C) | Cholesterol | 1070 |
| Alkyldiacylglycerol | 890 | | |
| Cod liver oil | 930 | | |
| Triacylglycerol | 930 | | |

Their density increases with a decrease in temperature and an increase in pressure, both of which occur with an increase in depth. A marked change in density occurs at the liquid–solid phase transition (freezing point) in some naturally occurring wax esters. However, reflecting the mixture of similar compounds present, these transitions are spread over several degree Celsius and are not characteristic of a specific compound.

The numbers in the left-hand column of Table 10.3 look rather similar but it is a feature of buoyancy that very small differences in density are important, especially in achieving neutrally buoyancy. Priede and colleagues (2020) have compared the compressibility of squalene and liver oils in Chondrichthyes with that of seawater. The small differential effects of pressure and temperature are sufficient to be significant in the energetics of the deeper living animals. Examples from engineering are illuminating. In a paper on whale buoyancy, Clarke pointed out that for a submarine to be neutrally buoyant it has to adjust its displacement weight (about 2000 tons) to within 0.002–0.003%, or, more easily imagined, to within 40 litres of water. Another good illustration is the way the Bathysphere Trieste (displacement 50 tons) could passively float just above the ocean floor. The slowly descending vessel suspended a rope, heavier than water, beneath it. As the rope contacted the floor its weight was reduced, slowing the rate of descent until the reduced weight brought the vessel to neutral buoyancy.

Many animals benefit from the energy saved in powered swimming if they are close to being buoyant, but a surprising number achieve neutral buoyancy for other reasons. A clear example of buoyancy, which is, after all, outside our everyday experience unless we are keen swimmers or astronauts, is to be seen in those public aquaria which are so big you can walk beneath them. Sharks are seen, continually swimming (hydroplaning), because they cannot achieve neutral buoyancy. Deep-sea sharks are closer to neutral buoyancy. They swim slowly and have less hydroplaning capability. In contrast, typical reef fish achieve neutral buoyancy and are suspended in the water, floating apparently without effort. Their tranquillity is put to good use in "tropical" fish tanks often found in dentist waiting rooms.

## 10.5.1 The Sperm Whale

The Sperm whale, *Physeter macrocephalus*, has remarkable buoyant properties. Adults weigh 30,000 kg or more, in air. Males can reach 16 m in length. In the early days of whaling, sperm whales had the convenient feature of being buoyant when killed by harpoon. As the scientific name suggests, the head is very big, contributing a third of the body weight. In addition to the usual features of a vertebrate head, it contains a thin clear oil, spermaceti, held in the spermaceti organ (Clarke 1978a). This has internal compartments holding spermaceti and a more solid basal section called a junk. In a 30,000 kg whale, there might be 7000 kg of spermaceti, which contributes to the buoyancy of the animal. In the nineteenth century, the spermaceti, a mixture of wax esters and triglycerides, made good candles and later was found to be useful in making cosmetics. The different oil obtained from whale blubber was used in oil lamps. The Sperm whale dives to great depth in pursuit of its main food, giant squid, 1000 kg of which an adult needs to consume each day (Vecchione et al. 2003; Kubodera and Mori 2005; Winkelmann et al. 2013). One idea is that the whale floats silently in the depths, waiting for a squid to come within range. The head of the sperm whale is thought to have at least two other specialised functions. The behaviour of males is, at times, aggressive, involving head butting (Carrier et al. 2002). There is no doubt that this behaviour has been responsible for the destruction of wooden sailing ships. Also, the structure of the spermaceti organ is consistent with focussing sound waves, and sound is certainly important in the life of the animal.

However, let us return to the problem of the animal's buoyancy at depth. Clarke (1978c) has made a study of sperm whales, in the flesh, so to speak, and has meticulously constructed a "buoyancy balance" sheet of the animal. He has postulated that the spermaceti enables the animal to become neutrally buoyant at depths greater than 200 m, important in its predation of squid. As the animal dives the gas in the lungs collapses, but dives are not regularly preceded by exhalation or inhalation, so the amount of residual compressed gas is variable. The tissues compress, which contributes to the reduced buoyancy. The density of the seawater into which the animal dives increases slightly with pressure and the decrease in temperature, which also favour buoyancy.

Casual observation shows that spermaceti above 30 C is a clear liquid which becomes cloudy when cooled a few degrees, evidence that some components of the mixture are solidifying. During a dive, the compression of the spermaceti and the very thick layer of blubber which covers the whole body, causes an initial rise in temperature (the heat of compression) which initially opposes the increase in density due to compression. The cooling of the spermaceti occurs through the head and in particular through the different nasal passages. The effect of pressure on spermaceti is influenced by its temperature. At 32 C, the plot of density against pressure shows a marked inflexion at 10 MPa, which is interpreted as a phase transition taking place in some of the constituents. At 31 C, the transition occurs at a lower pressure. The liquid to solid transition means the density changes markedly, making it difficult to predict its buoyancy contribution without knowing its precise temperature. The quantitative argument presented by Clarke (1978b) is that neutral buoyancy could be achieved at depths below 200 m, 2 MPa. The reduction in the buoyancy provided by the spermaceti as it compresses, cools and some of it freezes, is similar to the way wax esters behave in *Calanus acutus* (typically a few millimetres long), rendering it neutrally buoyant.

The hypothesis focuses on neutral buoyancy, which is theoretically possible if certain stringent conditions are met. In view of the variability in the composition of spermaceti (the degree of unsaturation and mixture of wax esters varies with age) and uncertainties about the animal's buoyancy, including the thermal equilibration of the spermaceti and residual gas in the lungs, it is easy to imagine that those conditions rarely prevail. Probably capturing squid is assisted merely by being able to passively hang in the water, rising or sinking slowly, when neutral buoyancy is only approximately achieved. Some aspects of the behaviour of Sperm whales seem at odds with the idea that near-neutral buoyancy is important in predation. Their entanglement with cables on the ocean floor is one example and the idea that Sperm whales normally hunt by direct attack and not by stealth is another. There is an annual gathering of Sperm whales off the Ogasawara Islands in the Pacific where they feed on giant squid at depths of 400–1000 m, so they are not particularly inaccessible. Following telemeter-tagged animals and monitoring the water in which they swim would appear to be essential to clarify their buoyancy and behaviour at depth.

## 10.5.2 Fish Buoyed by Oil and Lipids

For present purposes, there are three main classes of fish. The Class Elasmobranchii, also known as the Chondrichthyes, include sharks and smaller versions such as dogfish, and rays. They lack hard bones and are generally known as cartilaginous fish. They are also distinctive in lacking a swim bladder and their blood has a high concentration of urea. The class Actinopterygii comprises three important orders, Chondrostei, Holostei and Telostei. These are bony fish and usually have a swim bladder, but the bladder of some is filled with lipids rather than gas and in others, it has disappeared. The third class is the Choanichthyes, the lung fishes, which includes *Latimeria*, "the living fossil", of the Order Coalacanthinii.

### 10.5.2.1 Cartilaginous Fish

They have evolved various degrees of buoyancy using lipids, and in particular, squalene, stored in the enlarged liver (Corner et al. 1969; Bone and Roberts 1969; Van Vleet et al. 1984). Sharks and their smaller versions also generate uplift by "hydroplaning" on their pectoral fins, propelled by powerful tail and body movement.

Although not a deep water fish, the spiny dogfish, *Squalus acanthias*, is instructive. Its buoyancy is largely due to alkyldiacylglycerols and triacylglycerols which are stored in the liver. When weights are attached to the fish, the metabolism of the lipids changes. The alkyldiacylglycerol provides more buoyancy per gram than the latter, and its synthesis is increased in the weighted animals (Malins and Barone 1970). It would be possible to extend this type of experiment to deep-sea fish, using in situ benthic techniques.

Cartilaginous fish are significantly absent from very deep water. Only a few species occupy depths greater than 3000 m and none deeper than 4000 m (Priede et al. 2006; Musick and Cotton 2015). There is no reason to suppose this depth limitation is due to high hydrostatic pressure. It may, in part, be related to the energy cost of synthesising squalene and the need to keep up oil reserves in the liver. These fish have a limited capacity to store energy before buoyancy suffers, thus restricting the animals to depths where meals are frequently available. It is interesting that buoyancy based on oil is more energy demanding than the cost of using a swim bladder at depth, but that does not explain the limited depth distribution of Elasmobranchs.

Nevertheless, the lipid buoyancy system is well suited to the pressures experienced. The density of squalene (0.86) requires an animal to contain about 250 g of squalene per 1000 kg body weight to be neutrally buoyant, and as noted, its compressibility is similar to that of seawater.

Why squalene? It is an intermediary in the synthesis of cholesterol and therefore is part of the biochemical repertoire of most organisms. Its natural selection as a buoyant compound in the Chondrichthyes is partly understandable as it is light, metabolically inert and readily accommodated in large quantities in the liver, where it is synthesised. Generally, deep water sharks have large livers comprising up to 20–30% of the body weight. Up to 80% of the volume of the liver can be lipid (triglycerides and wax esters), with squalene a major constituent and very small quantities of pristane are also present. It used to be thought that cartilaginous fish could not achieve positive buoyancy but recently two species of deep water sharks have been tracked at depth and seem to "glide uphill for extended periods". The genera are *Hexanchus* and *Echinorhinus*. Some deep water sharks float when dead, for example *Centroscymnus coelepis* caught by long line from around 1500 m depths. The buoyancy analysis by Corner et al. (1969), now extended by Priede et al. (2020), showed that this animal and similar deep water sharks are close to neutral buoyancy in normal life. The general conclusion is that the quantity of liver oil, mainly squalene, see above, has to be regulated to within 2% to achieve this.

The presence of urea and other osmolytes in the plasma of Elasmobranchs (and Coelacanthini) has a positive buoyant effect (Table 10.2) but it is not clear that they are particularly significant in deep water species.

### 10.5.2.2  Bony Fish

Only one species of bony fish has a significant quantity of squalene in its liver (*Thaleichthyes pacificus*) but many contain various lipids in quantities sufficient to offset their tendency to sink. Furthermore, bony fish have evolved to live in greater depths than the Chondrichthyes and lipids clearly play an important role in these very deep living fish.

Regressed swim bladders contain wax between the layers of the bladder wall and they presumably provide some buoyancy and also serve as an energy store. Examples are *Lampanyctus*, the lantern fish, and *Latimeria chalumnae*. Many bony fish have lipids in their bones, but it is not clear if this is particularly common in deep water fish. Unlike squalene in cartilaginous fish, the lipids of bony fish are metabolically active and trying to distinguish buoyancy from nutrition is difficult, and perhaps pointless. Living in the twilight zone (1000 m), there are numerous species of bony fish which have no swim bladder, a high water content, not much muscle or dense skeleton and certainly no obvious buoyancy lipids (Blaxter et al. 1971). More recent studies have brought to light the presence of gelatinous tissue in deep-sea fish which appears to contribute buoyancy and to mechanically facilitate swimming (Gerringer et al. 2017). These features seem to have evolved as contributions to the eternal problem of balancing the need to obtain energy, using it effectively and reproducing.

# References[1]

Alexander RM (1966) Physical aspects of swim bladder function. Biol Rev 41:141–176

Barthelemy L, Belaud A, Chastel C (1981) A comparative study of oxygen toxicity in vertebrates. Resp Physiol 44:261–268

Berenbrink M, Koldjaer P, Kepp O, Cossins AC (2005) Evolution of oxygen secretion in fishes and the emergence of a complex physiological system. Science 307:1752–1757

Bidighare RR, Biggs DC (1980) The role of sulphate exclusion in buoyancy maintenance by Siphonophores and other oceanic gelatinous zooplankton. Comp Biochem Physiol A 66:467–474

Blaxter JHS, Wardle CS, Roberts BL (1971) Aspects of the circulatory physiology and muscle systems of deep sea fish. J Mar Biol Ass UK 51:991–1006

Bone Q, Roberts BL (1969) The density of elasmobranchs. J Mar Biol Ass UK 49:913–937

Boutilier RG et al (1996) Nautilus and the art of metabolic maintenance. Nature 382:534–536. *

Boyd CM, Gradmann D (2002) Impact of osmolytes on buoyancy of marine animals. Mar Biol 141:605–618

Brtttain T (2005) Root effect hemoglobins. J Inorgan Biochem 99:120–129. *

Carrier DR, Deban SM, Otterstrom J (2002) The face that sank the Essex: potential function of the spermaceti organ in aggression. J Exp Biol 205:1755–1763

Chu Z, Jin X, Yang B, Zeng Q (2007) Buoyancy regulation of Microcystis flos-aquae during phosphorus – limited growth. J Plankton Res 29:739–745. *

Clark KAJ, Brierley AS, Pond DW (2012) Composition of wax esters is linked to diapause behaviour of *Calanus finmarchicus* in a sea loch environment. Limnol Oceanogr 57:65–57. *

Clarke MR (1978a) Structure and proportions of the spermaceti organ in the sperm whale. J Mar Biol Ass UK 58:1–27

Clarke MR (1978b) Physical properties of spermaceti oil in the sperm whale. J Mar Boil Ass UK 58:19–26

Clarke MR (1978c) Buoyancy control as a function of the spermaceti organ in the sperm whale. J Mar Boil Ass UK 58:27–71

Clarke MR, Denton EJ, Gilpin-Brown JB (1979) On the use of ammonium for buoyancy in squid. J Mar Biol Assoc UK 59:259–276

Corner EDS, Denton EJ, Forster GR (1969) On the buoyancy of some deep sea sharks. Proc Roy Soc B 171:415–429

D'Aoust BG (1969) Hyperbaric oxygen: toxicity to fish at pressures present in their swim bladders. Science 163:576–578. *

D'Or R (2002) Telemetered cephalopod energetics: swimming, soaring and blimping. Intgr Comp Biol 42:1065–1070. *

Denton EJ (1971) Examples of the use of active transport of salts and water to give buoyancy in the sea. Phil Trans Roy Soc Lond B 262:277–287. *

Denton EJ (1974) On buoyancy and the lives of modern and fossil cephalopods. Proc R Soc Lond B 185:279–299. *

Denton EJ, Gilpin-Brown JR, Shaw TL (1969) A buoyancy mechanism found in cranchid squid. Proc Roy Soc Lond B 174:271–279

Diamond JM, Bossert WM (1967) Standing-gradient osmotic flow. J Gen Physiol 50:2061–2083. *

Dunstan AJ, Ward PD, Marshall NJ (2011) Vertical distribution and migration patterns of *Nautilus pompilius*. PLoS One 6:e16311

Enns T, Scholander PF, Bradstreet ED (1965) Effect of hydrostatic pressure on gases dissolved in water. J Phys Chem 69:389–391. *

Enns T, Douglas E, Scholander PF (1967) The role of the swim bladder rete of fish in secretion of inert gas and oxygen. Adv Biol Med Phys 11:231–244

---

[1] Those marked * are further reading.

Finn JK, Norman MD (2010) The argonaut shell: gas-mediated buoyancy control in a pelagic octopus. Proc Roy Soc B 277:2967–2973

Gerth WA, Hemmingsen EA (1982) Limits of gas secretion by the salting out effect in the fish swim bladder rete. J Comp Physiol 146:129–136

Greenwald L, Ward PD (2010) Buoyancy in Nautilus. In: Saunders WB, Landman NH (eds) Nautilus. Springer

Holbrook RI, de Perera TB (2011) Fish navigation in the vertical dimension: can fish use hydrostatic pressure to determine depth? Fish Fish 12:170–179. *

Jordan M, Chamberlain JA, Chamberlain RB (1988) Response of nautilus to variation in ambient pressure. J Exp Biol 137:175–189. *

Kanie Y et al (1980) Implosion of living Nautilus under increased pressure. Paleobiology 6:44–47

Kobayashi H, Pelster B, Scheid P (1989) $CO_2$ back-diffusion in the rete aids $O_2$ secretion in the swimbladder of the eel. Respirat Physiol 79:231–242

Korsoen OJ et al (2010) Behavioural responses to pressure changes in cultured Atlantic cod (Gadus morhua): defining practical limits for submerging and liftingsea cages. Aquaculture 308:105–115

Kubodera T, Mori K (2005) First-ever observations of a live giant squid in the wild. Proc Roy Soc B 272:2563–2585

Kuhn W, Ramel A, Kuhn HJ, Marti E (1963) The filling mechanism of the swim bladder. Experientia 19:497–552

Lapennas GN, Schmidt-Nielsen K (1977) Swimbladder permeability to oxygen. J Exp Biol 67:175–196

Lewis RW (1970) The densities of three classes of marine lipids in relation to their possible role as buoyancy agents. Lipids 5:151–153. *

Libes SM (1992) Introduction to biogeochemistry. Wiley

Ludwig H, Macdonald AG (2005) The significance of the activity of dissolved oxygen, and other gases, enhanced by high hydrostatic pressure. Comp Biochem Physiol 140:387–395

Maitra A, eight co-authors (2017) Fast charging self powered….piezoelectric separator. Nano Energy 40:633–645

Malins DC, Barone A (1970) Glyceryl ether metabolism: regulation of buoyancy in dogfish *Squalus acanthias*. Science 167:79–80

Musick JA, Cotton CF (2015) Bathymetric limits of chondrichthyans in the deep sea: a re-evaluation. Deep Sea Res. II 115:73–80

Nakamura I, Meyer CG, Sato K (2015) Unexpected positive buoyancy in deep sea sharks, Hexanchus griseus, and Echinorhinus cookie. PLoS One 10:e0127667. *

Nielsen JG, Munk O (1964) A hadal fish (Bassogigas profundissimus) with a functional swimbladder. Nature 204:594–595

Noble RW, Russell RP, Riggs A (1975) Studies of the functional properties of the haemoglobin from the benthic fish Antimora rrostrata. Comp Biochem Physiol 52B:75–81. *

Okonjo KO (2018) Bohr effect and oxygen affinity of carp, eel and human haemoglobin: quantitative analyses provide rationale for the root effect. Biophys Chem 242:45–59. *

Packard A, Wurtz M (1994) An octopus, Ocythoe, with a swim bladder and triple jets. Phil Trans R Soc B 344:261–275

Pelster B (1997) Buoyancy at depth. Chapter 5, Deep Sea Fishes. (Randall DJ, Farrell AP (eds) Academic Press

Pelster B (2001) The generation of hyperbaric oxygen tensions in fish. News Physiol Sci 16:287–291

Pelster B (2015) Swimbladder function and the spawning migration of the European eel *Anguilla Anguilla*. Front Physiol 5:art 486. *

Phleger CF (1971) Pressure effects on cholesterol and lipid synthesis by the swimbladder of an abyssal Coryphaenoides species. Amer. Zoologist 11:559–570

Phleger CF (1998) Buoyancy in marine fishes: direct and indirect role of lipids. Amer Zoologist 38:321–330. *

Pond DW (2012) The physical properties of lipids and their role in controlling the distribution of zooplankton in the oceans. J Plankton Res 34:443–453. *

Pond DW, Tarling GA (2011) Phase transition of wax esters adjust the buoyancy in diapausing Calanus acutus. Limnol Oceanogr 56:1310–1318. *

Pond DW, Tarling GA, Ward P, Mayor DJ (2012) Wax ester composition influences the diapause patterns in the copepod Calanus acutus. Deep Sea Res II 59–60:93–104

Prem C, Salvenmoser W, Wurtz J, Pelster B (2000) Swim bladder gas gland cells produce surfactant in vivo and in culture. Am J Physiol Regulatory, Integrative Comp Physiol 279:R2336–R2343

Priede IG (2018) Buoyancy of gas filled bladders at great depth. Deep Sea Res I 132:1–5. *

Priede IG et al (2006) The absence of sharks from the abyssal regions of the world's oceans. Proc R Soc B 273:1435–1441

Priede IG et al (2020) Near-equal compressibility of liver oil and sea water minimises buoyancy changes in deep sea sharks and chimaeras. J Exp Biol 223:222943

Reeves RB, Morin RA (1986) Pressure increases oxygen affinity of whole blood and erythrocyte suspensions. J Appl Physiol 61:486–494. *

Ross LG (1979) The permeability to oxygen and the guanine content of the swimbladder of a physoclist fish Pollachius virens. J Mar Biol Assoc 59:437–441

Rummer JL, Brauner CJ (2015) Root effect haemoglobins in fish may greatly enhance general oxygen delivery relative to other vertebrates. PLoS One 10:e0139477

Sanders NK, Childress JJ (1988) Ion replacement as a buoyancy mechanism in a pelagic deep sea crustacean. J Exp Biol 138:333–343

Sartoris FJ, Thomas DN, Cornils A, Schnack-Schiel SB (2010) Buoyancy and diapause in Antarctic copepods: the role of ammonium accumulation. Limnol Oceanogr 55:1860–1864

Saunders WB (1984) The role and status of nautilus in its natural habitat: evidence from deep water remote camera photosequences. Paleobiology 10:469–486. *

Saunders WB, Wehman DA (1977) Shell strength of nautilus as a depth limiting factor. Paleobiology 3:83–89

Scholander PF, Van Dam L (1954) Secretion of gases against high pressures in the swim bladder of deep sea fishes. Biol Bull 107:247–259. *

Scolander PF (1954) Secretion of gases against high pressures in the swim bladder of deep sea fishes. II the rete mirabile. Biol Bull 107:260–277. *

Scolander PF, Van Dam L, Enns T (1956) Nitrogen secretion in the swim bladder of whitefish. Science 123:59–60. *

Sherrad M (2000) Cuttlebpne morphology limits habitat depth in eleven species of sepia (Cephalopda: Sepiidae). Biol Bull 198:404–414

Sund T (1977) A mathematical model for the counter current multiplication in the swim bladder. J Physiol 267:679–696

Talley LD, Pickard GL, Emery WJ, Swift JH (2011) Descriptive physical oceanography. An introduction, 6th edn. Elsevier. *

Van Vleet ES et al (1984) Neutral lipid components of eleven species of Caribbean shark. Comp Biochem Physiol 79B:549–554

Vecchione M et al (2003) World wide observations of remarkable deep sea squids. Science, 294:2505–2506

Visser AW, Jonasdottir SH (1999) Lipids, buoyancy and the seasonal vertical migration of Calanus finmarchicus. Fish Oceanogr 8(suppl 1):100–106. *

Ward PD (1987) The natural history of nautilus. Allen and Unwin. *

Ward PD, Bioletzky SV (1984) Shell implosion depth and implosion morphologies in three species of sepia (Cephalopoda) from the Mediterraneansea. J Mar Biol Ass UK 64:955–966. *

Ward P, Greenwald L, Greenwald OE (1980) The buoyancy of the chambered nautilus. Sci American 243:190–203. *

Webber DM, Aitken JP, O'Dor RK (2000) Costs of locomotion and vertic dynamics of cephalopods and fish. Physiol Biochem Zool Ecologic and Evolut Approaches 76:651–662

Winkelmann I et al (2013) Mitochondrial genome diversity and population structure of the giant squid Architeuthis: genetics sheds new light on one of the most enigmatic marine species. Proc Roy Soc B 280:0273

Withers PC, Morrison G, Guppy M (1994) Buoyancy role of urea and TMAO in an elasmobranch fish, the port Jackson shark, *Heterodontus portusjacksoni*. Physiol Zoology 67:693–705. *

Wittenberg JB, Copeland DE, Haedrich RL, Child JS (1980) The swimbladder of deep sea fish: the swimbladder wall is a lipid rich barrier to oxygen diffusion. J Mar Boil Ass 60:263–276

Yancey PH, Gerringer ME, Drazen JC, Rowden AA, Jamieson A (2014) Marine fish may be biochemically constrained from inhabiting the deepest ocean depths. Proc Natl Acad Sci U S A 111:4461–4465

Yayanos YY, Benson AA, Nevenzel JC (1978) The pressure-volume-temperature (PVT) properties of a lipid mixture from a marine copepod, *Calanus plumchrus*: implications for buoyancy and sound scattering. Deep Sea Res 25:257–268. *

# Divers: Air-Breathing Animals, Including Humans, at High Pressure    11

A wide variety of air-breathing animals temporarily suspend their normal breathing, and go under water, experiencing the hydrostatic pressure arising from depths of a few centimetres to several kilometres. Generally, zoologists apply the term diver to vertebrate specialists such as seals and penguins, so, for the sake of completeness, it is necessary to mention those to which the term, as a convention, does not apply.

We do not regard as divers those insects and spiders which temporarily go under water. Normally they breathe air, which penetrates their bodies through fine tubes (tracheae) connecting the air outside the body to the cells deep within. Oxygen is thus made available to the cells and carbon dioxide escapes. An example is the aquatic spider, *Argyroneta aquatica*, which, having spun a web underwater, fills it with air and uses it as a physical gill (Seymour and Hetz 2011; Seymour and Matthews 2013). Many insects have evolved a means of using their tracheae under water by means of a film of air trapped on the surface of their exoskeleton and in contact with tracheae. The most fully developed form is called a plastron and comprises microscopic hairs or spikes projecting from the hydrophobic cuticle. This structure supports trapped air and is sustained more or less permanently (Hinton 1976; Jones et al. 2018). The insect's metabolism consumes oxygen and releases carbon dioxide, the latter dissolving in the water in contact with the plastron gas film. Oxygen diffuses in the reverse direction, from the ambient water. The bulk of the air in the plastron is nitrogen which is in equilibrium with that dissolved in water. Whilst the plastron can withstand any very small pressure difference which arises between the gas pressure in the film and the ambient water, experimentally applied hydrostatic pressures of 0.2–0.3 MPa compresses plastrons and causes their gas to slowly dissolve. Plastrons are intriguing devices which have interested many respiratory physiologists, notably Rahn and Paganelli, who have also been active in human diving research. The hydrophobic properties of plastrons have also interested materials scientists (Parra and Perez-Gill 2015).

A group of insects which have evolved into an aquatic environment which poses problems for their tracheal respiration comprises parasites living on or in diving vertebrates. The spectacular case is a louse, *Antarctophthirus microchir*, which lives

© Springer Nature Switzerland AG 2021
A. Macdonald, *Life at High Pressure*, https://doi.org/10.1007/978-3-030-67587-5_11

on the skin of the Sea Lion *Otaria flavescens*. Leonardi and Lazzari (2014) have described how this animal dives deep, so its parasitic louse experiences a high hydrostatic pressure, rare for an insect. Tests show that *A. microchir* survives immersion for 7 days but its pressure tolerance has not been examined. Significantly it seems to adopt a quiescent state on immersion, so perhaps it is able to acquire sufficient oxygen by diffusion through its tissues. Another louse, *Lepidophthirus macrorhini*, which infects Southern Elephant seals, survives its host's dive pattern of 15 MPa for 2 h.

A second group of animals, which are regarded as divers but are not dealt with here, include amphibia, reptiles and shallow diving birds. Only one species, the Leatherback turtle (*Dermochelys coriacea*) is a deep-diving specialist, capable of reaching a depth of more than 1200 m, comparable to the special birds and mammals which are discussed below (Fossette et al. 2010). However, the shallow dives of reptiles and birds are certainly not without interest. The latter, for example, have a diving reflex (see below) which has been much studied and certain sea snakes can obtain oxygen from the water through their skin.

So here we are primarily concerned with highly evolved specialist birds and mammals which dive to extraordinary depths and experience a significant increase and then a decrease in hydrostatic pressure. They are not deep-sea animals in the usual sense but they experience significant hydrostatic pressure, made all the more interesting by its rapid increase and decrease. They all dive whilst holding their breath, and human breath-hold divers are included here. Mention has been made (Chapter 1; Section 1.2.1) of whales tangling with cables on the seafloor at depths of 1000 m (Heezen 1957), but now modern research has provided many examples of the range of deep-diving mammals. An extreme case is a recorded dive of the beaked whale, *Ziphius cavirostris*, during which a depth of 2992 m was reached during a dive lasting 137.5 min. A great many physiological and biochemical questions arise from this sort of data. Many involve adaptations to the familiar homeostatic mechanisms common to vertebrates, such as control of the cardiovascular system and cellular respiration, and hydrostatic pressure effects have only recently been recognised and are the main concern here.

This chapter also deals with human divers who, because they are equipped with a gas supply, dive free of the peculiar stress of breath-holding, and experience significant hydrostatic pressures for longer periods than animal divers.

## 11.1  Breath-Hold Diving

### 11.1.1  Humans

To start thinking about hydrostatic pressure effects in these divers we begin with the traditional human diver, who simply holds his or her breath, when diving for food or pearls, as in Western Australia, Korea and Japan. Such activity has taken place for over 2000 years and has changed with technological developments, starting with the use of air pumps and diving helmets by the Australian pearl divers in the late nineteenth century (Lindholm and Lundgren 2009). In Japan and Korea, the mostly female divers, known as "Ama" and "Haenyeo", respectively, have retained a more traditional working practice. Competitive breath-hold diving, called Freediving, surely one of the most exotic of sports, is organised by the International Association for Developing Freediving. Athletes train for competitions in several types of dive and there is a growing body of knowledge, as in many modern sports, which provides a rational basis for training (Ostrowoski et al. 2012). In recent times the world depth record for a free breath-hold dive was more than 200 m. Breath-holding elicits a number of homeostatic adjustments in the diver but does hydrostatic pressure have specific effects? It probably does, by the now-familiar effect on chemical reactions in solution, and by acting mechanically by compressing the air held in the lungs and elsewhere. This latter aspect will be tackled first, and although some of the points have been mentioned in Chapter 4, they are mentioned here for convenience.

The volume of a gas is halved when its pressure is doubled. This means the volume of the gas in a diver's lungs is greatly affected by the ambient pressure. The gas of main interest is air which, as before, we will assume consists of 80% nitrogen and 20% oxygen. Thus, air at normal atmospheric pressure (100 kPa) contains nitrogen generating a partial pressure 80 kPa and oxygen 20 kPa. As we have previously seen, the amount of nitrogen and oxygen which dissolves in water is proportional to their partial pressures multiplied by their solubility in water. Oxygen is about twice as soluble as nitrogen in water, so when exerting the same partial pressure, more oxygen dissolves in water than nitrogen. In the case of air, the greater partial pressure of nitrogen offsets its lesser solubility, so more nitrogen than oxygen gas dissolves.

So far in this book pressure has been expressed as pressure in excess of atmospheric pressure (Chapter 1; Box 1.1). Thus, in the above example of doubling the pressure and halving the volume of lung gas, the final pressure can be expressed as 100 kPa above normal. This is also known as gauge pressure because historically, at atmospheric pressure gauges read zero. Or the pressure can be expressed as 200 kPa absolute or total pressure (100 kPa comes from the depth of water and 100 kPa from normal atmospheric pressure).

Lungs are large and quite delicate organs with distinct mechanical properties. A breath-hold dive to 10 m depth will subject the gas in the lungs to twice their normal pressure and, as we have seen, will therefore halve the volume of the soft spongy lung tissue will halve. The human rib cage is not very compressible so other soft tissue, mostly blood vessels and their contents, shift to occupy the space created by the compression of gas. In extreme breath-hold dives to depths of 50 m or more, the volume shift can be about a litre, which is drastic. It exposes the diver to the risk of tissue damage. Specialist diving animals have evolved anatomical features which avoid this problem, typically having compressible rib cages. Humans have other gas-filled cavities such as sinuses and around the middle ear, both of which can be a source of damage and pain if soft tissue cannot take up the extra space vacated by gas compression.

Compression of air increases the partial pressure of the constituent nitrogen ($PN_2$) and oxygen ($PO_2$). Oxygen diffuses from the minute air sacs in the lung (alveoli) into the blood flowing through the blood capillaries lining the alveoli. Conversely, carbon dioxide diffuses in the other direction. In steady-state breathing, the $PO_2$ and $PCO_2$ in the alveoli are kept constant by fresh air entering the lungs during rhythmic breathing. During a breath-hold at the surface, diffusion continues so the alveolar $PO_2$ declines and the $PCO_2$ increases. In a breath-hold at a depth of 10 m, carbon dioxide in the blood continues to diffuse onto the alveoli. Its partial pressure in the blood is the result of cellular metabolism and as time passes it increases during a breath-hold. It affects the brain, intensifying the need to resume breathing. Our familiar breathing rhythm is automatically generated by the brain and the act of breath-holding creates a novel conflict between this automatic process and the will power to oppose it. Experiments show the build-up in the $PCO_2$ in the blood correlates closely with the overwhelming need to resume breathing, the breakpoint. Our normal steady-state rhythmic breathing seems designed to maintain a low $PCO_2$ in the brain. If the $PCO_2$ increases, then breathing speeds up. In contrast, oxygen provides no such graded control over our breathing. If the $PO_2$ in the brain is very low we may start to breathe faster but soon become unconscious.

Now consider the breath-hold dive as it approaches its conclusion. The alveolar $PCO_2$ has steadily increased and now makes the diver feel the urgent need to resume breathing. So, as the diver ascends, the ambient pressure decreases, so the $PCO_2$ in the alveolar gas decreases, but in the brain, it continues to drive the need to breathe. At the same time the $PO_2$ in the alveoli, which has decreased during the course of the dive as oxygen continued to dissolve in the blood, abruptly decreases even further as the diver ascends and the pressure decreases. That low $PO_2$ soon affects the brain. In a safe breath-hold dive the diver surfaces with a strong need to breathe and with a low alveolar $PO_2$, but one which is above the level which causes loss of consciousness.

A trick to increase the duration of a breath-hold dive is to start it with a depleted $PCO_2$ in the lungs. This is easily produced by vigorously "over-breathing" or hyperventilating the lungs. The flux of carbon dioxide out of the body and into the lung gas continues normally but the lung gas is ventilated with fresh air much more, so the normal steady-state $PCO_2$ is reduced at the start of the breath-hold. This delays the onset of the increase in $PCO_2$ which makes the diver want to resume breathing, creating a dangerous situation. The $PO_2$ in the lung can be so reduced during the prolonged breath-hold that during the ascent it drops sufficiently to cause the diver to lose consciousness. Invariably that is fatal. Traditional breath-hold divers use moderate hyperventilation prior to a dive, and certainly not vigorous hyperventilation.

In the laboratory, I have seen medical students hold their breath for amazingly long times after prior hyperventilation. Involuntary contractions of the diaphragm are seen during the latter stages of the breath-hold, called the struggle phase, for good reason (Batinic et al. 2016). Healthy people cannot harm themselves holding their breath in the air, at least in the short term. Doing it often does introduce some physical stress which, as in any sport, can be regarded as harmful.

Physical activity during a breath-hold dive accelerates the reduction of the $PO_2$ in the lungs and increases the flux of carbon dioxide out of the body and into the lungs. This largely explains how it is that the world record breath-hold, carried out resting with only the face immersed in water, is 10.12 min whereas the equivalent record for a deep breath-hold dive to a depth of 214 m (using weights to descend and a flotation bag to ascend) lasted 4.24 min. These data will soon need revision!

What happens to the nitrogen, which comprises the bulk of the gas in the lungs? Its increased partial pressure enhances its solution in blood and on surfacing (decompression) the nitrogen has to come out of solution. In the human breath-hold diver it was once thought that the dissolved nitrogen was innocuous but we now know that it is not. Symptoms related to decompression sickness (the damage done when nitrogen comes out of solution as bubbles, see below) are now recognised. In a recent study of Japanese Ama, mature ladies with extensive diving experience, Kohshi et al. (2014), found a significant number had clearly experienced decompression symptoms and some had sub-symptomatic brain damage.

Although it is not obvious to the person undertaking a breath-hold dive, a set of reflexes play an important role in the dive. They are jointly referred to as the diving reflex and it is present in all air-breathing vertebrates and particularly clear in mammals (Panneton 2013). In birds, it is present in a slightly modified form. The cessation of breathing is accompanied by a decrease in heart rate and the blood circulation is modified, favouring the heart muscle and brain over the peripheral organs and muscles. The diving reflex is a defence against hypoxia, the lack of oxygen (Lee et al. 2016). When breathing stops, the body's aerobic metabolism is limited by the store of oxygen available. When the oxygen is depleted, anaerobic metabolism proceeds, lactic acid accumulates and so do oxygen free radicals, the latter being harmful (see below). The brain of mammals is dependent on the aerobic production of ATP to sustain the electrical gradients across the cell membranes. If they run down, the brain self-destructs. Understandably these changes have attracted a great deal of research, not just confined to the specialist divers but including other animals such as hibernators which also run the risk of hypoxia or anoxia. Cardiovascular conditions in humans can lead to hypoxia in the brain (a stroke) and irreversible brain damage. However, that said, we now have to leave the subject as hydrostatic pressure has not yet been shown to play a significant role in the hypoxia defence reactions of divers.

## 11.1.2 Specialised Deep-Diving Birds and Mammals

Impressive as the traditional and competitive human breath-hold divers are, the performance of the specialist diving birds and mammals, like the whale mentioned earlier, is in a different league.

The deepest diving bird is the Emperor penguin (*Aptenodytes forsteri*), Emperor of the Antarctic so to speak, with recorded dives lasting over 26 min and shorter, deep dives to over 560 m depths (Castellini 2010). The King penguin achieves depths of over 300 m and many sea birds dive to many tens of metres, such as the

Sooty shearwater (*Puffinus griseus*) and the Blue-eyed shag (*Phalacrocorax atriceps*).

Penguins have evolved with an interesting liability, although like other diving birds, they lack air spaces in their bones. Their "problem" is their rib cage is not compressible and so the orderly collapse of their lungs is not an option. In the case of the King penguin, its alveolar blood vessels become engorged, thus filling the space created by lung compression. Its gaseous store of oxygen is small, its tissue stores are large and alveolar gas exchange is minimised by the reduced blood flow in the lungs following engorgement of the alveolar blood vessels. Penguins are strong swimmers, using their wings vigorously at the start of a dive, apparently to reduce the buoyancy of the air in their lungs and feathers. That buoyancy rapidly diminishes as pressure reduces the volume of gas but vigorous exercise persists, as it must in all breath-hold dives. The so-called aerobic limit is often exceeded and the accumulation of lactic acid signals the need to surface soon.

The diving mammals are superior performers, and their phylogenetic classification is important to know.

Within the Class Mammalia specialist divers have independently evolved in two Orders, the Carnivores and the Whales. Within the latter, the Suborder Odontoceti—toothed whales, comprise three families: (1) The Physeteroidea includes the Sperm whale *Physeter* and (2) the Delphinoidea includes the narwhal (*Monodon*), dolphins (*Dolphinus*), porpoises (*Phocaena*), the Killer whale (*Orcinus*) and the Bottlenose whale, (*Hyperodon*). (3) For completion the third family, the Platanistoidea, includes river dolphins, which are not deep divers.

The Whale Suborder Mystceti includes the whalebone whales such as the giant Blue whale (Balaenoptera) and the Right whale (*Balaena*), once hunted for its oil. The big whales can dive to depths of several hundreds of metres, such as the Sperm whale (1800 m), the Bottlenose whale (*Hyperodon ampullatus*, 1400 m) and *Mesoplodon densirostris*, a beaked whale different to the one mentioned earlier, 800 m.

In the Order Carnivora, the Suborder Pinnipedia contains the deep divers, in two Families; Phocidae (seals) and Otariidae (eared seals, sea lions). Other Suborders are the Canoides (otters) and Odobenidae (walruses). Many of the Pinnipedia are born on dry land or beaches, and as juveniles have a terrestrial upbringing. In contrast, whales are born in water where they suckle their young.

The champion deep diver is a beaked whale, *Ziphius cavirostris*, whose record dive to a depth of 2992 m (i.e. to a pressure of 30 MPa) was mentioned earlier (Schorr et al. 2014; Tyack et al. 2015; Lopez et al. 2015). It surfaced after a dive time of 137 min, then the deepest and longest breath-hold dive ever recorded. It has been reported in the Times (September 2020) that a dive of 222 min has been recorded. The animal was wearing a recording "tag" which transmitted daily to a satellite link and by this means a great deal of information was gathered over many days. The record depth was just one of many dives to 1500 m or more. It is not clear if the deepest dives were a response to Naval sonar or a normal part of their behaviour (Southall et al. 2019). The deep dive profiles appear symmetrical suggesting a typical rate of descent (compression) of approximately 0.5 MPa per minute. Two champions of the Pinnipedia are the Northern Elephant seal (*Mirounga angustirostris*) and the Southern Elephant seal (*Mirounga leonine*) which achieve depths of 1400 m or more. (The latter has been used in Antarctic waters, fitted with sensors to measure salinity and temperature to a depth of 300 m. Each time the seal surfaced the sensors transmitted data to a satellite, Williams et al. 2016.) The California sea lion (*Zalophus californianus*) routinely dives to 1–200 m depths over 1–5 min and has been recorded at 420 m. The Weddell seal (*Leptonychotes weddellii*) can descend deeper and breath-hold for over an hour. It is a well-studied diver which lives on Antarctic ice, where one of the early leading workers in the field, Kooyman, of the Scripps Institution of Oceanography, California, spent a lot of time. One of his predecessors, Scholander, is regarded as a founding father of diving physiology through a paper published in 1940.

Emphasis is given here to the maximum depths achieved by these animals because of the high pressure they experience and to which they have become adapted, evolving from different ancestries. In the wild, they are primarily intent on feeding and their dive regimes generally diverge into long-duration exploratory dives or short deep dives.

## 11.2   Deep-Diving Lungs Under Pressure

As with human breath-hold dives, hydrostatic pressure acts mechanically on gas phases in these animal divers and, at sufficient depth by the familiar molecular mechanisms, on their cell physiology and biochemistry.

A factor important to all these deep divers is the store of air, oxygen and nitrogen, possessed by the animal. The greater the store of oxygen the longer the breath-hold time can be. Generally, large animals have a slower oxygen consumption and a large oxygen store and so can dive for longer than small animals. Oxygen is stored in combination with haemoglobin in the blood (a significant store of which is in the spleen which releases it when required), in combination with myoglobin in muscle and as air in the lungs. The air in the latter comprises alveolar gas which can diffuse into the blood circulation and gas in the "dead" space of the main airways, the trachea and bronchi. Diving mammals generally have collapsible rib cages and rigid airways, so that on compression the alveoli collapse and the gas moves into the rigid airways. In some animals, the volume occupied by alveolar and bronchial gas is taken up by engorgement of parabronchial blood vessels. The airways, in turn, enable compressed gas to re-expand during ascent without the risk of becoming trapped and rupturing soft tissue. Small Pinnipeds, which are practical experimental animals, unlike large whales, have been intensively studied and they show that air in the alveoli effectively disappears due to compression. The diffusion of gas between the alveoli and the blood perfusing the lungs stops when the alveoli collapse. This is judged to be advantageous as it reduces the quantity of nitrogen which dissolves in the blood during the dive and hence the amount which comes out of solution when the animal surfaces, i.e. is decompressed. Re-opening the alveoli during the ascent is an important and interesting process.

Pinnipeds vary in the emphasis placed on the store of oxygen in the air at the start of the dive. Some, the Elephant and the Weddell seals, for example, dive after exhaling, whilst others, the Fur seals, for example, inhale prior to a dive. The air store also affects the buoyancy of the animals, so it is a complicated problem to understand the advantages and disadvantages of the detailed techniques these animals adopt. A common feature of divers is their rapid breathing after a dive, clearly a way of restoring their depleted oxygen stores and releasing the accumulated carbon dioxide. Rapid breathing is itself problematic. When we, or a seal, breathes in, we do so by expanding the volume of the chest cavity in which the lungs are suspended. This causes the pressure in the cavity to drop below the pressure in the environment, so air moves through the mouth and large airways, into the lungs which expand. In quiet breathing, exhalation is largely due to the elastic recoil of the lungs. Rapid breathing is limited by the resistance to gas flow in the airways, especially when exhaling, as airway diameters can be squeezed (airway collapse).

The extent to which compressed gas diffuses into the blood prior to alveolar collapse depends on a number of variables and estimates of the pressure at which it ceases (i.e. the alveoli collapse) have been made for several species. The alveolar collapse pressure in the Weddell seal is somewhat less than 0.4 MPa (about 40 m depth). This has been directly monitored in simulated dives by Kooyman et al. (1970), using a pressure vessel equipped with X-ray equipment. The collapse pressure in the dolphin is around 0.8 MPa, and in the Harbour seal it is greater than 1.4 MPa. In the Weddell seal undergoing a simulated dive, the trachea was seen to decrease in volume at pressures of around 0.6 MPa but were not fully collapsed at 3 MPa. The volume of the bronchi and bronchioles were unaltered even at 3 MPa, so the gas within is a highly compressed reservoir. As pressure reduces during the ascent, the collapsed alveoli open. Other workers have used a pressure vessel made of fibreglass in conjunction with X-ray computer tomography to measure the airway collapse in dead seals whose lung volume could be inflated prior to the "dive". They found, for example, that the lungs of a Gray seal 50% inflated at the start of the dive were collapsed at a depth of 84 m. Generally, the results from these direct measurements support the idea that many diving mammals are at risk of decompression problems because sufficient nitrogen dissolves in their blood before all the alveoli collapse. This is consistent with the autopsies carried out on whales and seals which have become beached or trapped in fishing nets which often reveal evidence of gas bubbles in their tissues (Section 11.2.1). Some of these fatal episodes may be the result of animals becoming disorientated by military sonar during a dive (Talpalar and Grossman 2005).

During the ascent and conclusion of a normal dive the trapped gas expands. To understand the reinflation of the collapsed alveoli by the expanding gas we have to go back to the basic anatomy of the alveoli. The best-understood mammalian lungs are, of course, human lungs and their structure is similar to that of diving mammals (The lungs of birds are rather different and are not considered here.) Alveoli are approximately spherical and 300 µ in diameter. Their wall is made of a cell wrapped around a void, and a small number of a different cell, the Type Two cell, is found scattered in the wall. Closely associated with the alveolar wall are capillaries whose diameter is just sufficient to allow blood cells to circulate. Alveoli are clustered in their millions such that the total surface area for gas exchange in a typical human adult is 100–75 square metres.

Alveoli behave like bubbles and there is some interesting physics in bubbles. The spherical wall is lined with a liquid whose molecules are strongly attracted to each other. This creates a surface tension and the tendency for this tension to squeeze and collapse the bubble is inversely proportional to its radius. This is the Law of Laplace. The smaller the radius the greater the collapsing pressure. Breathing in (inspiration) requires the lungs to expand which is opposed by the surface tension, along with elastic fibres in the wall of the alveoli. We will soon return to the surfacing diving mammal but at present it is instructive to think about a newborn human baby taking its first breath, starting with collapsed alveoli. Considerable energy is required to overcome the surface tension to open the alveoli. In the last few weeks of pregnancy, the Type II cells in the alveoli secret chemicals which dissolve in the liquid lining the alveoli and reduce the intermolecular forces which create the surface tension. Healthy babies manage their first breath well enough, but premature babies may lack sufficient tension-reducing chemicals and manifest symptoms, the Newborn respiratory distress syndrome. Their breathing can be assisted by applying positive air pressure or by delivering the appropriate chemicals into the lungs through the airways. This is why respiratory physiologists are so impressed with the ability of diving mammals to routinely inflate their collapsed alveoli after repetitive dives. In the human lung, the surfactant in the alveoli is a mixture of chemicals whose concentration in the lining fluid is such that when the alveoli are small, potentially creating a high surface tension, its concentration in the reduced area which the liquid occupies, produces the required low surface tension. In normal breathing, the volume of the alveoli changes by +/− 10% so their surface area changes by rather less. As the alveoli expand, the potential for a reduced surface tension is compensated by the reduction in the concentration of surfactant. Thus, both the lung's elastic recoil and its resistance to inflation are reduced. Can this system, which has evolved for continuous breathing in terrestrial animals, work well in the discontinuous breathing of diving mammals? And do those animal divers which descend routinely to pressures sufficient to collapse their alveoli, require special surfactant? As we have seen, the collapse of alveoli confers the advantage of reducing the amount of gas which enters the bloodstream, so there is selective pressure in having efficient lung surfactants.

To understand the problem faced by these deeper divers we need more detailed information on the composition of the surfactant and any possible effects of high pressure on it. In a typical, i.e. non-diving, mammal, the surfactant consists mainly of lipids. They are: (1) 40% dipalmitoylphosphatidylcholine (DPPC), a saturated phospholipid which we have met before in cell membranes, (2) 25% unsaturated phosphatidylcholine, (3) phospholipids with a negatively charged headgroup such as phosphatidylglycerol (up to 15%), and (4) neutral lipids, mostly cholesterol (up to 13%). The rest of the surfactant is a mixture of four proteins, the major one being Surfactant Protein (SP) A. The others are SP-B, C and D. The surfactant can be collected by washing out the lungs, technically known as pulmonary lavage (which featured in Chapter 4; Section 4.3), and detailed analyses show that the composition of surfactant is characteristic of a species, and in humans, it can reveal pathophysiological states.

The surfactant is secreted by the Type II cells as compacted membranes (Lamellar bodies) which, when they come into contact with the liquid–air interface in the alveoli, spontaneously spread to form patches and then a continuous monolayer. When the area of the alveolar liquid decreases, the monolayer of surfactant kinks and collapses to form vesicles and bilayers below the surface. They are a reservoir, cyclically resupplying the monolayer as the area of the liquid–air interface expands (Duncan and Larson 2010). A variety of techniques have been used to study this surfactant behaviour, including fluorescence microscopy and ultra-delicate physical measurements, leading to the idea that the surfactant layer is a composite of patches of condensed lipids and more fluid phases. DPPC alone is effective at lowering the surface tension. The other components of the surfactant mix are thought to have specific functions. For example, the protein SP A does two things. It binds to bacteria, viruses and allergens in the first stage of protecting the lung from infection. It also enhances the adsorption of the surfactant mix onto the liquid surface. Protein SP B speeds the spreading of the lamellar bodies onto the liquid surface and if it is absent in humans a serious pulmonary condition arises. SP C assists in the spreading and, interestingly, is composed of only 35 amino acids linearly arranged to provide a hydrophobic domain which is buried in the bilayer formed by lipid and a more polar domain at the N-terminal end of the protein which lies in the aqueous phase. SP C is encoded by the gene sp-c and is chemically modified after initial assembly (Foot et al. 2007; Hobi et al. 2016). The structure and function of SP D are unknown.

A group at the University of Adelaide has studied the proteins from both divers (Blue whale, Southern Elephant seal, Californian Sea lion and other shallow divers) and from non-divers to see if differences in their amino acid sequence can be related to the particular function the protein performs. Generally, among mammals, the sequence in the hydrophobic transmembrane domain is highly conserved, which is understandable as it is the part of the protein which anchors it securely in the lipid layer. The N terminal domain, however, displays the most amino acid substitutions, suggesting that natural selection is at work in fine-tuning its function. In Whales and Pinnipeds, different substitutions achieve the common feature of increasing polarity which enhances the binding of the protein to the anionic head groups of the unsaturated phospholipids. This is thought to be advantageous for deep divers as it would minimise the extent to which they are squeezed out of the surface layer as the alveoli shrink and then collapse. The epithelial walls of collapsed alveoli tend to adhere to each other, and anionic head groups provide an "anti-glue" which facilitates separation and opening. There is a trade-off between reducing surface tension and reducing the adhesion of collapsed alveoli. Deep divers benefit from a certain balance between these two whilst shallow divers have no problem of adhesion to overcome. Indeed Pinnipeds, in general, have a surfactant which does not reduce surface tension as much as that of terrestrial mammals.

Remember, Pinnipeds, unlike Whales, give birth out of the water and their young grow from a terrestrial lifestyle to become proficient divers. Is this reflected in their lung surfactant? According to the Adelaide group, the answer is yes. The surfactant of the adult Californian sea lion (*Zalophus californianus*) has greater anti-adhesive properties than that of the young, having more anionic lipids and SP B protein. The surfactant of the young is typical of terrestrial mammals.

What about the direct effects of pressure on the surfactant? We can distinguish its synthesis and secretion from its surface properties. The Adelaide group (Miller et al. 2004) has measured the effect of pressure on the secretion of surfactant by Type II cells cultured after isolation from the lung of the Californian sea lion. The cells were pressurised in air for 30 or 120 min, simulating a dive. The "depths" of the dives were 5, 2.5, and 0.5 MPa absolute. Thus, 0.5 MPa absolute corresponds to a depth of 40 m which is a shallow dive. Five megapascals absolute corresponds to a depth of 490 m, which is a deep dive. The amount of phosphatidylcholine in the secreted surfactant was measured after the "dive" and compared with that from similar cells isolated from sheep (*Ovis aries*). The lowest pressure had no effect on either of the two groups of cells. Longer exposures, 120 min at 5 MPa, were more interesting.

The sheep cells showed increased secretion at pressure, which is surprising, and the increase was sustained for an hour after decompression. This suggests that pressure was interfering with the secretory process in some unknown way. The same dive had no effect on the sea lion cells. Interesting though this difference is, particularly as the sea lion cells seem to be pressure-resistant, the presence of both hydrostatic pressure and gas partial pressure makes it difficult to separate the effect of each factor.

The effect of pressure on surface tension has been measured in an experiment carried out in a rather different context by a distinguished trio. We have already met Bangham and Bennett, and the third is Paphadjopoulos, an expert in lipid bilayers (Chapter 4). They subjected a monolayer of phosphatidylcholine to a high pressure of a selection of gases. The monolayer was spread on an aqueous salt solution contained in a plastic trough, the whole assembly being ultra-clean as is required in this work. The monolayer surface tension was measured by a conventional micro-force balance and the entire apparatus was mounted in a pressure vessel. Some gases increased the surface tension. For example, 4 MPa nitrogen increased the tension by approximately 6% and oxygen by three times that amount, presumably by being more soluble in the monolayer. It is not clear how this effect might contribute to the balance of factors in a deep diver's alveoli, and I am unaware that it has even been considered. Helium, a highly insoluble and inert gas, had the effect of slightly reducing the surface tension, by 3% at a helium pressure of 4 MPa. Can this helium effect be interpreted as due to hydrostatic pressure, offsetting any trivial effect of dissolved helium, primarily ordering the phospholipid monolayer and somehow reducing the surface tension? Hydrostatic pressure may well have an effect on the collapsed alveoli and their subsequent opening and therefore might influence the "design" of the surfactants.

## 11.2.1 Consequences of Gas Dissolved in Tissues Under Pressure

### 11.2.1.1 Nitrogen

Alveolar gas exchange during a dive is often minimised but the gas exchange which does occur in a dive has a number of consequences. One is that nitrogen dissolves in the blood and gradually equilibrates throughout the body. The dissolved nitrogen is not involved in metabolism but is relatively soluble in fatty tissue, so typical pinnipeds, with their huge insulating layer of fat, possess a "sink" which soaks up a lot of nitrogen.

Remarkably, although nitrogen is metabolically inert it does possess anaesthetic properties, mentioned in Chapter 4; Section 4.3. They were discovered when humans dived deeper, breathing externally supplied air. This incipient anaesthesia, called nitrogen narcosis (a term which has nothing to do with psychoactive narcotics), poses a significant problem in human diving, of which more in Section 11.4.3. In deep-diving mammals, although any anaesthetic effect of nitrogen seems unlikely, there may be effects on individual excitable components of the nervous system.

A second consequence of the gas dissolved in the diving animal manifests itself during the ascent. As the pressure decreases so the $pN_2$ in the blood exceeds that in the lung, creating a supersaturated state. Nitrogen comes out of solution as very small bubbles in the blood and other tissues, and as pressure decreases further, the bubbles' size increases. They are normally filtered by capillaries in the lung and their gas eventually diffuses into the alveoli and is exhaled. After a normal dive, the "dose" of bubbles is likely to be sub-symptomatic but a severe dose of bubbles can cause acute symptoms such as blocked blood vessels or impaired nerve function (Hooker et al. 2012). This is, of course, classic decompression sickness similarly experienced by human divers (Section 11.6).

### 11.2.1.2 Oxygen

The partial pressure of oxygen in the alveoli may transiently increase as descent takes place but it presents no problem to diving birds and mammals (as it does in human SCUBA diving, see below). However, many of the divers' tissues experience a prolonged lack of oxygen (hypoxia). In typical mammals, and particularly in humans, the brain is particularly sensitive to hypoxia, which is why strokes can cause such devastating permanent damage (Joulia et al. 2003; Larson et al. 2014; Geiseler et al. 2016). In diving mammals, the brain has evolved molecular mechanisms of tolerating hypoxia, and these are more fully developed in deep-diving animals. There might well be a significant pressure effect in their function. Separately, when the supply of oxygen is restored by the circulation after a dive, a surge of free radicals and reactive oxygen species is produced, of which more below. This potential for intense oxidative damage to lipids and proteins is reinforced by other features of the lifestyle of diving animals and their evolution of compensatory antioxidant systems has been essential (Castellini 2010; Cantu-Medellin et al. 2011; Vasquez-Medina et al. 2012; Il'ina et al. 2017 and others listed at the end of the chapter).

## 11.3 Pressure Effects on Some Cells in Diving Animals

### 11.3.1 Muscle Enzymes

When the store of oxygen is depleted, the diver's muscles continue to generate energy anaerobically, a process in which the end product of glycolysis is converted to lactate, with the simultaneous regeneration of the important co-factor NAD from NADH. The enzyme which catalyses the reaction is lactate dehydrogenase, LDH, which we have met before. In shallow water fish, it is sensitive to high pressure, but in deep-sea fish, it is adapted to function at high pressure. Accordingly, LDH from the skeletal muscles of the Elephant seal, Sperm whale, Emperor penguin and for the purpose of comparison, the rabbit, have been investigated. The experiments measured the effect of 20 MPa on the LDH Km for NADH, at the animals' body temperature of 37 °C. No effect was detected, meaning the affinity of NADH for the enzyme (i.e. its binding) was unaffected by a pressure somewhat greater than normally experienced by these deep divers. The same enzyme in the rabbit was also unaffected. Perhaps the mammalian LDH, functioning at 37 °C, is fortuitously insensitive to pressure, which in the case of the divers appears as a fortuitous preadaptation (Croll et al. 1992).

### 11.3.2 Red Cells

The rate of glycolysis, i.e. the rate of glucose consumption or lactate production, in the red cells of mammals which only dive to a shallow depth, such as the Harbour seal, has been shown to be accelerated by 13 MPa (over 2 h at 37 °C). In the case of deep-diving species, the Weddell seal, for example the effect of pressure was much less and similar to that seen in human red cells. The significance of this curious finding is unclear. Glycolysis in these red cells mainly provides the energy for transporting solutes across the membrane (Castellini et al. 2001).

The structure and composition of the red cells of a variety of diving mammals have been investigated by Williams et al. (2001), with commendable thoroughness which was not rewarded with particularly clear-cut conclusions. Nevertheless, the results are important as they do not match our expectations. In the deepest diving examples, the Northern Elephant seal (*Mirounga angustirostris*) and the Harbour seal (*Phoca vitulina*) the red cell membranes were found to contain little cholesterol, a lot of unsaturated fatty acids and other components which would be expected to produce a fluid bilayer. Yet, as measured by DPH anisotropy, they were not very fluid. Furthermore, their fluidity decreased markedly with an increase in pressure, more so than the red cell membranes from the shallow diving Northern fur seal (*Callorhinus ursinus*) and non-diving mammals. These had a lipid composition which contrasted with that of the deep divers. There seems to be no obvious connexion between the structure of the red cell membranes and frequent exposure to a significant pressure (16 MPa or so), in which homeoviscous adaptation might have had a small role to play. Presumably, other factors are more important.

### 11.3.3 Platelets

Other blood cells, called platelets, have also been investigated and they are included in a following section on decompression in human diving.

### 11.3.4 Excitable Cells

We have seen (Chapter 4) that the normal activity of shallow water animals is increased and disorganised by moderate pressure, giving way to spasms and convulsions at high pressure. Are diving birds and mammals similarly affected by pressure? This question has not attracted much interest in the subject of diving animals but in human diving, its importance was recognised in the 1960s (Chapter 4; Sections 4.3, 4.4). Accordingly, what little we know about the effects of hydrostatic pressure on the activity and the nervous system of deep-diving animals will be discussed further in Section 11.5, on human diving. The adaptation of the nervous system of deep-diving animals to high pressure is insufficiently recognised as a major problem by many workers in comparative and environmental physiology, Castellini being a notable exception. Rapid compression and decompression are probably more challenging stress than the relatively constant high pressure experienced by deep-sea animals.

## 11.4   Human Diving Supported by a Self-Contained Gas Supply

A human diver can be supplied with air through a hose, as developed in the nineteenth century, or with special gases via an umbilical supply from a diving bell, as in modern deep oil well diving, or from pressurised tanks strapped to the diver's back, self-contained underwater breathing apparatus—SCUBA. The modern SCUBA diver, thus equipped, is a unique creature, with obvious similarities to animal divers, but independent of the constraints of breath-holding. However, special problems arise from the extended period of time spent breathing gas at high pressure, from hydrostatic pressure and from exposure to other conditions alien to human physiology.

The deepest SCUBA dives have reached depths in excess of 300 m but however deep or shallow the dive, the gas breathed is at the same pressure as the ambient water. The divers' lung volume is normal but the gas in it is compressed. This is quite unlike a breath-hold dive. The accumulation of gas dissolved in a diver's tissues creates difficulties when the pressure is decreased. This was discovered in an apparently separate activity, the construction of tunnels. Construction engineers developed ways of using air pressure to prevent water seepage. The pressure is confined by a structure called a caisson, usually made of steel, and caisson workers are "locked" in and out of the caisson. Whilst at work in the caisson they are subjected to significant air pressure. Experience gained in the nineteenth-century tunnel projects in New York and London paved the way to a realisation that gas dissolved in the workers' tissues created decompression sickness which has to be dealt with properly to avoid serious injury. It is an important issue for tunnel workers just as it is for divers.

### 11.4.1 Gases Breathed at High Pressure

The $PO_2$ in air becomes potentially toxic at a pressure of, say 0.2 MPa. Toxicity is time-dependent so there is no simple figure to quote, although diving codes specify safe limits. At a depth of 20–30 m, the $PN_2$ in air exerts incipient behavioural changes, usually starting with euphoria. This is nitrogen narcosis, mentioned earlier in Chapter 4. A $PN_2$ of 0.16 MPa can be equated with a dram, or a martini, informally perhaps, but as we shall (Section 11.4.3) see it is scientifically justified. Simultaneously, there is sufficient gas to dissolve in the divers' bloodstream, eventually reaching the limits set by the solubility of the gas in the different solvents present, water and fat. When the limits are reached, the tissues are said to be saturated. This is unlike the deep-diving animals which curtail the process. Finally, we should note that compressed gas has an increased density, which impairs its flow in and out of the lungs, changes its heat transfer properties and particularly in the case of helium, distorts the divers' speech.

### 11.4.2 Oxygen

The $PO_2$ in the air which we normally breathe is entirely safe, but a small increase, usually referred to as hyperbaric oxygen, soon overwhelms our natural antioxidant defence mechanisms. These exist in all organisms as mentioned previously. In the ancient evolutionary past, when photosynthesis evolved, it generated oxygen and hence reactive oxygen radicals, which damage macromolecules. In present-day organisms, the main reactive radicals are the superoxide anion and hydrogen peroxide, both of which arise from oxygen, in trace amounts, but nevertheless have to be kept at low concentration. The most toxic compound is the hydroxyl radical which is formed from the two previously mentioned. In healthy tissues, a very small fraction of the oxygen consumed generates these radicals, largely but not exclusively, in mitochondria. The evolution by natural selection of anti-oxidative mechanisms was fundamental to the subsequent evolution of life as we know it.

In humans the effects of oxygen toxicity are apparent in the lungs where the endothelium and alveolar epithelium are destroyed, resulting in fluid accumulation and impaired gas exchange. A $PO_2$ of 200 kPa (absolute) pressure, for about 3 h slightly reduces the vital capacity of the lungs, which is the volume of air which can be forcefully exhaled after taking a deep breath. It is a convenient measure of the health of the lungs, indicating the onset of Pulmonary oxygen toxicity. Higher doses of oxygen for many minutes or perhaps an hour, initiate Neurological oxygen toxicity. It starts with muscular twitching, often in the face, and spreads to the limbs and trunk as muscles strongly contract and alternate with rigid phases, a form of grand mal seizure. Any loss of consciousness is reversible in controlled conditions. Often there are few progressive warning signs. Paradoxically, mild hyperbaric oxygen is clinically useful in treating a number of conditions, for example, poisoning with cyanide or carbon monoxide, so safe therapeutic partial pressures are well established and treatment lasts only an hour or so. Clearly, oxygen is a potentially dangerous gas and, as with patients, the partial pressure to which divers and caisson workers should be exposed are carefully limited.

### 11.4.3 Nitrogen and Other Anaesthetic Inert Gases

Nitrogen narcosis was first observed in the early nineteenth century, but its true cause was only established by the American physician-scientist Behnke in 1935 (Grover and Grover 2014). The susceptibility of individuals varies greatly but the list of symptoms reveals its potential danger; euphoria (at around 300 kPa $N_2$), spatial disorientation, impaired reasoning and reckless conduct, increased reaction times, mental confusion and memory difficulties (at higher $PN_2$) culminating, during exposures to a very high $PN_2$, in stupefaction and eventually death. Behnke made two seminal points. First, the problem of nitrogen narcosis can be avoided by using artificial air, namely a gas mixture which is safe to breathe, providing the diver with a low or zero $PN_2$. Second, although nitrogen is chemically unreactive in the body and does not enter into any biochemical reactions, its potency might be attributed to its solubility in lipids. This brought it into line with the earlier ideas of Meyer and Overton concerning the action of anaesthetics.

This second point is that nitrogen's effect is similar to that of other inert gases such as xenon, nitrous oxide, methane and cyclopropane, described in Chapter 8 (Section 8.7.3) and Chapter 4 (Section 4.3). Volatile compounds such as chloroform and diethyl ether, alcohols and many other chemically diverse substances are included. These unreactive compounds all exert narcotic effects on biological systems, from cells to whole organisms, an effect which is highly reversible and involves little or no biochemistry or chemical change to the compound in question (Cullen and Gross 1951; Macdonald and Wann 1978). Nitrogen narcosis is such a narcotic effect on the central nervous system. General anaesthesia is a clinically useful version of a comparable effect, in which the loss of both consciousness and the sensation of pain takes place without serious effect on the heart or breathing (Franks 2008; Koch and Greenfield 2007). Local anaesthesia is quite separate, being achieved by a localised blockade of the conduction of impulses in sensory nerves.

The remarkable properties of anaesthetics were familiar to many people in the late eighteenth century, when laughing gas (nitrous oxide) was inhaled for fun, making people fall over without hurt. In 1799, the English physician-chemist Humphry Davy wrote of its possible use to suppress pain in surgery when he was Superintendent of the Medical Pneumatic Institute in Bristol. Many years elapsed before that happened. The chaotic emergence of clinical anaesthetics as safe and acceptable is well described by Stephanie Snow in "Operations without Pain". Here I will simply note the famous year of 1846 when a somewhat mysterious dentist by the name of William Morton demonstrated the effective use of diethyl ether as an anaesthetic. He extracted a tooth to the satisfaction of his patient in the presence of the great and the good at Massachusetts General Hospital. Within a few weeks, the news was exciting the London Establishment and in a year the use of diethyl ether had spread across the world. Chloroform and then nitrous oxide soon became widely used and with diethyl ether paved the way for modern surgery. Construction workers in caissons in the late nineteenth and early twentieth century who experienced mild nitrogen narcosis would have been bemused to learn that they were, in fact, experiencing incipient anaesthesia!

As Behnke thought, the narcotic potency of nitrogen is closely related to its solubility in hydrophobic solvents, a property it shares with many general anaesthetics. This important generalisation is based on experimental evidence, for example the righting reflex of the laboratory mouse. This has been mentioned earlier (Chapter 4). It is the reflex which is normally triggered if the animal is dropped upside down and is impaired by a dose of anaesthetic comparable to light anaesthesia in humans. The test animal is confined in a vessel which is rotated and the success of the animal in regaining its upright posture is scored as a percentage of responses. A gaseous anaesthetic whose partial pressure can be accurately measured is injected into the chamber and reduces the success with which the righting reflex is performed. The partial pressure which reduces the score to 50% is a measure of the anaesthetic's potency. Using inhalational anaesthetics and those which are administered intravenously, like the commonly used anaesthetic propofol, it is clear that their potency correlates very closely with their solubility in olive oil, whose solvent properties closely match those of cell membranes. It is a most remarkable correlation, extending over a 100,000-fold range of solubilities and has been studied for many years. There is nothing unique about nitrogen which is merely one of the inert gases which have an anaesthetic effect proportional to their solubility in olive oil. They are, in decreasing order of potency, xenon, krypton, argon, nitrogen, hydrogen and neon. Helium has a marginal anaesthetic effect, discussed below. They and the more potent clinical anaesthetics are thought to act on molecular sites in the animal's nervous system, with solvent properties similar to a membrane. Or, more likely, they bind to a similarly mildly hydrophobic region of a protein. Such a binding equilibrium would be readily reversible. A mouse anaesthetised with nitrous oxide recovers almost immediately when allowed to breath pure air, consistent with the idea that no biochemistry occurs in the anaesthetic process (and confirmed by other direct experiments). The inert gas xenon is an excellent general anaesthetic and would be widely used if it was less expensive, of which more soon. Along with other inert gases, including helium, it exerts surprising effects on cellular biochemistry which belies the historical term inert.

Divers had long been aware that nitrogen narcosis could be promptly reduced by ascending (decompressing). Nitrogen differs from nitrous oxide in being much less soluble in olive oil, so a high partial pressure is required for it to have any effect. Nitrous oxide's potency makes it a convenient anaesthetic: it can be breathed in air without affecting the $PO_2$. Like every anaesthetic nitrous oxide has its own clinical characteristics, the main one being its strong analgesic effect. It was developed by Tunstall at the Aberdeen University Medical School as a self-administered anaesthetic for use by women during the painful stages of childbirth and in mountain rescue (Tunstall 1961).

### 11.4.3.1 Anaesthesia and Inert Gas Narcosis Are Reversed by Pressure

The readily reversible nature of anaesthesia is also apparent in the way it is abolished by high pressure. The classic experiment reporting this was published in 1950, and it requires some introduction. A proper anaesthetic was not used. Instead, ethyl alcohol (the familiar stuff in a dram) was used to narcotise tadpoles. Alcohol, along with many other substances, has a narcotic potency which is closely related to its solubility in olive oil, but it is not biochemically inert. It deviates from the ideal anaesthetic in this regard, but then so do a number of very effective clinical anaesthetics, especially those with a halogen group attached, such as Halothane. Although they are biochemically processed in the liver this has nothing to do with their effect on the nervous system. The experiment consisted of narcotising tadpoles (of the frog Rana sylvatica) by immersing them in a solution of ethyl alcohol at a concentration of 3–6% (more like beer than whisky) filling a pressure vessel. The animals lay unresponsive on the bottom. The vessel was equipped with a window and on applying a hydrostatic pressure of 13–34 MPa the narcotised tadpoles were seen to swim up from the floor of the vessel and behave normally. In the absence of alcohol 13 MPa caused hyperactivity and 34 MPa immobilisation, as seen in most shallow-water animals under pressure. This "pressure reversal of anaesthesia" experiment stimulated a great deal of research. In its day it was an imaginative extension of the biochemical studies using high pressure then being carried out by Johnson and colleagues at Princeton University, USA.

The solubility of the inert gases neon and helium is so low that it implies that they may not act as anaesthetics and may simply transmit pressure. This was the reason why helium gas was suggested, by Behnke and others, as a safe gas for divers to breathe at high pressure, neon being too dense to breathe easily. Mixed with a small amount of oxygen to give a safe $PO_2$, helium is safe for divers to breath and eliminates the possibility of nitrogen narcosis. The mixture is called Heliox and is used in all deep diving. As discussed in Chapter 4, Heliox is also used to subject air-breathing animals to high hydrostatic pressure, acting to transmit pressure just as water does for aquatic animals. A good example of this is the modern equivalent of the tadpole experiment described above. Using a similar pressure vessel and procedures, a mouse whose righting reflex had been reduced to a low percentage by nitrous oxide, for example, is restored to normal by the application of the appropriate helium pressure. Similar experiments have shown that nitrogen narcosis can also be reversed by pressure. High-pressure helium can affect certain cellular processes, opposing the effects of hydrostatic pressure, a phenomenon described as pressure protection in Chapter 6. Quite separately helium, at close to normal atmospheric pressure, has neuroprotective effects on stressed cells, i.e. it affects their biochemistry by unknown mechanisms (Oei et al. 2010; Smit et al. 2015; Chapter 4).

However, convincing experiments demonstrating that helium pressure can exert a very weak anaesthetic-like effect in an animal were reported by Miller and his colleagues, in 1985. The experiments used tadpoles of *Rana pipiens*, a lively wriggly aquatic animal with a righting reflex. The tadpoles were exposed to water pressure, i.e. pure hydrostatic pressure or to water equilibrated with a matching partial pressure of helium. The tadpoles were paralysed by a mean hydrostatic pressure of 13.7 MPa but a higher pressure, 17.2 MPa partial pressure of helium-equilibrated water, was needed to achieve the same paralysis. It appears that helium opposes hydrostatic pressure. A second line of evidence is based on the use of the animals' righting response. When the performance of the response was reduced by an anaesthetic dissolved in the water in which the animals were swimming, hydrostatic pressure reversed the effect much more than the pressure applied with water equilibrated with same pressure of helium. That means helium was contributing to the anaesthetic dose. Despite this result, in practical diving and much experimental work using Heliox, it is assumed that helium is simply transmitting hydrostatic pressure and any anaesthetic effect is trivial, and in most cases undetectable.

Explaining the "pressure reversal" of narcosis and anaesthesia is just part of the problem of explaining how these chemically diverse compounds interact with components critical to the function of the nervous system. It is really very odd, for example, that the inert gas xenon and the quite complicated molecules of cyclopropane or Halothane act in the same way. For many years the problem has attracted physical chemists as well as physiologists. A group of the former from Oxford University produced the Critical Volume Hypothesis, to explain this feature, and to provide a general mechanism of anaesthesia. It proposed that a hydrophobic phase (membrane or region of a protein) would increase in volume as anaesthetic molecules dissolved in it. Hydrostatic pressure would compress the phase and restore its original volume. Thus, the anaesthetic state is associated with a critical volume of some undefined phase, which affects the normal function of the brain. In detail, the argument is quantitatively sound and thermodynamically elegant, but many physiologists think it explains nothing. The loss of consciousness and analgesia have to be explained in terms of brain circuitry, neurotransmitters and excitability. The different depths of anaesthesia which are managed in the operating theatre need an explanation, as does the clinically convenient fact that essential functions of the heart and breathing are little affected. If the latter was not the case it is difficult to see how anaesthesia could have developed in the nineteenth century.

The pressure reversal of anaesthesia can be explained as a case of antagonism, which is a pharmacologists' term. It could be a form of molecular antagonism in which, to imagine a plausible example, the anaesthetic molecules are dissociated from their molecular target by high pressure. Such a shift in a binding equilibrium is an effect of pressure we have met before and one which the thermodynamics of the Critical Volume Hypothesis describes. It is established that anaesthetics (xenon and nitrous oxide for example) can bind to the hydrophobic pockets of proteins (Quillin et al. 2000) perhaps involved in synaptic transmission, either enhancing their inhibitory action or reducing excitation. Alternatively, the pressure reversal of anaesthesia might be a case of physiological antagonism, in which the hyperexcitable effects of pressure over ride the depressive effects of the anaesthetic. This is probably how pressure reverses the anaesthetic effect of barbiturates, which are intravenous drugs, not general anaesthetics comparable to nitrous oxide, xenon or other classical inhalational anaesthetics. Antagonism between hydrostatic pressure and anaesthetics in the intact animal should also be apparent in isolated components of the nervous system. There are examples of the latter (Kendig 1984a, b; Wann and Macdonald 1988), but they are confounded by conflicting results. A particularly interesting experiment would be to target an anaesthetic bound to a specific molecule involved in excitability and then subject both to high pressure. An example is given in Box 11.1.

---

**Box 11.1**

The glutamate-gated ion channel in the leg muscle of the locust *Schistocerca gregaria* provides a suitable excitable molecule. This probably seems a strange choice, but it is similar to channels which occur in the vertebrate nervous system. It is also a good example of the Krogh Principle, which states that for certain experiments there is an animal (or part of) on which it can be best be carried out.

The experiment tested the idea that the effect of a general anaesthetic on an ion channel can be reversed (offset, antagonised) by high pressure. The anaesthetic dose should be in the clinical range and the pressure similar to that used for reversing anaesthesia in animals. It is assumed that the anaesthetic can only affect the function of an ion channel by binding to it, and if its effect is countered by pressure then it can be argued that pressure dissociates the anaesthetic from the channel protein.

The glutamate-gated channel in the locust muscle is amenable to the mega-ohm seal patch clamp recording technique (Chapter 4) for several reasons. Patlack, from the Göttingen Laboratory, Germany, mentioned before, and Usherwood and colleagues at Nottingham University, England, had established that the channel could be treated to prevent desensitisation. That meant a patch pipette containing glutamate in solution, gently pressed on the muscle surface to form a mega-ohm seal could record single-channel activity for a prolonged period and could then be moved to record from another site. A vast amount of channel open/closed data could be accumulated, which when analysed by computer, reveals a distinct population of open and closed times. They arise as the molecules of glutamate bind to and dissociate from the channel receptor in random fashion. Remember, this technique records the activity of a single protein molecule. Other channels desensitise after a short exposure to their transmitter and then fail to respond so they are useless for the intended experiment.

Recording channel currents with the mega-ohm seal technique at high pressure required a vertically mounted cylindrical pressure vessel of 15 cm internal diameter (Chapter 13; Section 13.5.2). It was constructed by the

(continued)

**Box 11.1** (continued)

collaboration of the author with Wann and colleagues at Aberdeen University and with Usherwood and his colleagues. A window in the top closure enabled a stereomicroscope to provide a view of the muscle, bathed in saline. An electrically driven micromanipulator inside the vessel positioned the patch pipette onto the muscle surface when the system was pressurised with helium. This method of applying high pressure was appropriate for combining with anaesthetics, mimicking the situation in animal experiments.

First, the effect of helium pressure on the channel gating kinetics was established. 10 and 30 MPa helium "on top of" the air in the vessel had no effect. But 50 MPa prolonged the mean closed time and halved the mean open time, without affecting the channel conductance. Only the lower pressures were used to examine the anaesthetic effect.

A low dose of nitrous oxide, a partial pressure of 50 kPa added to the air in the pressure vessel, had no effect. 150 kPa $N_2O$, which reduces the righting reflex in mice by 50% and can therefore be regarded as an anaesthetic dose, decreased the probability of the channel opening and the mean open time. It also increased the mean closed time. The channels were then subjected to 10 MPa helium pressure (actually 9850 kPa) combined with 150 kPa $N_2O$ "on top of" the air present in the vessel. The result was that pressure restored the channel parameters to normal. The quite different anaesthetic ketamine also had its effect on the channel kinetics similarly reversed by pressure. The result is consistent with pressure causing the anaesthetics to dissociate from their binding site. The binding equilibrium thus involves a significant volume increase. The alternative explanation that the pressure squashes the anaesthetic binding site back to its normal volume seems implausible. The result may be of mere chemical interest, but it is also consistent with pressure antagonising anaesthetic binding sites involved in anaesthesia by dissociating the anaesthetic. Applying Occam's razor, the idea of the physiological antagonism of anaesthesia is redundant, and molecular antagonism is probable.

This demonstration of the pressure reversal of the anaesthetic effect on an ion channel was carried out a long time ago (Macdonald et al. 1993; Macdonald and Ramsey 1995). But even then there was evidence that xenon, which is very similar to nitrous oxide in its anaesthetic properties, binds to proteins. Xenon is close to being the ideal general anaesthetic and our understanding of its protein binding has grown in recent years. X-ray crystallography has shown that xenon and nitrous oxide bind to the same cavity in proteins, for example the enzyme urate oxidase and the membrane-bound protein annexin V. (Ion channels, like most membrane-bound proteins, are difficult to crystallise.) An interesting feature is the increase in the volume of the anaesthetic-binding site, but it is not clear how such a volume, measured in a dry protein crystal, relates to the molar volume change which is expected as the anaesthetic binds to a hydrated protein. Another interesting outcome of the crystallographic work is that anaesthetic binding does not cause much structural change in the protein, but by filling a specific cavity it probably reduces the flexibility of the protein. Anaesthetic binding is allosteric, meaning action at a distance. General anaesthetics do not universally damage channel function; for example Franks and Lieb discovered that the once-popular clinical agent halothane activates a particular type of K channel. Xenon and nitrous oxide's ability to block various glutamate-gated ion channels is not yet understood with precision. Finally, it is worth noting that studying the effect of hydrostatic pressure on anaesthetic-protein binding seems to have been overlooked. The pressure reversal of anaesthesia in animals (and humans) surely means that the critical molecular sites of action must also be antagonised by pressure. An effect of an anaesthetic on an isolated excitable cell which is not antagonised by pressure is presumably not closely involved in anaesthesia.

Returning to nitrogen narcosis, it is important to remember that an effective dose of nitrogen entails several processes at the molecular level. Pressure drives the solution of nitrogen, getting a sufficient number of molecules to dissolve in, or bind to, the target molecule(s) whilst at the same time the hydrostatic pressure is possibly affecting the target (slightly) and probably opposing (slightly) its molecular interaction with nitrogen. Argon is more narcotic than nitrogen, requiring less partial pressure to achieve the effective dose. The inert gas hydrogen differs in the opposite direction. Its narcotic potency is slightly greater than that of helium. Compared with argon and nitrogen, it has a lower solubility in olive oil and is, therefore, less narcotic. It is easy to breathe as Hydrox, a hydrogen–oxygen mixture which the French diving company Comex has put to good use. An inappropriate hydrogen–oxygen mixture is potentially explosive, but that is a manageable issue. There are other inert gases, not used in diving, such as cyclopropane and ethylene, which have been used clinically. The former is unfortunately inflammable and the latter's potency is similar to that of xenon. Remarkably ethylene, at very low dose, is a plant hormone!

Inert gases have been used to control the excitable effects of hydrostatic pressure in human diving, HPNS (Chapter 4). They could be introduced relatively easily because, like clinical drugs being used to treat conditions they were not originally intended for, these gases were well known. An advantage of the hydrogen in Hydrox was that it ameliorated the HPNS, and the work of Bennett using nitrogen showed the same effect. These mixtures were used in numerous simulated dives with humans at a pressure of 5 MPa and more. Good accounts of these remarkable experiments are available in the readily available multi-author book edited by Brubakk and Neumann (2003). In particular, the chapter by Bennett and Rostain provides some of the details of the deepest simulated dives with human subjects, including the Comex Hydra 10 "dive" to 701 m depth, 7 MPa, in 1992. This dive employed helium and hydrogen in a complicated way, but to very good effect. Simulated dives with the monkey (*Macaca fascicularis*) attained a pressure of 12 MPa and showed that the beneficial effects of hydrogen were limited to 10 MPa, i.e. 1000 m depths (Rostain et al. 1994). However, for commercial ultra-deep work, perhaps 300 m depths, hydrogen provided other practical advantages, in decompression and thermal balance for example.

At the time, it was also speculated, by a few of us, that inert gases could be used by deep-diving mammals. Might they be intelligent enough to use up their oxygen to prevent oxygen toxicity and allow sufficient nitrogen to leak from their airways to suppress HPNS?

## 11.5  The High-Pressure Neurological Syndrome, HPNS

Having established that helium gas enables divers and other air-breathing animals to be pressurised in much the same way as aquatic animals in water, it was discovered that hyperexcitability, in contrast to narcosis, was a new symptom elicited in very deep dives, the HPNS. This has been discussed in Chapter 4, but some further comment is needed here. The HPNS is a multi-layered complex disturbance of the entire nervous system (Naquet 1988; Rostain et al. 2010; Gronning and Aarli 2011). The simplest manifestations in humans, mentioned in Chapter 4, but also seen in baboons (*Papio papio*), are reflexes mediated at the spinal level, probably involving synaptic transmission by acetylcholine. At the other end of the scale of complexity, seizures seen in the baboon implicate, among other factors, GABA (gamma aminobutyric acid) transmission in cortical pathways. And there is plenty of evidence that other neurotransmitters are involved in other parts of the brain. Perhaps it is not surprising that hydrostatic pressure affects so many components of the nervous system, apparently directly, and we cannot exclude the possibility that secondary, indirect effect also play a part. Counteracting the HPNS in humans with the inert gases nitrogen and hydrogen and in experimental animals with general anaesthetics and other compounds is remarkable. Does it provide a useful model for investigating the tolerance of deep-diving animals to pressure?

The stability of the nervous system of deep-sea animals at high pressure is a major facet of their evolution which awaits proper study. But air-breathing divers face a dramatic version of a similar problem. The rapid deep dives expose the animals to HPNS, so we must assume that countermeasures have evolved. The spectacular dive of the Beaked whale already mentioned involved a rate of descent of 0.4 MPa per minute to an eventual 30 MPa. Other species dive at similar rates to lesser depths but which are, nevertheless, sufficient to cause the HPNS in mammals. If we assume the onset pressure for tremors and involuntary muscle jerks is a reasonable threshold of impaired motor control, then 4 MPa or 400 m or perhaps somewhat less, is a critical pressure/depth at which an animal's hunting performance might suffer. This ignores effects on the heart and other muscles. A reasonable prediction is that factors counteracting the HPNS have probably evolved in those species which dive deeper than 400 m depths, such as the Weddell seal (*Leptonychotes weddellii*), the Hooded seal (*Ommatophoca rossii*), the Elephant seals, both Southern and Northern and the large whales. One possibility, already mentioned, is that a suitable $PN_2$ is present in the animal's brain, the nitrogen coming from the pulmonary air present at the start of the dive. Again, making a plausible extrapolation from simulated dives in which nitrogen offsets HPNS, an effective $PN_2$ would be less than 400 kPa. As pointed out by Ponganis et al. (2003), such a $PN_2$ is within the range known to occur in diving mammals and has, in the case of the Weddell Seal, been directly measured in experiments. The idea is thus entirely plausible and a good hypothesis to test. Much as I relish the idea as the simplest way to account for the extraordinary performance of the deep-diving specialists, I imagine that nature can do better than using nitrogen to control HPNS. Furthermore, using hyperbaric nitrogen would create a liability during ascent, decompression, which is our next topic. Using metabolites to stabilise pressure-sensitive macromolecules in the nervous system is an obvious extrapolation from what we know of the pressure adaptation of enzymes. Does TMAO have a role in the deep-diving CNS?

## 11.6    Decompression

Scuba diving and even breath-hold diving lead to gas dissolving under pressure and being distributed to the tissues by the blood circulation. When the pressure is decreased, during ascent, the body fluids become supersaturated with gas, which comes out of solution as bubbles. The commonly used illustration is of a bottle of "fizzy" lemonade being opened, releasing bubbles as the contents experience a decrease in pressure. This is a valid if rather dramatic illustration but a better one is a glass of tap water, which has been used to illustrate solubility before. When freshly filled with cold tap water and allowed to stand in a warm room, a small number of air bubbles will gradually form on the sides of the glass. The solubility of gases in water decreases with rise in temperature so the warmer water achieves a supersaturated state. The gas comes out of solution as bubbles but the important point of the illustration here is, why do they appear separately on the glass surface, and not as one large bubble? The process of bubble formation involves an energetically favourable site, probably a micro crevasse on the glass surface, which enable the bubble to nucleate and then grow by diffusion. Similarly, in the divers' body fluids, bubbles form on special micro-surfaces and micro-particulate material (of which more below), and they do so because of the supersaturated state. Any practical dive will result in bubbles arising during the most gradual decompression. Haldane's decompression theory assumed it was possible to avoid the appearance of bubbles but that is not the case. Bubbles formed on decompression can be directly detected by ultrasonic methods and the more bubbles there are the greater is the risk of pathological consequences. Modern decompression theory is directed at managing the dose of bubbles the diver experiences. Decompression sickness embraces a variety of effects, ranging from the simple obstruction to blood flow caused by bubbles in small vessels, pain in the joints (which caused early sufferers to walk with a bent posture, hence the traditional divers' term "bends"), to necrotic (dead) patches in load-bearing bones being discovered months or years after the decompression. This is a serious condition and can render the diver unfit to continue diving. The skeletons of Sperm whales have been discovered with similar lesions. In at least one case the stranded carcase revealed bone ulceration, degeneration and remodelling of cartilage, which, in the absence of infection, was described as osteonecrosis, similar to that seen in human divers. Whilst pressure is the primary cause of these problems it is not a factor in the unfolding consequences of decompression.

Microparticles, less than 1 um in size, mentioned earlier, are found in the circulation of decompressing human divers and appear to be formed of cellular protein fragments enclosed by a membrane (Thom et al. 2014). Moderate pressures of inert gases, including xenon, nitrogen (<200 kPa) and helium, interact with oxygen to generate reactive oxygen compounds which, in turn, cause damage to neutrophils (also called polymorphonuclear leukocytes) and the release of particles. Cytoskeletal actin is one source of particulate protein. Microparticles are thought to initiate an inflammatory response. This complicated sequence does not seem to be a specifically hydrostatic phenomenon.

There are, however, some interesting effects of hydrostatic pressure on other pathological processes in decompression which involve blood platelets.

Platelets are blood cells. They are formed by pinching off from bone marrow cells called megakaryocytes, and although rather simplified cells, lacking a nucleus, they are crucial in haemostasis. This is the process by which potential leaks in the blood circulatory system are plugged. Platelets are activated by specific physical and chemical stimuli. For example, platelets are activated by contact with collagen, which is liable to be exposed in a mechanically damaged blood vessel. The platelets respond with a change in shape, from small smooth-surfaced discs to become spherical cells from which fine processes extend. The platelets thus entangle and adhere to others, and then release chemicals by exocytosis which drives aggregation further to create a plug. In the presence of fibrin molecules, it becomes a more durable blood clot, often coloured red as erythrocytes become trapped in it. Platelet activation and blood clotting involve a cascade of reactions, bewildering in their complexity, which both reinforce and control the reaction such that unrequired clots are not produced. The point of interest here is that divers often have a low platelet count, immediately after a dive and in subsequent days, even after an approved decompression procedure. This has been found in a variety of real wet dives and in carefully controlled simulated dives. So, what is going on?

When gas bubbles form in blood during decompression, the blood plasma–gas interface acquires a film of denatured protein which, it was thought, might trigger platelet activation. Careful experiments show that platelets are not that easily activated, but the flow of plasma carrying the platelet past the bubble in a confined space could create a shear force which triggers activation, which in turn would take platelets out of circulation and reduce the post-dive platelet count.

Are human platelets affected by hydrostatic pressure? They are not irreversibly affected by, for example, 27.3 MPa for 40 min, because after being decompressed they aggregate normally when exposed to an aggregating chemical such as ADP. It is assumed that exposure to lower diving pressures for longer would also have no effect. So hydrostatic pressure is probably not responsible for the drop in the platelet count, which leaves decompression a likely candidate.

Platelet aggregation, however, is affected in vitro by diving pressures. For example, Pickles et al. (1989, 1990) used methods which applied purely hydrostatic pressure or a high partial pressure of inert gases in solution (Chapter 13; Fig. 13.7).

When a suspension of platelets aggregates in response to a specific chemical trigger (ADP in the following), the clear suspension becomes turbid so the reaction can be followed optically. The initial shape change and adhesion, mentioned above, was shown to be unaffected by hydrostatic pressure (5 MPa) but the second stage is inhibited. The exocytosis of ATP, one of several aggregating substances released by the platelet, is inhibited by 3–5 MPa. The effects of nitrogen and other inert gases were studied by Pickles et al. (1989, 1990) and by McIver et al. (1990) and their results were not inconsistent, but complicated to compare and summarise. Essentially, high partial pressures of nitrogen inhibit ADP-induced aggregation, but not the initial stage. It is difficult to separate the hydrostatic and partial pressure contributions. However, both hydrostatic pressure and nitrogen have no effect on the binding of ADP to the platelet surface.

As deep-diving mammals have bubbles in their blood and tissues during decompression (ascent), are they adapted to cope with a vastly more frequent exposure to bubbles than humans experience.? Is their platelet count reduced, post dive? Such an unhealthy outcome seems unlikely, but we need more basic information. Some preliminary work on seal platelets has proved both interesting and encouraging. Platelets from the Northern Elephant seal (*Mirounga angustirostris*) have been compared with human platelets (Field and Tablin 2012). This seal dives to a depth of 1600 m and thus experiences significant pressure (16 MPa). Fortunately, cooperative specimens were available on a Californian beach where they provided blood samples. One property was common to both seal and human platelets, namely activation by cooling to 4 °C from a body temperature of 37 °C. This is surprising as the peripheral circulation of the seal presumably experiences such low temperatures frequently, so presumably, other factors offset this obvious health risk. Some effects of pressure were studied in vitro by holding a platelet suspension at 19 MPa for 55 min at 37 °C, with 5 min for compression and decompression, respectively. This corresponds the pressure excursion experienced in a very deep dive. A high proportion of human platelets underwent shape change, but the seal platelets were unaffected. This is a very nice example of pressure tolerance. The human platelets showed a marked increase in fibrinogen binding some 30 min after decompression (in the absence of gas bubbles), which is further evidence of activation, but the seal platelets were unaffected.

In the same study, the membrane of the seals' platelets was investigated. They have a very high cholesterol content, about three times that of the human platelet, and thus a highly ordered membrane structure. How this relates to the platelets' insensitivity to pressure is not clear, but it would be surprising if it was not involved and perhaps, in some way, it offsets cold activation at high pressure.

The Killer whale, *Orcinus orca*, actually a deep-diving porpoise, is another diver whose platelets have been studied, but only at normal atmospheric pressure. According to Patterson et al. (1993), Orcinus can dive to 700 m depths without any apparent effect on the platelets in circulation, but no evidence was cited. Platelet aggregation in vitro following exposure to ADP or arachidonate is limited and reversed but collagen causes the normal full aggregation, unlike the platelets of the Northern Elephant seal (but just as in human platelets!). Mammalian platelets show a lot of variation in activation mechanisms so to relate individual patterns to environmental adaptation requires a great deal of evidence. Access to diving mammals in the wild and in aquaria is such that securing blood samples and hence platelets is not an insuperable difficulty, so that a full story of platelet adaptation to compression and decompression in diving mammals might eventually be achieved.

## 11.7   Conclusion

Historically, deep-diving mammals have attracted the interest of respiratory physiologists and the expectation was that neurophysiologists, familiar with the HPNS in humans, would take up the challenge of studying the divers' central nervous system. To some extent, this is taking place. The comparative physiology/biochemistry of the high-pressure adaptation of their nervous system is just one of several aspects of the brain of these remarkable animals which await concerted investigation. The cellular physiology/biochemistry of their non-nervous tissues has special interest too.

## References[1]

Baj Z, Olszanski R, Majewska E, Konarski M (2000) The effect of air and nitrox diving on platelet activation tested by flow cytometry. Aviat Space Environ Med 71:925–928. *

Balmert A, Bohn HF, Ditsche-Kuru P, Barthlott W (2011) Dry under water: comparative morphology and functional aspects of air-retaining insect surfaces. J Morphol 272:443–451. *

Barnard EEP (1971) Submarine escape from 600 feet (183 metres). Proc Roy Soc Med 64:1271–1273. *

Barthelemy L, Belaud A, Salou A (1981) A study of the specific action of "per se" hydrostatic pressure on fish considered as a physiological model. Underwater Physiology VII pp. 641–649. Undersea Medical Society Bethesda. USA *

Batinic T et al (2016) Dynamic diaphragmatic MRI during apnea struggle phase in breath-hold divers. Resp Physiol Neurobiol. 222:55–62

Bennett PB, Rostain JC (2003) The high pressure nervous syndrome. Chap. 9 in Brubaak AO and Neuman TS (eds) Physiology and medicine of diving. 5th edn. Saunders *

Bennett PB, Papahadjopoulos D, Bangham AD (1967) The effect of raised pressure of inert gases on phospholipid membranes. Life Sci 6:2527–2533. *

Bennett PB, Blenkarn GD, Roby J, Youngblood A (1974) Suppression of the high pressure nervous syndrome in human deep dives by he-N2-O2. Undersea Biomed Res 1:221–237. *

Bitterman N (2004) CNS oxygen toxicity. Undersea Hyperbar Med 31:63–72. *

Blix AS, Walloe L, Messelt EB (2013) On how whales avoid decompression sickness and why thay sometimes strand. J Exptl Biol 216:3385–3397. *

Brauer RW, Sheean ME (1974) N2, H2 and N2O antagonism of high pressure nervous syndrome in mice. Undersea Biomed Res 1:59–72. *

Brauer RW, Beaver RW, Hogue CD, Ford B, Goldman S, M., and Venters, R.T. (1974) Intra- and inter species variability of vertebrate high pressure neurological syndrome. J Appl Physiol 37:844–851. *

---

[1]Those marked * are further reading.

Brauer RW, Beaver RW, Mansfield WM, O'Connor F, White LW (1975) Rate factors in the development of the high pressure neurological syndrome. J Appl Physiol 38:220–227. *

Brauer RW, Mansfield WM, Beaver RW, Gillen H, W. (1979) Stages in the development of the HPNS in the mouse. J Appl Physiol 46:756–765. *

Brubakk AO, Neumann TS (2003) Physiology and medicine of divine, 5th edn. Saunders

Cantu-Medellin N et al (2011) Differential antioxidant protection in tissues from marine mammals with distinct diving capacities. Shallow / short vs. deep/ long divers. Comp Biochem Physiol A 158:438–443

Carpenter FG (1954) Anesthetic action of inert and unreactive gases on intact animals and isolated tissues. Am J Phys 176:505–509. *

Castellini M (2010) Pressure tolerance in diving mammals and birds. In: Sebert P (ed) Comparative high pressure biology. Science Publishers, pp 379–398

Castellini MA, Castellini JM, Rivera PM (2001) Adaptations to pressure in the RBC metabolism of diving mammals. Comp Biochem Physiol A 129:751–757

Clark JM, Thom SR (2003) Oxygen under pressure. In: Brubak AO, Neuman TS (eds) The physiology and medicine of diving. Saunders. *

Colloc'h N et al (2007) Protein crystallography under xenon and nitrous oxide pressure: comparison with in vivo pharmacology studies and implications for the mechanism of inhaled anesthetic action. Biophys J 92:217–224. *

Croll DA, Nishiguchi MK, Kaupp S (1992) Pressure and lactate dehydrogenase function in diving mammals and birds. Physiol Zool 65:1022–1027

Cullen SC, Gross EG (1951) The anaesthetic properties of xenon in animals and human beings, with additional observations on krypton. Science 113:580–582

Demchenko IT et al (2017) Antiepileptic drugs prevent seizures in hyperbaric oxygen: a novel model of epileptiform activity. Brain Res 1657:347–354. *

Dinse A et al (2005) Xenon reduces glutamate-, AMPA-, and kainite -induced membrane currents in cortical neurones. Brit. J Anaesthesia 94:479–485. *

Duncan SL, Larson RG (2010) Folding of lipid monolayers containing lung surfactant proteins SP-B 1-25 and SB-C studied via coarse grained molecular dynamics simulations. Biochim Biophys Acta 1798:1632–1650

Field CL, Tablin F (2012) Response of northern elephant seal platelets to pressure and temperature changes: a comparison with human platelets. Comp Biochem Physiol A 162:289–295

Foot NJ, Orgeig S, Daniels CB (2006) The evolution of a physiological system: the pulmonary surfactant system in diving mammals. Resp Physiol Neurobiol 154:118–138. *

Foot NJ et al (2007) Positive selection in the N-terminal extramembrane domain of lung surfactant protein C (SP-C) in marine mammals. J Molec Evolut 65:12–22

Fossette S et al (2010) Behaviour and buoyancy regulation in the the deepest diving reptile: the leathertback turtle. J Exptl Biol 213:4074–4083

Franks NP (2008) General anaesthesia: from molecular targets to neuronal pathways of sleep and arousal. Nat Rev Neurosci 9:370–386

Franks NP, Lieb WR (1988) Volatile general anaesthetics activate a novel neuronal K current. Nature 333:662–664. *

Franks NP, Lieb WR (1994) Molecular and cellular mechanisms of general anaesthesia. Nature 367:607–613. *

Friedrich O (2010) Pressure, the nervous system and ion channels: are humans too complicated? Undersea Hyperbar Med 37:241–244. *

Geiseler SJ, Larson J, Folkow LP (2016) Synaptic transmission despite severe hypoxia in hippocampal slices of the deep-diving hooded seal. Neuorscience 334:39–46

Gronning M, Aarli JA (2011) Neurological effects of deep diving. J Neurol Sci 304:17–21

Grover CA, Grover DH (2014) Albert Behnke: Nitrogen narcosis. J Emerg Med 46:225–227

Gutierrez DB et al (2015) Phosphatidylcholine composition of pulmonary surfactant from terrestrial and marine diving mammals. Resp Physiol Neurobiol 211:29–36. *

Hall AC, Pickles DM, Macdonald AG (1993) Aspects of eukaryote cells. In: Macdonald AG (ed) Comp. and environ. Physiol. 17. Effects of high pressure on biological systems. Springer, pp 30–85. *

Harris DJ, Coggin R, Roby G, Turner G, Bennett PB (1982) Slowing of S.E.P. late waves in gas breathing and liquid breathing dogs compressed up to 101 bars. Undersea Biomed Res 9 (abstract 1):7. *

Heezen BC (1957) Whales entangled in deep sea cables. Deep Sea Res 4:105–115

Hinton HE (1976) Plastron respiration in bugs and beetles. J Insect Physiol 22:1529–1550

Hobbs M, Kneller W (2015) Inert gas narcosis disrupts encoding but not retrieval of long term memory. Physiol Behaviour 144:46–51. *

Hobi N et al (2016) A small key unlocks a heavy door: the essential function of the small hydrophobic proteins SP-B and SP- C to trigger adsorption of pulmonary surfactant lamellar bodies. Biochim Biophys Acta 1862:2124–2134

Hooker SK et al (2012) Deadly diving? Physiological and behavioural management of decompression stress in diving mammals. Proc R Soc B 279:1041–1050

Il'ina TN et al (2017) Antioxidant defense system in tissues of semi aquatic mammals. J Evolut Biochem Physiol 53:282–288

Johnson FH, Flagler EA (1950) Hydrostatic pressure reversal of narcosis in tadpoles. Science 112:91–92. *

Jones KK, Hetz SK, Seymour RS (2018) The effects of temperature, activity and convection on the plastron Po2 of the aquatic bug Aphelocheirus aestivalis (Hemiptera; Aphelochieridae). J Insect Physiol 106:155–162

Joulia F et al (2003) Breath-hold training of humans reduces oxidative stress and blood acidosis after static and dynamic apnea. Respirat Physiol Neurobiol 137:19–27

Kendig JJ (1984a) Ionic currents in vertebrate myelinated nerve at hyperbaric pressure. Am J Phys 246C:84–90. *

Kendig JJ (1984b) Nitrogen narcosis and pressure reversal of anesthetic effects in node of Ranvier. Am J Phys 246C:91–95. *

Koch C, Greenfield S (2007) How does consciousness happen? Sci Am 297:76–83

Kohshi K et al (2014) Brain damage in commercial breath-hold divers. PLoS One 9:e105006

Kooyman GL, Hammond DD, Schroeder JP (1970) Bronchograms and tracheograms of seals under pressure. Science 169:82–84

Kooyman GL, Schroder JP, Denison DM, Hammond DD, Wright JJ, Bergman WP (1972) Blood nitrogen tensions of seals during simulated dives. Am J Phys 223:1016–1020. *

Kylstra JA, Nantz R, Crowe J, Wagner W, Salzman HA (1967) Hydraulic compression of mice to 166 atmospheres. Science 158:793–794. *

Larson J, Drew KL, Folkow LP, Milton SL, Park TJ (2014) No oxygen? No problem! Intrinsic brain tolerance to hypoxia in vertebrates. J Exptl Biol 217:1024–1039

Lee J-L, Lee H-H, Kim S, Jang Y-J, Baek Y-J, Kang K-Y (2016) Diving bradycardia of elderly Korean women divers, Haenyeo, in cold seawater: a field report. Ind Health 54:183–190

Leonardi MS, Lazzari C (2014) Uncovering deep mysteries: the underwater life of an amphibious louse. J Insect Physiol 71:164–169

Lever MJ, Miller KW, Paton WDM, Smith EB (1971) Pressure reversal of anaesthesia. Nature 231:368–371

Lindholm P, Lundgren CEG (2009) The physiology and pathophysiology of human breath-hold diving. J Appl Physiol 106:284–292

Lonati GL, Westgate AJ, Pabst DA, Koopman HN (2015) Nitrogen solubility in odontocete blubber and mandibular fats in relation to lipid composition. J Exptl Bio 91(218):2620–2630. *

Lopez LMM, Miller PJO, de Soto NA, Johnson M (2015) Gait switches in deep diving beaked whales: biomechanical strategies for long-duration dives. J Exp Biol 218:1325–1338

Lopez-Rodriguez E, Perez-Gil J (2014) Structure -function relationships in pulmonary membranes: from biophysics to therapy. Biochim Biophys Acta 1838:1568–1585. *

Lundgren CEG, Ornhagen HC (1976) Hydrostatic pressure tolerance in liquid breathing mice. Proc. fifth symposium on underwater physiology. FASEB Bethsda, USA, p 397–404 *

Macdonald S, Kitching JA (1976) The effect of partial pressures of inert gases on the behaviour and survival of the heterotrich ciliate *Spirostomum ambiguum*. J Exp Biol 64:615–627. *

Macdonald AG, Ramsey RL (1995) The effects of nitrous oxide on a glutamate gated ion channel and their reversal by high pressure: a single channel analysis. Biochim Biophys Acta 1236:135–141

Macdonald AG, Wann KT (1978) Physiological aspects of anaesthetics and inert gases. Academic Press, London. *

Macdonald AG, Ramsey RL, Drewry J, Usherwood PNR (1993) Effects of high pressure on the channel gated by the quisqualate sensitive glutamate receptor of the locust muscle and its blockade by ketamoine: a single channel analysis. Bichimica et Biophysica Acta 1151:13–20

Marassio G et al (2011) Pressure response analysis of anesthetic gases xenon and nitrous oxide on urate oxidase: a crystallographic study. FASEB J 25:2266–2275. *

McIver DJL, Fields ND, Philp RB (1990) Isolated human blood platelets discriminate between anaesthetic and non-anaesthetic gases at high pressures. Brit J Anaesthes 64:77–84

Miller KW (2002) The nature of sites of anaesthetic action. Brit J Anaesthesia 89:17–31. *

Miller KW, WDM P, Smith EB, Smith RA (1972) Physicochemical approaches to the mode of action of general anaesthetics. Anesth 36:339–351. *

Miller KW, Paton WDM, Smith EB, Smith RA (1973) The pressure reversal of anaesthesia and the critical volume hypothesis. Molec Pharmacol 9:131–143. *

Miller JB, Aidley JS, Kitching JA (1975) Effects of helium and other inert gases on Echinosphaerium nucleofilum (Protozoa, Helizoa). J Exp Biol 63:467–481. *

Miller NJ et al (2004) Control of pulmonary surfactant secretion in adult California Sea lions. Biochem Biophys Res Commun 313:727–732

Miller AJ, Postle AD, Orgeig S, Koster G, Daniels CB (2006) The composition of pulmonary surfactant from diving mammals. Resp Physiol Neurobiol 152:152–168. *

Moore Mj, Early GA (2004) Cumulative sperm whale bone damage and the bends. Science 306:2215. *

Moore MJ et al (2011) Hyperbaric computed tomographic measurements of lung compression in seals and dolphins. J Exptl Biol 214:2390–2397

Naquet R (1988) In: Lemaire C, Rostain J-C, Octares M (eds) The high pressure nervous syndrome and performance. France

Oei GTML, Weber NC, Hollmann MW, Preckel B (2010) Cellular effects of helium in different organs. Anesthesiol 112:1503–1510

Ostrowski A, five authors (2012) The role of training in the development of adaptive mechanisms in Freedivers. J Human Kinetics 32:197–210

Panneton WM (2013) The mammalian diving response: an enigmatic reflex to preserve life? Physiology 28:284–297

Parra E, Perez-Gill J (2015) Composition, structure and mechanical properties define performance of pulmonary surfactant membranes and films. Chem Phys Lipids 185:153–175

Patlack JB, Gration KAF, Usherwood PNR (1979) Single glutamate-activated channels in locust muscle. Nature 278:643–645. *

Patterson WR, Dalton LM, McGlasson DL, Cissik JH (1993) Aggregation of killer whale platelets. Thrombosis Res 70:225–231

Philp RB (1990) Pharmacological studies on the mechanism of pressure inhibition of human platelet aggregation. Aviat Space Environ Med 61:333–337. *

Pickles DM, Ogston D, Macdonald AG (1989) Effects of gas bubbling and other forms of convection on platelets in vitro. J Appl Physiol 76:1250–12355

Pickles DM, Ogston D, Macdonald AG (1990) Effects of hydrostatic pressure and inert gases on platelet aggregation in vitro. J Appl Physiol 69:2239–2247

Ponganis PJ, Kooyman GL, Ridgway SH (2003) Comparative diving physiology. Chap. 7 in Brubakk AO, Neuman TS (eds) The Physiology and medicine of diviing. Saunders

Price RA, Ogston D, Macdonald AG (1986) Effects of high hydrostatic pressure on platelet aggregation vitro. Undersea Biomed Res 13:63–75. *

Quillin ML, Breyer WA, Griswold IJ, Matthews BW (2000) Size versus polarizability: binding of noble gases within engineered cavities in phage T$ lysozyme. J Molec Biol 302:955–977

Rahn H, Paganelli CV (1968) Gas exchange in gas gills of diving insects. Resp Physiol 5:145–164. *

Reisinger RY, Keith M, Andrews RD, de Bruyn PJN (2015) Movement and diving of killer whales (*Orcinus orca*) at a Southern Ocean archipelago. J Exptl Marine Biol Ecol 473:90–102. *

Righetti BPH, Simoes-Lopes PC, Uhart MM, Filho DW (2014) Relating diving behaviour and antioxidant status: insights from oxidative stress biomarkers in the blood of two distinct divers, Mirounga leonine and *Arctocephalus australis*. Comp Biopchem Physiol A 173:1–6. *

Rostain J-C, Dumas JC, Gardette B, Imbert JP, Lemaire C (1984) Effects of addition of nitrogen during rapid compression of baboons. J Appl Physiol 57:332–340. *

Rostain JC, Gardette-Chauffure MC, Gardette B (1994) Neurophysiological studies in *Macaca fascicularis* during exposures with breathing mixtures containing hydrogen up to 12000 metres sea water. In: Bennett PB, Marquis RE (eds) Basic and applied high pressure biology. University of Rochester Press, pp 243–252

Rostain JC, Risso JJ, Abraini JH (2010) The high pressure nervous syndrome. In: Sebert P (ed) Comparative high pressure biology. Science Publishers, pp 431–460

Schaefer KE et al (1968) Pulmonary and circulatory adjustments determining the limits of depths in breathold diving. Science 162:1020–1023. *

Schlamm NA, Perry JE, Wild JR (1974) Effect of helium at elevated pressure on iron transport and growth of *Escherichia coli*. J Bact 117:170–174. *

Scholander PF (1940) Experimental investigations on the respiratory function in diving mammals and birds. Hvalradtes Skrifter 22:1–131. *

Schorr GS, Falcone EA, Moretti DJ, Andrews RD (2014) First long-term behavioural records from Cuvier's beaked whales (*Ziphius cavirostris*) reveal record breaking dives. PLoS One 9(3): e92633

Seymour RS, Hetz SK (2011) The diving bell and the spider: the physical gill of Argyroneta aquatic. J Exptl Biol 214:2175–2181

Seymour RS, Matthews PGD (2013) Physical gills in diving insects and spiders: theory and experiment. J Exptl Biol 216:164–170

Smit KF et al (2015) Effects of helium on inflammatory and oxidative stress-induced endothelial cell damage. Exptl Cell Res 337:37–43. *

Smith EB et al (1984) Species variation and their mechanism of pressure-anaesthetic interactions. Nature 311:36–37. *

Snow SJ (2006) Operations without pain. Palgrave Macmillan, Basingstoke. *

Southall BL et al (2019) Sonar disturbs blue whales when feeding. J Exp Biol 222:1. *

Spragg RG et al (2004) Surfactant from diving aquatic mammals. J Appl Physiol 96:1626–1632. *

Talpalar AE, Grossman Y (2005) Sonar versus whales: noise may disrupt neural activity in deep diving cetaceans. Undersea Hyperbaric Med 32:135–139

Taylor CD (1979) Growth of a bacterium under a high-pressure oxy-helium atmosphere. Appl Environ Microbiol 37:42–49. *

Thom SR, Marquis RE (1981) Contrasting actions of hydrostatic pressure and helium pressure on growth of Saccharomyces cerevisae. Underwater Physiol VII:667–673. Bachrach AJ, Matzen MM (eds) Undersea Medical Soc., Bethesda *

Thom SR, Bhopale VM, Yang M (2014) Neutrophils generate microparticles during exposure to inert gases due to cytoskeletal oxidative stress. J Biol Chem 289:18831–18845

Trudell JR, Koblin DD, Eger EI (1987) A molecular description of how noble gases and nitrogen bind to a site of anesthetic action. Anesth Analgesia 87:411–418. *

Tunstall ME (1961) Obstetric analgesia. Lancet 2:964

Tyack PL et al (2015) Formal comment on Schorr GS et al (2014) "first long term behavioural records from Cuvier's beaked whales (*Ziphius cavirostris*) reveal record breaking dives". PLoS One 9:e92633

Udyawer V, Simpfendorfer CA, Heupel MR, Clark TD (2016) Coming up for air: thermal dependence of dive behaviours and metabolism in sea snakes. J Exptl Biol 219:3447–3454. *

Valeri CR et al (1974) Effects of hyperbaric exposures on platelets. Aerosp Med 45:610–616. *

Vasquez-Medina JP, Zenteno-Savin T, Elsner R, Ortiz RM (2012) Coping with physiological oxidative stress: a review of antioxidant strategies in seals. J Comp Physiol B 182:741–750

Wann KT, Macdonald AG (1988) Actions and interactions of high pressure and general anaesthetics. Prog Neurobiol 30:271–307

Williams EE et al (2001) Hydrostatic pressure and temperature effects on the molecular order of erythrocyte membranes from deep-, shallow-, and non-diving mammals. Can J Zool 79:888–894

Williams GD, et al. (2016) The suppression of Antarctic bottom water formation by melting ice shelves in Prydz Bay. Nature Communications, p 1

Winkler DA, Thorntom A, Farjot G, Katz I (2016) The diverse biological properties of chemically inert noble gases. Pharmacol Ther 160:44–64. *

Winter PM, Smith RA, Smith M, Eger EI (1976) Pressure antagonism of barbiturate anesthesia. Anesthesiology 44:416–419. *

# Adaptation to High Pressure in the Laboratory

**12**

Can microorganisms and animals be adapted to high pressure in controlled experimental conditions? The susceptibility of organisms to pressure is called tolerance or resistance and increasing it by exposure to high pressure is a form of adaptation, which in animal physiology, is called acclimation or acclimatisation, of which more in Section 12.2. Adaptation to high pressure by other means is also possible. The converse adaptation of a high-pressure organism to a lower pressure can also occur. The formal term for these changes is the plasticity of the phenotype, (Hendry 2016; Ghalambor et al. 2015). How they relate to evolution is not simple, and well beyond the scope of this book.

## 12.1  Microorganisms

Microorganisms can undergo adaptive changes in their tolerance to high pressure, or to lower than normal pressure which, in practice are detected in a population of cells. Improved growth or survival in the face of an initially adverse effect of pressure is a common type of adaptation. Its limits are set by the initial genome, but such microorganisms have a short generation time and can undergo mutations and thus change their genome. Laboratory evolution is a powerful method of studying the molecular processes underlying the natural selection for specific traits.

The resistance of bacteria to antibiotics is a well-known medical crisis and in the present context, it illustrates the extraordinary ability of bacteria to cope with short-term challenges to their survival. Antibiotics were introduced 70 years ago, to great effect, but we now know that the spread of resistance was inevitable. The intrinsic ability of bacteria to cope with foreign toxic chemicals has been demonstrated in DNA extracted from permafrost formed in the Late Pleistocene and in the current prevalence of antibiotic resistance in ordinary bacteria distributed across our planet. The essential step in developing antibiotic resistance in a population is, of course, the survival of a small number of inherently resistant individuals. The initial, fortuitous mechanism of resistance is not important but cells possessing it prosper. However, it

© Springer Nature Switzerland AG 2021                                           327
A. Macdonald, *Life at High Pressure*, https://doi.org/10.1007/978-3-030-67587-5_12

leads by mutations and horizontal gene transfer (involving plasmids, bacteriophages, and the uptake of DNA and other genetic tricks), to populations of cells becoming well equipped to deal with antibiotics which were initially lethal to the vast majority of their kind. The antibiotics may be prevented from entering the cell, inactivated, or sequestered, whilst at the same time their presence, even at low concentration, can enhance the rate of mutation and generally favour diversification, which in turn favours survival (D'Costa et al. 2011; Mazel and Mobashery 2012).

Three species of microorganisms are selected here to illustrate some examples of adaptation to high pressure in the laboratory; two species of bacteria and one Eukaryote species, yeast

### 12.1.1 *Escherichia coli*

In High Pressure Biology the bacterium *E. coli* is the cell that we understand most fully, a statement that probably applies to Biology in general. It qualifies as a Mesophile, that is, it is capable of growing at intermediate pressures. How is it that an organism, which normally lives in the gut of mammals and never experiences high pressure, can grow at pressures found at significant ocean depths? How does natural selection achieve such a state of affairs? One naïve possibility is that the thermodynamics underlying its metabolic reactions are simply unaffected by pressure, but we can discount that on experimental and theoretical grounds. A clue lies in the finding that prior exposure to a high temperature increases the pressure at which *E. coli* can grow. The heat stress response (Chapter 2), causes proteins to be produced which can handle those damaged by high temperature. It is thought that a transient increase in the concentration of a molecule that promotes transcription leads to the synthesis of the heat shock proteins, which can also deal with those damaged, unfolded perhaps, by high pressure. (This is also seen in yeasts, see below). Natural selection results in broad-acting stress responses.

Although *E. coli* may not encounter high pressure in Nature it most certainly does in the food processing industry. High pressure is used to Pasteurise (Pascalise in the case of pressure) and sterilise foods and the survival of *E. coli* and other potentially dangerous bacteria is a matter of great concern. Changes in the pressure tolerance of *E. coli* and the multiplicity of factors that can influence it are important to understand.

Most of *E. coli* cellular processes are affected by intermediate pressures. Bartlett (2002), has helpfully summarised the effects of pressure, with motility being the most sensitive. 10 MPa inhibits the activity of the driving flagellum, and cell division is the next most sensitive process, being affected by about 20 MPa. Transmembrane transport follows, with gene transcription susceptible to 30 MPa, followed by DNA synthesis (50 MPa), translation (60 MPa) and RNA transcription (77 MPa). Its stress response may also be adversely affected by high pressure because, like the previously mentioned processes, it involves a multiplicity of non-covalent bonds that are susceptible to pressure, as are the resultant oligomeric stress proteins.

Can *E. coli* adapt to high pressure? For example can it become capable of growing at previously inhibitory pressures and become more resistant to initially lethal pressures? Here are some examples of experiments that attempt to tackle these questions.

The adaptive laboratory evolution of *E. coli*, over 505 generations, was achieved by Bartlett and his colleagues (Marietou et al. 2015), by initially growing cells at 41 MPa and then, after dilution, at progressively higher pressures. This strategy culminated in growth at 62 MPa, and compensatory changes in the cells' membrane fatty acids. An isolated clone, AN62, proved able to grow at the astonishing high pressure of 62 MPa with a doubling time of 70 mins. The parent cells were unable to grow at that pressure. The ratio of unsaturated: saturated fatty acids in the membrane of AN62 increased considerably when it grew at high pressure, largely through the increase in the synthesis of the 18 carbon cis-vaccenic acid. This outcome was attributed to a mutation that occurred in one of the 505 generations, which led to greater synthesis of fatty acids. Doubtless additional changes are also required to account for the growth performance. AN62 grown at normal atmospheric pressure was, like the parent strain, typically rod shaped, but when grown at 62 MPa it became filamentous, an indication that its division mechanism was still impaired by pressure (Marietou et al. 2015).

The *E. coli* strain NCTC8164 was grown at different temperatures by Casadei and colleague, to produce cells with membranes of different fluidities, an example of homeoviscous adaptation (Abe and Usui 2013). For example cells grown at 10 °C had a high proportion of unsaturated fatty acids in their membrane lipids, which, when extracted, were found to have a lower phase transition temperature than lipids from cells grown at higher temperatures. That implies the 10 °C cells had a more fluid membrane bilayer than cells grown at 37 °C. So far, so good—it is always worthwhile to confirm (or refute!) previous work. Then the cells with fluid membrane bilayers were tested for their resistance to a 5-minute exposure to a very high, lethal, pressure (100–600 MPa). The 10 °C cells proved to be more resistant than cells grown at higher temperatures, with less fluid bilayers. Some years later an interesting parallel finding emerged. A mutant *E. coli* strain K1060, which is defective in its membrane lipid metabolism, can be grown in culture medium enriched with specific fatty acids with the result that cells can be grown at a common temperature with membrane bilayers containing different fatty acid proportions and hence different fluidities. Such cells with fluid membrane bilayers proved to be more resistant to very high pressure than those with less fluid membranes. How is it that a fluid bilayer is linked to high-pressure resistance? One irreversible effect of a very high pressure, as used in these experiments, is to severely order the bilayer whose lateral compressibility is considerable. The result is those membrane proteins are extruded from the bilayer, an irreversible effect. The more fluid bilayer would be less susceptible to this than an ordered bilayer. Another possibility is the denaturation of membrane proteins.

A growing culture of *E. coli* subjected to a pressure of, say 55 MPa, undergoes a reduction in growth rate and protein synthesis is inhibited. However, special stress proteins, pressure-induced proteins, PIP, are a notable exception. Using a method of radiolabelling the cells at high pressure, Welch et al. (1993) showed the proteins are induced within 60–90 mins of exposure to pressure. The population of cells resumed growth in mass after a lag period. This is a much slower response than seen when the cells are heat shocked but it is an adaptation to pressure. It is not known what the stress proteins actually do to adapt the cells to pressure. In the case of heat shock the stress proteins are thought to restore denatured proteins or help dispose of them. 55 MPa is not much pressure to denature proteins. E. coli also has a separate stress response, called SOS, which is primarily involved with the repair of DNA damaged by, for example toxins or U V radiation. Very high pressure, 100 MPa, appears to stimulate this response in a rather indirect and complicated way as it does not damage the DNA directly. This is significant because the SOS response also plays a role in creating genetic changes that serve to enhance the cells' ability to adapt to changed environmental conditions.

The general stress response can determine the resistance of E. coli to pressure (Charoenwong et al. 2011). The RpoS regulon plays an important role in the magnitude of the stress response and in other aspects of the cells' physiology, including membrane lipids. It is the basis of stationary phase cells (strain BW2952), being more resistant to high pressure than cells in the exponentially growing phase of a culture. A very high pressure (100–400 MPa for 8 mins), was sufficient to damage the cell membrane by extruding proteins. Stationary phase cells (BW2952) are less affected in this regard than exponentially growing cells and a mutant strain (BW3709) which has a defective rpos gene, suffered the most damage.

E. coli and other bacteria of interest in food preservation are found to be more pressure resistant in high concentrations of salt or sucrose. This is due to the cells accumulating compatible solutes which stabilise proteins against the effects of high pressure (Chapter 8).

In very high-pressure adaptation experiments (300 MPa) with the extremely important food pathogen E. coli 0157:H7, which can cause haemorrhagic colitis in humans, the cells with a marked increase in tolerance to pressure and to high temperature appeared. (Vanlint et al. 2013). In some particularly careful pressure tolerance studies of the E. coli strain MG1655, Griffin and colleagues, applying an idea long established in microbiology, suggested that within an apparently homogeneous population there exists a subpopulation of resistant cells, in this case, pressure resistant, which exist with no prior selection by pressure. A plausible explanation involves the selection of RpoS, which might be present in a mutated or other highly potent form.

Laboratory evolution experiments have been carried out with E. coli to study rates of mutation in various conditions. The 500 generations mentioned earlier are greatly exceeded in such studies, which can achieve many thousands of generations. However, as is thought that mutations often arise during cell division it has been argued that the cumulative number of cell divisions is important. Accordingly, experiments have been carried out with E. coli, not at high pressure, in which many billions of cell divisions have taken place, and the large number of mutations were recorded by sequencing the genomes. Such experiments involve the serial inoculation of batch cultures but adaptation to pressure has not been tested (Lee et al. 2011).

Black and colleagues (2013) adopted a different approach, screening a collection of single gene mutants (i.e. cells lacking a specific gene) using a high throughput, high-pressure cell culture method. It reveals what genes are required for the growth of *E. coli* K-12 at a given pressure. For example it established that for growth at 30 MPa six specific genes were required and for growth at 40 MPa a further 3 were needed. The elegance of the approach is that the "need" can be confirmed by re-introducing the absent gene, to restore growth. This is done via a plasmid containing the gene. The role of a gene required for growth at high pressure may be direct or indirect. For example the gene *dedD* codes for a protein which stabilises the septal ring critical in cell division mentioned earlier, and is essential at pressures that would otherwise destabilise its structure. This is a direct effect. Others might enable essential molecules to enter the cell from the external environment, a more indirect role. Such genes can be said to contribute to adaptation to high pressure, as in their absence the cells perform less well.

A simple selection procedure was adopted by Hauben and colleagues (1997), in experiments with *E. coli* strain MG1655 which examined the resistance of the cells to the lethal effects of very high pressure, so-called inactivation. It consisted of exposing cultures to a lethal pressure (280 MPa, 15 mins, room temperature) followed by a period at normal atmospheric pressure during which time the survivors would grow (separately at 30, 37 and 42 °C) and provide cultures for exposure to another lethal dose of pressure. A total of 18 cycles of such treatment, with the lethal dose of pressure being increased, produced three strains of *E. coli* from the three temperatures used, each with an increased resistance to pressure inactivation. For example the resistance to an exposure to 800 MPa was hugely increased over that of the parent strain. The increased resistance to pressure inactivation was permanent, persisting after growth at atmospheric pressure and was taken to reflect multiple mutations. It had nothing to do with any subpopulation of resistant cells which may have existed in the parent culture. Most interesting of all, the resistance to inactivation was not linked to an increase in the ability to actually grow at high pressure. In fact, growth at 50 MPa was slightly diminished. This result was significant in the design of high-pressure sterilisation procedures used in the food industry, and indeed the work was the result of a collaboration between food microbiologists, in Belgium, and marine microbiologists, in the USA.

For the very high-pressure experiments carried out by Sharma and his colleagues a diamond anvil cell was used (Chapter 13). *E. coli* and *Shewanella oneidensis* MR1 were separately pressurised in a micro-chamber mounted in the anvil. A spectroscopic technique was used to monitor the anaerobic oxidation of formate at pressure, and 25 °C. The concentration of formate was reduced markedly over the first 300 MPa but a low rate of oxidation persisted up to 1060 MPa. This evidence of metabolic activity at such a high pressure is astonishing. It was supported by control experiments in which, for example dead cells or cells poisoned with cyanide, failed to oxidise the formate, but extracellular formate oxidation cannot be excluded. Microscopic examination confirmed the presence of motile cells, with the implication that motility was evidence of life. Experiments at higher pressures incurred the formation of a form of Ice, called Ice VI, at 1250 MPa and room temperature. Both *Shewanella* and *E. coli* showed a decline in motility and formate oxidation at pressures up to 1200 MPa. At higher pressure, up to 1600 MPa, ice VI formed with fluid persisting in interstitial cracks in the mosaic of solid ice. Decompression melted Ice VI and cell counts revealed a decrease in the number of cells present, i.e. surviving, with some larger than normal. This was interpreted as the result of cell division being blocked. Clearly this is a classic case of provocative experiments producing more questions than answers. The paper was discussed in the journal Science (2002), in which Yayanos stressed the lack of proof that the cells in question could grow in the classical way, by forming colonies on an agar plate. Do the results seriously imply that life might exist at pressure 16 times greater than that at the bottom of the Mariana trench, or some 50 Km down in the earth's crust? Sharma's experiment stirred things up and claiming evidence of living processes at pressures in the GPa range broke the mental barrier many had, that 1000 MPa, or thereabouts, was a pressure limit for life.

Subsequent experiments with *E. coli* have demonstrated remarkable adaptation to high pressure, the criterion being the survival of pressures well into the GPa range. Vanlint et al. (2011a, b) used a simple repetitive process of exposing growing cells of strain K-12MG1655 of *E. coli* to high pressure for 15 mins (initially 100 MPa) during which time a proportion of the cells were killed. The "kill" was quantified by the reduced number of cells able to from colonies on agar, at normal atmospheric pressure. However, the survivors included pressure tolerant cells which, when subjected to the next high-pressure test, increased their presence. Cycles of 15-minute long pressure "shocks," incrementally increased by 100 MPa, followed by recovery, growth and then another higher pressure exposure, produced cells whose resistance to 2000 MPa (2 GPa) was similar to that originally seen after exposure to 600 MPa. An interesting detail is that the "kill," the rate of the inactivation measured over 45 mins was slower at 2 GPa than at 800 MPa. This apparent increased resistance in the surviving cells might, in part, arise from the peculiar state of water, which forms a special form of Ice at 2 GPa. Another interesting point is that equivalent experiments attempting to increase the heat tolerance of the same strain of cells, achieved only a very small increase, whereas, by any standards the increase in the pressure tolerance was huge.

The immediate questions arising from these results focus on the potential damage to proteins and membranes. Do they occur and/or, are there protective or repair mechanisms at work? We do not know. It is also unclear how surviving very high pressure for 15 mins relates to the ability to grow at an equivalent pressure. However, the important point is that the bacterium has the genetic capability to acquire extreme levels of high-pressure resistance when selected. Surprisingly there seem to be no follow-up experiments to report.

## 12.1.2 Other Examples of Organisms Surviving Very High Pressures

Various organisms, prokaryotes and eukaryotes, which have evolved a strategy for surviving desiccation, can survive prolonged exposure to GPa range pressures in their desiccated state. Examples are the eggs and cysts of a crustacean *Artemia* and the Tardigrade, *Milnesium tardigradum* (Ono et al. 2012; Hazael et al. 2019). It seems the absence of water confers this high degree of pressure resistance, providing us with an unexpected demonstration of the role of water in the properties of macromolecules.

The tolerance of bacteria to ultra-high pressures has also been studied in the distinct stress of shock compression, which simulates the fleeting GPa pressures generated when a meteorite hits the earth (Leighs et al. 2012). This subject is concerned with the idea that simple forms of life might survive being transferred from one planetary body to another in this way. Experiments have been carried out with bacteria and bacterial spores in these ultra-extreme conditions. For example Hazael et al. (2017), have shown that *Shewanella* subjected to a transient, less than a second, exposure to 2.5 GPa are not all killed, leaving some survivors able to grow colonies on agar plates, Shock compression subjects cells to something more complicated than a very brief exposure to a high hydrostatic pressure, but nonetheless there is plenty of evidence to support the idea that simple forms of life could survive arriving by meteor.

## 12.1.3 *Photobacterium profundum*

The piezophile *Photobacterium profundum* has almost become the *E. coli* of the deep sea. It belongs in the family Vibrionaceae, and the strain SS9 was originally isolated from Amphipods collected at a depth of 2500 m in the Sulu Trough. Its optimum growth is at 28 MPa and 15 °C, but growth both at normal atmospheric pressure and up to 90 MPa over the temperature range of 2 to 20 °C, is sufficient to support experiments. And a great number of interesting experiments have been reported in the literature. To avoid confusion note that another strain, DSJ4 was also isolated from deep-sea sediments from a depth of 5700 m in the Ryukyu Trench, (Nogi et al. 1998). Many of the experiments with *P. profundum* involve growing the cells and conducting various genetic manipulations. One disconcerting aspect is that certain cold-sensitive mutants were manifest only when grown on agar but not in liquid cultures. This point reminds us that adaptation experiments carried out in laboratory conditions have limitations that we should not overlook when we try to transfer the implications of the results to Nature. Despite this reservation, *P. profundum* SS9 has been a gold mine for microbiologists. Its genome comprises two circular chromosomes and a plasmid. The genome sequence was published by Vezzi et al in a 2005 paper, in which its status as a piezophilic bacterium was underscored by the fact that stress response genes are activated when the cell is grown at normal atmospheric pressure. *E. coli* shows the converse, with its stress response appearing at high pressure.

The second strain of *P. profundum*, designated strain DSJ4, mentioned above, has its optimum growth is at 10 MPa and 10 °C, but it grows well at pressures up to 50 MPa. A third strain of *P. profundum* (strain 3TCK) was isolated from shallow water in San Diego Bay. Its optimum growth is at normal atmospheric pressure at 2–4 °C (Campanaro et al. 2005; Chapter 8, Section 8.9.1). In case you are wondering how these three strains can belong to the same species, the evidence is strong, being based on Campanaro's analysis of the sequence of the 16sRNA. Strain 3TCK is 97.7% identical with SS9 and 98.7% identical to DSJ4.

To quote a paper on the growth of SS9 at various pressure, it undergoes "heavy reorganisation in gene expression between atmospheric pressure and 28 MPa, while this is not seen moving from 28 MPa to 54 MPa." A distinct feature of adaptation experiments on SS9 is that when grown at normal atmospheric pressure it is being "challenged" to adapt to low pressure and when grown at pressures higher than its optimum it is also being "challenged," by high pressure. The idea that a piezophile might adapt to a lower pressure was, at one time, novel, and now it is not, but nevertheless it is instructive. As was mentioned earlier SS9 grown at atmospheric pressure manifests a stress response although it is not clear what molecular changes caused by decreasing pressure from 28 MPa might trigger the response. In fact, four genes which are upregulated are involved in protein folding and the same genes are upregulated in *E. coli* when subjected to high pressure. Additionally other genes in SS9 which are involved in the repair of DNA are also upregulated, mirroring the response of *E. coli* to high pressure. In the latter case, the pressure was insufficient to directly damage the DNA, so indirect damage was assumed, perhaps by enzymes getting out of control. This would also account for the response seen in SS9.

What other adaptational changes can be elicited in SS9 ? There are many examples of genes, other than those involved in the stress response, becoming activated or repressed by a change in pressure. An important discovery by Welch and Bartlett was that a protein situated in the inner membrane of the bacterium plays a role in the mechanism by which pressure is "sensed." It is called Tox R because it is closely related to a protein in the intensively studied bacterium which causes cholera, *Vibrio cholerae*, where it regulates membrane structure when environmental conditions change. The expression of OMP H in the membrane of *P. profundum* grown at high pressure is a form of adaptation to pressure, which has been described in Chapter 8.

The homeoviscous adaptation to high pressure of SS9 was mentioned earlier (it is partial, Chapter 7, Section 7.3). The mono-unsaturated fatty acid C 18: 1 in the membranes is required for growth at high pressure. This was shown by a mutant unable to synthesise it, which fails to grow at pressure, an outcome reversed by the provision of exogenous 18:1. The strain DS14 grown at 10 °C at 27 MPa and at normal atmospheric pressure for 3 days had fatty acids with a higher unsaturation index in the cells grown at high pressure, consistent with homeoviscous adaptation (Wannicke et al. 2015).

*P. profundum*, grown at 28 MPa, accumulates pressure protecting piezolytes, which is a form of extrinsic adaptation to high pressure (Chapter 6). However, it does this in a very flexible way. When grown at atmospheric pressure the accumulation of piezolytes is much less than when grown at 20–30 MPa, during which familiar protective solutes, betaine and glutamate, are significant. This is an excellent example of laboratory adaptation to high pressure. The experiments were very simple and direct; cells were grown in standard conditions to early stationary phase and the solutes they contained were analysed by nuclear magnetic resonance, a technique well suited to the range of solutes of interest. The "solute cocktail," was complicated but its broad effect quite clear (Martin, et al. 2002).

P. profundum has also been the subject of a high-pressure proteomics study, that is, a screening of the proteins which are differentially expressed when the cells are grown at 28 MPa versus atmospheric pressure. This is a powerful approach which certainly reveals the scale of the response of the cells to a high pressure which is also optimum for growth. Of a total of 966 proteins identified, 213 were upregulated at 28 MPa, i.e. their presence increased, and 168 were down regulated. Some of the upregulated proteins were heat shock proteins, for example Gro EL, which implies, but does not prove, a direct effect of pressure. Other, "knock on" effects of disrupted metabolism may occur.

## 12.1.4  Other Microorganisms

P. profundum is not unique in its ability to grow over a wide pressure range. Shewanella violacea, strain DSS12, isolated from sediments at a depth of 5110 m, is another example. It grows optimally at 30 MPa and 8 °C, but adequately at atmospheric pressure, and has been studied by Sato and colleagues. It appears that the stress response is elicited by pressures similar to those of its natural environment, suggesting that their chaperone function may continuously sustain proteins in good shape. This appears to be achieved by the sustained activity of promoters and can be regarded as a dynamic, extrinsic, adaptation to high pressure.

It is not clear if Shewenella manifests a stress response at normal atmospheric pressure but the piezo hyperthermophile Archaeon Thermococcus barophilus appears to do so (Martiensson et al. 1999). This organism was isolated from a depth of 3500 m and grows well at the corresponding pressure. When subjected to low pressure (0.3 MPa) a distinct protein, P60, was significantly induced over the temperature range 75–90 °C. At 40 MPa and "normal" high temperatures, it is only present at low concentration in the cells. Raising the temperature to 98 °C slightly induces it. P60 resembles a heat shock chaperonin protein found in another Archaeon, Pyrococcus sp. Strain KOD1. The idea is that P60 could protect the cells when they are swept up into shallow water, but exactly from what is obscure.

## 12.1.5  Yeast, Saccharomyces sp., a Eukaryote Microorganism

As with Prokaryotes, creating numerous mutants at random in yeasts will produce a number of pressure tolerant strains, by means unknown. The Eukaryote yeasts have proved attractive and rewarding cells in adaptation studies and in other experiments in which high pressure is used to activate gene expression with other goals in mind. For example *Saccharomyces cerevisiae* exposed to 50 MPa for 30 mins at room temperature manifested numerous activated genes, one of which enhanced ethanol production, clearly of interest to the drinks and energy industries (Bravim et al. 2016). In the matter of adaptation to high pressure in this cell we need to distinguish between exposures to very high pressures, e.g. 150–200 MPa, which activates nearly three hundred genes in *S. cerevisiae*, and exposures to sub-lethal pressures, say 25 MPa. The latter has little unfolding effect on soluble proteins and probably acts via membrane proteins. The stress responses which these two pressures elicit, presumably differ.

In *S. cerevisiae* the heat stress response to a moderate temperature enhances the tolerance to a subsequent high temperature. Following exposure to the sub-lethal temperature of 42 °C, the ability of cells to survive short exposures to 150–180 MPa is increased. The heat shock protein 104, induced by thermal stress, appears to limit the damage caused by the high pressure. In strain BY4742 growth at a sub-lethal pressure, 25 MPa, increases the expression of a number of heat shock proteins, including 104, but only one seems to be essential for sustained growth, namely Hsp 31. Mutants which lack this protein in their stress response fail to adapt to pressure (Miura et al. 2006). Thus, the apparently odd ability of a cell-like yeast to adapt to high pressure, which is not present in its normal environment, is not directly selected for in evolution. The adaptation comes about as a fortuitous consequence of a broadly acting stress response to temperature and, for example exposure to hydrogen peroxide.

Iwahashi et al. (2005) and Iwahashi (2015), promoted the idea that *S. cerevisiae* can temporarily adapt to 30 MPa concomitantly activating 366 of its total of 6000 genes. These include stress response genes. There are technical limitations in many experiments in the pressure—stress field which need to be addressed, and which should simplify the interpretation of much of the data. This point is made in the context of animal experiments (Section 12.2 below), to which it is certainly not confined.

The adaptation of yeast to high pressure in the form of acquiring the ability to grow at pressures which are initially inhibitory involves more than stress proteins. Growth depends on the uptake of the amino acid tryptophan and this is inhibited by pressure. Raising the concentration of tryptophan can offset the failing transport system and growth will pick up. Two distinct tryptophan transport proteins called permeases are involved, each the product of separate genes. Introducing one of the genes via a plasmid enables the cells to grow at previously inhibitory pressures. One transporter protein, whose gene is Tat 2, is situated in ordered regions (lipid raft) of the membranes of the Endoplasmic reticulum, Golgi Apparatus and to a limited extent the cell membrane. The other, the product of the gene Tat 1, lies in more fluid bilayer regions of the cell membrane and is a transporter with a low affinity for tryptophan. Its transport of tryptophan is less sensitive to pressure than that of the higher affinity transporter incorporated in ordered bilayers. Transporter (Tat 1) can also benefit from homeoviscous adaptation. The control and the properties of these transporters is a complicated story which has been worked out in detail by Abe and

Iida (2003), in a most impressive way. Another contribution to yeast's adaptation to high pressure comes from trehalose which, as we have seen previously, protects proteins from high pressure (Fujii et al. 1996a, b).

## 12.2 Animals

Adaptation within the lifetime of an animal usually means compensation for the initial effect caused by some external factor, or an increase in tolerance to it. Such adaptations can either be a resistance adaptation, that is a change in lethal limits or a capacity adaptation, a change in the rate of a specific function. Acclimation is the term usually reserved for adaptations to a specific factor in an experiment whilst acclimatisation is used for similar changes taking place in Nature, where a single variable cannot be assumed to be at work.

Within a multiplicity of physiological changes which take place in acclimatisation, some may be more obviously beneficial for function or survival than others. The classic case is high altitude acclimatisation in humans. Some changes are beneficial, for example the increase in the number of red cells in circulation, which tends to offset the reduced partial pressure of oxygen prevalent at high altitude. However, this change is only beneficial within certain limits, because of the increase in blood viscosity, which slows blood flow, and hence the delivery of oxygen. The human body is not particularly efficient at acclimatising to high altitude, probably because the stimulus of the low oxygen pressure acts on a mechanism primarily effective at countering anaemia. The point is acclimation experiments might produce plausible results but not necessarily ideal compensation or adaptation. And they might produce inconclusive, non-lethal pathological effects, which should not be regarded as acclimation. A final point to stress is that the state of acclimatisation is not inherited; rather it is the capacity to acclimate, or not, which is inherited.

Adaptation to temperature change is a much more developed subject than adaptation to high pressure (Somero 2010). We have met temperature adaptation in freshwater fish, in the case of the homeoviscous adaptation of cell membranes (Chapter 7; Section 7.2.2). Can animals acclimate to high pressure? We start with experiments on animals which do not experience pressure changes in their normal life. If they can adapt to high pressure, the interesting question is, how?

### 12.2.1 Shallow Water Animals

One of the first experiments to test an animal for its ability to acclimate to the pressure used the North American fiddler crab, Uca pugilator, and was carried out by Avent (1975). The crab lives on coastal and estuarine mudflats and never experiences pressures higher than those created by the tides. Male animals were kept in 30% seawater at 24 °C in a high-pressure aquarium at 3.4 MPa for 5 or 10 day periods. Then they were decompressed and transferred to a small high-pressure vessel in which they were tested for their response to a stepwise compression schedule of 1.3 MPa per minute. Control experiments showed the untreated crabs responded with agitated motor activity at an average pressure of 14 MPa and became immobilised (paralysed), at 20.9 MPa. Holding the animals at high pressure did not affect the pressure at which agitated activity appeared, but it did increase the pressure which caused paralysis. The 5-day treatment increased the mean pressure causing paralysis by 1.33 MPa (i.e. more than a third of the holding pressure) and the 10-day treatment had only a slightly greater effect. Paralysis describes a state in which the walking legs are curled up tightly into the body and it generally corresponds to the immobilised state seen in many normally active animals at high pressure, subsequent to the seizure stage (Chapter 4). It is interesting and probably significant that the agitated activity did not acclimate to pressure. It might well be a natural behavioural response, attempting to swim to a lower pressure. Paralysis is presumably caused by the direct effect of pressure on excitable cells, in other words it is pathological. Although we do not understand the detailed mechanisms behind the two types of behaviour they are clearly affected by pressure in different ways. The effectiveness of the first five days' exposure to high pressure indicates a fairly rapid acclimation process and one which is clearly limited, judging from the diminished effect of the longer exposure.

More recently the European salt marsh shrimp *Palaemonetes varians* has been the subject of pressure acclimation experiments. It has been shown to survive 10 MPa for seven days surprisingly well. Its general motor activity was no different to that of the controls. In a follow-up study New et al. (2014), tested the animals' acclimation to pressure by applying higher pressures in 1 MPa steps at 5-minute intervals. The animals' response was categorised as follows: motionless, active movement (proba-bly normal behaviour in response to pressure) and, most instructive in the present context, loss of equilibrium, which was defined as the animal lying on its side, or upside down, for more than two seconds at the end of a pressure step. This is probably a pathological, direct effect of pressure on the balance mechanism. This provides a critical pressure (CP), defined as the pressure at which 50% of a group of animals exhibited loss of equilibrium. An example will clarify this. After 7 days of confinement at 5 °C and normal pressure, the pressure test was applied, producing a CP value of 11.1 MPa. After 7 days at 10 MPa the CP value increased to 13.3 MPa, which is a significant change. The animals had become more pressure tolerant, but not greatly so. Given the criteria, loss of equilibrium, this might arise from a compensatory, regulatory, change in the balance mechanism, partially offsetting the direct effect of pressure.

Molecular biology techniques have been used to monitor gene expression in studies of acclimation to high pressure. Gene expression can be part of homeostasis, the routine regulation which underlies a healthy animal, but it is also involved in more a drastic stress response, which initiates damage repair, and which appears to have a role in acclimation. The experiment summarised below illustrates some of the problems which pressure acclimation experiments with animals face. The shallow water amphipod *Eogammarus possjeticus* was kept at 20 MPa for 16 hrs to see what changes in gene expression were elicited during acclimation (Chen et al. 2019). The amphipods were not held in an observation pressure vessel, so their motor activity was unknown. Their general pressure tolerance is similar to that of *M. marinus*, which is immobilised at 20 MPa. Consider then the acclimation conditions of *Eogammarus*. It was probably paralysed for the duration and survived because, by chance, like *M marinus*, it tolerates much reduced respiration. Is it being tested for its ability to adapt to pressure, or depressed respiration, or some other linked factor?

### 12.2.1.1 The Design of Experiments Eliciting Stress Responses and the Expression of Other Genes

The above is a good example of the problem of introducing a single independent variable in stress experiments. A further problem in the design of such experiments is that after an animal (or cell suspension) has been held at high pressure for a time it is rapidly decompressed and then frozen and processed to assess gene expression. How much change can occur during the short but critical post decompression period? The problem was avoided by Welch et al. (Section 12.1.1 above). Certain circumstances call for a method of "fixing" or freezing the animal (or cell suspension) at high pressure (as in Ishii et al. 2004, Chapter 2; Lerch et al. 2014, Chapter 13; Studer et al. 2008, this chapter). Furthermore, if high pressure increases gene expression, as it often does, is this caused by pressure driving a series of biochemical processes (a pathological effect) or by pressure changing the structure of certain proteins which, in turn, elicit a normal stress response? Often the pressures involved seem rather small to denature proteins. This latter point also applies to experiments with microorganisms. It is plausible to imagine pressure dissociating a protein to subunits, with no effect on the stress sensing mechanism, but on decompression the subunits mis-assemble and trigger the response. In such a hypothetical case the cause of the stress response would be decompression, not pressure. Finally, pressure may also affect the oligomeric state of the shock proteins, making their functional significance suspect. These problems of experimental design, elegantly avoided in the simple Ananthan experiment described in Chapter 2, do not seem to be discussed in the literature.

The shrimp *Palaemonetes* mentioned above, has been subjected to high-pressure acclimation experiments after which gene expression was measured at normal pressure, i.e. post decompression (Morris et al. 2015a, b). Exposure to 10 MPa at 15 °C for two hours was used as the standard treatment after which the animals were frozen in liquid nitrogen at time intervals post decompression, starting at five minutes. One of the most interesting genes to be examined was narg, closely similar to a gene in the mosquito that codes for a special membrane protein, the NMDA receptor-regulated protein. NMDA (N Methyl-D-aspartate) is a neurotransmitter with a multiplicity of roles in major groups of animals, including Crustacea. Its receptor, a membrane protein, is an ion channel and like many such molecules is composed of subunits. The transcription of the subunits was increased by pressure treatment, which is also known to elicit increased motor activity dependent on neural control. It is proposed that the enhanced gene expression is part of a process which regulates the molecular components whose activity and turnover are increased by pressure. Is this idea tied to pressure driven activity? In other Crustacea stress proteins are synthesised in response to a visual stimulus or to increased motor activity. The expression of other genes was also simultaneously increased by pressure-treating Palaemonetes. They include the heat shock gene hsp 70 and the expression of the gene gapdh which codes for a glycolytic enzyme and the beta-actin gene, which codes for the protein actin.

The rapid synthesis of heat shock proteins following a heat shock in *E. coli* is not matched by the synthesis of pressure-induced proteins, which takes 60–90 mins. The evidence from heat shock experiments with human osteosarcoma cells is that general protein synthesis is promptly arrested (as in *E. coli*), but heat shock proteins increase slowly, within 8–10 mins of the shock, which is broadly similar to the timescale of the cytoskeletal changes caused by a pressure pulse of 4 MPa (Haskin et al. 1993). An increased presence of heat shock proteins was apparent on decompression after 20 mins at 4 MPa pressure, at 37 °C. Not unexpectedly, the heat shock response measured in marine snails is much slower. Tomanek and Somero found the intertidal *Tegula brunnea* started to produce heat shock proteins 2–14 hrs after the shock. Synthesis peaked over 15–30 hrs post-shock. The kinetics of the expression of heat shock proteins does not give us a reliable guide on how to design the equivalent pressure experiments. It is likely the data obtained in the above experiments with Palaemonetes is approximately what the experimenters hoped for but fixing/freezing the animals at pressure would be preferable. Special equipment would be needed to do this and already exists in one form or another (Section 12.2.1.1 and Chapter 13).

## 12.2.2  Animals from Deeper Water: Acclimation to Atmospheric Pressure

*Gnathophausia ingens*, mentioned earlier (Chapter 5; Section 5.4.1), is an active bathypelagic crustacean which normally lives at several hundred metres depth. It appears healthy at normal atmospheric pressure and in the experiments of interest here animals were kept at 7.5 MPa for 30–45 days at 5 °C, without feeding. This ensured the animals became acclimated to a pressure they normally experience. Control animals were similarly kept (acclimated) at atmospheric pressure. Their steady state oxygen consumption was unaffected by pressures of up to 25 MPa in both groups. At pressures of up to 22.5 MPa the beat frequency of the pleopods was also very similar in the two groups. It seems that prolonged exposure to atmospheric pressure or to 7.5 MPa has no effect on the respiration or the pleopod activity of these animals. Another crustacean, the amphipod *Stephonyx biscayensis*, also shows remarkable pressure tolerance (its vertical rage is 500–2000 m). Experimental animals were collected from bottom depths of 1530–1765 m in a baited trap, Brown and Thatje (2011). They appeared to remain healthy for two months at atmospheric pressure, 8 °C and can be regarded as acclimated. The length of time kept at atmospheric pressure had no effect on the oxygen consumption measured at selected pressures, up to 30 MPa. This is a benthic animal and presumably experiences lesser changes in depth/pressure than the planktonic *Gnathophausea*. Both have a remarkable disregard for the pressures they normally inhabit, which means they do not acclimate (change) when shifted from one pressure to another (Mickel and Childress 1982).

The King crab *Lithodes maja* has been kept at high pressure for various lengths of time in experiments which produced some interesting results but raise the question of the length of time which pressure acclimation experiments should involve. The animals were collected from 60 m depths and kept at normal atmospheric pressure for more than two months to ensure temperature acclimation at 6 °C. A team of scientists (Brown et al. 2017) measured the oxygen consumption and heart rate of adult males during exposure to selected pressures, 7.5, 10 or 12.5 MPa for 216 hrs, using the IPOCAMP apparatus (Chapter 13). Oxygen consumption increased abruptly after compression and up to 12.5 MPa showed a sustained increase. At higher pressures, it was somewhat reduced but still above control the level. The heart rate decreased at 12.5 MPa and remained constant. At lesser pressures it was also reduced, but less clearly. During this time the concentration of lactate in the blood (haemolymph) remained constant. Other measurements were carried out, of gene expression and of the fatty acid composition of whole body phospholipids. The latter showed no change in the 9 day exposure, but is 9 days long enough to conclude that homeoviscous adaptation of membrane and other lipids does not occur? The authors make no conclusion either way, but many experiments acclimating invertebrates and fish to temperature involve longer times (Williams and Hazel 1994). Decompression from 12.5 MPa, but not lesser pressures, resulted in an increase in the expression of several genes, but is this physiological or the recovery from a disrupted system? Given the depth range of Lithodes, from 60 to well over 600 m, it would be of interest to compare the genomes and the pressure tolerance of individuals collected by isobaric means from the 600 m with the shallow water individuals.

Acclimation of a deep living animal to a lower pressure appears to take place in the case of the sessile hydrothermal vent mussel, *Bathymodiolus azoricus* (also mentioned in Chapter 9). Specimens collected with decompression from 850 m depths on the Menez Gwen site on the Mid Atlantic Ridge were kept at normal atmospheric pressure in seawater supplied with methane and sulphide to sustain the animals' symbionts. Martins et al. (2016), were interested in the antioxidant defence enzymes, with the idea that exposure to atmospheric pressure may stimulate a change in them. The physiological situation is complicated, but there appeared to be evidence for the induction of the antioxidant enzyme catalase at atmospheric pressure. In separate studies using animals from the same source also kept at atmospheric pressure, Barros et al. (2015), followed gene expression in both the symbionts and the host cells for three weeks. Rather like the expression of stress or heat shock proteins by other deep-sea organisms subjected to lower than normal pressure, *Bathymodiolus* demonstrates a positive adaptive kind of response to atmospheric pressure. Given that the same species occupies depths of 800 to 3000 m on the Mid Atlantic Ridge, it would be interesting to measure how well animals from 3000 m cope at normal atmospheric pressure. According to Martins et al. (2017), survival is no longer than a few days. It is, of course, debatable whether a few weeks of survival constitutes acclimation. It is of interest that the common seashore mussel, *Mytilus edulis*, has been kept in cages at 650 m depths for two months and showed little or no signs of stress.

Pressure acclimation experiments, meaning exposing animals to pressure for a significant length of time in healthy conditions and monitoring function, reveals something of the plasticity of the phenotype, but if they are to reveal a feature of adaptive significance then it would appear that mature animals may not always be the best choice. In the case of *Lithodes* for example it is the pelagic larvae that are most likely to possess a significant pressure tolerance and the capacity to adapt to pressure (Chapter 9, Section. 9.7.2).

### 12.2.3  Genetic Adaptation

In Chapter 4 the susceptibility of mice to compression in *Heliox* was described. Breeding experiments have demonstrated that the Type I seizure is largely genetically determined. The mouse strain DBA/2 undergoes seizures at approximately 6 MPa when pressurised at a rate of 1.7 MPa per hour whereas strain C57BL/6 shows seizures at 8 MPa. This line of research suggests that in mammals a safe depth limit can be extended some 200 m, not by acclimation but by genetic changes similar to bacterial adaptation. In the evolution of diving mammals, many other aspects of the animal's physiology have to co-evolve to enable the animal to dive deeper, but the mouse model is relevant and instructive. Quite separately, it is very likely that within the life of diving mammals, individual breath-hold performance improves, i e acclimates with training, as it does in humans, (Chapter 11).

### 12.2.4  The Eel, *Anguilla anguilla*

This extraordinary animal has a life history that presents biologists with a variety of interesting phenomena in osmoregulation, endocrinology, exercise physiology, navigation, and adaptation to moderate hydrostatic pressures. Its place in the history of Science is well summarised in the title of a 2012 review article by Righton and colleagues, " *The Anguilla sp.* migration problem: 40 million years of evolution and two millennia of speculation," but some substantial experimental information is now available.

Before discussing the role which high pressure might play in the life of Anguilla, let us briefly outline that life. The larvae, a centimetre in length and called leptocephali, emerge from the twilight zone of the Sargasso Sea and swim northeast, with the ocean current, towards the Atlantic coastline of North Africa and Western Europe, a distance of 5–6000 km (Van Ginneken and Maes 2005). Remarkably, from approximately the same region, leptocephali of the distinct North American eel, *A. rostrata*, swim west to the eastern coastline of the USA. In the Pacific Ocean the Japanese eel, *Anguilla japonica* similarly migrates to spawn in an area from which the north-flowing current will assist the leptocephali's migration to estuaries.

The pattern of breeding in the ocean and migrating to freshwater to feed and grow is called catadromous and is the converse of the anadromous pattern seen in salmon.

The migrating *A. anguilla* larvae, at the stage of entering estuaries and therefore at the transition from seawater to freshwater, have grown bigger and are called glass eels. As they migrate upstream they become fully adapted to life in freshwater and are called elvers. They feed and grow in streams and marshes, attaining a length of 30 cm or so, at least five times the size of the glass eel stage. These are Yellow eels, which might appear to be an adult but in fact are a long way from sexual maturity. Yellow eels live for many years, well into their mid-teens, as freshwater fish and build up energy reserves in the form of fat. Then, following some unknown cue, they begin to prepare for the long migration back to the Sargasso Sea to breed. They undergo a metamorphosis to the Silver stage, usually in the summer, which prepares them for life in seawater. They acquire a silvery underside and a dark topside, typical camouflage of pelagic fish, and their eyes get bigger (and more complicated within). The pectoral fins elongate and the skin gets thicker and tougher. Remarkably, Silver eels do not feed. The alimentary tract shrinks but continues to function in absorbing water and excreting salts. The gills and kidneys also discharge salts as part of the animals' ionic regulation in seawater. The swim bladder, an important buoyancy organ for any pelagic fish, develops its capacity and the lateral line sensory system also improves. All of this is preparation for a 5000–6000 km swim, taking several months, without feeding. The energy store needed for this feat of endurance has been shown to be present by laboratory swimming trials. A number of Silver Anguilla, fitted with sensors, have been tracked swimming along the migration route at depths of around 200 m in the night and 560 m in the day. This diel pattern is thought to minimise the risk of predation and keeps the animals in a preferred temperature range. There is an argument that a cool temperature will delay the onset of sexual maturity to match the anticipated time of arrival at the spawning zone. The red aerobic muscle powers this marathon swim, with slow steady contractions, probably assisted in certain circumstances by the white muscle. Fuel is provided by fat reserves, of which about 40% support muscle activity and the rest supports the development of gonads. A proportion of female eels from the Baltic sea are likely to run short of fuel by the time they arrive at the breeding zone. Full sexual maturity is only achieved "on site," and it may, in some way, be physiologically dependent on swimming and pressure.

The hormonal control of sexual development is nicely illustrated by some experiments carried out in the Mediterranean with the support of the Musee Oceanographique, Principaute de Monaco. Female silver eels were held in cages at a depth of 450 m (4.5 MPa) for three months. Compared to control animals, held in shallow water, the experimental group had a much higher level of pituitary gonado-trophin, indicating that pressure stimulates, at least in part, sexual development. This might be a case of pressure triggering gene expression, and it is interesting to note that it is much less pressure than the seashore mussel seems to tolerate (Section 12.2.2).

It is thought that much of the spawning takes place at a depth of at least 200–300 m, i.e. at a pressure of 2–3 MPa. This is within the depth range at which migration takes place and which probably has a maximum depth of 1000 m, 10 MPa.

A reasonable way for experimenters to proceed is to regard the Yellow eel as a shallow, freshwater fish that can adapt to seawater and the Silver eel as an oceanic fish whose pressure tolerance is uncertain. The extent and possible mechanisms of pressure adaptation have been investigated extensively. One of the difficulties researchers face is that a maximum depth range of 1000 m for the majority of migrants means the Silver eels are clearly not deep-sea fish and their pressure adaptations might not be different to those of many shallow water, open ocean fish (and which are still largely unknown). Intriguingly, there is evidence that catadromous eels such as Anguilla may have evolved from deep water ancestors. Nevertheless, some pressure adaptations in the Silver eel are likely to be present, but to less than 10 MPa and therefore difficult to establish. Compensating for this, the eel is a robust and convenient experimental animal.

Where to begin ? From the viewpoint of high-pressure physiology, if not eel physiology, motor activity at high pressure is of particular interest, especially the convulsion threshold pressure. As we have seen it is the pressure which, in a stepwise sequence (here 1 MPa each 5 mins) elicits seizures, writhing convulsions of the whole musculature of the animal. Convulsions persist in bursts as pressure continues to increase and then give way to a paralysed state. The reader may think these experiments are rather gruesome and indeed they are, but they have a serious purpose and in the UK they are closely controlled. Those involved would point out that the eels experience something less severe than that caused by electric fishing, which is used to clear waters such as canals for construction work and, of course, the sport fisherman inflicts severe stress on his catch, for fun.

Experiments using seizures as a criterion were carried out using a large pressure vessel as a rather primitive high-pressure aquarium. An individual eel was confined in the vessel for 72 hrs. at 9 °C, with fresh seawater circulated twice daily either at normal atmospheric pressure (control conditions) or at 4 MPa. To determine the animal's sensitivity, the pressure was increased in steps of 1 MPa each 5 mins, and the onset of seizures noted. Yellow eels, about 30 cm long and adapted to seawater, underwent seizures at a mean pressure of 9.1 MPa in the case of control treatment and 13.5 MPa after being held at 4 MPa (Johnstone et al. 1989). Thus, their onset pressure was increased by an amount similar to the holding pressure. This is evidence for a change in response to an absolute pressure, 9 MPa, but for no change in the differential (extra) pressure required to elicit seizures $(4 + 9 = 13)$.

Silver eels, collected as they migrated to the sea, underwent seizures at 14.2 MPa (control) and 19.1 MPa (held at 4 MPa). These fish are more pressure tolerant than the Yellow eels and they too show the same sort of potential adaptive shift in the mean onset pressure for seizures. An interesting addition to the experiments was the slow decompression of three animals held at 4 MPa for 72 hrs to normal atmospheric pressure. After an hour each was pressurised and manifested seizures at 13.3, 15.3 and 20.4 MPa. This seems to be telling us that two animals lost their acclimation in the hour at normal pressure. These experiments show that the neuromuscular system of Silver eels is more pressure tolerant than that of Yellow eels. It undergoes a reversible change in its sensitivity to compression when held at pressure, acclimated, a finding confirmed by Sebert and co-workers. The change is in the direction of increased tolerance to the pressure they might experience during their spawning migration. However, pressure acclimated Silver eels show a paradoxical change in the twitch tension the red muscle produces at normal pressure. It is significantly reduced, by means unknown (Sebert 2008).

As eels have a swim bladder that will be drastically compressed during these pressure experiments how do we know that the response of the eels is not a sensory response, in which the bladder acts as a sensor, rather than a direct effect on the neuromuscular system, as assumed above? There are several reasons, chief of which is that fish lacking a swim bladder respond in much the same way as the eel (Chapter 4; Section 4.2). The point is the neuromuscular system of the Yellow eel fails to demonstrate any prior tolerance to high pressure, and after 72 hrs at the equivalent of 400 m depth, it convulses at a higher pressure but still within the range seen in ordinary marine fish. The Silver eel is more tolerant prior to exposure to pressure and it probably acquires greater tolerance during its migration.

The swimming trials mentioned earlier, which demonstrated that the eels' food reserves were sufficient for their long migration, were carried out at normal pressure. They left unanswered the question, how might pressure affect the eels' performance? This is a difficult question to tackle experimentally. We do not know the condition of the eels at the end of their migration, which is, after all, a one-way journey. The adults do not return and they only have to arrive in a state sufficient to release their gametes. Their migration almost certainly entails programmed adaptive changes that might not respond to well-meaning experimental exposure to high pressure. And, on intuitive grounds, acclimation experiments at high pressure with an active animal are not best carried out in a confined space over a short time interval.

Despite these difficulties, Sebert and his colleagues have made impressive progress using a compact high-pressure aquarium in their laboratory in Brest (Sebert et al. 1990). A water flow device in which fish swim "upstream" has been constructed in the confines of a cylindrical pressure vessel. The oxygen consumption of individual male Silver eels has been measured at different swimming speeds at normal atmospheric pressure and at 10 MPa. Oxygen consumption increases as the fish swim faster, but at 10 MPa the oxygen consumption at a given speed was much lower than at normal pressure. The speed of swimming varied from zero to 0.9 Body Lengths per second. An average speed for the migration would be about 0.5 B L per second. This finding strongly suggests that 10 MPa affects metabolism in some advantageous way. In the Yellow eel the fluidity of the membranes in the gill cells is increased by a month's exposure to 10 MPa (Sebert et al. 1993). This is consistent with the finding of Theron et al. (2000), that acclimation to pressure enhances the efficiency of oxidative phosphorylation. The adaptive advantage of this in Nature is uncertain for two reasons; one is the experiments were short term (understandably) and the other is that 10 MPa pressure was used. The migrating eels are unlikely to spend much time at 10 MPa (1000 m depth). They probably spend most of their time at about half that pressure. Other metabolic changes might be induced by such pressures, especially in the longer term. The suggestion is made that the fluidity of mitochondrial membranes (presumably in the red muscle cells) is increased prior to migration, so that they support efficient oxidative metabolism at depth. It is noteworthy that the membrane fluidity of the mitochondria extracted from the red muscle of Silver eels is, in fact, greater than that of the equivalent membranes in Yellow eels, due to low cholesterol and high unsaturated fatty acid levels. This condition is present in Silver eels that have not experienced high pressure and is regarded as one of the numerous changes involved in the metamorphosis from Yellow to Silver states. If the fluidity of the Silver eel mitochondrial membranes enhances oxidative metabolism in some way, that would be interesting. The oxidative phosphorylation carried out by mitochondria extracted from the red muscle of Yellow eels exposed to 10 MPa for three weeks is enhanced (by means uncertain) and such an increase in efficiency seems important, and quite possibly an acclimated feature. In the case of the trout (*Oncorhynchus mykiss*), pressure inhibits the equivalent form of mitochondrial respiration, and therefore it is incapable of an eel-like migration (Sebert and Theron 2001). Perhaps the time has come when Silver eels have to become routine experimental animals. Catching them in the sea is probably an unrealistic approach so in situ acclimation, perhaps using moored cages, may be the best way to produce them. It may also be necessary to use isobaric techniques to examine the state of their cellular biochemistry, and their neurobiology, to gain a fuller understanding of their adaptation/acclimatisation to pressure.

The activity of the associated oxidative defence enzymes, superoxide dismutase and catalase, have also been measured in muscle. As Yellow eels have a higher rate of oxidative metabolism than Silver eels there is greater oxidative stress in Yellow eels but in both, 10 MPa exposure for three weeks, reduced the activity of the two enzymes, measured at normal atmospheric pressure. The significance of this is not clear as surely what matters is the rate at high pressure. Perhaps it relates to the high swimming efficiency of the eel mentioned earlier (Sebert et al. 2009). Not only is the Anguilliform style of swimming intrinsically efficient but the eel seems to be very good at it, about 4–6 times more efficient than other fish, according to Van Ginneken and Maes (2005) and Van Ginneken et al. (2005)

## 12.3    Conclusion

Prokaryote microorganisms and yeast can be experimentally adapted to high pressure by various routes. Aquatic animals manifest an adaptive shift in their motor activity after time at pressure. In certain mouse strains the pressure sensitivity of the Type I seizure is largely genetically determined and resistant strains have been bred. This is not acclimation but is a form of adaptation.

The eel Anguilla sp. migrates in ocean waters at depths of several hundred metres and probably benefits from adaptive changes to pressure which are apparent in experiments. These include adaptive responses in sexual maturation and mitochondrial respiration and other metabolic processes. Changes in the pressure sensitivity underlying seizures provide a hint of the importance of neurobiology, including the cardiovascular system, which are unknown territory.

Exactly how pressure affects the stress response is not clear. The pressure equivalent of the Anantham heat stress experiment is required (Chapter 2). Many acclimation experiments would benefit from better experimental design, including the use of isobaric methods to arrest biochemical reactions prior to decompression, and a longer time scale.

Adaptation to high pressure in the laboratory prompts speculation about the extent to which comparable changes are important in Nature. Animals that experience a very significant change in pressure in their normal life, such as the larvae of benthic animals (Chapter 9), are likely to possess sufficient pressure adaptation/acclimation (phenotypic plasticity) to be of survival value. However, linking acclimation to high pressure to the evolution of deep-sea animals is best avoided by reductionists, and left to experts on evolutionary processes.

# References[1]

Aarestrup K et al (2009) Oceanic spawning migration of the European eel (Anguilla anguilla). Science 325:1660. *

Abe F (2007) Exploration of the effects of high hydrostatic pressure on microbial growth, physiology and survival: perspectives from piezophysiology. Biosc Biotechol Biochem 71:2347–2357. *

Abe F, Iida H (2003) pressure-induced differenetial regulation of the two tryptophan permeases Tat 1 and Tat 2 by ubiquitin ligase Rsp5 and its binding proteins, Bul1 and Bul2. Mol Cell Biol 23:7566–7584. *

Abe F, Usui K (2013) Effects of high hydrostatic pressure on the dynamic structure of living Escherichia coli membrane: a study using high pressure time-resolved fluorescence anisotropy measurement. High Pressure Res 33:278–284

Aertsen A, Michiels C (2005) Mrr instigates the SOS response after high pressure stress in Escherichia coli. Mol Microbiol 58:1381–1391. *

Allen EE, Bartlett DH (2000) FabF is required for piezoregulation of cis-Vaccenic acid levels and piezophilic growth of the deep sea bacterium Photobacterium profundum SS9. J Bacteriol 182:1264–1271. *

Allen EE, Bartlett DH (2002) Structure and regulation of the omega-3 polyunsaturated fatty acid synthase genes from the deep sea bacterium Photobacterium profundum strain SS9. Microbiology 148:1903–1913. *

Allen EE, Facciotti D, Bartlett DH (1999) Monounsaturated but not polyunsaturated fatty acids are required for growth of the deep sea bacterium Photobacterium profundum SS9 at high pressure and low temperature. Appl Environ Microbiol 65:1710–1720. *

Amerand A, Vettier A, Sebert P, Moisan C (2006) Does hydrostatic pressure have an effect on reactive oxygen species in the eel? Undersea Hyperb Med 33:157–160. *

Avent RM (1975) Evidence for acclimation to hydrostatic pressure in Uca pugilator (Crustacea: Decapoda: Ocpodidae). Mar Biol 31:193–199

Barrick JE et al (2009) Genome evolution and adaptation in a long term experiment with Escherichia coli. Nature 461:1243–1249. *

---

[1]Those marked * are further reading.

Barros I et al (2015) Post-capture immune gene expression studies in the deep sea hydrothermal vent mussel Bathymodiolus azoricus acclimatized to atmospheric pressure. Fish Shellfish Immunol 42:159–170

Bartlett DH (1999) Microbial adaptations to the Psychrosphere/Piezosphere. J Mol Microbiol Biotechnol 1:93–100. *

Bartlett DH (2002) Pressure effects on in vivo microbial processes. Biochim Biophys Acta 1595:367–381

Black LS, Dawson A, Ward FB, Allen RJ (2013) Genes required for growth at high hydrostatic pressure in E. coli K-12 identified by genome-wide screening. PLoS One 8:e73995. *

Brauer RW et al (1974) Intra- and interspecies variability of vertebrate high pressure neurological syndrome. J Appl Physiol 37:844–851. *

Bravim F et al (2016) High hydrostatic pressure activates gene expression that leads to ethanol production enhancement in a Saccharomyces cerevisiae distillery strain. Appl Microbiol Biotechnol 97:2093–2107

Brown A, Thatje S (2011) Respiratory response of the deep sea Amphipod Stephonyx biscayensis indicates bathymetric range limitation by temperature and hydrostatic pressure. PLoS One 6: e28562. *

Brown A, Thatje S (2014) Explaining bathymetric diversity patterns in marine benthic invertebrates and demersal fishes: physiological contributions to adaptation of life at high pressure. Biol Rev 89:406–426. *

Brown A et al (2017) Metabolic costs imposed by hydrostatic pressure constrain bathymetric range in the lithodid crab Lithodes maja. J Exp Biol 220:3916–3926

Campanaro S et al (2005) Laterally transferred elements and high pressure adaptation in Photobacterium profundum strains. BMC Genomics 6:122–137

Campanaro S et al (2012) The transcriptional landscape of the deep sea bacterium Photobacterium profundum in both a toxR mutant and its parental strain. BMC Genomics 13:567–587. *

Casadei MA et al (2002) Role of membrane fluidity in pressure resistance of Escherichia coli NCTC 8164. Appl Environ Microbiol 68:5965–5972. *

Charoenwong D, Andrews S, Mackey B (2011) Role of rpoS in the development of cell envelope resilience and pressure resistance in stationary phase Escherichia coli. Appl Environ Microbiol 77:5220–5229

Chen J, Liu H, Cai S, Zhang H (2019) Comparative transcriptome analysis of Eogammarus possjeticus at different hydrostatic pressure and temperature exposures. Sci Rep 9:10

Chow S, Kurogi H, Mochioka N, Kaji S, Okazaki M, Tsukamoto K (2009) Discovery of mature freshwater eels in the open ocean. Fish Sci 75:257–259. *

Clevestam PD, Ogonowski M, Sjoberg NB, Wickstrom H (2011) Too short to spawn? Implications of small body size and swimming distance on successful migration and maturation of the European eel Anguilla. J Fish Biol 78:1073–1089. *

Cottin D et al (2012) Sustained hydrostatic pressure tolerance of the shallow water shrimp Palaemonetes varians at different temperatures: insights into the colonisation of the deep sea. Comp Biochem Physiol A 162:357–363. *

D'Costa VM et al (2011) Antibiotic resistance is ancient. Nature 477:457–461

El-Hajj ZW et al (2010) Insights into piezophily from genetic studies on the deep sea bacterium Photobacterium profundum SS9. Annals N Y Acad Sci 1189:143–148

Else PL, Hulbert AJ (2003) Membranes as metabolic pacemakers. Clin Exp Pharmacol Physiol 30:559–564. *

Fajardo-Cavazos P, Langenhorst F, Melosh HJ, Nicholoson WL (2009) Bacterial spores in granite survive hypervelocity launch by spallation: implications for lithopanspermia. Astrobiology 9:647–657. *

Fernandes PMB, Domitrovic T, Fao CM, Kurtenbach E (2004) Genomic expression pattern in Saccharomyces cerevisiae cells in response to high hydrostatic pressure. FEBS Lett 556(15):3–160. *

Fontaine Y-A, Dufour S, Alimat J, Fontaine M (1985) L'mmersion prolongee en profondeur stimule la function gonadotrope de l' Anguille europeenne (Angulla anguilla.) femelle. CR Acad ScParis Serie III(300):83–87. *

Forlin L et al (1996) Molecular investigations into pollutant impact on Roundnose Grenadier (C. rupestris) and translated common mussel (M.edulis) in Skagerrak, the North Sea. Mar Environ Res 42:209–212

Fujii S et al (1996a) Characterization of a barotolerant mutant of the yeast Saccharomyces cerevisiae: importance of trehalose content and membrane fluidity. FEMS Microbiol Lett 141:97–101. *

Fujii S et al (1996b) Saccharides that protect yeast against hydrostatic pressure stress correlated to the mean number of equatorial OH groups. Biosci Biotechnol Biochem 60:476–478. *

Ghalambor CK et al (2015) Non-adaptive plasticity potentiates rapid adaptive evolution of gene expression in nature. Nature 5625:372–375

Griffin PL, Kish A, Steele A, Hemley RJ (2011) Differential high pressure survival in stationary phase Escherichia coli MG1655. High Presssure Res 31:325–333. *

Haskin CL, Athanasiou K, Klebe R, Cameron IL (1993) A heat shock like response with cytoskeletal disruption occurs following hydrostatic pressure in MG-63 osteosarcoma cells. Biochem Cell Biol 71:361–371

Hauben KJA et al (1997) Escherichia coli mutants resistant to inactivation by high hydrostatic pressure. Appl Environ Microbiol 63:945–950. *

Hazael R, Meersman F, Ono F, McMillan PF (2016) Pressure as a limiting factor for life. Life 6:34. *

Hazael R et al (2017) Bacteria survival following shock compression in the GigaPascal range. Icarus 293:1–7

Hazael R et al (2019) Pressure tolerance of Artemia cysts compressed in water medium. High Pressure Res 39:293–300

Hendry, AP (2016) Key questions on the role of phenotypic plasticity in eco- evolutionary dynamics. J. Heredity, Symposium Am. Genetic Assoc. August, pp 25–41.

Inoue JG et al (2010) Deep ocean origin of the freshwater eels. Biol Lett 6:363–366. *

Ishii R, Aramaki M, Mochizuki T, Abe F (2019) Identification of a regulatory element for yeast tryptophan permease Tat2 ubiquitination using high hydrostatic pressure. High Pressure Res 39:258–266. *

Iwahashi H (2015) Pressure dependent gene activation in yeast cells. In: Akasaka K, Matsuki H (eds) High pressure bioscience, subcellular biochemistry. Springer Science+Business Media, Dordrecht

Iwahashi I, Nwaka S, Obuchi K (2000) Evidence for contribution of neutral trehalase in baraotolerance of Saccharomyces cerevisiae. Appl Environ Microbiol 66:5182–5185. *

Iwahashi H, Odani M, Ishidou E, Kitagawa E (2005) Adaptation of Saccharomyces cerevisiae to high hydrostatic pressure causing growth inhibition. FEBS Lett 579:2847–2852

Johnstone ADF, Macdonald AG, Mojsiewicz WR, Wardle CS (1989) Preliminary experiments in the adaptation of the European eel (Anguilla Anguilla) to high hydrostatic pressure. J Physiol 417:87

Kanda N, Abe F (2013) Structural and functional implications of the yeast high-affinity tryptophan permease Tat2. Biochemistry 52:4296–4307. *

Komatsu Y et al (1991) Deuterium oxide, dimethylsulfoxide and heat shock confer protection against hydrostatic pressure damage in Yeast. Biochem Biophys Res Commun 174:1141–1147. *

Kpecka-Pilarczyk J, Coimbra J (2010) The effect of elevated hydrostatic pressure upon selected biomarkers in juvenile blackspot seabream Pagellus bogaraveo in a 14 day long experiment. J Fish Biol 77:279–284. *

Kultz D (2005) Molecular and evolutionary basis of the cellular stress response. Annu Rev Physiol 76:225–257. *

Kurosaka G, Abe F (2018) The YPR153W gene is essential for the pressure tolerance of tryptophan permease Tat 2 in the yeast Saccharomyces cerevisiae. High Pressure Res 38:90–98. *

Le Bihan T, Rayner J, Roy MM, Spagnolo L (2013) Photobacterium profundum under pressure: a MS-based label -free quantitative proteomics study. PLoS One 8:e60897. *

Lee D-H, Feist AM, Barrett CL, Paisson BO (2011) Cumulative number of cell divisions as a meaningful timescale for adaptive laboratory evolution of *Escherichia coli*. PLoS One 6:e26172

Leighs JA et al (2012) A sealed capsule system for biological and liquid shock-recovery experiments. Rev Sci Instrum 83:115113

Linke K et al (2008) Influence of high pressure on the dimerization of ToxR, a protein involved in Bacterial signal transduction. Appl Environ Microbiol 74:7621–7623. *

Linke K et al (2009) Influence of membrane organization on the dimerization ability of ToxR from Photobacterium profundum under high hydrostatic pressure. High Pressure Res 29:431–442

Lorenz R, Molitoris HP (1997) Cultivation of fungi under simulated deep sea conditions. Mycol Res 101:1355–1365. *

Marietou A, Nguyen ATT, Allen EE, Bartlett DH (2015) Adaptive laboratory evolution of *Escherichia coli* K-12 MG1655 for growth at high hydrostatic pressure. Front Microbiol 5:749

Marti E, Variatza E, Balcazar JL (2014) The role of aquatic ecosystems as reservoirs of antibiotic resistance. Trends Microbiol 22:36–41. *

Martiensson VT, Reysenbach A-L, Birrien J-L, Prieur D (1999) A stress protein is induced in the deep sea barophilic hyperthermophile Thermococcus barophilus when grown under atmospheric pressure. Extremophiles 3:277–282

Martin DD, Bartlett DH, Roberts ME (2002) Solute accumulation in the deep sea bacterium Photobacterium profundum. Extremophiles 6:507–514

Martins I et al (2016) Activity of antioxidant enzymes in response to atmospheric pressure induced physiological stress in deep sea hydrothermal vent mussel Bathymodiolus azoricus. Mar Environ Res 114:65–73

Martins I et al (2017) Physiological impacts of acute Cu exposure on deep sea vent mussel Bathymodiolus azoricus under a deep mining scenario. Aquat Toxicol 193:40–49

Mazel D, Mobashery S (2012) Antibiotics as physiological stress inducers and bacterial response to the challenge. Curr Opin Microbiol 15:553–554

McCall RD (2011) HPNS seizure risk: a role for the Golgi-associated retrograde protein complex? Undersea Hyperbaric Med 38:3–9. *

Mickel TJ, Childress JJ (1982) Effects of pressure and pressure acclimation on activity and oxygen consumption in the bathypelagic mysid Gnathophausia ingens. Deep-Sea Res 29:1293–1301

Miura T, Minegishi H, Usami R, Abe F (2006) Systematic analysis of HSP gene expression and effects on cell growth and survival at high hydrostatic pressure in Saccharomyces cerevisiae. Extyremophiles 10:279–284

Molina-Hoppner A, Doster W, Vogel RF, Ganzle MG (2004) Protective effect of sucrose and sodium chloride for Lactococcus lactis during sublethal and lethal high pressure treatments. Appl Environ Microbiol 70:2013–2020. *

Morris JP et al (2015a) Acute combined pressure and temperature exposures on a shallow water crustacean: novel insights into the stress response and high pressure neurological syndrome. Comp Biochem Physiol A 181:9–17

Morris JP et al (2015b) Characterising multi-level effects of acute pressure exposure on a shallow water invertebrate: insights into the kinetics and hierarchy of the stress response. J Exp Biol 218:2594–2602

Munro C et al (2015) The role of ontogeny in physiological tolerance: decreasing hydrostatic pressure tolerance with development in the northern stone crab Lithodes maja. Proc R Soc B 282:20150577. *

New P et al (2014) The effects of temperature and pressure acclimation on the temperature and pressure tolerance of the shallow water shrimp Palaemonetes varians. Mar Biol 161:697–709

Nogi Y, Masui N, Kato C (1998) Photobacterium profundum sp. nov., a new, moderately barophilic bacterial species isolated from a deep sea sediment. Extremophiles 2:1–7

Ono F et al (2012) Effect of very high pressure on life of plants and animals. J Phys Conf Ser 377:012053

Palhano FL, Orland MTD, Fernandes PMB (2014) Induction of baroresistance by hydrogen peroxide, ethanol, and cold shock in Saccharomyces cerevisiae. FEMS Microbiol Lett 233:139–145. *

Pribyl AL, Schreck CB, Parker SJ, Weis VM (2012) Identification of biomarkers indicative of barotrauma and recovery in black rockfish Sebastes melanops. J Fish Biol 81:181–196. *

Righton D et al (2012) The Anguilla spp. migration problem: 40 million years of evolution and two millenia of speculation. J Fish Biol 81:365–386. *

Rossignol O, Sebert P, Simon B (2006) Effects of high pressure acclimation on silver eel (Anguilla anguilla, L.) slow muscle. Comp Biochem Physiol A 143:234–238. *

Sato H, Nakasone K, Yoshida T, Kato C, Maruyam T (2015) Increases of heat shock proteins and their mRNAs at high hydrostatic pressure in a deep sea piezophilic bacterium, Shewenella violacea. Extremophiles 19:751–762. *

Scaion D, Belhomme M, Sebert P (2008) Pressure and temperature interactions on aerobic metabolism of migrating European silver eel. Respir Physiol Neurobiol 164:319–322. *

Sebert P (2002) Fish at high pressure: a hundred year history. Comp Biochem Physiol A 131:575–585. *

Sebert P (2008) Fish muscle function and pressure. In: Sebert P, Onyango D, Kapoor BG (eds) Fish life in special environments. Science Publishers, Enfield, NH, USA

Sebert P, Macdonald AG (1993) Fish. In: Macdonald AG (ed) Effects of high pressure on biological systems. Springer Verlag. *

Sebert P, Theron M (2001) Why can the eel, unlike the trout, migrate under pressure. Mitochondrion 1:79–85

Sebert P, Barthelemy L, Simon B (1990) Laboratory system enabling long term exposure to hydrostatic pressure (101 atm) of fishes or other animals breathing water. Mar Biol 104:165–168

Sebert P, Cossins A, Simon B, Meskar A, Barthelemy L (1993) Membrane adaptations in pressure acclimated freshwater eels. Proceedings of the 32nd Congress Int. Union of Physiological Sciences, Glasgow. 177.2/P121

Sebert P, Meskara A, Simon B, Barthelemy L (1994) Pressure acclimation of the eel and liver membrane composition. Experientia 50:121–123. *

Sebert P, Simon B, Pequeux A (1997) Effects of hydrostatic pressure on energy metabolism and osmoregulation in crab and fish. Comp Biochem Physiol A 116:281–290. *

Sebert P, Scaion D, Belhomme M (2009a) High hydrostatic pressure improves the swimming efficiency of European migrating silver eel. Respir Physiol Neurobiol 165:112–114. *

Sebert P, Vettier A, Amerand A, Moisan C (2009b) High pressure resistance and adaptation of European Eels. In: den Thillart V et al (eds) Spawning migration of the European eel. Springer Science + Business Media, BV. *

Sharma A et al (2002) Microbial activity at Gigapascal pressures. Science 295:1514–1516. *

Simonato F et al (2006) Piezophilic adaptation: a genomic point of view. J Biotechnol 126:11–25. *

Somero GN (2010) The physiology of climate change: how potentials for acclimatization and genetic adaptation will determine "winners" and "losers". J Exp Biol 213:912–920

Strauss DB, Walter WA, Green CA (1987) The heat shock response of E. coli is regulated by changes in the concentration of Q (sigma) 32. Nature 329:348–351. *

Studer D, Humbel BM, Chiquet M (2008) Electron microscopy of high pressure frozen samples: bridging the gap between cellular ultrastructure and atomic resolution. Histochem Cell Biol 130:877–889

Suzuki A et al (2013) Pressure induced endocytic degradation of the Saccharomyces cerevisiae low-affinity tryptophan permease Tat1 is mediated by Rsp5 ubiquitin ligase and functionally redundant PPxY motif proteins. Eukaryot Cell 12:990–997. *

Tan M-L, Balabin I, Onuchi JN (2004) Dynamics of electron transfer pathways in cytochrome c oxidase. Biophys J 86:1813–1819. *

Theron M, Guerrero F, Sebert P (2000) Improvement in the efficiency of oxidative phosphorylation in the freshwater eel acclimated to 10.1 MPa hydrostatic pressure. J Exp Biol 203:3019–3023

Tolgyesi FG, Bode C (2010) Pressure and heat shock proteins. In: Sebert P (ed) Comparative high pressure biology. Science Publishers USA. *

Tomanek L, Somero GN (2000) Time course and magnitude of synthesis of heat shock proteins in congeneric marine snails (Genus Tegula) from different tidal heights. Physiol Biochem Zool 72:249–256. *

Tsukamoto K (1992) Discovery of the spawning area for the Japanese eel. Nature 356:789–791. *

Tsukamoto K (2009) Oceanic migration and spawning of anguillid eels. J Fish Biol 74:1833–1852. *

Van Ginnecken V et al (2005) Eel migration to the Sargasso: remarkably high swimming efficiency and low energy costs. J Exp Biol 208:13290–11335

Van Ginneken VJT, Maes GE (2005) The European eel (Anguilla Anguilla) its lifecycle, evolution and reproduction: a literature review. Rev Fish Biol Fish 15:367–398

Vanlint D et al (2011a) Rapid acquisition of Gigapascal high pressure resistance by *Escherichia coli*. MBio 2:e00130

Vanlint D, Michiels CW, Aertsen A (2011b) Piezophjysiology of the model bacterium *Escherichia coli*. In: Horikoshi K (ed) Extremophiles handbook. Springer, pp 670–686

Vanlint D et al (2013) Exposure to high hydrostatic pressure rapidly selects for increased RpoS activity and general stress resistance in *Escherichia coli* O157:H7. Int J Food Microbiol 163:28–33

Vettier A et al (2006) Hydrostatic pressure effects on eel mitochondrial functioning and membrane fluidity. Undersea Hyperb Med 33:149–156. *

Vezzi A et al (2005) Life at Depth: Photobacterium profundum genome sequence and expression analysis. Science 307:1459–1461. *

Vik SB, Capaldi RA (1980) Conditions for optimal electron transfer of cytochrome c oxidase isolated from beef heart mitochondria. Biochem Biophys Res Commun 94:348–354. *

Wannicke N et al (2015) Measuring bacterial activity and community composition at high hydrostatic pressure using a novel experimental approach: a pilot study. FEMS Microbiol Ecol 9: fiv036

Welch TJ, Bartlett DH (1998) Identification of a regulatory protein required for pressure responsive gene expression in the deep sea bacterium Photobacterium species strain SS9. Mol Microbiol 27:977–985. *

Welch TJ, Farewell A, Neidhardt FC, Bartlett DH (1993) Stress response of *Escherichia coli* to elevated hydrostatic pressure. J Bacteriol 175:7170–7177

Williams EE, Hazel JR (1994) Thermal adaptation in fish membranes:temporal resolution of adaptive mechanisms. In: Cossins AR (ed) Temperature adaptation of biological membranes. Portland Press Proceedings, pp 91–106

Willis MJ, Ahrens TJ, Bertani LE, Nash CZ (2006) Bugbuster—survivability of living bacteria upon shock compression. Earth Planet Sci Lett 247:185–196. *

Yayanos AA (2002) Are cells viable at gigapascal pressures? Science 297:295

Zobell CE, Johnson FH (1949) The influence of hydrostatic pressure on the growth and viability of terrestrial and marine bacteria. J Bacteriol 57:179–189. *

# High-Pressure Equipment for Use in the Laboratory, at Sea and at Depth

# 13

In Chapter 1, the point was made that the greatest pressure in the oceans, about 110 MPa, is not particularly challenging in engineering terms. It is at the high end of the pressure range used in much hydraulic equipment and industrial machinery. The humble hydraulic car jack and road digger employ pressures in the 35–100 MPa range. However, interesting engineering problems arise in high-pressure biology for lots of reasons. Measurements are required to investigate life processes at high pressure, organisms and cells have to be kept healthy at high pressure, and often they need to be examined with a microscope. Humans wish to explore the depths in submersible craft and see the environment for themselves, human divers need to be housed in hyperbaric chambers for weeks at a time, and in rare circumstances, submariners need to escape from their damaged boat.

We have seen, also in Chapter 1, that nineteenth-century engineering enabled the first observations to be made on organisms subjected to deep-sea pressure. In the modern era Bridgman, mentioned in Chapter 2 as a pioneer high-pressure scientist, can also be regarded as a founder of high-pressure engineering (Hemley 2010). He extended the pressure generating and containing capabilities of equipment from 400 MPa to 40,000 MPa. High-pressure technology was driven by the manufacture of guns, polyethylene and the use of isostatic pressing, which is a method of forming material in sheet or powder form into a solid material structure. The manufacture of diamonds and precious stones, which also uses very high pressures, was important too, and many other developments, in the food industry for example, are taking the technology in new directions.

This chapter provides some basic information about high-pressure equipment and describes some special devices and apparatus referred to earlier in the book. I am grateful to my engineer colleague Dr. Ian Gilchrist for commenting on the chapter. Its purpose is to render the descriptions given in earlier chapters more intelligible. In many cases, it is only possible to give the gist of how things work, but the baffled and enthusiastic reader can make use of the copious references. Textbooks on high-pressure techniques such as the book by Sherman and Stadmuller (1988), Fryer and Harvey (2012) or papers such as that by Chandra Shekar and Sahu (2007), and

© Springer Nature Switzerland AG 2021
A. Macdonald, *Life at High Pressure*, https://doi.org/10.1007/978-3-030-67587-5_13

Skelton and Crossland (1980), can be consulted for fuller information. Papers shown in bold in the list at the end of the chapter can be viewed on the Web without affiliation or payment. As with other chapters, the list of references includes extras, for further reading.

## 13.1   Generating High Pressure

### 13.1.1 Pumps

A simple hand-powered pump is often adequate to pressurise a small volume. A long lever magnifies the force applied to a piston whose displacement generates pressure in a fluid, water or oil, the pressure being "trapped" by a non-return valve. For very small pressure circuits, such as microscope pressure cells, high-pressure liquid chromatography pumps can be used, as they can also provide a continuous flow. However, Maltas et al. (2014) used a pair of piston screw pumps, one driving and the other a slave, in their high-pressure perfusion apparatus.

Reciprocating piston pumps are used in the chemical and food processing industries. These are designed to avoid contamination, but in general, they are most appropriate for use in situations where significant flow rates are required. A popular design for laboratory work uses a low-pressure air supply that acts on a large diameter piston (called a diaphragm) that drives a small diameter piston. The pressure in the small-diameter cylinder is the product of the ratios of the area of the diaphragm : small piston. Regulating the air pressure on the low-pressure side automatically regulates the pressure on the high-pressure side. The non-return valves on both the low-pressure and high-pressure sides of the pump allow cycling to occur. This type of pump can be used to run a high-pressure aquarium. Gas can be pumped by a similar mechanism, and the pumps can be described as intensifiers. Usually, the reservoir of low-pressure gas is a commercial cylinder which supplies gas at a moderate pressure, to the intensifier, which then increases it. A notable feature of pumping gas to high pressure is the heat generated on compression. When transferring a large volume of gas from one storage vessel to another, it is important to proceed slowly as considerable heat can be generated. Conversely, decompressing a gas cylinder should also be done slowly as the decrease in temperature at the point of decompression can cause condensation and then freezing, the basis of compressor-refrigeration. The low temperature might also change the properties of the metal fittings. The amount of heat generated as a fluid is compressed is given by the general formula: $dT/dP = a/p.C \times T$ in which a is the thermal expansion coefficient, p the density of the fluid and C the specific heat capacity and T and P are the temperature (degrees Kelvin) and pressure, respectively. The first set of parameters change with pressure and temperature, so the calculation gets complicated if it is to be used to predict how hot something may get when pressurised. This is of real concern in certain industries, for example in food technology, but in experimental work it is usual to simply measure and report the temperature changes in the site of interest, using a thermocouple or a thermistor. The latter can be sensitive to high pressures.

Sun et al. (2018) have provided some good data for the heat of compression of organic solvents for the food industry. Data for water is available in Lemmon et al. (2010). Water has a very small heat of compression. In electrophysiological work, it is often necessary to have inert medicinal liquid paraffin transmitting pressure to the preparation and providing electrical insulation, and as it has a high heat of compression, temperature changes are an important issue. For example, Harper et al. (1987, Chapter 5) had to wait for 2 h after the rapid compression the experiment needed to allow for temperature to equilibrate Schmalwasser et al. (1998) discussed the thermal features of their pressure apparatus, in which compression to 50 MPa over 3 min produced less than a 1-C temperature increase.

## 13.1.2  Generating Pressure by Centrifugation

Centrifugation is a mechanical process that utilises an applied centrifugal force field to separate the components of a mixture according to density. It is routinely used in many industries: mineral, petrochemical, chemical, medical, pharmaceutical, municipal/industrial waste, dairy, food, polymer, energy and agriculture.

Laboratory centrifuges usually work with 20–40 cm diameter rotors and 100–20,000 rotations per minute. Ultracentrifuges work at higher velocities, usually from 60,000 revolutions per minute, creating a gravitational acceleration of up to 1,000,000 g.

In biochemistry, centrifugation is one of many basic techniques, and the realisation that high pressure accompanies centrifugation has an interesting history, with lessons for us all. In 1966, Josephs and Harrington published a paper on the properties of myosin, a major muscle protein. They were using high-speed centrifugation, then a relatively novel technique. They described their equipment as follows: "The practical picture we need is of a small glass tube, say 5 mm internal diameter × 30 mm long, holding a carefully prepared solution. Several such tubes fit snugly in a solid metal holder, carefully balanced, and mounted on a vertical rotor drive shaft". They reported that, "unusual sedimentation behaviour appears to be the result of a shift in the monomer—polymer equilibrium at high rotor velocity which favours the monomer species and emphasises the delicate balance of the equilibrium state". In a later section, they state: "We are investigating the possibility that the shift in the equilibrium at high rotor velocity stems from the hydrostatic pressure gradient in the centrifuge cell. . . ." They confirmed this in a short paper in 1967. These very competent chemists had stumbled on something of considerable technical importance, and not well known among biochemists. In 1967, the prestigious journal, The Proceedings of the National Academy of Science, USA, published a paper by Kegeles, Rhodes and Bethune summarising the situation with high-speed centrifugation, pointed out that "the careful reinterpretation of already existing findings based upon high speed ultracentrifugation may lead to interesting conclusions". Is this statement commendable tact? Presumably, it reflected the then-current lack of knowledge in the biochemical community about centrifugation.

A paper by Heden, published in Bacteriological Reviews in 1964, contained a graph illustrating centrifugally generated pressures of up to 500 MPa generated by rotors of different diameters spinning at different speeds. Heden was based at the Karolinska Institute in Stockholm, Sweden, but had the previous year (1963) given a lecture at a major Microbiology Meeting in Cleveland, Ohio, USA, in which centrifugally generated pressure would have been discussed.

Subsequently, other biochemists found that ribosomes were dissociated by centrifugation, actually pressure (Hauge 1971) and in 1985 two papers were published describing the use of centrifuges to generate predictable pressures. Mertens-Strljthagen and colleagues used both centrifugation and a static pressure vessel to perturb rat mitochondrial membranes, and Wu et al. used centrifugally generated pressure to perturb the phase transition of liposome bilayers. In 1997, Halle and Yedgar used centrifugation to generate a relatively low hydrostatic pressure to affect the order of the bilayer in human blood cells.

The simple version of the equation used to calculate centrifugally generated hydrostatic pressure (P) is $P = \omega^2 \rho (x^2 - x_o^2)/2$, in which $\omega$ is the angular velocity of the spinning rotor, p the density of the solution, x and $x_o$ the distance from the centre of rotation to the bottom of the tube and the top of the solution, respectively. It is assumed that the density of the solution is constant throughout its depth, which is not strictly true. Centrifugation is only used to generate pressure in rare circumstances, mentioned earlier, as it is inconvenient and, of course, introduces high gravitational forces.

## 13.2   High-Pressure Plumbing

The term plumbing is an apt description but in no way is it intended to trivialise how a pump is connected to the vessel it serves. Pipes in the form of thick-walled steel tubes, couplings, tees, elbows, crosses and stop valves rated for use at certain pressures can be purchased. Similarly, rupture discs mounted in holders and flexible steel, high-pressure capillary tubing can also be purchased. The means of measuring pressure is usually a pressure gauge giving an analogue display on a large dial. This is the traditional Bourdon Gauge. It works by the pressurised fluid entering a coiled tube which has an oval cross-section. The coil tends to straighten in proportion to the contained pressure, and the movement is coupled to a delicate mechanism that actuates the needle on the dial. Transducers are also commonly used. They are essentially thick-walled tubes that expand slightly when pressurised. The expansion is detected in various ways, all of which provide an electrical output. The accuracy of the gauge or transducer is checked by means of a deadweight tester. These devices are commercially available and their accuracy is certificated by a nationally recognised laboratory. The tester comprises a vertically mounted cylinder with a piston on which weights are placed. The known weight acting on a defined area of the piston creates a corresponding hydrostatic pressure. The tester is coupled to the gauge or transducer which is then adjusted to give accurate readings. Optical methods are used in special circumstances in the physical sciences (Wahl 1913a, b), for example the ruby fluorescence technique, for pressures up to 1 GPa and within a defined temperature range (Grasset 2001).

## 13.3    Pressure Vessels

Vessels that are designed to contain high pressure are usually cylindrical. For the purpose of stress analysis, they can be divided into two types, those with either thin or thick walls. Well established equations describe the stresses in each of these (Sherman and Stadmuller 1988). A familiar thin-walled pressure vessel is a commercial gas cylinder containing oxygen, helium, etc., usually at a pressure of 15–19 MPA and rarely higher. Large hyperbaric chambers that accommodate human patients or divers are also thin-walled, with the complication, for the designer, of windows and a large door and perhaps a "lock-in system" for humans and separately for small samples. The upper safe working limit for specialised vessels used in diving research is around 5–10 MPa but plenty of good work has been carried out in vessels with a much lower rating. Experimental vessels used in laboratories are usually thick-walled cylinders with, if required, very high safe working pressures. Vessels for high-pressure Biological work might be rated for 100 MPa, which is not a particularly high pressure. For ultra high-pressure experiments in physical science, a device called a diamond anvil is used. This provides a very small cell at pressures up to 77,000 MPa.

Steel is the most common material used to make a pressure vessel and its exact properties are carefully chosen by the designer. Titanium has advantages but is expensive and for very low-pressure purposes, transparent material such as polymethylmethacrylate, PMMA (known as Perspex and Lucite), has been used (Quetin and Childress 1980). Sophisticated manufacture of large vessels, including the hyperbaric chambers just mentioned, can make use of carbon and glass fibre composite materials. The typical laboratory pressure vessel is usually machined from a piece of steel in a specialist workshop. It has to have at least one end closure, which is often a plug held in place by a thick-walled ring in which a thread is machined. The plug enters the bore of the pressure cylinder and supports a soft seal, an O-ring made of special rubber or other elastomers, which confines the pressurised fluid. To quote my engineer colleague, the O-ring is more clever than it looks. It fits in a groove cut in the plug, the width of the groove being slightly smaller than the diameter of the cross-section of the O-ring. When the plug slides into the vessel, the O-ring is forced into an oval shape, providing the initial seal between the plug and vessel wall. When pressure is applied, the O-ring is squashed into the groove, further sealing the pathway the pressurised fluid would otherwise take. The unsung O-ring self-seals. With repeated use at its maximum pressure, an O-ring can show a ridge of near extruded material, illustrating how it works and that it is time to be replaced. For use at very high pressure, O-rings can be "backed up" by a separate chamfered ring.

The ensuing thrust (force x area) on the end plug is taken by the retaining ring and distributed along the thread. Such end rings may be made of high strength aluminium bronze, which minimises the risk of thread-locking. The plug can be used to carry electrically insulated feed-throughs, which can be made from mineral-insulated electrical cable or purchased. A window can be accommodated, see below. A rotating shaft is often conveniently mounted in an end plug. It has to be specially machined and fitted with a thrust bearing (purchased) and a special type of soft seal. A rotating shaft is a versatile means of moving something within the vessel as its motion does not create a change in volume or pressure.

## 13.4   Safety

Pressure vessels are potentially dangerous. The energy in a compressed fluid is roughly proportional to its compressibility. Therefore, the energy stored in compressed gas is much greater than in compressed water, which is compressed by only 4% by 100 MPa. However, it is not the total amount of energy stored in the system which should be the only concern. The rupture of a capillary connection can produce a jet of water, especially serious when some compressed gas is in the system. The damage which occurs does not depend directly on the total stored energy but on the density of the energy in the jet that emerges. When the jet hits a person, it is focussed on a very small area, so guards that can withstand that energy should be fitted. Even in systems in which the total amount of stored energy is not great, i.e. water-filled vessels, safety precautions must be applied. For the professional engineer, failures are discussed in the paper by Skelton and Crossland (1980). The minimum factors of safety required for commercial compressed gas cylinders are between 2.5 and 3; however, the conditions for the use and storage of these cylinders are closely specified. The generally recommended factors of safety for pressure vessel design are between 3.5 and 6, although for certain materials and in certain circumstances much higher may be required.

The various codes and directives in high-pressure technology will be familiar to a professional engineer. The history of fatalities and accidents arising from the design and use of high-pressure equipment has led to it becoming one of the most regulated areas in engineering. All relevant codes and standards should be observed. Guidelines are provided by The United Kingdom Department for Business, Energy and Industrial Strategy: Pressure Equipment (Safety) Regulations 2016. The legally enforced EU Pressure Equipment Directive 2014/68/EU came into force in 2016. This directive applies to the design, manufacture and conformity assessment of pressure equipment and assemblies of pressure equipment with a maximum allowable pressure greater than 0.05 MPa. Designs that depart from these standards and codes of practice are not generally illegal but they have to be justified on their own terms. Equivalent regulations prevail in other parts of the world. Departure from any standard or code of practice may make insurance problematic and should only be considered if it can be shown that the regulations are inappropriate.

## 13.5   Examples of High-Pressure Experimental Apparatus

### 13.5.1  Experiments with Small Animals

The vessel shown in Fig. 13.1 has been used with small animals compressed with either water or gas (heliox).

The vessel was machined from stainless steel round Bar (Firth Vickers 520B equivalent to S143steel) which is a low carbon precipitation hardening steel, with good fracture toughness and highly resistant to stress corrosion cracking. The pressure rating for the vessel is much higher than is usually needed (110 MPa

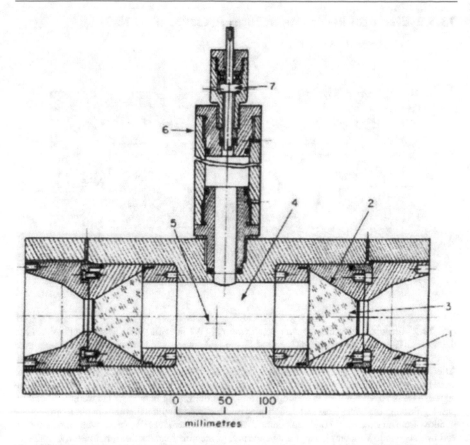

**Fig. 13.1** An observation vessel with high-pressure windows, a rotating shaft inlet and an additional instrument chamber. This vessel is fitted with two conical polymethacrylate (PMMA) windows, with a 60-degree included angle, and a viewing aperture of 3.8 cm (see Section 13.6.2). The vessel has an internal diameter of 7.6 cm and an internal chamber length of 17.8 cm. The pressurised instrument chamber on the top of the vessel has a 5-cm internal diameter and is 25 cm long. A through-flow of seawater at constant high pressure is achieved by an air-powered pump. Details: (1) end ring which supports the window mount (2). (3), widow; (4), main chamber; (5), transfer port, see below, out of plane; (6) instrument chamber; (7), rotating shaft. Not shown is the pressure inlet port which connects to the hydraulic circuit of the pump, gauge, control valves and rupture disc. The vessel fits in a steel jacket through which temperature-controlled water is circulated. From Gilchrist and Macdonald (1983)

routinely, 140 MPa maximum). The vessel can also connect with the isobaric collection vessel, described below, by way of a radial port, (5). Smaller versions of this general design have been used by the author as they are light enough to be air freighted to join research ships of opportunity. A long-range stereobinocular microscope is normally used with these vessels.

## 13.5.2 Electrical Recording at High Pressure (Fig. 13.2)

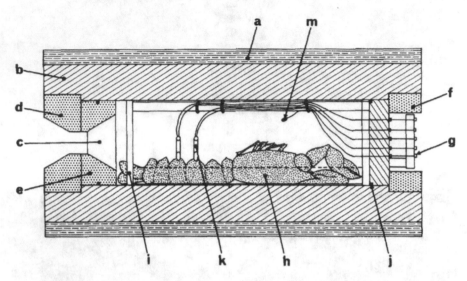

**Fig. 13.2** Electrical recording from an aquatic animal (Wann et al. 1979). The vessel is similar to that in Fig. 13.1, with an internal diameter of 12 cm and enclosed in a water jacket for temperature control. The experimental lobster, mounted on a cradle outside the vessel, is fitted with electrodes. After initial tests show all is working well, the apparatus is lowered vertically into the pressure vessel, displacing air from the seawater which will pressurise it. The safe working pressure of the apparatus was 50 MPa, set by the window design. The thermostat regulating the water temperature passing through the water jacket was duplicated and backed up with a thermal cut out to protect the window from an excessively high temperature. Details: a, water jacket, b, vessel wall, c, window, d, end ring, e, window mount, f, supporting end ring, g, electrical connections, h, lobster, i, cradle to which it is strapped, j, O-ring, k, support for electrodes, m, thermistor (Wann et al. 1979)

### 13.5.2.1 Patch Clamp Recording at High Pressure

Some electrophysiological techniques were mentioned in Chapter 4 and here methods of patch clamp recording at high pressure are described. The technique provides data on the kinetics of ion and other channels (Chapter 4, Section 4.5.3).

First, the mega-ohm seal patch clamp method, which is featured in experiments in Chapter 11, is described, without an illustration. A vertically mounted cylindrical vessel similar to that in Fig. 13.1 contains the recording apparatus. At the base, a plug replaces the window and accommodates mechanical drives, electrical connections, a light guide and a pressure inlet port. At the top, the window enables the experimenter to view the target cell and the electrode tip, using a long-range binocular microscope. The tip of the patch electrode/pipette has to be gently pressed on to the surface of a cell using an electrically powered micrometre drive. This is commercially available and can be used in a helium-filled pressure vessel, as here. The cell preparation is mounted in recording solution at the bottom of the vessel and its position is adjusted by direct drive mechanical linkages. The patch electrode is connected to its preamplifier, mounted in a pressure-resistant casing, supported by the micrometre drive. The experimenter, viewing the cell and the electrode tip, controls the micrometre drive and judges the contact and seal the electrode/pipette makes with the cell membrane, using electrical information displayed on an oscilloscope. After a good seal is formed, channel open/closed activity is recorded. The patch pipette can then be raised and a new seal formed on a different part of the cell membrane (Chapter 11, Box 11.1) (Macdonald et al. 1989).

The giga-ohm seal technique is the main, widely used patch clamp method, which Heinemann adapted for use at high pressure, devising the apparatus shown in Fig. 13.3.

The Heinemann design has been modified in various ways, making it more versatile (Fig. 13.4) (Macdonald and Martinac 1999). A version of this design has been used with high-pressure oxygen, to study oxygen toxicity (Macdonald and Vjotosh 1999).

Heinemann and colleagues also adapted a two electrode voltage clamp apparatus for use at high pressure, (Fig. 13.5), whose use was described in Chapter 4, Section 4.5.3.1. These three apparatuses are illustrated in the following pages 362–364.

### 13.5.3 Other High-Pressure Apparatus

#### 13.5.3.1 Stopping a Reaction at High Pressure

Various ways of stopping a reaction proceeding in a pressure vessel are described in the literature. The simplest design has a steel ball in a tube which is partitioned by a thin glass plate, the whole confined in a pressure vessel. When the vessel is inverted, the ball falls, it breaks the glass partition separating the fixing solution from, for example, a cell suspension, thereby fixing the cells (e.g. Murayama 1987; Landau and Thibodeau 1962).

A method of freezing a pressurised sample is illustrated in Fig. 13.6 page 365. It is designed to kinetically "trap" the high-pressure conformation adopted by a protein and is therefore a very fast process. The cartoon-style presentation is largely self-explanatory. The use of the old-fashioned term "pressure bomb" is incomprehensible to the author. It means, of course, the reverse of what is intended! Note that the pressurising fluid is ethanol and the silicon pistons are made in the laboratory. At stage C the sample, still pressurised, is moved to the bottom of the vessel where it cools very rapidly to 200 K ($-73$ °C). At stage D decompression takes place and in E the sample is transferred to liquid nitrogen, 77 K ($-196$ °C).

#### 13.5.3.2 Starting a Physiological Reaction at High Pressure

An example is shown in Fig. 13.7 page 366, a pressure vessel of 5.5 cm internal diameter, similar to that in Fig. 13.1, contains a PMMA cuvette which holds a suspension of platelet-rich plasma. This is equilibrated with the gas, helium, nitrogen, etc., in the pressure vessel by holding the pressure vessel vertically (window at the bottom), and gently rocking, see inset. Note the geometry of the cuvette. When the platelet suspension is equilibrated, the pressure vessel is returned to the horizontal position and the solution of the aggregating agent (ADP) is injected by the rotation of

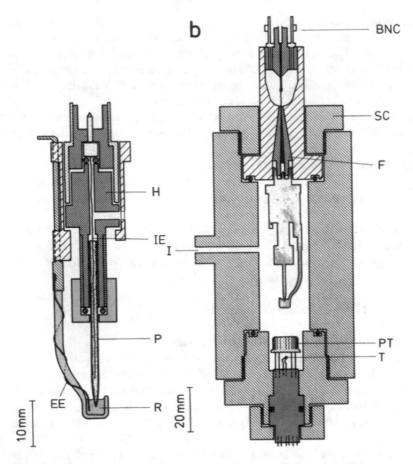

**Fig. 13.3** The giga-ohm patch clamp apparatus for use at high pressure (Heinemann et al. 1987). The patch electrode holder, shown on the left, is used to form a patch in the normal fashion, with the miniature bath swung out of the way. Then the minibath, containing recording solution (R), is then swung under the tip of the electrode providing a complete circuit, such that the apparatus can be transferred to the pressure vessel shown on the right (hence the term flying patch clamp). The pressure vessel is filled with medicinal liquid paraffin oil which floats on the recording solution, insulates the electrical components and transmits the hydrostatic pressure which is increased by pumping oil into the vessel through a side port, I. Pressure and temperature are measured by transducers at the bottom of the vessel, PT and T, respectively. Other components: H, electrode holder; IE, internal silver/silver chloride electrode; P patch pipette; BNC; co-axial electrical connection (note the preamplifier is not inside the vessel); SC, threaded steel retaining collar; and F, Teflon-coated electrical connector (see Chapter 4)

the screw cap which pushes the rod (PR) down. Not shown is a pair of small windows in the wall of the vessel, aligned to the cuvette and coupled, outside the vessel, to light guides that connect to a photometer. The light path through the cuvette is shown by a dotted circle (See Chapter 11, Section 11.6). Aggregation is measured by the change in optical density measured by the photometer. Details: W, window; EP, end plug; EC, electrical connection for the fan motor; CT, high-pressure capillary tube

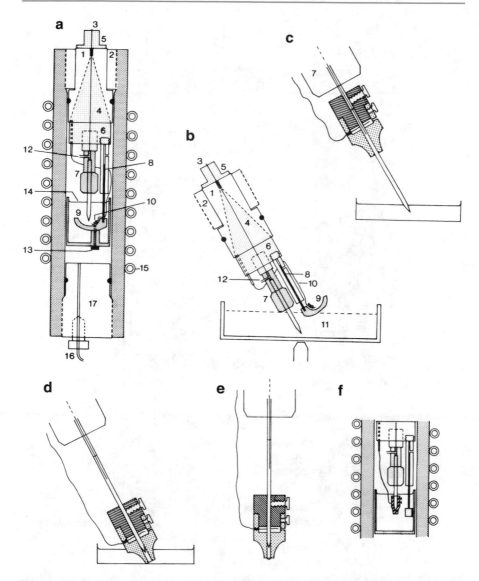

**Fig. 13.4** **A** shows the assembly fitted into a 29-mm diameter vessel, filled with paraffin oil. Its novel feature is the heat shield beaker of the recording solution, supported by a bolt (13). (14) is the oil–water interface. The large volume of water enables pressures of up to 90 MPa to be applied without excessive heating or equilibration time (Chapter 4). Details: 1, end closure; 2, retaining ring; 3, lead connecting the patch pipette to the amplifier; 4, Teflon insulation; 5, Teflon plug socket; 6, extension of 4, supporting 7, pipette holder and 8, mechanical sliding support for 9, the transfer dish; 10, bath electrode; 12, suction line used to form a seal and then disconnected; 15, copper tubing providing a circulation of thermostatted water; 16, pressure capillary; 17, lower plug. **B** shows how the initial seal is made. **B** to **F** shows the apparatus modified to patch clamp record from Prokaryote channels reconstituted in liposomes. The initial patch seal is made with liposomes in suspension, in the dish in C, using the technique described in Delcour et al. (1989), Chapter 4, Section 4.5.3.1. A PMMA collar, holding a small volume of recording solution by capillarity, is moved down the pipette and secured to enclose the patch at the tip of the pipette, and is then itself enclosed by the heat shield beaker and pressurised as in **A**

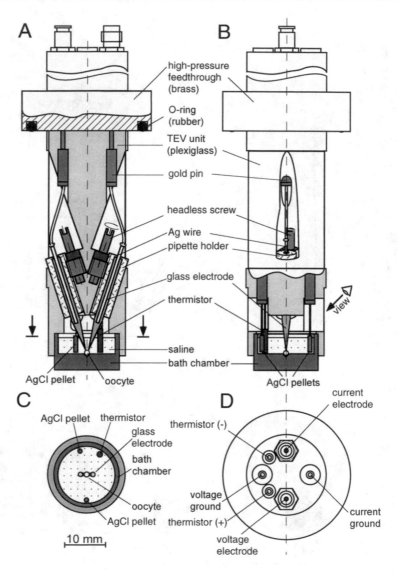

**Fig. 13.5** Two electrode voltage clamp high-pressure apparatus of Schmalwasser et al. (1998). The pressure vessel (not shown), which accommodates the apparatus, is of 2-cm internal diameter and 6-cm internal height. After the procedures which would be carried out with a normal benchtop apparatus have been completed, mounting the recording apparatus in the pressure vessel proceeds. The oocyte is positioned at the centre of the recording bath. The pipette holders, shown on the left with a view at right angles on the right, which support concentrically aligned glass pipettes, are linearly adjusted by screw threads to impale the oocytes, a procedure assisted by using a long-range microscope. The amplifier is switched to a voltage clamp setting and the oocyte is tested for channel activity. At that stage the apparatus in the figure is inserted into the pressure vessel, displacing the air such that compression is by oil. Other routine electrical checks are made before compression, which generates a temperature rise of less than 1 °C on pressurising to 50 MPa in 3 min (Chapter 4, Section 4.5.3.1)

**Fig. 13.6** Fast freezing of a sample at high pressure, from Lerch et al. (2014)

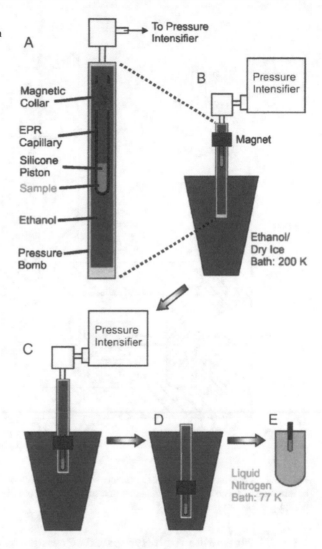

connection; H, humidifying solution in cotton wool; CV, cuvette containing 0.9 ml plasma, stirred by the motor SM; P1 is the pressure injection device; SC, screw cap; PR, pushrod.

The same vessel may be used to apply hydrostatic pressure. The vessel is used vertically, with the window at the top and the propellor removed. It is partially filled with water to exclude the pressurising gas from the sample, which in turn, is separated from the water (Fig. 13.7) (see Pickles et al. 1990, Fig. 1B).

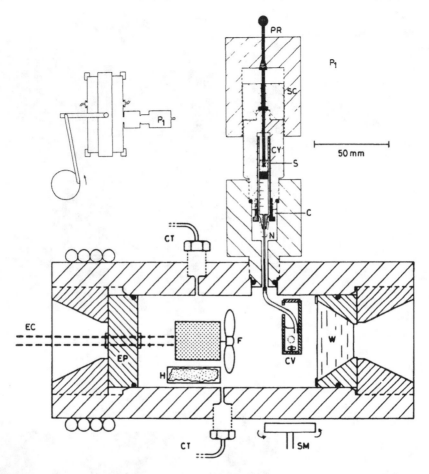

**Fig. 13.7** A pressure vessels for experiments with blood platelets (Pickles et al. 1990). A method of initiating and recording the aggregation of a suspension of blood platelets equilibrated with high partial pressures of inert gases is shown

### 13.5.3.3 Measuring High Hydrostatic Pressure in Load-Bearing Joints

The location of pressure transducers built into the artificial head of a femur is shown (Fig. 13.8). Each transducer lies behind a diaphragm which is continuous with the surface of the spherical head. A film of fluid transmits pressure between the head and the socket in which it fits. Beneath the diaphragm is a strain gauge sensor with a sensitivity of 0.28 um of deflection per MPa. Within the head electronic circuitry sequentially samples the transducer outputs and transmits a radio telemetry signal. Pseudofemoral heads implanted in a 73-year-old woman (A) and an 82-year-old man (B). The lower figure (C) shows the orientation of the head when facing a standing individual. The numbers relate to specific sensors whose output is recorded, see Fig. 1.4, Chapter 1.

**Fig. 13.8** Femoral head prosthesis for the human hip joint (Morrell et al. 2005)

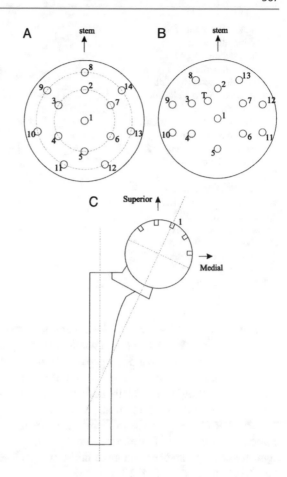

## 13.6 High-Pressure Windows

High-pressure windows are either large or small, which is not as trivial as it may seem, and the latter are commonly used with various physical instruments in addition to microscopes and the human eye. A glass capillary tube can serve both as a sample holder and as a pressure window.

### 13.6.1 Small High-Pressure Windows

In the late nineteenth century, Regnard used high-pressure windows to observe small aquatic animals at high pressure, and indeed to project their image onto a screen (Chapter 1, Section 1.2.1; Chapter 4). His glass windows, 1-cm diameter by 4-cm

**Fig. 13.9** The Poulter seal. (a) 12-mm diameter monocrystalline sapphire window is mounted on a polished steel seat; (b). The unsupported area is indicated by the diameter; (d). The window thickness was varied from 0.6 to 3 mm and subjected to tests, which required the holding-plug (c) to be screwed into a pressure vessel. The thickest windows were used in the PICCEL apparatus rated for 30 MPa, see below, described by Pradillon et al. (2004). An example of a Poulter seal is shown in Fig. 13.11A (Chervin et al. 1994)

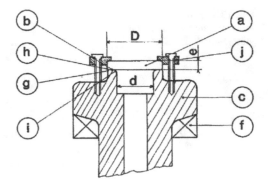

long, were capable for use to 104 MPa and larger versions were used up to 52 MPa. Amagat and then Wahl developed windows for higher pressure and Poulter, working in the 1930s, produced windows capable of use at more than 3000 MPa. The Poulter seal, which can be used for plugs as well as windows, has a cylinder with a perfectly smooth end sitting on an equally perfectly smooth supporting ring. The cylinder is in compression, and therefore strong, and the seal between the two smooth surfaces experiences very high pressure because the central area is unsupported, the unsupported area principle of Bridgman. Poulter-type windows are used in high-pressure light microscopy and, for example in the study of sapphire windows carried out by Chervin et al. (1994) (Fig. 13.9).

A cuvette (a sample cell) functions as a test tube but is designed for use in an optical instrument. Standard cuvettes are sometimes cylindrical but are usually square in cross-section to avoid refraction artefacts. Depending on what part of the spectrum is used, they may be made of quartz, sapphire or optical glass although plastic cuvettes do exist for less demanding measurements (as in Fig. 13.7). Most cuvettes have two transparent sides opposite one another so the spectrophotometer light can pass through, although some use reflection and only need a single transparent side. For fluorescence measurements, a transparent side at right angles to those needed for the excitation light can provide for the emission. A cuvette is often used within a small pressure vessel, aligned to its pressure windows, and immersed in the pressure transmitting fluid, as in Paladini and Weber (1981), who corrected for the birefringence induced by pressure straining the quartz windows (Chapter 2, Box 2.4).

The distinctive feature of high-pressure light microscopy is that the objective lens has a short focal length and thus the high-pressure window, through which the light passes, must be sufficiently thin yet capable of withstanding the pressure. On the underside of an orthodox upright microscope, the condenser lens, which collects light to pass through the objective, has a longer focal length and so the lower pressure window can be thicker than the upper window. High-pressure microscopy is carried out with high quality manufactured microscopes, upright or inverted, to which an optical pressure cell or chamber is fitted. Marsland made extensive use of high-pressure microscopy, starting in the 1930s (Marsland 1958; Chapter 2). The glass windows were 6-mm thick and used at up to 110 MPa. Kitching, in the UK (1954 and later, Miller et al. 1975) used similar tapered-cylindrical glass windows in a microscope optical chamber to observe single cells in an aqueous suspension. A drop of the suspension is placed on the underside of the upper window, the so-called hanging drop method. The chamber is filled with medicinal paraffin which does not disturb the hanging drop as it transmits pressure to it. The same liquid paraffin fills the hand-powered high-pressure pump connected to the chamber.

Salmon and Ellis' observations of the microtubules in the mitotic apparatus of cells dividing under high pressure required a more sophisticated design, which is thoroughly explained in their 1975 paper, and in Inoue et al. (1975). Their microscope provides phase-contrast microscopy at high pressure, capable of resolving otherwise transparent intracellular structures. It also provided polarising microscopy, capable of resolving and quantifying the birefringence retardation (see Physics textbook) of microtubular arrays. The papers give a detailed engineering account which cannot be summarised here, but some practical points are worth noting. The volume of the microscope chamber is small, 0.075 ml, but nevertheless sufficient to contain sufficient dissolved oxygen for the needs of the cells. It can be rapidly sealed by means of a bayonet-type closure. The high-pressure windows are made of strain-free optical glass "flats", 5 mm in diameter $\times$ 1.75 mm thick, mounted Poulter style and seated and sealed by the unsupported area principle and silicone cement. The window stress birefringence was analysed in detail by Salmon and was shown to be sufficiently low at 73 MPa so that measurements of birefringence retardation at high pressure were acceptable. That seems to be the factor limiting the working pressure indicated in Inoue et al. (1975). For phase-contrast microscopy, the working limit is twice that value. Finally, the temperature increase following a stepwise increase to 40 MPa was 2 °C when the inert fluorocarbon oil Kel-F10 filled the sample chamber, and it equilibrated in 35 s. With water in the chamber, the increase was much less.

The birefringence retardation (BR) is a measure of the polymerised tubulin in the microtubules. Rebhun et al. (1974) showed that by treating the spindle microtubules in dividing eggs with glycol, much bigger and more dense microtubule bundles were formed whose BR was similarly increased. In the present pressure experiments, the experimenter rotates a disc on the microscope assembly to measure the BR of a spindle in focus as the cell proceeds through its mitosis, or as pressure is varied. Salmon (1975a) shows results obtained with the spindle microtubules in the oocytes of the marine worm *Chaetopteris*. Application of 13.6 MPa causes the BR to decrease to an equilibrium value, and on decompression, it returns to the original value. This is interpreted as the depolymerisation of tubulin and subsequent polymerisation, manifesting the reversible equilibrium mentioned in Chapter 2. A similar effect of pressure is seen with tubulin extracted from the rabbit brain. When reconstituted as microtubules in solution the degree of polymerisation is measured by light absorption. The effect of pressure on the polymer–monomer equilibrium can be measured using a high-pressure light absorption cell (Salmon 1975d). And finally, Salmon also showed that the microtubules in cultured mammalian cells (Hela cells) undergo depolymerisation at high pressure (68 MPa at 37 °C), in this case, quantified by electron microscopy. This was done by fixing cells at high pressure. Subsequent electron microscopy showed the microtubules had disappeared at pressure and reappeared 10 min post decompression.

The outstanding quality of the high-pressure microscopy of Nishiyama and colleagues has been mentioned in Chapter 2, Section 2.5 (Fig. 13.10).

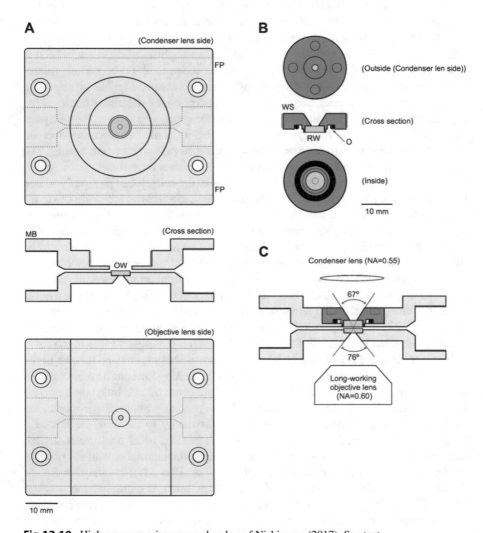

**Fig 13.10** High-pressure microscope chamber of Nishiyama (2017). See text

The chamber is mounted on an inverted microscope and a vibration-free table. The long-range viewing objective has a numerical aperture of 0.6. Note the viewing window is thinner than the window which serves the condenser lens (N A 0.55). The special glass used for the high-pressure windows, a borosilicate glass designated BK7, has low dispersion and a low refractive index, closely matching the refractive index of the glass used in commercial microscope objectives. Nishiyama et al. argue that alternative window materials, sapphire, diamond or quartz, provide a poorer match than the special glass. The pressure windows have a safe working pressure of more than 100 MPa. The optical performance is illustrated in Nishiyama's 2015 account of photographs of images of 1 um beads in four different modes: dark field, light field, phase contrast and fluorescence, at 100 MPa. The onset of pressure causes the image of the beads to be distorted, but focussing the objective restores sharpness. Images of single molecules obtained with the equipment were reported by Ishii, Nishiyama and Yanagida in 2004.

The "multi-purpose modular system" produced by Vass and his group at Edinburgh University, UK, is notable for being used with an argon laser in a study of the effects of pressure on the calcium-sensitive fluorescent dyes Fluo-4 and Fura red (Schneidereit et al. 2016), following earlier experiments by Friedrich et al. (2006). Figure 13.11 shows the design for a microscope pressure cell which is also capable of imaging of single fluorescent molecules.

Other examples of high-pressure microscopy equipment are described in Frey et al. (2006), Raber et al. (2006) and by Deguchi and Tsujii (2002) whose supercritical state flow cell is mentioned in Chapter 9 (Fig. 9.1). Fluorescence, both polarised and depolarised, is used in high-pressure instruments (Chapter 2; Tekmen and Muller 2004). A method of measuring the concentration, or partial pressure, of oxygen dissolved in water under high pressure based on fluorescence is described by Stokes and Somero (1999).

Readers might wish to examine more thoroughly than is possible here the work on sapphire windows mentioned earlier (Chervin et al. 1994), and the variety of equipment used by the following: Ranatunga et al. (1990), a miniature flat PMMA window in a pressure cell used for measuring the effect of pressure, up to 10 MPa, on muscle fibres; Pagliaro et al. (1995), for low pressure (6.7 MPa), observation of cultured mammalian cells; Mizuno (2005), microscopic-pressure cell for determining intracellular calcium in chondrocytes (Chapter 3); Haver et al. (2006), fluorescence imaging in a capillary; and Maltas et al. (2014), optical microscopy perfusion apparatus. A high pressure microscopy chamber for growing cells illustrated in Fig. 13.12.

The term window is also used for components that withstand high pressure and transmit parts of the electromagnetic spectrum beyond visible light, for example X-rays used in X-ray scattering, and neutrons used for neutron scattering, described by Winter (2002) (Fig. 13.13) and more recently by Testemale et al. (2016).

Nuclear magnetic resonance spectroscopy (NMR) is nowadays conducted by pressurising the sample in a sapphire or a zirconia ceramic cell as Fig. 13.14.

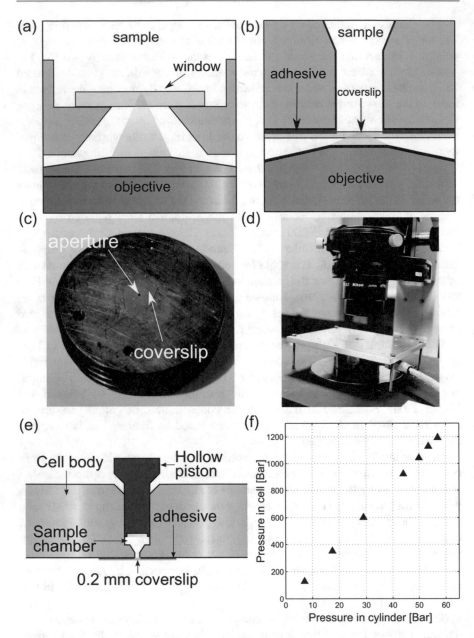

**Fig. 13.11** High-pressure microscope chamber of Vass et al. (2013). The microscope is inverted, with the viewing objective beneath the cell. It incorporates a flat quartz "coverslip" window 0.2 mm thick, glued to the pressure cell (b) and has a 100 MPa capability. This is achieved because the diameter of the unsupported area of the window is very small, as shown in c. In contrast, the Poulter seal is shown in a. Pressure is applied by a hollow piston applying force downwards, e and f. Light passes through a diamond window in the piston. This remarkable design includes interesting practical features, such as the coverslip windows are disposable and readily replaced

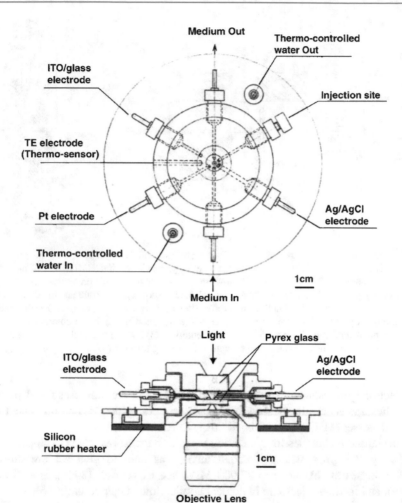

**Fig 13.12** High-pressure microscope chamber for growing cells (Koyama et al. 2001). The significant feature of this microscope chamber is that it is temperature-controlled (2 °C to 80 °C) and coupled to a pressure-controlled perfusion system that supplies nutrients and dissolved oxygen. An inverted differential interference contrast microscope is used. The upper (condenser) high-pressure window is an 8-mm-thick Pyrex glass. The lower (objective) window is made of 2-mm-thick Pyrex glass with a 2-mm viewing diameter and the chamber is rated for pressures up to 100 MPa. The plan view of the apparatus shows radial connecting ports. Starting at six o'clock and proceeding clockwise they are: culture medium inlet, Pt electrode, indium tin oxide electrode, culture medium outlet, pressure injection port, connected to a hand pump, and reference silver/silver chloride electrode. Other connections control the temperature of the chamber

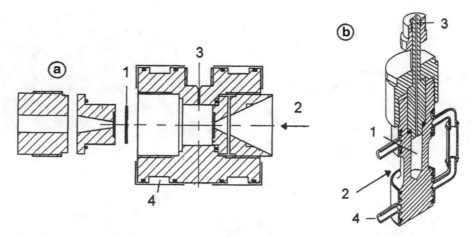

**Fig. 13.13** (a) High-pressure cells for X-Ray scattering and Neutron scattering measurements (Winter 2002). The cell is made of stainless steel or a special alloy (NIMONIC 90) and it has beryllium windows 1–2 mm thick (1). (2) shows the beam path and (3) the pressure connection port. Safe working pressure is 200 MPa. The assembled state is shown on the right and the disassembled state on the left. (**b**) A pressure cell for neutron scattering measurements, up to 250 MPa, made of aluminium alloy, AlMgSi1. The arrow (2) shows the direction of the beam entering the sample chamber, diameter 3 mm. Water at high pressure enters at (3) and is separated from the sample by PTFE. Temperature control is achieved by water circulating through (4)

Electron spin resonance spectroscopy (ESR) at high pressure uses fused silica or borosilicate glass capillary tubes to hold the sample, which is then immersed in a pressure vessel (McCoy and Hubbell 2011; Lerch et al. 2014).

Specialised techniques for generating very high pressures were developed in the 1930s by Bridgman whose diamond anvil was able to generate pressures of 40,000 MPa, now extended to 77,000 MPa, see Lipp et al. (2005), Haberl et al. (2019), and for lower pressure biological applications, Oger et al. (2006).

**Fig. 13.14** Sample cell for
NMR measurements at high
pressure (Peterson and Wand
2005). The NMR cell is
shown at the bottom of the
assembly. It connects via a
cell base to the main body,
which, at its top, is sealed by a
plug. The safe working
pressure is 70 MPa. The NMR
cell is a tube made of a
non-magnetic, ceramic
material, alumina-toughened
zirconia, with an outside
diameter of 5 mm and a
working volume of 200 ul. It
is resistant to the short-chain
alkanes used to dissolve
encapsulated proteins, which
are delivered, at high pressure,
through the side port in the
main body

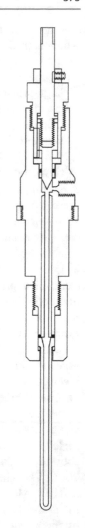

## 13.6.2 Large High-Pressure Windows

Large windows are fitted in human scale hyperbaric chambers used at low pressure
in clinical work, and at higher pressures, up to 10 MPa, in specialised diving research
chambers. Manned submersibles also require large windows and as most such
vehicles are limited to depths of no more than 7000 m, no deeper depth rating for
the widows has been required. Few manned vessels have been capable of descending
to the greatest ocean depths, namely the bathyscaphe developed by Piccard, and two
others, of which more soon. Large windows are also required for observational
laboratory pressure vessels, and their design followed on from Piccard's early work
on bathyscaphe windows using polymethyl methacrylate, PMMA. Earlier experience with glass windows showed that compression could cause cracks at
imperfections and in local stresses arising from the movement between the window

and the supporting mount. The brittle nature of glass leads to the cracks developing which immediately leads to window collapse. In contrast, surface cracks in PMMA windows do not cause the same failure modes as the window plastically deforms when compressed.

High-pressure PMMA windows are usually truncated cones, with the pressure acting directly on the larger diameter (Stachiw 1966, 1970).

Gilchrist (1972) designed PMMA windows for observing small aquatic animals and an isobaric receiving and observation vessel with working pressures of 110 MPa shown in Fig. 13.1. The cone shape has a 60-degree included angle and a thickness to viewing diameter ratio of 1.57. The windows studied had a viewing diameter of 3.96 cm, giving a clear view of the interior of the cylindrical vessel they were to serve. The window is supported in a matching conical mount with sufficient depth to accommodate any slight displacement which might occur. Gilchrist's comprehensive programme of tests was carried out at 20 °C in a test pit 2 m deep for safety, using water as the pressurising fluid. The results are discussed in Gilchrist and Macdonald (1983) and exhaustively in Gilchrist (1972). The tests comprised: (a), static high pressures, (b) cyclic pressures and (c) creep measurements of long duration (1000 h) at 138 MPa, i.e. at a pressure 1.4 times the designed maximum working pressure. A separate test programme was also conducted by Gilchrist to examine how these windows fail. Pressure tests up to 345 MPa were carried out (Gilchrist and Macdonald 1983). We can skip the more demanding engineering aspects of the work, such as the photoelastic analysis, the detailed analysis of failure mechanisms and the effects of fast decompression from very high pressure, but take on board three practical points: (a) lubricating the conical face of the window sometimes led to surface cracks forming and so was avoided; (b) no advantage was accrued by annealing the window after machining and (c) the window can be scaled down to suit smaller vessels whilst retaining the key proportions. The tests showed the design is safe for the stated pressure at temperatures between 0 and 20 °C. The windows are ready for use after machining from stock PMMA round bar and have a long life if kept clean of particles (Fig. 13.15). Nevertheless, research into the design and properties of conical PMMA windows continues, stimulated in part by the recently built manned submersibles, Deep Sea Challenger and the Limiting Factor which have the capability of descending to the greatest ocean depths (Jamieson 2020; Lutz and Falkowski 2012; Stokstad 2018 all in Chapter 1). See also papers by Wang et al. (2019), Kemper and Kemper (2019), Liu et al. (2018) and Snoey and Katona (1970). Latterly questions have been expressed in these papers about the safety of the windows when they are used at pressures above 70 MPa, but those issues had already been addressed in Gilchrist's research programme (see Gilchrist and Macdonald 1983). The programme showed that these windows were suitable for repeated long-term use in laboratory situations at pressures of up to 110 MPa, and with regular inspection, pressures of up to 140 MPa could be employed.

**Fig. 13.15** High-pressure PMMA windows, freshly machined. Smaller diameter 39.6 mm (Gilchrist 1972)

## 13.7   The Isobaric Collection and Transfer of Organisms from High-Pressure Environments

The collection and handling of organisms from a high-pressure environment without loss of pressure, isobaric collection, is one of the most distinctive problems in the study of life at high pressure.

---

**Box 13.1 Isobaric Collection of Small Deep-Sea Animals**
A brief biographical background to the project to develop isobaric equipment provides another illustration of the way scientists go about their business. As a cell physiologist and post-doctoral research worker involved in high-pressure experiments, the author was attracted to work on deep-sea animals, for many reasons. One was the intriguing problem of the isobaric collection of deep-sea, preferably small, animals. Why had no one accomplished it? With the backing of Professor J A Kitching, the author's former Ph.D. supervisor, collaboration was arranged with Professor B Crossland, head of the Engineering Department at the Queen's University of Belfast, and research funds obtained from the Natural Environment Research Council, UK to produce equipment for that purpose. That was followed by the appointment of Ian Gilchrist, previously employed by the shipbuilders Harland and Wolff, and doing research at Queen's, to carry out what the author regarded as the difficult stuff, the engineering. As part of the application for research funds, the principles of

(continued)

**Box 13.1** (continued)

the passive isobaric collection of deepwater had been argued, but Gilchrist reworked the case during the design of practical equipment, taking into account the advantage of incorporating a gas-filled component (accumulator). This was important, especially to the author who had been exposed to counter-arguments, that isobaric collection would require an active and simultaneous input of energy, making the task more difficult. The accumulator is a simple and passive source.

Another interesting aspect of the project was that it took place at a stage when the tribal organisation of biological research in the UK was changing. The new Natural Environment Research Council was, eventually, to provide ship time for scientists from many disciplines and not just those employed in Oceanographic or Marine Laboratories. Other countries went through similar changes, led by the USA.

Currently, there is a variety of methods for collecting samples of water (and gas), microorganisms in suspension or in sediments and macroscopic animals, from their ambient high pressure without significant change in that pressure. Some of these methods provide for the onward transfer to experimental equipment, again with no change in pressure. Techniques of isobaric collection and transfer are essential for certain investigations, but generally, they are not used sufficiently often, probably because of the expense and difficulty in obtaining and using the equipment.

However, the principles basic to the technique of isobaric collection and transfer are simple enough and were first clearly stated by Gilchrist (1972). When an open-ended, thick-walled steel cylinder is lowered several kilometres into the sea, its wall is slightly compressed. If the ends are rigidly closed by some hypothetical means, and the cylinder is brought back to the surface with no change in temperature, the pressure of the water trapped inside will be close to 5% less than the pressure at which closure took place. This decrease arises from the elasticity of the steel and its very small incompressibility, acting in conjunction with the compressibility of the water. In the case of a steel pressure cylinder with a ratio of outside diameter : inside diameter of 1.5, the pressure loss is 8%. A hypothetical vessel with an infinitely thick wall would undergo a loss of 4.9% (Gilchrist and Macdonald 1983).

Gilchrist's analysis showed that the trapped pressure in a practical vessel will also undergo a decrease due to soft seal displacement, which he quantified, and there will also be a pressure change due to temperature variation. These were expressed mathematically by Gilchrist and compared to values obtained from laboratory tests. A large water-filled pressure vessel was used to simulate the deep sea, and within it, the recovery vessel was pressurised. The recovery vessel was a thick wall cylindrical vessel connected to an accumulator, similar to the vessel in Fig. 13.17 with one end sealed by a plug and a globe valve at the other. This type of valve, better seen in Fig. 13.19, works by the 90-degree rotation of a sphere, through which a hole has been drilled. No volume change is involved. The valve was closed at a

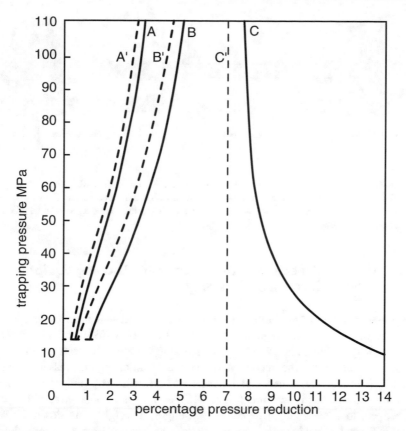

**Fig. 13.16** Experimentally measured reduction in the "trapped" pressure in a recovery vessel similar to that in Figs. 13.17 and 13.19 (Gilchrist and Macdonald 1983). Vertical axis: pressure at which the globe valve was closed. Horizontal axis: the trapped pressure, as a percent of the closing pressure. Curve A: accumulator volume 590 ml nitrogen gas charged at 13.79 MPa. Curve B: accumulator volume 160 ml nitrogen gas charged at 13.79 MPa. Curve C: No accumulator present The dotted lines show the drop in pressure if no seal displacement takes place

selected pressure and on decompressing the large vessel, the trapped pressure in the recovery vessel was measured and compared to the closing pressure. The data were consistent with the mathematically predicted values (Fig. 13.16).

The results show the value of an accumulator and also exemplify the engineering practice of "belt and braces". Thus, the passive isobaric collection of deep sea water is possible, a point which was of some concern to the author, see Box 13.1.

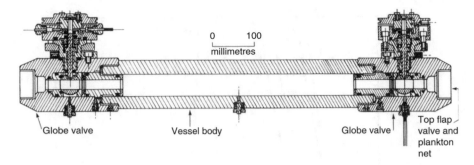

**Fig. 13.17** Isobaric recovery vessel with accumulator. (Gilchrist and Macdonald 1983). The internal diameter of the vessel is 5 cm and of the hole through the globe valves, 2.5 cm. The tests described (Fig. 13.16) used a similar vessel sealed by a plug at one end and a globe at the other. The transfer piston was not present

### 13.7.1  Isobaric Collection of Deep-Sea Animals Using Equipment Operated from a Surface Ship

The pressure retaining properties of the recovery vessel had been proved in tests (Fig. 13.16), so the next stage was to demonstrate its performance at sea. HMS Hydra, a Royal Naval survey ship, provided ship time for the tests when making passage to Gibraltar. The recovery vessel was similar to the vessel shown in Fig. 13.17, except that it had a globe valve at one end and a plug at the other. It was mounted in a handling frame (Fig. 13.18), and lowered on a wire, with a globe valve open, to a prearranged depth. A messenger, a weight, sent down the wire, actuated the valve closing mechanism. There were two pressure measuring devices connected to the vessel. The pressure (depth) recorder provided the closing pressure (Table 13.1) and the subsequent pressures. The transducer confirmed the recorder's pressure and the fluctuations during transfer which were +/−2% (not shown in the table). The slight increase in pressure seen in casts 5 and 6 was probably due to the vessel warming as the package was not thermally insulated. The first cast was a valve closing practice run and the second cast incurred a trivial problem that was promptly resolved. The subsequent four casts produced isobaric samples from depths of 1400–2000 m and on two occasions the sample was smoothly transferred to the experimental chamber (Fig. 13.1, Section 13.5).

The original idea was to attach the recovery pressure vessel to a large plankton net, so that small animals would be filtered into the vessel. It would be closed at depth and when recovered on board the ship, the high-pressure suspension of semi-microscopic and somewhat larger animals could be flushed into an observational experimental vessel with no change in pressure. Observations and experiments would follow.

Subsequent ship time on the Plymouth Marine Biological laboratory vessel Sarsia, in 1969, allowed us to collect a not very deep isobaric plankton sample (from 530 to 660 m) and transfer it to the experimental vessel at a pressure that fluctuated between 4.66 and 5.3 MPa. The vessel was thermally insulated. The hydraulic circuit used for this involved an air-powered reciprocating pressure

**Fig. 13.18** Test assembly of the Isobaric Collecting Vessel (Macdonald and Gilchrist 1969). The recovery vessel (5 cm internal diameter and 55 cm internal length) is mounted in a handling frame. At the top, there is a globe valve and at the bottom a plug is connected to a stop valve. (a) globe valve with rotatory, spring-powered closing mechanism, (b) body of vessel, (c) handling frame, (d) end plug, (e) depth recorder pressure housing, (f) accumulator, (g) pressure transducer housing. The depth gauge and pressure transducer connect to the vessel mid-way along its length

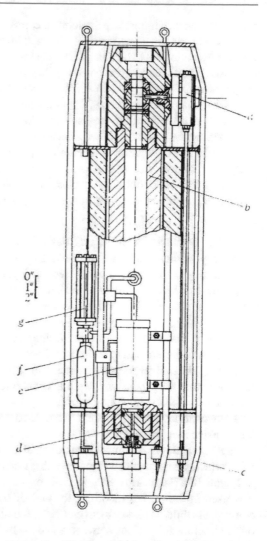

**Table 13.1** Isobaric collection of deep sea water, simplified from Macdonald and Gilchrist (1969)

| Depth of recovery | | Pressure at which valve closed; MPa | | Pressure in vessel on recovery MPa | |
|---|---|---|---|---|---|
| From wire paid out | Pressure recorder | | Depth recorder | Depth recorder | Transducer |
| 1400 m | 14 MPa | | 14.4 | 14.4 | – |
| 1400 m | 14 | | 14.4 | 14.4 | 14.4 |
| 2000 m | 20 | | 20.06 | 21.1 | – |
| 2000 m | 20 | | 20.06 | 21.1 | 21.5 |

pump and an accumulator to smooth the pressure pulses. Coupling the recovery vessel via its globe valve to the experimental vessel was made easier when the ship's motion was minimal (Macdonald and Gilchrist 1972). This was merely a technical exercise as bad weather prevented a fuller programme of work, but two important points were learned: (1) The globe valves could seal in the presence of dense particulate material (plankton). This is a valuable feature and arises from the wiping action of the rotation of the steel sphere (globe). (2) Plankton was not a good source of experimental animals. It was decided that it would be better to convert the recovery vessel to function as a benthic trap and collect larger animals. So, unknowingly, we joined the international fraternity of deep-sea biologists who were collecting that most obliging of creatures, benthic amphipods, from considerable depths, using simple traps and in two cases, described below, isobaric traps.

### 13.7.1.1 Isobaric Trapping of Benthic Animals

Feasibility studies were first carried out using nonpressure-retaining traps and the results are described in Chapter 5, Section 5.1.1.

In the isobaric benthic trap, the globe valve originally used in the mid-water trapping equipment is fitted to a new sampling vessel with a large window (Fig. 13.19a).

This amphipod isobaric trap was designed to be operated by conventional surface ships, but as we shall see, isobaric collecting vessels have been produced which require a submersible to put specimens in the collecting pressure vessel and close it. That makes for expensive operating costs but the isobaric collection of fish much more practical than attempting it from a surface ship (see Section 13.7.2).

Two methods of deployment were used with the original isobaric trap. An inexpensive mooring method was at first used to place the traps, mounted on sledges, on the ocean floor. A 6-mm-diameter steel cable was used for the main mooring wire, which was suspended from inflatable buoys at the surface. Moorings were laid to a depth of 2700 m. However, to avoid excessive cable weight, polypropylene rope was used in the deeper moorings for part of the main mooring wire. Usually, the equipment could be cast or recovered within one and a quarter hours. Traps were left on the seafloor for periods between 8 and 24 h. The traps generally remained where they were cast, although those moored at depths of 2700 m could be dragged by the currents.

The superior free-fall ("pop up") method was adopted when funds allowed (Fig. 13.20). This can be used to trap and recover animals at any ocean depth. Although its use is a routine procedure the following description of its design and deployment is interesting. The apparatus was designed by Gilchrist to sink at about 45 m per minute. Twelve buoyancy spheres, widely used in oceanographic work, were incorporated. Each unit comprised a matching pair of glass hemispheres, lightly clamped to form a sphere capable of withstanding any ocean pressure. Each sphere was mounted in a protective, tough plastic casing. In total, they provided 320 kg of buoyancy (lift). Negative buoyancy of 220 kg came from the

**a**

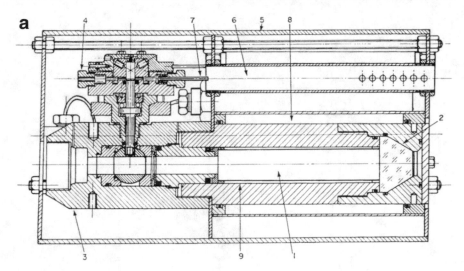

**b**

**Fig. 13.19** (**a**) The trap comprises a thick wall cylinder of 7.5 cm internal diameter, with a globe valve (5 cm diameter hole) at one end and a PMMA window at the other. This was a truncated cone of 60 degrees included angle, described previously, with a safe working pressure of 140 MPa. Additionally, there is an accumulator, a pressure transducer and a rotary valve closing mechanism powered by a spring, all mounted within a robust handling frame. Details: 1, body of the vessel, 2, window with an external plate retaining atmospheric pressure on its outer face. An "O" ring seal fitted to the cylindrical section of the window mount prevents any ingress of water as the window moves into its sealing position during the vessel's descent to the ocean floor. 3, globe valve, 4, valve closing mechanism, 5, handling frame, 6, spring housing for the closing mechanism, 7, trigger for closing mechanism, actuated by electronic timer, not shown, 8, water jacket for cooling the vessel after recovery (from Macdonald and Gilchrist 1978). (**b**) Photograph of the isobaric benthic trap (from Gilchrist 1972)

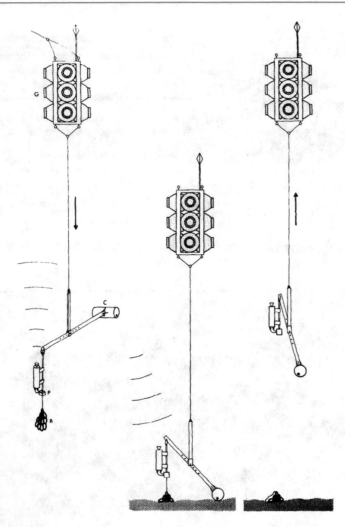

**Fig. 13.20** Gently placing the trap on the ocean floor. This is an example of the widely used "pop up" technique (Macdonald and Gilchrist 1982)

rest of the apparatus (mainly the pressure trap) plus 130 kg from disposable ballast. Thus $220 + 130 = 350$ kg. minus 320 kg of lift left 30 kg of negative buoyancy, causing the whole apparatus to sink steadily. When the device reached the ocean floor the extended arm carrying the pressure trap gently settled to put the trap on the floor with a load of 2–8 kg (Fig. 13.20). The apparatus included an acoustic "pinger" which enabled us to track its descent. Conventional navigation was used to fix its location on the seafloor. After many hours on the seafloor, the globe valve was closed by a spring-powered rotatory mechanism, triggered by an electronic timer. (Yes, it was important to send the trap down with the globe valve open; we used a

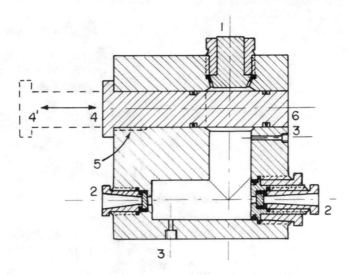

**Fig. 13.21** Isobaric trap employing a sleeve valve (Yayanos 1977). The vessel is made of a titanium alloy and the internal diameter of the sleeve valve and holding chamber is 5.08 cm. (1), plug sealing a machined hole, (2), quartz windows, (3), high-pressure connection, (4), the sleeve valve is a sliding piston which closes the entrance to the trap. The figure shows the piston in the closed position and the open position is shown by dashed lines, (5), a relieving groove, (6), trap entrance, 5 cm diameter. An accumulator, not shown, is needed to ensure full retention of the trapped pressure

detailed checklist!) The whole apparatus was recovered by first sending an acoustic command from the ship to fire a pyrotechnic (explosive) device, releasing 130 kg of ballast. Thus $350 - 130 = 220$ kg which was offset by the 320 kg lift from the glass spheres. The resulting 100 kg of positive buoyancy "floated" the apparatus up through the water column at about 75 m per minute. So, the ascent from a modest 2000 m depth took half an hour and was also followed acoustically. When the apparatus reached the surface a visual search followed, sometimes a frantic one. However, we never lost our gear.

The results are summarised in Chapter 5, Section 5.2.1.

Yayanos (1977) produced a different design for an isobaric benthic trap which was also deployed by a free-fall method. It collected amphipods from the greatest depths in the Pacific (see Chapter 5, Section 5.2.2; and Fig. 13.21). Only limited observations were made on the amphipods collected; their main contribution to science has been the bacteria they provided (Chapter 8).

The abundance of the deep sea, including hadal, amphipods, is exemplified by the collections made by nets, operated by a robot system in Puerto Rico, at more than 8000 m depths (Lacey et al. 2013). Conversion to isobaric collecting would be particularly rewarding.

**Fig. 13.22** Isobaric fish trap
(Phleger et al. 1979)

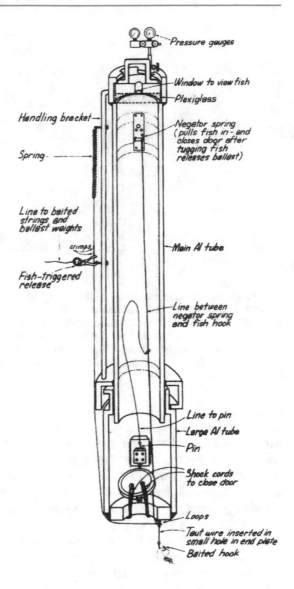

## 13.7.2 Isobaric Collection of Fish from a Surface Ship

The isobaric collection of large fish is a difficult proposition. Attempts to make isobaric fish-collecting equipment work from a surface ship are interesting and well worth noting. Phleger et al. (1979) produced a trap with a safe working pressure of 15 MPa which could be deployed mid-water or on the seafloor (Fig. 13.22). The authors report collecting *Coryphaenoides acrolepis* from a depth of 1200 m and holding it at 10 MPa. The simplicity of this device makes it inexpensive to operate

**a**    **b**

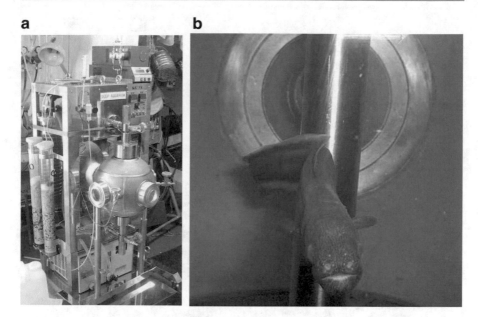

**Fig. 13.23** (a) The "Pressure-stat" vessel mounted in a laboratory (Koyama et al. 2002). (b) The deep-sea fish *Simenchelys parasiticus* in the pressure-stat aquarium at 10 MPa and 5 °C. It was collected at a depth of 1162 m and has a vertical range of 366–2630 m. Pectoral fin cells were cultured from this particular specimen (from Koyama 2007)

and it is surprising that no experimental work seems to have been achieved with it. Some years later, Wilson and Smith constructed a pressure-retaining fish trap which was operated as a free-fall vehicle, returning to the surface rapidly where it was connected to high-pressure seawater circulation system, and functioned as an aquarium (Fig. 13.22).

Brown (1975) and Drazen et al. (2005) have also produced isobaric fish traps. The former is limited to 0.6 MPa but the latter, which can be viewed online, has collected *Coryphaenoides acrolepis* from 1450 m depths with a pressure drop of only 5%. The animals were used in respiration experiments and were probably in a sufficiently good condition for neurobiological and/or swim bladder experiments.

## 13.7.3   The Isobaric Collection of Animals Using a Submersible

In 2002, Koyama et al. described a pressure vessel called a "Pressure stat-aquarium" which can be operated by the crew of a submersible. Using a suction device, specimens can be drawn into the vessel and the doors closed to trap the animal at the ambient pressure. Figure 13.23 shows the vessel transferred to a laboratory and Fig. 13.24 the life support system and mode of operation.

The vessel is a stainless steel sphere 36 cm in diameter, with a volume of 20 l. The windows are of two different sizes in the shape of truncated cones, with inner and

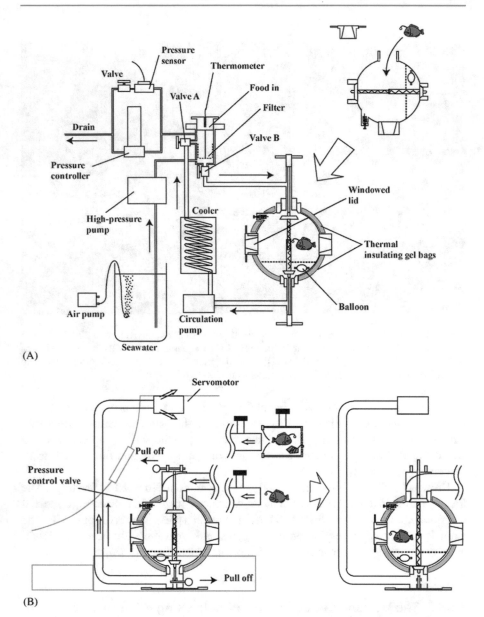

(A)

(B)

**Fig 13.24** The operation of the "Pressure-stat" system at depth (Koyama et al. 2002)

outer diameters of 11 and 8 cm and 9 and 6 cm, respectively. They are made of acrylic resin and bonded to their steel support. Various connections are made to the vessel for life support water circulation and of course, its temperature is controlled.

At depth, the submersible operator uses a suction motor to draw a selected animal, a shrimp or small fish, into the sphere through a 6-cm entrance port. The operator closes the two "doors", which are internally mounted, by triggering a spring-loaded mechanism. This traps the animal at the ambient pressure and the submersible ascends. The pressure within the sphere is retained at 20 MPa if closure took place at a higher pressure. If closure takes place at a lesser pressure, then that is retained. In either case, a gas-filled balloon (shown in the figure) acts as an accumulator. The design ensures there is no need to transfer the collected animal to a long-term holding vessel.

One specimen of *Zoarcidae sp* was collected at 1171 m depth by the submersible Shinkai 2000 and maintained at 10 MPa and 5 °C for 3 days. Other fish, *Ebinania brephocephala*, trawled from 500 m depths was transferred to the aquarium with decompression. It was kept at 5 MPa for 64 days during which time its oxygen consumption was measured, by means of a polarographic sensor. The equipment was also used by Koyama et al. (2005) to isobarically collect shrimps (*Alvinocaris*) from a depth of 1157 m at a cold seep (see Chapter 9, Section 9.7.3).

An isobaric recovery apparatus which also requires a submersible, or a ROV (Remotely Operated Vehicle), was described by Shillito et al. (2008). Here the design concept is quite different to that of the Jamstec "Pressure stat" system. The submersible, manned or ROV, collects the specimen in a sampling cell (Fig. 13.25)

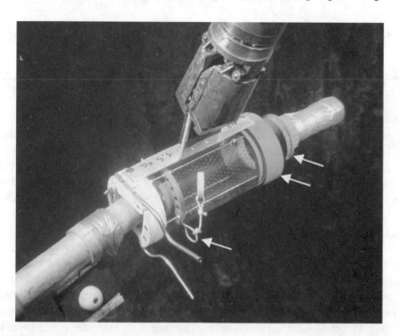

**Fig. 13.25** The sampling cell used in PERISCOP (Shillito et al. 2008). The submersible's manipulator arm (top, grey) holds the sampling container which is connected to a suction device (lower left, white). It comprises concentric cylinders, the inner one of which is 10 cm in diameter with holes in the wall. Suction draws water and a specimen into the cylinder, from right to left in the picture. The suction device is then disconnected and the inner cylinder, the sampling cell, with its captive specimen, is transferred to the PRD (Fig. 13.26)

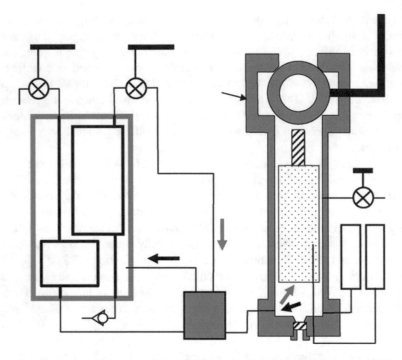

**Fig. 13.26** The Pressure Retaining Device, PRD, used in PERISCOP (Shillito et al. 2008)

and places it in a pressure-retaining vessel (a pressurised recovery device, PRD) which is then sent, pop up style, to the mother ship. Thus, a single submersible dive can operate an isobaric shuttle service for scientists on the mother ship. The Projet d'Enciente de Remontee Isobare Servant la Capture d'Organismes Profonds, mercifully abbreviated to PERISCOP, thus comprises two components: a sample cell (Fig. 13.25) which is then transferred to the PRD, for transport to the surface mother ship, shown in Fig. 13.26.

The PRD has a volume of 6.6 L and is rated for use at 30 MPa. There is a 26-mm diameter window at the bottom and at the top there is a globe valve, through which a 10 cm diameter sampling cell may pass, containing the specimen. The globe valve is closed and the free ascent to the mother ship begins. It subjects the PRD to changes in ambient temperature (minimised by buoyant thermal insulation) and a decrease in pressure. The latter is "felt" and is compensated for (+/−1 MPa) by a complicated system shown on the left of the figure whose details we can skip. Once onboard the mother ship, the specimen can be visually inspected. Temperature and pressure probes are shown on the right. Connection to a second pressure vessel uses collars that clamp and mate with the globe valve. The idea is that transferring the specimen in the sampling cell allows the PRD to be returned to the submersible, which is still at depth.

A shipboard pressure-retaining experimental chamber called BALIST can receive the isobarically transferred specimen. As originally designed, the maximum depth rating for PERISCOP was 3000 m (30 MPa) but this has been increased.

Three hydrothermal vent shrimp species were sampled on the Mid-Atlantic Ridge, from depths of 1700 and 2300 m. In addition, a fish caught at 2300 m depth reached the surface in very good condition. A high-pressure experimental vessel and incubator called IPPOCAMP is used with the PERISCOP system (Shillito et al. 2001, 2014), It is an orthodox experimental pressure vessel, similar to many, and has an internal diameter of 20 cm and a length of 60 cm, a working pressure of 30 MPa and is fitted with an endoscope and light guides and a through-flow system. It has been adapted to record from the heart of a shallow-water crab at pressures up to 15 MPa by Robinson et al. (2009).

The use of the PERISCOP shuttle service to the surface is illustrated by the work of Ravaux et al. (2013) in collecting the Pompeii worm from 2500 m depths on the East Pacific Rise, in order to study its heat tolerance (Fig. 13.27; Chapter 9, Section 9.3.1).

Ravaux et al. (2019) also used the ROV Victor 6000 and the PERISCOP system to collect the hydrothermal vent shrimp *Rimicaris exoculata* from the chimney walls of the hydrothermal vent Tag, depth 3630 m. Pressure was maintained to within 70% and 84% of the trapping pressure. The thermal insulation fitted to the pressure recovery system was sufficient to prevent the temperature from rising above 15 °C. Some animals underwent isobaric transfer to a receiving vessel at 30 MPa and 10 °C. Technically, this was a very satisfactory isobaric and isothermal operation (Chapter 9, Section 9.3).

Non-isobaric experimental pressure vessels: IPOCAMP and PICCEL.

These acronyms relate to descriptions, in French, of experimental pressure vessels which have been successfully used with hydrothermal vent and other animals. The IPOCAMP vessel is described by Shillito et al. (2014). The PICCEL system, described by Pradillon et al. (2004), is designed for use with small animals, larvae and egg cells. It has a working pressure of 30 MPa at temperatures of up to 100 °C, and a pair of 100 ml chambers each fitted with a cylindrical sapphire window 3 mm thick, suitable for use with a long-range objective. See Fig. 13.9.

## 13.8 The Isobaric Collection of Microorganisms

The need for the isobaric collection of deep-sea bacteria was intuitively obvious to many workers in the mid-twentieth century, and probably before, but is intuition enough to justify the considerable expense and effort involved in developing and using such equipment?

The isobaric collection and onward transfer of microorganisms gives experimental access with minimal disturbance to the cells' normal physiological state. In specific cases, it has been shown to avoid both structural changes and impairing the cells' viability (Chastain and Yayanos 1991). It can, in other cases, avoid significant inactivation which would distort community composition and it even

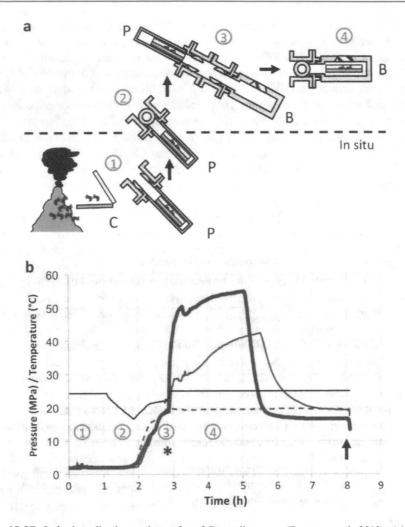

**Fig. 13.27** Isobaric collection and transfer of Pompeii worms (Ravaux et al. 2013). (**a**) The cartoon (upper) summarises the progress of the PRD. from a vent; (1) sampling; (2) ascent; (3) onboard ship the sample is transferred to an experimental pressure vessel called BALIST; (4) experimental treatments. (**b**) The PRD is equipped with pressure and temperature recording devices, whose output is shown in the graph. Pressure is shown by a blackline and the numbers relate to the stages in collecting and transfer. The red lines relate to experimental temperature changes starting at the asterisk and ending at the arrow (Chapter 9, Section 9.3)

avoids affecting gene expression. On the other hand, it has been argued that piezophiles survive decompression and can be subsequently grown at their appropriate pressure for investigation.

An interesting example arose in the use of PERISCOP in collecting the mussel *Bathymodiolus* (above). Whereas the symbiotic bacteria living in its gills seem to be recovered adequately by either isobaric or non-isobaric methods, the animal itself is in better condition for physiological and host–symbiont experiments when collected by isobaric means (Szafrinski et al. 2015). Even the hadal MT-41 retains some viability after a decompression for an hour or two (Yayanos and Dietz 1983), so why bother with isobaric collection? As with deep-sea animals, the recovery of microorganisms with decompression still enables some good experiments to be carried out, of the right sort, but isobaric techniques are an essential part of the experimenter's repertoire. This applies to both the Deep Sea and the Deep Biosphere.

We are primarily concerned with Prokaryote microorganisms which are distributed, mid water, in suspension and usually attached to particles, in sediments, variously attached to particulate substrates, on the surface of, and within, animals and as biofilms, on solid wet substrates. Furthermore, it is sometimes necessary, after collecting a sample, to isolate individual species or strains without changing pressure, following procedures that microbiologists have established over the years. Collecting is the relatively easy part. Any toxic effects of the materials used in the equipment is avoided, usually by coating metal surfaces with inert Teflon or polyether ether ketone (PEEK) (Remember that stainless steel inhibits the growth of Archaea but glass does not, Chapter 8, Section 8.3.) Maintaining sterile procedures and preventing contamination is manageable, and of course, essential.

A pressure-retaining microbiological sampler was proposed by Gibbon and is on record as a patent application in 1968. It was not unlike an oil well logging sampler. In 1972, Gundersen and Mountain, working at the University of Hawaii, published a description of isobaric equipment for collecting pelagic microorganisms and transferring them to a second vessel. The description was detailed and clearly, some practical tests had been carried out, but like the Gibbon device, there seems to be no record of any work carried out with it. After these false starts several laboratories have produced microbiological isobaric collecting and isolating equipment.

## 13.8.1 Meso and Bathy Pelagic Microorganisms

The first successful isobaric microbiological vessel was a thin-walled cylinder, rated for use up to 20 MPa. It collected a sample of seawater, avoiding shear forces that can damage cells, by permitting only a slow entry of water. The equipment was developed in stages by Jannasch and colleagues working at the Woods Hole Oceanographic Institution, USA (Jannasch et al. 1973, 1976; Jannasch and Wirsen 1977), culminating in an isobaric isolation apparatus to which the isobaric sampler connects (Jannasch et al. 1982).

The following account provides an outline of how the early version of the equipment was used. It was designed to be operated by a submersible or from a surface ship. In the latter case, mechanical triggers at the end of the wire used to lower it, closed the vessel at a prearranged depth (Fig. 13.28, page 395).

The sampler can also be used as an incubator and it can receive or deliver experimental samples with minimal pressure change. Stirring is provided by a magnetic stirring bar (g) set in a recess in the top endplate. The transfer device (h) provides these options when autoclaved, filled with sterile water and connected to the top endplate. Valves (i) and (j) in the top endplate and the screw-driven piston (k) in the transfer device are used for this. To assist the accumulator in maintaining constant pressure, sterile distilled water can simultaneously be injected or withdrawn using d and f.

Experience has shown that a brief decompression followed by recompression is tolerated by many species of Prokaryote but it is clearly important to avoid the risk of failing to isolate a sensitive species by subjecting them to potentially lethal decompression. Accordingly, the isobaric method was developed to carry out the isolating procedures, known in conventional microbiological practice as "plating out".

The ordinary laboratory procedure consists of transferring a drop of liquid culture (or seawater sample), the inoculum, using a fine platinum wire with a loop at its end, traditionally sterilised before use by heating it in the flame of a Bunsen burner. The inoculum is deposited on to the surface of nutrient gel (agar) covering the base of a transparent plastic dish (called a plate), fitted with a detached transparent lid. Again traditionally, the liquid is streaked over the gel surface. The dish is then incubated and after a time creamy white spots appear. They are colonies of bacteria. If a very dilute suspension of bacteria is plated out in this fashion, then a colony can be assumed to have arisen from an individual cell, a procedure which can provide a statistically rigorous measure of the number of cells per unit volume of the original source.

The Woods Hole group developed equipment called an isolation chamber, a special vessel containing an apparatus for isolating deep-sea bacteria in the absence of decompression. The isobaric sampler (Fig. 13.28) connects to the isolation chamber, enabling a sample, concentrated by filtering 3 litres down to 13 ml, to be transferred to it at constant pressure. Unfortunately, a formal diagram of this crucial piece of equipment does not appear to be available.

The normal atmosphere of the laboratory is replaced by an oxygen–helium gas mixture. The pO2 is controlled within safe limits. (The partial pressure of oxygen can be selected to suit individual samples: the inert nature of helium used in this way is considered in Chapter 11, Section 11.4.3.1.) A platinum wire loop is sterilised by electrical heating and transfers a drop of the sample to the plates which are organised on a conveyer belt. The manipulation of the loop and the conveyer belt is by a direct mechanical coupling and electrical power provides lighting and heat. The whole assembly fits snugly in a 10-cm-diameter high-pressure vessel and is viewed through a window. After colonies have grown on the plates the constituent bacteria, ideally a single species, are transferred by the wire loop to a volume of liquid culture housed in a small vessel, then isobarically transferred to a one-litre culture vessel. The system is designed for use up to 100 MPa at a controlled (low) temperature and at the time of publication had been used at 60 MPa.

A way of using agar gel to support the separate growth of bacterial colonies under high hydrostatic pressure has been devised by Masui and Kato of the JAMSTEC Deep Star group (1999). Essentially a type of agar inoculated with bacteria is enclosed in a sterile film pouch which is heat-sealed free of trapped air. In this

**Fig. 13.28** Isobaric microbiological sampler (Jannasch et al. 1973). Two stainless steel cylinders are clamped by six longitudinal tie bars (not shown) to end plates and a central dividing plate. The two free pistons, a and b, create three compartments, A, C and B, the latter being divided by the dividing plate and is filled with sterile distilled water. The sampling chamber, A, holds one litre and is lined with Teflon. Its endplate, top, contains the sample inlet valve (e) on the right. Chamber C is filled with air at a pressure selected to match the pressure anticipated at the sampling depth. At sampling depth the valve (e) is opened, allowing water to slowly enter at a speed set by the valved orifice (c) in the dividing plate. The normal filling time is 15 min, after which valves (e) and (c) are closed. The figure shows the filled sampler with the air in compartment C acting as an accumulator which is essential for isobaric collection

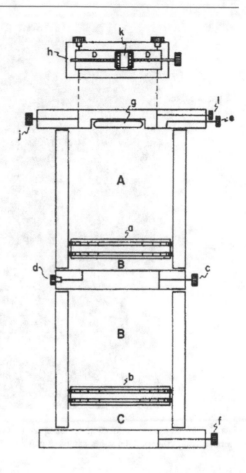

state, the pouch can be pressurised in a water-filled pressure vessel and separate colonies identified on decompression. However, this convenient technique does not allow subsequent isobaric plating out. The Deep Star group has also developed an isolating procedure appropriately called underwater isolation, of which more soon.

Other, non-isobaric methods of plating out bacteria have used various gels and growth substrates and media to good effect (Dietz and Yayanos 1978; Zhang et al. 2018).

Isobaric microbiological sampling from deep water was also developed by Tabor et al. (1981) at the University of Maryland, USA at about the same time as the Woods Hole group. The apparatus is complicated, providing for isobaric transfers, but for reasons of space is not shown here. In 1988, Deming et al. reported results of samples collected at in situ pressure in the Puerto Rico Trench.

Inlet valve

Carousel trigger                                                     Outlet valve

                                                                     500 ml sampler

Exhaust tank

                                                                     Pressure sensor

                                                                     Polypropylene board

                                                                     Pressure accumulator

                                                                     CTD

**Fig. 13.29** Carousel supporting six sampler units (Bianchi et al. 1999). A cluster of isobaric samplers and conventional water samplers can be operated during one deepwater cast

The isobaric collecting equipment described by Bianchi and Garcin in 1991 then more fully, Bianchi et al. (1999), comprises eight 500 ml collecting vessels mounted in a carousel (Fig. 13.29).

The isobaric sampler has numerous features that are similar to those of the Jannasch sampler, so the following description deals with the developments and differences. A single floating piston moves down the vessel as it takes in a sample through the inlet valve (Fig. 13.30), simultaneously ejecting sterile distilled water from the lower chamber. All surfaces in contact with the sample are coated with PEEK (Fig. 13.31).

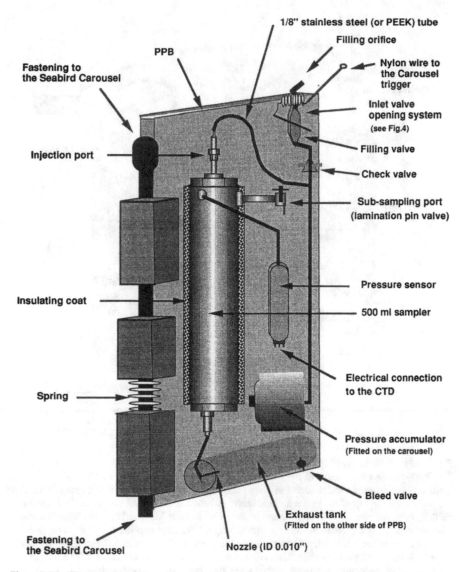

**Fig. 13.30** The sampler and associated components mounted on a board which attaches to the carousel (Bianchi et al. 1999). The sampler has, at its top, the inlet (injection) connection and at its bottom, the ejection valve connected to the exhaust chamber. A separate accumulator is shown connected to the inlet tube. The seawater passes through a throttling valve into a separate exhaust chamber. This means that the piston can move downwards until it is stopped by reaching the bottom of the sample chamber cylinder. Thus, the pressure of any gas in the exhaust chamber never reaches the sampling pressure, so that the system cannot act as a hydraulic accumulator to maintain pressure during recovery. However, data show that constant pressure is maintained during recovery by the connection to a separate gas accumulator shared with other samplers mounted alongside it

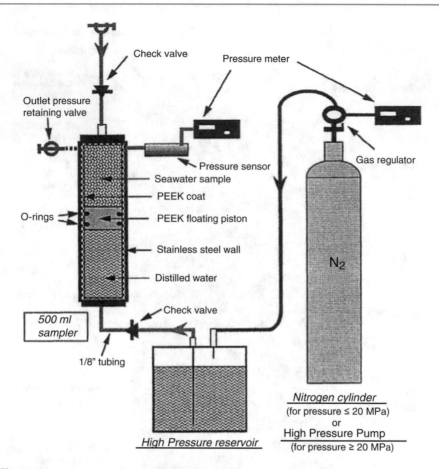

**Fig. 13.31** Isobaric sampling and transfer (Bianchi et al. 1999). The collecting vessel is on the left. The high-pressure reservoir and the nitrogen gas cylinder remain in the ship's laboratory. The sterilising and assembly which precedes using the sampler are ignored here. When ready for casting the piston is at the top of the vessel and the chamber below is filled with sterile distilled water and chilled. Each sampler on the carousel can be triggered to open independently at different depths. The ambient pressure forces the sample into the upper chamber and the distilled water into the exhaust chamber, at a rate determined by the exhaust nozzle. The pressure is recorded. When the sampler is back on board the ship, it, and its connections, are disconnected from the carousel and reconnected to onboard equipment, notably the nitrogen cylinder or high-pressure gas pump. Sub-samples are isobarically (+/−10%) withdrawn through the port at the top of the sample chamber. The 1999 paper includes impressive data showing the variations in pressure which the sampler experiences in use. Garel et al. (2019) designed a development of the Bianchi model, explicitly for widespread use. An orthodox floating piston is used and the inflow rate during sampling is controlled by constricting the flow of sterile water from the section of the chamber below the free-floating piston. The pressure vessels were made either from stainless steel coated with PEEK plastic or from titanium to avoid corrosion. The equipment is rated for pressures of 60 MPa and samples have been recovered with less than 2% variation from the pressure at depth. The published description can be seen on the web

Also, in 2019, Peoples and a multidisciplinary team published an account of a full ocean depth rated modular lander which could accommodate an isobaric water (bacteria) sampler.

## 13.8.2 DEEPBATH

In 1990, JAMSTEC, the Japan Marine Science and Technology Center, set up the "Deep Star" programme which uses manned and unmanned submersibles to search for and study deep-sea microorganisms living at extreme depths, particularly in the Japan Trench. The DEEPBATH programme emerged from this. It was described by Kyo et al. in 1991 and later by others, including Kato et al. (2008), and featured in Chapter 8, Section 8.2.3.

The system consists of four parts, each performing a distinct function: (i) a sediment sampler and pressure retaining vessel, (ii) a dilution device, (iii) an isolation apparatus and (iv) a culture facility. It is designed to operate at up 68 MPa and 300 °C. The collecting operation begins with the manipulator arm of an ROV or submersible putting 5 ml of sediment from the deep seafloor into the sterilised tubular pressure-retaining vessel. The ball or globe valves are then closed to maintain pressure, assisted by an accumulator and the vessel is separately sealed in a second container to retain the temperature during the time taken to surface. On board the mother ship the pressure retaining sample vessel is connected to pressure circuitry and refrigerated for onward transport to a land laboratory, where it is connected to the dilution device. This is a complicated isothermal and isobaric setup, which injects sterile seawater to dilute samples. Sediment samples can be treated much like water samples. However, the isolation apparatus is the centre piece of the operation, and here we see the industrial scale thinking behind the DEEPBATH concept, which is primarily about exploring the bacteriological potential of the depths for useful organisms. The DEEPBATH isolation technique is unlike that of Jannasch et al. (1982) which uses a high-pressure helium atmosphere in which operations are carried out. Instead, the DEEP Bath system uses a hydraulic, i.e. a fluid-based method and is called underwater isolation. Multiple prediluted samples are simultaneously processed. It does this with a large, water-filled isobaric pressure vessel, the base of which has multiple pressure connecting points. To each of these (there are 36 in all), sterile tubes are connected, like inverted test tubes, projecting into the water-filled pressurised vessel. Each tube is filled with 20 ml of nutrient solution and is now called a cell. Successive serially diluted samples, for example diluted by a factor of 10 6 or 10 7, etc., are injected into successive cells. Each cell (like an inverted test tube) contains some silicone oil above which the sample is injected. Thus, the oil is displaced down and seals the cell. The growth of a population of bacteria in each cell is monitored by the scattering of a laser light beam, a novel method developed for DEEPBATH. The population which grows from the most diluted sample is judged to be a pure culture, comprising one species, probably of unknown identity. (Samples subjected to greater dilution will be "diluted to extinction" and will fail to produce cells.) The chosen sample is then transferred to be cultured in a one-litre vessel. This was the procedure used by Yanagibayashi et al.

(1999) (Chapter 8). Unfortunately, the DEEP BATH equipment is not well illustrated in accessible sources but using the Web can be informative.

### 13.8.3  Microorganisms from Deep-Sea Animals

We should not overlook the use of deep-sea animals in providing deep-sea bacteria, but not necessarily in an isobaric way. A famous example is the isolation of the obligate piezophile MT-41 from the hadal amphipod Hirondellea, collected with decompression from 10,473 m depths in the Mariana trench by Yayanos et al. (1981). Another is *Colwellia marinimaniae*, which was extracted from an Amphipod collected from a depth of 10,918 m in the Challenger Deep, Mariana Trench, during the Deepsea Challenge expedition (Kusube et al. 2017). It was collected with decompression but restored to 110 MPa for transport to the Scripps Institution of Oceanography, California, USA. It grows over the pressure rang 80–140 MPa (at 6 °C), with its optimum at 120 MPa, the highest pressure at which any organism can grow. Other animals have been used; the deep-sea cucumber and deep-sea fish from more than 4000 m depths, (Nakayama et al. 1994) and symbiotic mycoplasmas have been recovered from an Isopod, Bathynomous, whose depth range extends to more than 2000 m (Wang et al. 2016).

### 13.8.4  Fluid and Dissolved Gas Samplers

An interesting example is that devised by Seewald et al. (2001) which collects fluid samples containing a lot of gas from hydrothermal vents. It operates at pressures of up to 45 MPa and temperatures up to 400 °C. It has been developed to serve as an experimental vessel when recovered on board a ship. A paper by McNichol et al. (2016) explains how it is used (Fig. 13.32).

The Crab Spa vent on the East Pacific Rise was sampled by the equipment shown in the figure, operated by the ROV Jason II and its mother ship R V Atlantis. The vent water was at a temperature of 24 °C and a depth of 2506 m. The object was to obtain water samples containing microorganisms, dissolved gases and salts, and to incubate the microorganisms in defined conditions to study their metabolism, all without significant change in temperature or pressure.

The ROV placed the sampler in the main orifice of the vent and opened the sampling inlet valve. We can skip the detailed preparations required prior to this, except the charging of the left-hand section (B) with nitrogen gas (to 10% of the anticipated sampling pressure) and the filling of the middle section with filtered bottom water. The figure shows the sample chamber (A), 150 ml volume, has drawn in vent fluid and compressed the nitrogen gas accumulator. The sampling valve permits a slow entry of vent fluid, and 2 min is required to fill the sampler. The valve is then closed, the equipment is returned to the mother ship's laboratory where the sampler serves as an experimental incubator. Various options are available. For example, 10 ml samples can be withdrawn from the system through the inlet valve, without pressure loss, by the compensatory injection of water by the high-pressure

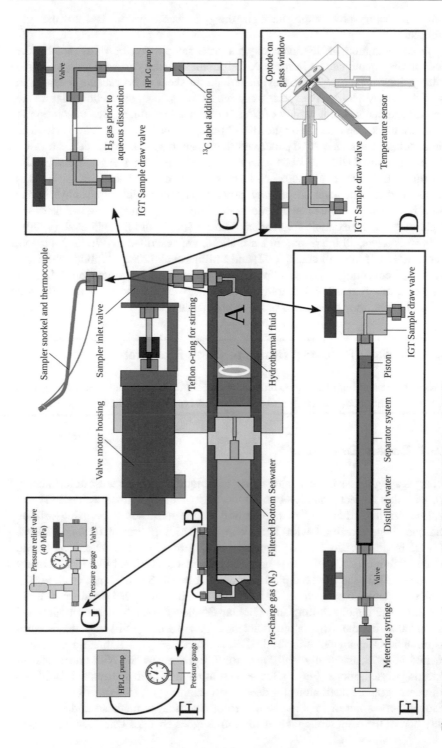

**Fig. 13.32** Isobaric water/gas sampler and incubator (McNichol et al. 2016)

liquid chromatography pump through the pre-charge valve B into the gas accumulator.

The concentration of dissolved oxygen in a sample can be measured by the optode in the circuit, shown in inset D. Cell counts can be carried out on a sample similarly drawn into a separation cylinder without shear stress causing damage. In this field of study additions of specific chemicals to a growth or reaction mixture are called amendments, and a variety of such additions, including gases in solution at high pressure, can be injected by the HPLC pump connected to the inlet valve. The main conclusion from the 2016 paper was that many reactions proceeded at a faster rate when measured with this isobaric technique than had been measured previously, at normal atmospheric pressure. A further paper by McNichol et al. (2018) demonstrated a high level of primary production (chemolithoautotrophy) in the subsea biosphere, measured at in situ conditions. The same is true of deep-sea pelagic bacteria, for example Bianchi and Garcin (1993) and Bianchi et al. (1999).

Other samplers, all interesting to learn about, are described by Wu et al. (2018), Jin et al. (2014), Malahoff et al. (2002) and Phillips et al. (2003). Mobile automated methods of collecting organisms from deep water have been developed by, for example, Evans et al. 2019; Humphris and Soule 2019; and Brandt et al. 2016, but as far as I know, they have yet to produce biological results.

## 13.9    Isobaric Recovery from the Deep Biosphere

Microorganisms living in the deep biosphere are obtained from fluids emerging from seeps, just mentioned, and from and sediment and rock cores obtained by drilling.

### 13.9.1 Isobaric Drill Cores

Drilling is a highly developed engineering technique, which we must unfortunately skip, with due respect, and simply note the various organisations which carry out scientific drilling. This kind of exploration of the deep sea floor using drilling techniques began in 1968 with The Deep Sea Drilling Project (DSDP). It was followed by the Ocean Drilling Program (ODP) in 1985, the Integrated Ocean Drilling Program (IODP) in 2004 and the present International Ocean Discovery Program in 2013 (see for example Huey and Storms 1985; D'Hondt et al. 2007). National organisations such as JAMSTEC are also involved in deep drilling operations. The famous drilling ship Glomar Challenger was launched in 1968 and was operated by the US National Science Foundation (NSF) and the Scripps Institution of Oceanography in the DSDP. Its successor, the JOIDES Resolution, launched in 1978, is equally well known and operated by the NSF and a host of supporting organisations (Fig. 13.36b shown at the end of this chapter) (JOIDES – Joint oceanographic institutions for deep earth sampling or, in short, JR).

Isobaric cores are of importance to geologists because methane hydrate, commonly part of the deep sediment structure, undergoes changes when decompressed.

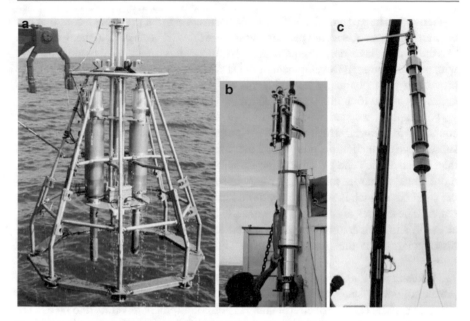

**Fig. 13.33** The Multiple Autoclave Corer of Abegg et al. (2008). (**a**) shows the equipment about to be deployed, fitted with two laboratory transfer pressure vessels (chambers). An individual chamber is shown in (**b**) and an individual piston corer in (**c**)

Free gas fizzes out of solution and methane escapes from the hydrate more slowly with the result that the fabric of the core is changed. Isobaric sediment coring systems have been developed to preserve the integrity of the core.

The Multiple Autoclave Corer (Fig. 13.33) and the Dynamic Autoclave Piston Corer described by Abegg et al. (2008) permit the pressurised gas hydrate-bearing core to be examined by X-ray methods, within the pressure retaining vessel (autoclave). They are designed for use up to 14 MPa but not for isobaric transfer. Engineers from Chanchun (Li et al. 2016) have published a technique for improving the efficiency of the pressure-retaining performance of isobaric corers by freezing the core mouth and valves.

The lighter system of Case et al. (2017) can be operated by ROVs, which have a limited payload. It is able to produce a 10 cm core and transport it to the surface. The published report notes that unfortunately only 20% of the trapped pressure was retained because grit damaged the soft seal, a technically trivial problem with important consequences. Unlike the previously mentioned samplers which can only cope with chemical studies, the case sampler can receive isobaric injections of substrates and is thus capable of microbiological experiments. The corer described by Jackson et al. (2017) is designed for isobaric retrieval from 3500 m depths, and onward transfer, but has yet to be taken beyond its initial sea trials stage. For isobaric experiments with deep sedimentary microorganisms, it is a very promising development.

Geologists have devised a method of isobaric coring that produces a core that can be transferred to a second pressure vessel—HYACE, translating the French as "Hydrate autoclave coring equipment" (Autoclave is often used synonymously with pressure vessel). Development of the HYACE system led to the comprehensive procedure for isobarically accessing microorganisms from deep cores, which is described below (Schultheiss et al. 2006; Parkes et al. 2009).

To simplify, we start with the arrival of a pressurised core at the surface. It is contained in a barrel, which was driven, hammered, into the sediment/substrate and subsequently pulled into a thin-walled pressure vessel, called an autoclave, which is then sealed at each end by valves, thus trapping the pressure. An alternative method, more suited to harder rock, uses a rotary cutting device, a shoe, which cuts around the solid core which is subsequently pulled into a barrel, then into the autoclave and sealed. These types of core are approximately 50 mm in diameter and have to be kept at the in situ temperature, and not allowed to warm, to prevent the methane from dissociating from the hydrate. Pressure relief valves are consequently integral fittings.

Parkes et al. (2009), and other workers, developed an impressive and complete procedure which starts with their HYACINTH drilling system producing isobaric cores, HYACINTH is derived from the HYACE equipment developed to be used in New Tests on Hydrate. Cores a metre long and 51–57 mm in diameter are produced by the drilling technique. They are kept at 2–6 °C for transport to a land laboratory where the core processing usually takes place. The core is handled by PRESS, an isobaric apparatus which (a) couples to the core and then (b) provides for a central inner 20 cm diameter of the primary core to be pushed through globe valves into a second chamber, the axial sub-core extrusion chamber. Tests have shown that an inner 20-mm diameter of the primary core is free of contaminants.

The receiving vessel contains a solution of salts and nutrients which changes the sediment into a slurry which is then passed to the isolation chamber. This is the equipment that enables bacterial suspensions to be diluted and "plated out". The one used is essentially the design of Jannasch et al., previously described, and from it pure cultures (i.e. one species) are obtained, at the pressure from whence the sediment came. The maintenance of pressure in the complete system, given the name DEEPisoBUG, is ensured by using pure nitrogen gas, and the initial core, held at the trapped pressure, is supported by water pressure. The details are not clear and records of pressure during the various stages are not provided, but no matter. The system is remarkable, providing biologists with mixed populations and single species from the most inaccessible environment. (However, the use of nitrogen gas pressure raises obvious questions about narcotic and toxic effects, which I assume have been duly considered.) A diagrammatic summary is freely available via the Web from Parkes et al. (2009).

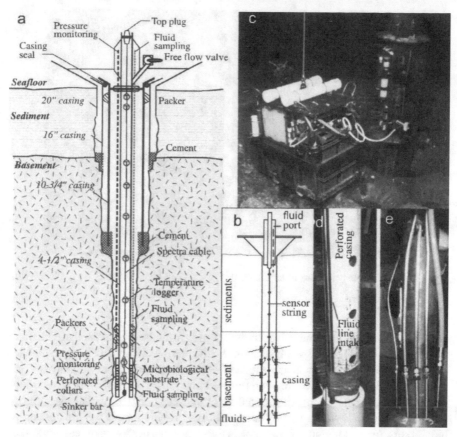

**Fig. 13.34** CORK II (Cowen et al. 2012). The first design is shown in (b) and the revised, functional design in (a) which shows continuous fluid delivery lines, as dashed lines. (c) shows the GeoMICROBE instrument sled. (d) and (e) show details of the delivery lines. The size of the CORK is apparent from Fig. 13.35, which shows the upper two metres

## 13.9.2  CORKS and FLOCS: Drill Hole Sampling

A CORK (Circulation Obviation Retrofit Kit) is a device for sampling fluid from a drill hole in the ocean crust, creating a sampling point or observatory for chemical and microbiological markers. An early example, described by Cowen et al. (2003), was used on the Juan de Fuca Ridge in the Pacific, on borehole 1026B (ODP leg 186) from which flowed water from basalt 500 m below the seafloor. The flow rate was fast but contamination, both chemical and microbiological, had to be eliminated, hence the development of the CORK II, as shown in Fig. 13.34.

The system is controlled by a Geo MICROBE instrument sled (upper right) Cork II draws geothermal fluid some 400 m through individual narrow bore pipes, lined with PTFE. The sled provides battery power and its computer is programmed to

**Fig. 13.35** Scientists pose prior to launching their CORK on board JOIDES Resolution (Mayer 2012)

switch on and off pumps and valves to draw fluids from the crustal depths. The fluids pass through sensors and analysers on board the sled. An ROV or a manned submersible is required to install the sled which has operated continuously for a year. CORKS have become more complicated and joined by FLOCS (Flow-through Osmo Colonization system) (illustrated in Orcutt et al. 2010), which use osmotic pumps to draw fluid into a narrow tube and thus provide a fluid sample containing biomarkers, over a long time. Important as these systems are for chemical work, they do not provide cellular samples for experiments.

Figure 13.35 shows the working deck of a drillship and the top 2 m of the CORK about to be launched by the assembled scientists. Figures 13.36 and 13.37, illustrating some of the workhorses of deep-sea research, conclude this account, but we should not forget their counterparts in the basic sciences of chemistry and biology.

**Fig. 13.36**  Three veterans of deep-ocean research. (**a**) The Lander ROBIO, Robust Biodiversity lander, equipped with camera and strobe system for use to depths of 6000 m. It floats above the ocean floor in a tension mooring, tethered by a line and ballast (visible), buoyed by floats, (not shown) and operated by a team from Oceanlab, University of Aberdeen, UK. Photograph courtesy of I.G. Priede, University of Aberdeen, UK (see Chapter 5, Fig 5.3). (**b**) The manned submersible ALVIN, crewed by scientists from the Woods Hole Oceanographic Institution, working on the Juan de Furca Ridge (Ben-Ari 2002). (**c**) The drilling ship JOIDES Resolution (Mayer 2012)

Pictures of recently developed vehicles and equipment are readily available on the Web, including the unfortunately named long-range autonomous vehicle "Boaty McBoatface".

**Fig. 13.37** Veterans of another sort. JAMSTEC hospitality for visiting scientists, a long time ago. Background: early version of Shinkai. Foreground, from left to right; C. Kato, the author, D. Bartlett, F. Abe and M. Ganzer

# References[1]

Abegg F, Hohnberg H-J, Pape T, Bohrmann G, Freitag J (2008) Development and application of pressure-core-sampling systems for the investigation of gas and gas-hydrate – bearing sediments. Deep-Sea Res I 55:1590–1159

AD 2000-Merkblatt (2019) Technical rules for pressure vessels, Section N4: pressure vessels made of glass. https://www.ad-2000-online.de/en. Accessed Sep 2019 *

Agamat E (1893) Reports on the elasticity and expansibility of liquids at very high pressures. Ann Chim Phys Series 6 29:68–96. *

Agamat E (1896) Reports on the elasticity and expansibility of liquids at very high pressures. Notice sur les travaux scientifques de M. E. H. Agamat, Paris *

ASME (2015) Boiler and pressure vessel code, Section VIII Rules for construction of pressure vessels, Division 1. American Society of Mechanical Engineers *

ASME (2016a) Safety standard for pressure vessels for human occupancy: ASME PVHO-1-2016. American Society of Mechanical Engineers *

ASME (2016b) Safety standard for pressure vessels for human occupancy: in-service guidelines ASME PVHO-2-2016. American Society of Mechanical Engineers *

Ben-Ari E (2002) Intimate connections: geomicrobiologists explore the interactions between biosphere and geosphere. Bioscience 52:326–331

Bianchi A, Garcin J (1991) In stratified waters the metabolic rate of deep sea bacteria decreases with decompression. Deep-Sea Res I 40:1703–1710

---

[1]Those in bold can be seen on the web without fee or affiliation. Those marked * are further reading.

Bianchi A, Garcin J (1993) In stratified waters the metabolic rate of deep sea bacteria decreases with decompression. Deep-Sea Res I 40:1703–1710

Bianchi A, Gardin J, Tholosan O (1999) A high pressure serial sampler to measure microbial activity in the deep sea. Deep-Sea Res I 46:2129–2142

Biggers RE, Chilton JM (1959) Review and bibliography on design and use of windows for optical measurements at elevated temperatures and pressures 1881–1959. Oak Ridge National Laboratory. *

Braeuer A (2015) In situ spectroscopic techniques at high pressure. Elsevier, ISBN 0444634207, 9780444634207. *

**Brandt A et al (2016) Cutting the umbilical: new technological perspectives in benthic deep sea research. J Mar Sci Eng 4(2):36**

Bridgman PW (1931) The physics of high pressure. G. Bell and Sons, Ltd. *

Brown DM (1975) Four biological samplers: opening-closing midwater trawl, closing vertical tow net pressure fish trap, free-vehicle drop camera. Deep-Sea Res 22:565–567

Case DH et al (2017) Aerobic and anaerobic methanotropic communities associated with methane hydrates exposed on the seafloor: a high pressure sampling and stable isotope-incubation experiment. Front Microbiol 8:art 2569

Chandra Shekar NV, Sahu PC (2007) High pressure research on materials: 2. Experimental techniques to study the behaviour of materials under high pressure. Resonance 12. https://doi.org/10.1007/s12045-007-0082-6

Chastain RA, Yayanos AA (1991) Ultrastructural changes in an obligately barophilic marine bacterium after decompression. Appl Environ Microbiol 57:1489–1497

Chervin JC, Syfosse G, Besson JM (1994) Mechanical strength of sapphire windows under pressure. Rev Sci Instrum 65:2719–2725

Cowen JP et al (2003) Fluids from aging ocean crust that support microbial life. Science 299:120–123

Cowen JP et al (2012) Advanced instrument system for real-time and time-series microbial geochemical sampling of the deep (basaltic) crustal biosphere. Deep Sea Res 61:43–56 *

Cui W (2018) An overview of submersible research and development in China. J Marine Sci Appl 17(4):459–470. https://doi.org/10.1007/s11804-018-00062-6. *

Cui WC, Hu Y, Guo W, Pan BB, Wang F (2014) A preliminary design of a movable laboratory for hadal trenches. Methods Oceanogr 9:1–16. *

Cui WC, Wang F, Pan BB, Hu Y, Du QH (2016) Issues to be solved in the design, manufacture and maintenance of full ocean depth manned cabin. Adv Eng Res 11:1–29. *

D'Hondt S, Inagaki F, Ferdelman T, Jørgensen BB, Kato K, Kemp P, Sobecky P, Sogin M, Takai K (2007) Exploring subseafloor life with the integrated ocean drilling program. Scientific Drilling 5:26–37

Damasceno-Oliviera A, Goncalves J, Silva J, Fernandez-Duran B, Coimbra J (2004) A pressurising system for long term study of marine or freshwater organisms enabling the simulation of cyclic vertical migrations. Scientia Marina 68:615–619. *

Das P (2002) Redesign of NEMO-type spherical acrylic submersible for manned operation to 3000 feet (914 m) ocean depth. J Pres Ves Technol 124(1):97. https://doi.org/10.1115/1.1428746. *

Deguchi S, Tsujii K (2002) Flow cells for in situ optical microscopy in water at temperatures and pressures up to supercritical state. Rev Sci Instrum 73:3938–3941

Delcour AH, Martinac B, Adler J, Kung C (1989) Modified reconstitution method used in patch clamp studies of Escherichia coli ion channels. Biophys J 56:631–636

Deming JW, Somers LK, Straube EL, Swartz DG, Macdonell MT (1988) Isolation of an obligately barophilic bacterium and description of a new genus, Colwellia gen. nov. Syst Appl Microbiol 10:152–160

Dietz AS, Yayanos AA (1978) Silica gel media for isolating and studying bacteria under hydrostatic pressure. Appl Environ Microbiol 36:966–968

Downs JL, Payne RT (1969) A review of electrical feed-through techniques for high pressure gas systems. Rev Sci Instrum 40. *

**Drazen JC, Bird LE, Barry JP (2005) Development of a hyperbaric trap-respirometer for the capture and maintenance of live deep-sea organisms. Limnol Oceanogr Methods 3 (2005):488–498**

Evans S, Birch J, Brier J, Jakuba M, Salto M, Robidart J (2019) Ocean robots uncover microbial secrets. Microbiol Society Oceans 5 Feb

Faes, W., Lecompte, S., Van Bael, J., Salenbien, R., Bäßler, R., Bellemans, I., Cools, P., De Geyter, N., Morent, R., Verbeken, K., De Paepe, M. (2019, November) Corrosion behaviour of different steel types in artificial geothermal fluids Geothermics Volume 82, 182–189. *

**Frey B et al (2006) Microscopy under pressure – an optical chamber system for fluorescence microscopic analysis of living cells under high hydrostatic pressure. Microsc Res Tech 69:65–72**

Friedrich O, Wegner FV, Hartmann M, Frey B, Sommer K, Ludwig H, Fink RHA (2006) In situ high pressure confocal $Ca^{++}$ fluorescence microscopy in skeletal muscles: a new method to study pressure limits in mammalian cells. Undersea Hyperbaric Med 33:181–195

Fryer DM, Harvey JF (2012) High pressure vessels. Springer Science & Business Media

**Garel M et al (2019) Pressure-retaining sampler and high pressure systems to study deep sea microbes under in situ conditions. Front Microbiol 10:art 453**

Gibbon HA (1968) Pressure liquid sampling system and apparatus. U.S. Patent No 3,379,065, 1968 *

Gilchrist I (1972) Equipment for the recovery and study of deep sea animals. PhD thesis, The Queens University of Belfast, UK

Gilchrist I, Macdonald AG (1983) Techniques for experiments with deep sea animals. In: Macdonald AG, Priede IG (eds) Experimental biology at sea. Academic Press, London

**Giovanelli D, Lawrence NS, Compton RG (2004) Electrochemistry at high pressures: a review. Electroanalysis 16:789–810**

Grasset O (2001) Calibration of the R ruby fluorescence lines in the pressure range (0–1 GPa) and the temperature range (250–300 K). High Pressure Res 21:139–157

Gundersen KR, Mountain CW (1972) A pressure and temperature preserving sampling, transfer, and incubation system for deep sea microbiological and chemical research. In: Brauer RW (ed) Barobiology and the experimental biology of the deep sea. University of Northe Carolina, USA. *

Guorui L et al (2021) Self-powered soft robot in the Mariana Trench. Nature 591:66–71 *

**Haberl B, Molaison JJ, Neuefeind JC, Luke L, Daemen LL, Boehler R (2019) Modified Bridgman anvils for high pressure synthesis and neutron scattering. High Pressure Res 39:426–437**

Halle D, Yedgar S (1988) Mild pressure induces resistance of erythrocytes to hemolysis by snake venom phospholipase A2. Biophys J 54:393–396. *

Harper AA, Macdonald AG, Wardle CS, Pennec J-P (1987) The pressure tolerance of deep sea fish axons; Results of Challenger Cruise 6/b85. Comp

Hauge JG (1971) Pressure induced dissociation of ribosomes during ultracentrifugation. FEBS Lett 17:168–172

Haver T, Raber EC, Urayama P (2006) An application of spatial deconvolution to a capillary high pressure chamber for fluorescence microscopy microscope imaging. J Microsc 230:363–371

Heden CG (1964) Effects of hydrostatic pressure on microbial systems. Bacteriol Rev 28:14–29. *

**Heinemann SH, Stuhmer W, Conti F (1987) Si9ngle acetylcholine receptor channel currents recorded at high hydrostatic pressures. Proc Natl Acad Sci U S A 84:3229–3233**

Hemley R (2010) Percy W. Bridgman's second century. High Pressure Res 30:581–689

**Hodge A, Fujan RS, Carlson KL, Burgess RG, Harris WH, Mann RW (1986) Contact pressures in the human hip joint measured in vivo. Proc Natl Acad Sci U S A 83:2879–2883**

Holtgrewe N, Greenberg E, Prescher C, Prakapenka VB, Goncharov AF (2019) Advanced integrated optical spectroscopy system for diamond anvil cell studies at GSECARS. High Pressure Res 39(3):457–470. *

Huey D, Storms M (1985) The ocean drilling program IV: deep water coring technology-past, present, and future. Published in: OCEANS '85 – Ocean Engineering and the Environment, 12–14 Nov

Humphris S, Soule A (2019) Vehicles for deep sea exploration. Encyclopedia of Ocean Sciences, Academic Press

Inoue S, Fuseler J, Salmon ED, Ellis GW (1975) Functional organization of mitotic microtubules. Bioophys J 15:725–744

Ishii A, Nishiyama M, Yanagida T (2004) Mechano-chemical coupling of molecular motors revealed by single molecule measurements. Curr Protein Pept Sci 5:81–87

**Jackson K, Witte U, Chalmers S, Anders E, Parkes J (2017) A system for retrieval and incubation of benthic sediment cores at in situ ambient pressure and under controlled or manipulated environmental conditions. J Atmos Oceanic Tech 34:983–1000**

**Jannasch HW, Wirsen CO (1977) Retrieval of concentrated and undecompressed microbial populations from the deep sea. Appl Environ Microbiol 33:642–646**

Jannasch HW, Wirsen CO, Winget CL (1973) Bacteriological pressure-retaining deep-sea sampler and culture vessel. Deep-Sea Res 20:661–662

**Jannasch HW, Wirsen CO, Taylor CD (1976) Undecompressed microbial populations from the deep sea. Appl Environ Microbiol 32:360–367**

Jannasch HW, Wirsen CO, Taylor CD (1982) Deep sea bacteria: isolation in the absence of decompression. Science 216:1315–1316

Jin Y, Konno Y, Nagao J (2014) Pressurised subsampling system for pressurised gas-hydrate-bearing sediment: microscale imaging using X-ray computed tomography. Rev Sci Instrum 85:094502

**Jose A, Caro A, Wand J (2018) Practical aspects of high pressure NMR spectroscopy and its applications in protein biophysics and structural biology. Methods 148:67–80**

Josephs R, Harrington WF (1966) Studies on the formation and physical chemical properties of synthetic myosin filaments. Biochemistry 5:3474–3487. *

Josephs R, Harrington WF (1967) Sedimentary studies on interacting systems: pressure dependence of the polymerization of myosin. Fed Proc 26:abstract 2630. *

Kato C, Nogi Y, Arakawa S (2008) Isolation, cultivation and diversity of deep sea piezophiles. Chapter 12. In: Michiels C, Bartlett DH, Aertsen A (eds) High pressure microbiology. ASM Press, Washington, DC

Kegeles G, Rhodes L, Bethune JL (1967) Sedimentation behaviour of chemically reacting systems. Proc Natl Acad Sci U S A 58:45–51. *

Kemper B, Kemper K (2019) NEMO joint design in the ASME PVHO Code. Presented at 16th Annual Manned Underwater Vehicles Symposium (Marine Technology Society)

**Kitching JA (1954) The effects of high pressure on a suctorian. J Exp Biol 31:56**

Koyama S (2007) Cell biology of deep sae multicellular organisms. Cytotechnology 55:125–133

**Koyama S, Miwa T, Sato T, Aizawa M (2001) Optical chamber system designed for micro-scopic observation of living cells under extremely high hydrostatic pressure. Extremophiles 5:409–415**

Koyama S et al (2002) Pressure-stat aquarium system designed for capturing and maintaining deep sea organisms. Deep-Sea Res I 49:2095–2102

**Koyama S et al (2005) Survival of deep sea shrimp (Alvinocaris sp.) during decompression and larval hatching at atmospheric pressure. Marine Biotechnol 7:272–278**

Kusube M et al (2017) Colwellia marinimaniae sp. nov., a hyperpiezophilic species isolated from an amphipod within the Challenger Deep, Mariana Trench. Int J Syst Evol Microbiol 67:824–831

Kyo M, Tuji T, Usui H, Itoh T (1991) Collection, isolation and cultivation system for deep-sea microbes study: concept and design. Oceans 1:419–423

Landau JV, Thibodeau L (1962) The micromorphology of Amoeba proteus during pressure-induced changes in the sol-gel cycle. Exp Cell Res 27:591–594

Lemmon EW, Hubner ML, McLinden MO (2010) NIST standard reference database 23: reference fluid thermodynamic and transport properties-REFPROP.9.0

**Lerch MT, Yang Z, Brocks EK, Hubbell WL (2014) Mapping protein conformational heterogeneity under pressure with site directed spin labelling and double electron-electron resonance. Proc Natl Acad Sci:E1201–E1210**

Li L et al (2016) Pressure retaining method based onphase change forcoring of gas hydrate-bearing sediments in offshore drilling. Appl Thermal Eng 107:633–641

Li Y, Peng J, Huang C, Wang M (2019) Experimental study on a sampling technique based on a freeze-sediments valve for deep sea microorganisms. Appl Ocean Res 82:470–477. *

**Lipp MJ, Evans WJ, Yoo CS (2005) Hybrid Bridgman anvil design: an optical window for in situ spectroscopy in large volume presses. High Pressure Res 25:205–210**

Liu PF, Wang SB, Li XKJ (2018) Finite element analysis of failure behaviors of the inspection window of the deep-sea submersible. Fail Anal Preven 18:1198

**Macdonald AG (1997) Effects of high hydrostatic pressure on the BK channel in bovine chromaffin cells. Biophys J 73:1866–1873**

Macdonald AG, Gilchrist I (1969) Recovery of deep sea water at constant pressure. Nature 222:71–72

Macdonald AG, Gilchrist I (1972) An apparatus for the recovery and study of deep sea plankton at constant temperature and pressure. In: Brauer RW (ed) Baro-biology and the experimental biology of the deep sea. University of North Carolina Press, pp 394–407

Macdonald AG, Gilchrist I (1978) Further studies on the pressure tolerance of deep sea crustacea with observations using a new high pressure trap. Mar Biol 45:9–21

Macdonald AG, Gilchrist I (1982a) The pressure tolerance of deep sea Amphipods collected at their ambient high pressure. Comp Biochem Physiol 71A:348–352. *

Macdonald AG, Gilchrist I (1982b) The pressure tolerance of Amphipods collected at their ambient high pressure. Comp Biochem Physiol 71A:349–352. *

**Macdonald AG, Martinac B (1999) Effect of high hydrostatic pressure on the porin Omp C from *Escherichia coli*. FEMS Microbiol Lett 173:329–334**

Macdonald AG, Vjotosh AN (1999) Patch-clamp recording of BK (Ca) channels in hyperbaric oxygen. J Physiol 51P:111P

Macdonald AG, Gilchrist I, Wann KT, Wilcock SE (1979) The tolerance of Animals to pressure. In: Gilles R (ed) Animals and environmental fitness, vol 1. Pergamon Press. *

Macdonald AG, Ramsey RL, Shelton CJ, Usherwood PNR (1989) An apparatus for single channel patch recording at high pressure. J Physiol 409:2P

Malahoff A, Gregory T, Bossuyt A, Donachie S, Alam M (2002) A seamless system for the collection and cultivation of extremophiles from the deep ocean. IEEE J Ocean Eng 27:862–869

Maltas J, Long Z, Huff A, Maloney R, Ryan J, Urayama P (2014) A micro-perfusion system for use during real-time physiological studies under high pressure. Rev Sci Instrum 85:106106

**Marsland D (1958) Cells at high pressure. Sci Am 199:36–43**

Marsland D (1970) Pressure–temperature studies on the mechanisms of cell division. In: Zimmerman A (ed) High pressure effects on cellular processes. Academic Press, pp 259–312. *

Masui N, Kao C (1999) New method of screening for pressure sensitive mutants at high hydrostatic pressure. Biosci Biotechnol Biochem 63:235–237

Mavor JW (1965) Observation windows of the deep submersible, ALVIN (AD0629592). Technical Report 65–62. Woods Hole Oceanographic Institution, Woods Hole (MA). *

Mayer A (2012) Life in the deep biosphere. Bioscience 62:453–457

**McCoy J, Hubbell WL (2011) High pressure EPR reveals conformational equilibria and volumetric properties of spin-labeled proteins. Proc Natl Acad Sci 108:1331–1336**

McNamee MJ, Wawersik WR, Shields ME, Holcomb DJ (1991) A compact coaxial electrical feedthrough for use up to 400 MPa. Rev Sci Instrum 62:1662. *

McNichol J et al (2016) Assessing microbial processes in deep-sea hydrothermal systems via incubations at in situ temperature and pressure. Deep-Sea Res I 115:221–231

McNichol J et al (2018) Primary productivity below the sea floor at deep sea hot springs. Proc Natl Acad Sci U S A 115:1–6. *

Mertens-Strljthagen J, De Schryver C, Wattiaux-De Conninck S, Wattiaux R (1985) The effect of pressure on fetal and neonatal liver mitochondrial membranes. Arch Biochem Biophys 236:825–831. *

Mickel TJ, Childress JJ (1982) Effects of pressure and pressure acclimation on activity and oxygen consumption in the bathypelagic mysid Gnathophausia ingens. Deep-Sea Res 29:1293–1301. *

**Miller JB, Aidley JS, Kitching JA (1975) Effects of helium and other inert gases on Echinosphaerium nucleofilum (Protozoa, Heliozoa). J Exp Biol 63:467–481**

Mizuno S (2005) A novel method for assessing effects of hydrostatic fluid pressure on intracellular calcium: a study with bovine articular chondrocytes. Am J Physiol Cell Physiol 288:C329–C337

**Morrell KC, Hodge WA, Krebs DE, Mann RW (2005) Corroboration of in vivo cartilage pressure with implications for synovial joint tribology and oateoarthritis causation. Proc Natl Acad Sci U S A 102:14819–14824**

Murayama M (1987) Compression inhibits aggregation of human platelets under high hydraulic pressure. Thromb Res 45:729–738

Nakayama A, Yano Y, Yoshida K (1994) New method for isolating barophiles from intestinal contents of deep sea fishes retrieved from the abyssal zone. Appl Environ Microbiol 60:4210–4212

Nishiyama M (2015) High pressure microscopy for studying molecular motors. Chapter 27. In: Akasaka K, Matsuki H (eds) High pressure bioscience. Springer Science+Business Media Dordrecht

**Nishiyama M (2017) High pressure microscopy for tracking dynamic properties of molecular machines. Biophys Chem 231:71–78**

Nishiyama M, Kojima S (2012) Bacterial motility measured by a miniature chamber for high pressure microscopy. Int J Mol Sci 13:9225–9239. *

Nishiyama M, Sowa Y (2012) Microscopic analysis of bacterial motility at high pressure. Biophys J 102:1872–1880. *

Nishiyama M, Kimura Y, Nishiyama Y, Terazima M (2009) Pressure-induced changes in the structure and function of the kinesin-microtubule complex. Biophys J 96:1142. *

Nishiyama M, Kato C, Abe F (2012) Intracellular pressure measurement by using pressure sensitive Yellow fluorescent protein. Biophys J 102:419a. *

**Noji H, Yasuda R, Yoshida M, Kinosita K (1997) Direct observation of the rotation of F1-ATPase. Nature 386:299–302**

Oger PM, Daniel I, Piccard A (2006) Development of a low pressure diamond anvil cell and analytical tools to monitor microbial activities in situ under controlled P and T. Biochim Biophys Acta 1764:434–442

**Orcutt B, Wheat CG, Edwards KJ (2010) Subseafloor ocean crust microbial observatories: development of FLOCS (Flow through Osmo Colonization System) and evaluation of borehole construction materials. Geomicrobiol J 27:143–157**

Pagliaro L, Reitz F, Wang J (1995) An optical pressure chamber designed for high numerical aperture studies on adherent living cells. Undersea Hyperbaric Med 22:171–181

Paladini AA, Weber G (1981) Absolute measurements of fluorescence polarisation at high pressures. Rev Sci Instrum 52:419–427

**Parkes RJ et al (2009) Culturable prokaryotic diversity of deep, gas hydrate sediments: first use of a continuous high-pressure, anaerobic, enrichment and isolation system for subseafloor sediments (DeepIsoBUG). Environ Microbiol 11:3140–3153**

Peoples LM et al (2019) A full ocean depth rated modular lander and pressure retaining sampler capable of collecting hadal-endemic microbes under in situ conditions. Deep-Sea Res 143:50–57. *

Peterson RW, Wand AJ (2005) Self contained high pressure cell, apparatus and procedure for the preparation of encapsulated proteins dissolved in low viscosity fluids for NMR spectroscopy. Rev Sci Instrum 76:094101

Phillips H, Wells LE, Johnson RV, Elliott S, Deming JW (2003) LAREDO: a new instrument for sampling and in situ incubation of deep sea hydrothermal vent fluids. Deep-Sea Res I 50:1375–1387

Phleger CF, McConnaughey RR, Grill P (1979) Hyperbaric fish trap operation and deployment in the deep sea. Deep-Sea Res 26A:1405–1409

Piccard J, Dietz RS (1957) Oceanographic observations by the Bathyscaph Trieste (1953–1956). Deep-Sea Res 4:221–229. *

**Pickles DM, Ogston D, Macdonald AG (1990) Effects of hydrostatic pressure and inert gases on platelet aggregation in vitro. J Appl Physiol 69:2239–2247**

Poulter TC (1930) A glass window mounting for withstanding pressures of 30,000 atmospheres. Phys Rev 35:297. *

Poulter TC (1932) Apparatus for optical studies at high pressures. Phys Rev 40:860–871. *

Poulter TC, Benz CA (1932) The lens effect of pressure windows. Phys Rev 40:872–876. *

Poulter TC, Buckley F (1932) Diamond windows for withstanding very high pressures. Phys Rev 41:364–365. *

Poulter TC, Wilson RO (1930) Some recent developments in high pressure windows. Proc Iowa Acad Sci 37:259–302. *

Poulter TC, Wilson RO (1932) The permeability of glass and fused quartz to ether, alcohol and water at high pressures. Phys Rev 40:877–880. *

Pradillon F, Shillito B, Chervin J-C, Hamel G, Gaill F (2004) Pressure vessels for in vivo studies of deep sea fauna. High Pressure Res 24:237–246

Qinghai D, Yong H, Weicheng C (2017) Safety assessment of the acrylic conical frustum viewport structure for a deep-sea manned submersible. Ships Offshore Struct 12(sup1):S221–S229. *

**Quetin LB, Childress JJ (1980) Observations on the swimming activity of two bathypelagic mysid species maintained at high hydrostatic pressure. Deep-Sea Res 27:383–391**

Raber EC, Dudley JA, Salerno M, Urayama P (2006) Capillary based, high pressure chamber for fluorescence microscopy imaging. Rev Sci Instrum 77:096106

Rabinowitz S, Ward IM, Parry JSC (1970) The effect of hydrostatic pressure on the shear yield of polymers. J Mater Sci 5:29–39. *

**Ranatunga KW, Fortune NS, Geeves MA (1990) Hydrostatic compression in glycinerated rabbit muscle fibers. Biophys J 58:1401–1410**

Ravaux J et al (2013) Thermal limit for metazoan life in question: in vivo heat tolerance of the Pompeii worm. Plos One 8(e):64074

Ravaux J, Léger N, Hamel G, Shillito B (2019) Assessing a species thermal tolerance through a multiparameter approach: the case study of the deep-sea hydrothermal vent shrimp Rimicaris exoculate. Cell Stress Chaperones 24:647–659

Rebhun LI, Mellon M, Jemiolo D, Nath J, Ivy N (1974) Regulation of size and biretringence of the in vivo mitotic apparatus. J Supramol Struct 2:466–485

Regnard P (1884a) Effekts des hautes pressions sur les animaux marins. Compt Rend Soc Biol 36:394–395. *

Regnard P (1884b) Recherches experimentales sur l'influence de tres haute pressions sur les organisimes vivantes. Comptes Rendues Acad Sci Paris 98:735–747. *

Regnard P (1884c) Note sur les conditions de la vie dans les profondeurs de la mer Comptes. Rend Soc Biol Paris 36:164–168. *

Regnard P (1891) Recherches Experimentales sur les conditions physiques de la Vie dans les Eaux. Masson, Paris. *

Robinson NJ, Thatje S, Osseforth C (2009) Heartbeat sensors under pressure: a new method for assessing hyperbaric physiology. High Pressure Res 29:422–430

Roche J, Royer CA, Roumestand C (2017) Monitoring protein folding through high pressure NMR spectroscopy. Prog Nucl Magn Reson Spectrosc 102–103:15–31. *

Salmon ED (1975a) Pressure induced depolymerization of spindle microtubules I. J Cell Biol 65:603–614. *

Salmon ED (1975b) Pressure induced depolymerization of spindle microtubules II. J Cell Biol 66:114–127. *

Salmon ED (1975c) Spindle microtubules: thermodynamics of in vivo assembly and role in chromosome movement. Ann N Y Acad Sci 253:383–406. *

Salmon ED (1975d) Pressure induced depolymerization of brain microtubules in vitro. Science 189:884–886. *

**Salmon ED, Ellis GW (1975) A new miniature hydrostatic pressure chamber for microscopy. J Cell Biol 65:587–602**

Salmon ED, Goode D, Maugel TK, Bonar DB (1976) Pressure induced depolymerization of spindle microtubules III. J Cell Biol 69:443–454. *

Sauer P, Glombitza C, Kallmeyer J (2012) A system for incubations at high gas partial pressure. Front Microbiol 3:art 25. *

Schmalwasser H, Neef A, Elliott AA, Heinemann SF (1998) Two electrode voltage clamp of Xenopus oocytes under high hydrostatic pressure. J Neurosci Methods 81:1–7

**Schneidereit D, Vass H, Reischi B, Allen RJ, Friedrich O (2016) Calcium sensitive fluorescent dyes Fluo 4 and Fura Red under pressure: behaviour of fluorescence and buffer properties under hydrostatic pressures up to 200 MPa. Plos One 11:0164509**

Schultheiss PJ, Francis TJG, Holland M, Jackson PD et al (2006) Pressure coring, logging and subsampling with the HYACINTH system. Geol Soc Lond Spec Publ 267:151–163

Seewald JS, Doherty KW, Hammar TR, Liberatore SP (2001) A new gas tight isobaric sampler for hydrothermal fluids. Deep-Sea Res I 49:189–196

Shelton DP (1992) Lens induced by stress in optical windows for high-pressure cells. Rev Sci Instrum 63:3978. https://doi.org/10.1063/1.1143248. *

Sherman WF, Stadmuller AA (1988) Experimental techniques in high pressure research, 1st edn. Wiley. ISBN-10: 0471103136 ISBN-13: 978-0471103134

Shillito B et al (2001) Temperature resistance of Hesiolyra bergi, a polychaetous annelid living on deep sea vent smoker walls. Mar Ecol Prog Ser 216:141–149

Shillito B, Hamel G, Duchi C, Cottin D, Sarrazin J, Sarradin P-M, Ravaux J, Galli F (2008) Live capture of megafauna from 2300 m depth, using a newly designed pressurized recovery device. Deep-Sea Res I 55:881–889

Shillito B, Gaill F, Ravaux J (2014) The IPOCAMP pressure incubator for deep-sea fauna. J Mar Sci Technol 22:97–102

Skelton WJ, Crossland B (1980) High pressure safety: a review of high pressure equipment development at the Queen's University of Belfast. Int J Pres Ves Piping 8(1980):377–404

Smiley JE, Drawbridge MA (2007) Techniques for live capture of deepwater fishes with special emphasis on the design and application of a low-cost hyperbaric chamber. J Fish Biol 70:867–878. *

Snoey MR, Katona MG (1970) Structural design of conical acrylic viewports. Technical Report No.: R-686. Naval Civil Engineering Laboratory, Port Hueneme (CA)

Stachiw JD (1966) Critical pressure of conical acrylic windows under short term hydrostatic loading A.S.M.E. Paper 66-WA/UNT.2

Stachiw JD (1967) Windows for external or internal hydrostatic pressure vessels. Part 2: Flat acrylic windows. Technical Report R527 U.S. Naval Civil Engineering Laboratory *

Stachiw JD (1970) Conical; acrylic windows under long term hydrostatic pressure of 20,000 lbf/in2 Trans A.S.M.E. J Eng Ind 92:237–256

Stachiw JD, Gray KO (1967) Windows for external or internal hydrostatic pressure vessels. Part 1: Conical acrylic windows. Technical Report R512. U.S. Naval Civil Engineering Laboratory *

Stokes MD, Somero GN (1999) An optical oxygen sensor and reaction vessel for high pressure applications. Limnol Oceanogr 44:189–195

Sun W et al (2018) Adiabatic compression heating of selected organic solvents under high pressure processing. High Pressure Res 38:325–336

Szafrinski KM, Piquet B, Shillito B, Lallier F, Duperron S (2015) Relative abundance of methane- and sulfur-oxidizing symbionts in gills of the deep sea hydrothermal vent mussel Bathymodiolus azoricus under pressure. Deep-Sea Res I 101:7–13

Tabor PS et al (1981) A pressure-retaining deep ocean sampler and transfer system for measurement of microbial activity in the deep sea. Microb Ecol 7:51–65

Taylor CD et al (2006) Autonomous Microbial Sampler (AMS), a device for the uncontaminated collection of multiple microbial samples from submarine hydrothermal vents and other aquatic environments. Deep-Sea Res I 53:894–916. *

Tekmen M, Muller JD (2004) High pressure cell for fluorescence fluctuation spectroscopy. Rev Sci Instrum 75:5143–5148

Testemale D, Prat A, Lahera E, Hazemann J-L (2016) Novel high-pressure windows made of glasslike carbon for X-ray analysis. Rev Sci Instrum 87:075115

Tian CL, Hu Y, Liu DQ, Cui WC (2010) Creep analysis on deep-sea structure's viewport windows. J Ship Mech 14(5):526–532. *

Turner A (2016) The evolution of thicker & more complex acrylic windows for manned submersibles. In: Proceedings of the 13th annual manned underwater vehicles symposium (Marine Technology Society). New Orleans, LA. http://www.underwaterintervention.com. Accessed Sep 2019 *

Undersea and Hyperbaric Medical Society (2004) Hyperbaric facility design guidelines version 1.0 July 2004. UHMS Associates. https://www.uhms.org/images/Safety-Articles/aia_facility_design_1_2.pdf. Accessed Sep 2019 *

Undersea and Hyperbaric Medical Society (2012) Safe design and operation of hyperbaric chambers. Reviewed by the UHMS Hyperbaric Oxygen Safety Committee. https://www.uhms.org/images/Safety-Articles/safe_design_and_operation_of.pdf. Accessed Sep 2019 *

Undersea and Hyperbaric Medical Society (2019) Safety references. https://www.uhms.org/publications/safety-documents/safety-references.html. Accessed Sep 2019 *

United Kingdom Department for Business, Energy and Industrial Strategy (2016) Pressure equipment (safety) regulations 2016. https://assets.publishing.service.gov.uk/government/uploads/system/uploads/attachment_data/file/640795/nlf-pressure-equipment-regulations-2016-guidance.pdf. Accessed Sep 2019

Usui K, Hiraki T, Kwamoto J, Kurihara T, Nogi Y (2012) Eicosapentaenoic id plays a role in stabilising dynamic membrane structure in the seep-sea piezophile Shewanella violacea: a study employing high-pressure time-resolved fluorescence anisotropy measurement. Biochim Biophys Acta 1818:574–583. *

Vass H, Lucas-Black SL, Herzig EM, Ward FB, Clegg PS, Allen RJ (2010) A multipurpose modular system for high-resolution microscopy at high hydrostatic pressure. Rev Sci Instrum 81:053710. *

Vass H et al (2013) Single molecule imaging at high hydrostatic pressure. Appl Phys Lett 102:154103

Wahl W (1913a) Physio-chemical determination at high pressure by optical methods, I. Apparatus for optical determinations at high pressures. Trans Roy Soc A 212:117–148. *

Wahl W (1913b) Optische Untersuchung verfestigter Gase. II. Zeitschrift für Physikalische Chemie 84U(1):112–122. *

Wang Y et al (2016) Genomic characteriuzation of symbiotic mycoplasmas from the stomach of deep sea isopod Bathynomus sp. Environ Microbiol 18:2646–2659

Wang F, Wang W, Zhang Y, Du Q, Jiang Z, Cui W (2019) Effect of temperature and nonlinearity of P.M.M.A. material in the design of observation windows for a full ocean depth manned submersible. Mar Technol Soc J 53:27–36

Wann KT, Macdonald AG, Harper AA, Wilcock SE (1979) Electrophysiological measurements at high hydrostatic pressure: methods for intracellular recording from isolated ganglia and for extracellular recording in vivo. Comp Biochem Physiol 64A:141–147

Wilcock SE (1979) The effects of high hydrostatic pressure on a marine crustacean, on the crustacean abdominal nerve cord and another nerve bundle. PhD thesis, Aberdeen University, UK *

Wilcock SE, Wann KT, Macdonald AG (1978) The motor activity of Crangon crangon subjected to high hydrostatic pressure. Mar Biol 45:1–7. *

Wilson RR, Smith KL (1985) Live capture, maintenance and partial decompression of a deep sea grenadier fish (Coryphaenoides acrolepis) in a hyperbaric trap-aquarium. Deep-Sea Res 32:1571–1582. *

Winter R (2002) Synchrotron X-ray and neutron small-angle scattering of lyotropic lipid mesophases, model biomembranes and proteins in solution at high pressure. Biochim Biophys Acta 1595:160–184

Wu W, Chong PL-G, Huang CH (1985) Pressure effect on the rate of crystalline phase formation of L-a-dipalmitoyl phosphatidylcholine in multilamellar dispersions. Biophys J 27:327–342. *

Wu S-J, Wang S, Yang C-J (2018) An isobaric sample transfer apparatus for deep-sea pressurized fluid sample. Rev Sci Instrum 89:086110

Yayanos AA (1977) Simply actuated closure for a pressure vessel: design for use to trap deep sea animals. Rev Sci Instrum 48:786–789

Yayanos AA (1978) Recovery and maintenance of live amphipods at a pressure of 580 bars from an ocean depth of 5700 meters. Science 200:1056. *

Yayanos AA (1995) Microbiology to 10,500 meters in the deep sea. Annu Rev Microbiol 49:777–805. *

Yayanos AA, DeLong EF (1987) Deep sea bacterial fitness to environmental temperatures and pressures. In: Jannasch HW, Marquis RE, Zimmerman AM (eds) Current perspectives in high pressure biology. Academic Press, Orlando, FL. *

Yayanos AA, Dietz AS (1983) Death of a hadal deep sea bacterium after decompression. Science 220:497–498

Yayanos AA, Dietz AS, Von Boxtell R (1981) Obligately barophilic bacterium from the Mariana trench. Proc Natl Acad Sci U S A 78:5212–5215

Yoder P, Vukobratovich D (eds) (2017) Opto-mechanical systems design, Volume 1: Design and analysis of opto-mechanical assemblies. CRC Press, ISBN 1482257718, 9781482257717. *

Zhang J, Zuo X, Wang W, Tang W (2014) Overviews of investigation on submersible pressure hulls. Adv Nat Sci 7(4):1–8. https://doi.org/10.3968/6129. ISSN 1715-7870 *

Zhang Z, Wu Y, Zhang X-H (2018) Cultivation of microbes from the deep sea environments. Deep-Sea Res II 155:34–43

Zhu Y, Liang W, Zhao X, Wang X, Xia J (2019, Sep) Strength and stability of spherical pressure hulls with different viewport structures. Int J Pres Ves Piping 176. https://doi.org/10.1016/j.ijpvp.2019.103951. *

# Index

Printed in the United States
by Baker & Taylor Publisher Services